THE AVIAN BROOD PARASITES

Reed warbler tending a nestling common cuckoo. After a photo by Wyllie (1981).

THE AVIAN BROOD PARASITES

DECEPTION AT THE NEST

Paul A. Johnsgard

New York Oxford
Oxford University Press
1997

Oxford University Press

Oxford New York
Athens Auckland Bangkok Bogota Bombay Buenos Aires
Calcutta Cape Town Dar es Salaam Delhi Florence Hong Kong
Istanbul Karachi Kuala Lumpur Madras Madrid Melbourne
Mexico City Nairobi Paris Singapore Taipei Tokyo Toronto Warsaw

and associated companies in
Berlin Ibadan

Published by Oxford University Press, Inc.
198 Madison Avenue, New York, New York 10016

Oxford is a registered trademark of Oxford University Press

Library of Congress Cataloging-in-Publication Data
Johnsgard, Paul A.
The avian brood parasites : deception at the nest / by Paul A.
Johnsgard.
p. cm.
Includes bibliographical references and index.
ISBN 0-19-511042-0
1. Parasitic birds. 2. Brood parasites. I. Title.
QL698.3.J68 1997
598.2556—dc20 96-8884

9 8 7 6 5 4 3 2 1
Printed in the United States of America
on acid-free paper

CONTENTS

PREFACE

> The cuckoe is a idle and lazy birde, never buildinge herselfe a nest, but layinge her eggs in the nests of other birdes as in wood-pigeons, hedge-sparrowes, wagtayles or such other like.
>
> E. Topsell, *The Fowles of Heaven, or History of Birds* (translation of Harrison & Hoeniger, 1972).

I must confess at the outset that my decision to write this book, rather than being the direct result of sudden inspiration, came about via a brainstorming session. Specifically, Charles Brown, Josef Kren, and I were sitting around a table at the University of Nebraska's Cedar Point Biological Station in western Nebraska one long summer evening in 1993, and the subject of desirable but not yet available ornithological books came up. After discussing several areas of current interest to ornithologists, the idea of a book dealing with avian brood parasitism rose to the top of potentially valuable subjects. No world-comprehensive English-language book exists on brood parasitism, and comprehensive foreign-language books on the subject are outdated or inaccessible (Makatsch, 1955; Mal'chevsky, 1987), despite the fact that the evolutionary, ecological, and behavioral questions posed by obligate brood parasites are among the most intriguing contemporary ornithological topics. Soon thereafter I decided to consider writing such a book, and began seriously gathering references and reviewing the large and greatly scattered world literature on the subject.

Originally I had planned to restrict my coverage to the sufficiently daunting task of reviewing the nearly one hundred species of obligate interspecific avian brood parasites in the world. However, it soon became evident that I would also have to consider some parallel intraspecific brood parasites, which exhibit such behaviors as nest-sharing and facultative conspecific brood parasitism, at least in terms of their possible relevance to the evolutionary history of obligatory brood parasitism. Such hypothetical and other peripheral topics of brood parasitism are dealt with in Chapter 1. I have tried to cover the other major evolutionary and comparative aspects of brood parasitism in Chapters 2–5. The second and major part of the book is devoted to 94 individual species accounts that include all of the world's known obligatory avian brood parasites as well as some others that, because of their close relationships to known parasites, are almost certain to fall into that category once their breeding biologies have been better studied. In all these accounts, emphasis is placed on field and in-hand species identification and on those aspects of breeding biology that are related specifically to brood parasitism, rather than on summarizing their overall ecologies and entire life histories, which would obviously be impossible in a single-volume work.

The terminology associated with discussions of brood parasitism and related aspects of social parasitism is still somewhat unsettled and, indeed, rather unsatisfactory. Even the term "brood parasitism" is somewhat inaccurate, since in many host species the "brood" in a parasitized nest consists of only the parasite chick, at least after it has eliminated the host's own eggs or chicks. "Egg parasitism" seems slightly better semantically, but this term has had little use except in reference to intraspecific egg dumping. "Prehatching brood amalgamation" is much too cumbersome to be useful, and better fits the phenomenon of intraspecific egg dumping than that of surreptitious introduction of alien eggs into a host species' nest. "Clutch parasitism" probably offers the fewest semantic difficulties in describing the phenomenon, but to my knowledge it has never been used by ornithologists. I leave it to other writers to suggest better solutions to this semantic problem, although I have devised a few new terms (e.g., host-tolerant vs. host-intolerant species) that seemed to be useful descriptors. For such reasons, and to make the text more accessible to nonornithologists, I have included a fairly extended glossary, which defines technical terms that appear in the text, especially those that relate specifically to social parasitism or are more generally relevant to behavior, genetics, and ecology.

The taxonomic sequence in the species accounts, and my choices of appropriate nomenclature for species and higher-level taxonomic groups, are based on the world checklist of Sibley and Monroe (1990). The listed subspecies are those that appear to be the most widely accepted, judging from recent technical literature. In each species account, all such geographic races are initially listed in logical geographic order, but subsequent listings of multiple subspecies (as, for example, under their measurements and masses) are alphabetic. I have also provided an appendix consisting of an alphabetic list of the English vernacular and corresponding Latin names of the approximately 1000 avian species (other than the individually described brood parasites) that are mentioned in the text. With a few minor exceptions, these vernacular names are also those recommended by Sibley and Monroe.

As usual in any world survey of bird groups, one is faced with an overwhelming amount of information on a few well-studied species (such as the common cuckoo and brown-headed cowbird), but relatively little reliable information on the majority of other parasitic species. For example, Wyllie (1981) listed nearly 200 citations relevant to the common cuckoo, and nearly 1000 citations concerning the five parasitic cowbirds were collectively provided by Friedmann (1963), Friedmann et al. (1977), and Friedmann and Kiff (1985). In other multispecies monographic treatments, Friedman (1960) listed nearly 500 citations on the African parasitic finches, and more than 200 on the honeyguides (1955).

Yet, for most brood parasites, the literature is surprisingly sparse, and often much of it is unreliable. In part this is because brood parasites are generally harder to study than any other categories of birds; they have survived and successfully adapted largely as a result of their deceptive behavior, elusiveness, and apparent cunning. It should thus not be surprising that almost as many false beliefs about them exist as factual knowledge. As a result, discerning truth from mythology and folklore, from contrived misinformation, and from honest but erroneous conclusions was a major problem in assembling materials for this book.

In spite of such problems, few groups of birds provide such intellectual appeal or provide so many opportunities to learn evolutionary lessons. It is easy to become emotional and judgmental when discussing social parasites; like other parasites they are best regarded simply as organisms that have managed to survive and thrive by exploiting the readily available energies provided by others. Many humans regularly use the same survival strategy without having universal condemnation heaped upon them—capitalists and welfare recipients provide obvious examples of opposite extremes of our own social spectrum that exhibit widely differing but nevertheless acceptable and individually adaptive behaviors.

Great advances have been made in the theoretical basis of exploitative social interactions such as social parasitism, and many well-documented field studies of this phenomenon in birds have been performed. Indeed, so many studies are being published on topics related to avian brood parasitism that some of the information in this book will likely become dated soon after publication. With the advent of modern biochemical analysis techniques, it is possible to identify eggs of unknown brood parasites, not only as to their species, but also at a level that permits identification of eggs laid by individual females. Thereby we may begin to verify previously speculative parasite:host combinations, get a grasp of the existence and significance of host-specific female "gentes," obtain better measurements of female egg-laying ranges during the breeding season, and help determine annual female egg production. Yet, part of the appeal of the avian brood parasites is that even such well-studied species as the brown-headed cowbird still offer fertile areas for behavioral and ecologic study using innovative field studies or sophisticated laboratory techniques. Additionally, the majority of the parasitic species lack simple field observations that would help fill in some of the all-too-frequent "no information" statements that are abundantly sprinkled through the species accounts of this book. For example, host species for less than half of the world's honeyguides have so far been documented, and the same is true of the bronze cuckoos. Almost nothing is known of the biologies of the hawk cuckoos, the several genera and species of endemic New Guinea cuckoos, or the New World pheasant cuckoos. Furthermore, we have almost no information on the actual costs (in terms of their reduced productivity) of brood parasitism for most host species, as well as the nature and effectiveness of their possible antiparasitism defense systems.

Many persons helped me during the preparation of this book. I must especially thank Josef Kren, who helped me in many ways, especially in reading various manuscript versions. Parts of the manuscript were also read and constructively critiqued by William Scharf and other associates and friends. Dr. Karin Johnsgard located many obscure references for me at the Cornell University libraries, and Dr. Lloyd Kiff photocopied other references from the Western Foundation of vertebrate Zoology library. The librarians at the University of Nebraska-Lincoln, the University of Kansas, the University of Michigan (Van Tyne Memorial Library), and at several other libraries also provided valuable assistance. I especially appreciate the help of the curatorial staff of the American Museum of Natural History (especially Dr. Lester Short, Jr.) and of the National Museum of Natural History (especially Dr. Richard Banks) in obtaining and providing specimen data, and I thank the National Museum of Natural History for allowing me personal access to their specimen collection and library facilities.

Lincoln, Nebraska P.A.J.
June 1996

Part I

COMPARATIVE BIOLOGY

Olive-backed tailorbird host feeding a nestling plaintive cuckoo. After a photo by J. Koolman (in Becking, 1981).

1

AN OVERVIEW OF BROOD PARASITISM

Evolutionary Pathways to Avian Brood Parasitism

Laying eggs in the nest of another individual, and allowing or tricking the nest owner to rear such "parasitic" young rather than, or in addition to, its own, is one of the rarest forms of reproduction known. Except for the social insects, in which intraspecific brood parasitism is sometimes well developed (Wilson, 1971), such social parasitism is almost unknown in animal groups other than birds. Brood parasitism occurring within members of the same species (intraspecific or conspecific brood parasitism) has been reported for fewer than 100 species of birds (Yom Tov, 1980), and possibly occurs among other vertebrates only in a few fish. Likewise, fewer than 100 species of birds are known to be obligatory brood parasites (Payne, 1977b). Sporadically interspecific egg dumping, or "facultative" brood parasitism, occurs among even fewer species, based on the available evidence. The rarity of brood parasitism as a reproductive strategy is rather surprising, considering that exploiting the energy of another species through true external or internal parasitism is extremely common among animals. Indeed, some phyla of animals are predominantly or even exclusively parasitic and, at least in terms of actual numbers of individuals, there are probably far more parasitic than free-living animals alive in the world.

Vertebrates and other chordates differ from many invertebrate groups in that there are no true internal parasites represented in this phylum, and only a few external parasites (e.g., the lampreys) exist. To be sure, there is a substantial number of vertebrate scavengers, commensals, and other forms of social exploitation, but few brood parasites. The explanation for this lies in the fact that among most fish, amphibians, and reptiles, little or no parental care for the eggs and young occurs, and so no benefits would likely accrue by the surreptitious introduction of eggs into another's nest. Among placental mammals, the fertilization of eggs and the most vulnerable stages of development occur within the female's body. Thus, no opportunities for interspecific social parasitism exist in mammals, although fertilization by a conspecific male other than a "mate" is possible. However, few mammals maintain definite, extended pair-bonds, so even this potential for intraspecific sexual exploitation is limited.

It is only among birds, whose fertilized eggs are tended in a variably exposed nest throughout the entire developmental period, and in which intense parental care is often extended well beyond hatching, that almost unlimited opportunities exist for various kinds of reproductive exploitation, both within and between species. These opportunities may include, for example, the takeover of a nest that has been built by other birds of the same or a different species ("nest parasitism"), either by forceful eviction (nest supplanting) or by simple, uncontested replacement such as by an occupation of completed but unoccupied nests (nest takeovers).

Intraspecifically, reproductive exploitation sometimes occurs in the form of fertilization of a female by a male other than her pair-bonded mate, either through noncoercive or forced "extra-pair copulations" of already mated females (McKinney, et al. 1983; McKinney, 1985). Such copulation-stealing behavior, at least in humans, is often called cuckoldry. This emotionally laden term has only limited scientific usefulness: copulation-stealing among birds and other nonhumans is probably better referred to as kleptogamy (Gowaty, 1984, 1985). Indeed, the term *cuckoldry* is derived from the Norman French *cucuald,* and refers to the common cuckoo's behavior of insinuating its eggs into the nests of other species. It is that aspect of sexual exploitation, interspecific brood parasitism, that is the subject of this book, rather than intraspecific sexual promiscuity, even though somewhat similar reproductive benefits may be derived from both behaviors. However, in the case of kleptogamy, only the promiscuous males obtain obvious benefits (females may also potentially benefit if the genes they receive from the successfully inseminating male result in better overall selective advantage than those available from their mate), whereas the care-giving males incur corresponding reproductive costs.

The occasional or "chance" dropping of fertile eggs in another bird's nest by a female is sometimes called "egg dumping" or "dump nesting"; the latter term is especially used to describe nests with large numbers of eggs that are usually laid by several females but sometimes incubated by none. In such cases of occasional fortuitous egg-dumping behavior or more frequent facultative brood parasitism, as well as in cases of more "purposeful" or obligatory brood parasitism, the care-giving birds of both sexes ("hosts") are victimized and are presumed to be evolutionary losers (Gowaty, 1985). In contrast, both sexes of a brood parasite can benefit to the degree that their reproductive potential might be variably enhanced, either by spreading their risks of egg or chick predation over a larger number of nests than they could protect individually, or by generating and disseminating a larger number of potential offspring than they could care for and rear by themselves.

A kind of intermediate situation occurs among species that engage in "cooperative" or communal breeding, in which one or more nonbreeding "helpers" participate to varying degrees in chick rearing and sometimes also help in incubation or nest protection activities with breeding birds of their own species (Brown, 1978). Such helpers at the nest (Skutch, 1987) gain no immediate benefit from their participation, but often are close relatives of the breeders and thus may benefit indirectly through kin selection. In addition, the active participation of a young male in assisting an established breeding pair may increase the chances of the helper male eventually acquiring an adjoining breeding territory as he matures.

A special case of communal nesting occurs in the groove-billed and smooth-billed anis, in which up to four monogamous pairs share a common nest. All the females deposit their eggs in this common nest, producing clutches of as many as 20 (groove-billed) to 29 (smooth-billed) eggs, although no more than 13 eggs have been known to hatch from any single nest (Skutch, 1987). Some of these excess eggs are buried under leaves during incubation, and others may fall from or even be thrown out of the nest by participating females. In one study, the oldest and dominant female was the last to lay, and she ejected all the eggs present in the nest before she began to lay her own clutch. Some of these lost eggs of other females were replaced, but few if any of the nondominant females contributed as many eggs to the final clutch as did the dominant female. However, incubation and other posthatching parental duties are subsequently shared by all the participants, and the removal of some of the early-laid eggs may help assure that most of the nestlings will hatch at about the same time and thus more of the chicks will have a greater probability of surviving to fledging (Verhrencamp, 1976, 1977).

Clearly, this "selfish" egg-removal behavior among otherwise cooperative breeders suggests similarities with the egg-removal behavior of many obligate brood parasites and suggests that intermediate conditions exist between cooperative and exploitive breeding interactions and suggest a possible evolutionary route leading toward intraspecific brood parasitism. As a result, it is desirable to distinguish here between such communal but not necessarily entirely cooperative nesting behavior and more clearly functionally cooperative types of nesting interactions. An outline of these various "nonparasitic" interactions as well as various categories of actual brood parasitism is presented in table 1.

Brood parasitism among birds has been the subject of a vast ornithological literature, but until now only Makatsch (1955) has attempted to survey this subject from a worldwide perspective. Other authors (e.g., Chance, 1922; Friedmann, 1929, 1955, 1960; Baker, 1942; Wyllie, 1981) have described various species, genera, or even families of brood parasites, and Payne (1973b, 1977b, 1982) has contributed greatly to an overall understanding of the ecology and evolution of brood parasitism among birds.

Evolutionary Aspects of Avian Brood Parasitism

The evolutionary origins of avian brood parasitism provide one of the most interesting unsolved questions in contemporary ornithology. After reviewing some historical and obviously sometimes far-fetched notions on the origins of brood parasitism in the Old World cuckoos, Friedmann (1929) advanced three hypotheses. The first of these might be identified as the reproductive asynchrony model. In this model, originally advanced by Herricks (1910) for the common cuckoo, an asynchrony inexplicably develops between the egg-laying and nest-building phases of the breeding cycle, leading to deposition of eggs before the nest is com-

TABLE 1 Types of Cooperative and Exploitive Reproductive Interactions Occurring among Birds

A. Conspecific Exploitive/Cooperative Interactions
  1. Stealing of food and/or nest materials ("kleptoparasitism")
  2. Copulation stealing: Includes all extra-pair copulations (EPCs), including forced copulations ("kleptogamy")
  3. Nest renovation and nest supplanting ("nest parasitism")
    a. Breeders claim inactive nests (nest renovation)
    b. Breeders expel other conspecifics from active nests and use them for breeding (true nest parasitism).
  4. Brood-sharing and intraspecific brood parasitism
    a. Dump-nesting (egg-dumping): Not all egg layers can, or sometimes even attempt, to participate in parental behavior. Reduced hatching success is common, often owing to nest abandonment, conflitcts over incubation participation, or excessive clutch sizes. Benefits to participants may vary greatly, depending on the hatching success of their own eggs relative to their individual degree of parental involvement.
    b. Cooperative nesting: Egg-layers as well as other auxillary nonbreeders ("helpers") all participate in providing care to the eggs or young. Presumably all participatns eventually benefit, either through individual selection (e.g., improved chances of becoming breeders later) or by kin selection (improved survival of near relatives).
    c. Communal nesting: Two or more females lay in the same nest, and all remain to participate in parental care giving. However, some cheating may occur, through removal or burial of other participating females' eggs, and thus differential benefits may accrue to each of the individually involved females.
    d. Conspecific brood parasitism: Parasitic females do not participate in care giving after egg laying but may benefit regardless, usually at host's expense.

B. Interspecific Exploitive/Cooperative Interactions
  1. Interspecific piracy: Food stealing or prey stealing from other species.
  2. Interspecific nest parasitism: As in intraspecific nest parasitism, but exploiting other species' nests.
  3. Fortuitous or facultative brood parasitism: In addition to normally caring for their own eggs and offspring, females may also deposit eggs in other species' active nests.
  4. Obligate brood parasitism: Females regularly deposit eggs in other species' nests, but do not perform any nest building, incubation or parental behavior.
    a. Nonexploitive brood parasitism: Host species does not suffer from the parasitic interactions (host breeding success maintained; possibly rarely improved).
      (i) Parasitic young do not strongly compete with host young and may improve brooding/foraging efficiency of host parents.
      (ii) Parasitic young may improve survival rate of host young by elimination of ectoparasites.
    b. Exploitive brood parasitism: Parasite benefits but host suffers from their breeding interactions (host breeding success variably reduced).
      (i) Host-tolerant parasites: Those whose young are reared in the nest with host chicks and compete with them for attention; some host offspring might also fledge, depending on competition levels.
        Host-generalist parasites: Those parasites that exploit many host species, with little or no evolved host mimicry.
        Host-specific parasites: Those parasites that exploit a single host species, usually with evolved host mimicry at the species level.
      (ii) Host-intolerant parasites: Those species whose chicks expel host eggs or kill their young soon after hatching; thus no host young normally survive.
        Host-generalist parasites. As described above.
        Host-specific parasites. As above, but individual females may collectively make up subpopulations (gentes) having individual host specificity. Host mimicry may occur within gentes, although the parasitic species often collectively exploits many host species.

pleted. A similar hypothesis might be called the visual stimulus model, in which the female parasite is stimulated to lay eggs by the sight of a nest containing eggs similar to its own. This idea had been advanced by Chance (1922) for the common cuckoo, and Friedmann suggested that although it seemed to fit some of the parasitic cowbirds, it could not serve as a general model unless a corresponding reduction in attachment of the female to its own nest occurred concurrently. A third, related explanation is that some species had developed habits of breeding in abandoned nests and sometimes failed to discriminate between such nests and newly completed, active ones. Friedmann doubted the likelihood of this scenario, suggesting that because of the time lag between discovery of a nest and the laying of an egg in the nest, the female would likely discover that the nest was already occupied and would probably abandon efforts to use it.

Friedmann did agree that at least the parasitic cowbirds, and presumably also other brood parasites, evolved from species that originally nested in a normal fashion of biparental or uniparental care and that their brood parasitism was secondarily acquired. Within the cowbirds, he noted that all the relatives of the parasitic species are nest builders, and that the seemingly most behaviorally "primitive" cowbird species, the bay-winged cowbird, is nonparasitic, but uses the abandoned nests of other birds for breeding more often than it builds its own nest. Friedmann also noted that the shiny cowbird rarely attempt to build a nest, but is never successful.

The monogamous and territorial tendencies of some cowbirds, as seen in the bay-winged cowbird, compared to the weak development of these traits in most of the obligatorily parasitic cowbirds, support the predication that a breakdown in territoriality and monogamous pair-bonding should be expected corollaries of evolving brood parasitism. In Friedmann's hypothetical scenario, the screaming cowbird, with its relative host specificity, represents an advanced stage in brood parasitism, whereas the shiny cowbird, with a broad host diversity, represents an earlier stage. This progression seems to make logical sense, although a recent biochemical study (Lanyon, 1992) suggests a reverse behavioral sequence, with host specificity representing the primitive parasitic state and host diversity representing the advanced condition.

Friedmann (1968) suggested that brood parasitism may be an older behavior in honeyguides than it is in any of the other families of birds exhibiting the trait, based on the observations that all of the species are parasitic, none exhibits strongly developed pair-bonds or apparent territoriality, and newly hatched chicks of at least some species have special structural modifications for killing host nest-mates. Friedmann suggested that brood parasitism evolved independently in each of the groups exhibiting this behavior and hypothesized that brood parasitism is more common in the cuckoos because they may have been able to "let go" their rather weak nest-building tendencies more easily than passerine groups which have strong nest-building instincts and complex nests, such as the weavers and icterines. In addition to the obligate brood parasitic cuckoos, facultative brood parasitism or egg dumping occurs in various nonparasitic cuckoos; Wyllie (1981) lists 11 species that have been reported as recipients of eggs of yellow-billed or black-billed cuckoos in North America.

Davis (1940b) has proposed that brood parasitism in birds arose from nest parasitism, egg parasitism, or both. Nest parasitism (nest rehabilitation or nest takeover) might result in the gradual loss of the nest-building instinct, in a similar manner to that suggested by Friedmann. Jourdain (1925) similarly believed that nest parasitism, which was followed by the dropping of eggs in the occupied nests of other species, represented the first evolutionary step toward brood parasitism. Egg parasitism, involving the initial chance laying of eggs in the nests of other birds (dump nesting), might produce a gradual loss of the species' ability to build a nest, especially as such parasitism became more effective. Davis suggested that this latter route most likely accounted for the unusual breeding behavior of the communally nesting New World ground cuckoos, including the anis; in other words, interspecific egg parasitism may have resulted in communal nesting evolution, although communal nesting as a behavioral stepping stone between normal parental nesting and brood parasitism has also been hypothesized. More recently, a similar argument to that of Davis has been advanced: parasitic egg laying may have been the evolutionary precursor to communal nesting in some ratites (Handford & Mares, 1985). In such colonially nesting species, the "indolent" behavior of the participants might stimulate some females, having lost their ability to build individual nests, to nevertheless deposit their eggs in a communal nest and thus potentially become parasitic. Davis (1940a) suggested that the guira cuckoo, in which most pairs nest separately within colonial territories but with some nest-sharing, is the most generalized pattern of coloniality in this group. In the greater and groove-billed anis, which breed in colonies composed of colonial pairs that lay in a single common nest, a second stage is reached (Davis, 1941). Finally, in the smooth-billed ani, the colony consists of promiscuous breeders that strongly defend their common nest against intruders (Davis, 1942).

A third possible hypothetical route to brood parasitism in Davis's (1940b) view involves a progressive loss of the brooding instinct, perhaps owing to pituitary changes associated with the loss of prolactin production or a possible loss of target sensitivity to prolactin relative to its associated control of broodiness. Höhn (1959) and Dufty et al. (1987) have since determined that reduced levels of prolactin are not typical of the brown-headed cowbird, and thus it is more likely that reduced target-organ sensitivity may be responsible for the absence of brood patches and broody behavior in this species.

The next important review of the evolutionary route to brood parasitism in altricial birds was that of Hamilton and Orians (1965). They reviewed the two existing major hypotheses discussed previously, which they respectively labeled "progressive degeneration of nesting instincts" and the "failure to synchronize nest-building and egg laying" models, and found both to be faulty in various ways. Among these perceived faults is the implicit view that brood parasitism is a linear (orthogenetic) degenerative evolutionary process, rather than the result of positive selection pressures. They also faulted the implied supposition that such a deterioration of nesting or brooding tendencies or asynchronous laying tendencies must have occurred in all of the members of the species simultaneously as the species became parasitic. Hamilton and Orians proposed an alternative hypothesis based on the idea that brood parasitism may instead arise through the potential benefits of genetically controlled tendencies of some species to deposit their eggs in the nests of others, possibly as a proximate result of nest destruction, accidental egg deposition, and inappropriate temporal asynchronies occurring between nest building and egg laying. The chances of the survival of introduced eggs would depend on such factors as incubation periods, nestling food requirements, and nestling begging effectiveness, plus fortuitous similarities to host-species characteristics. If a sufficient percentage of parasitic eggs are successful in producing fledged young, the genes facilitating parasitic tendencies should spread through the parasite's population. In such a case, additional adaptive modifications favoring egg acceptance and nestling survival are likely to occur, such as improved mimicry of eggs or chicks, adjustments to match host incubation periods, and modifications in nestling behavior patterns that might improve their fledging rates.

Payne (1977b) also reviewed the evolution of avian brood parasitism, both intraspecific and interspecific, and pointed out some behavioral similarities between the two types. Females of some species may occasionally lay eggs in the nests of conspecifics ("facultative intraspecific parasitism or dump nesting"), with little or no further efforts toward parental care. As noted earlier, somewhat similar behavior occurs in some communally nesting birds such as anis, in which shared incubation behavior regularly occurs, but in which some cheating behavior (removal of other females' eggs from the nest) might also occur before the onset of incubation.

Additionally, Payne (1977b) noted that some species of birds renovate abandoned nests of other species and lay eggs in them, but they also remain at the nests to hatch and rear their own young. The active takeover of still-occupied and possibly defended nests of another species for the purpose of egg deposition provides a more direct potential precursor to brood parasitism, especially if the original owners are still so attached to their nest as to remain and accept any alien eggs. Obviously, to the extent that such parasitically deposited eggs result in fledged young, a single egg-dumping female can supplement the offspring she is able to rear by herself with those reared by others. If more offspring survive as a result of her incipient or facultative parasitic behavior than through her own incubation and rearing efforts, obligatory rather than facultative brood parasitism is likely to evolve. Payne further calculated that at least one egg of any three-egg clutch has a greater statistical chance of surviving predation if each of the eggs is laid in a different nest than if all are laid in the same nest, which provides an additional potential selective advantage that might facilitate a previously nonparasitic species to become a brood parasite. However, the selection pressures from this source are probably quite weak in most cases (Petrie & Möller, 1991).

Dump-nesting and Intraspecific Brood Parasitism in Birds

Because dump nesting or occasional facultative intraspecific brood parasitism seems to be a potential route of interspecific avian brood parasitism, a review of some of the available information on its occurrence and effectiveness as a reproductive strategy is warranted.

Although largely ignored in the past, the occurrence and possible biological significance of intraspecific brood parasitism has been recently discussed and reviewed by several authors (Yom-Tov, 1980; Andersson, 1984; Petrie & Möller, 1991; Yamauchi, 1995). The related phenomenon of dump nesting, or "prehatch brood amalgamation," has been discussed by Eadie et al. (1988), at least with regard to its occurrence in waterfowl, the group in which this behavior is best documented. Yom-Tov (1980) documented 53 species as engaging in intraspecific facultative parasitism, including 32 anseriforms, 6 passerines, 5 galliforms, 4 columbiforms, and a few representatives of four additional orders. Rohwer and Freeman (1989) increased this list to include 61 precocial species (out of 103 total North American and Western Palearctic birds) and 28 (of 825) altricial species.

Clearly, precocial species, especially waterfowl, are the primary players in this activity (see tables 2 and 3), and Yom-Tov judged that the parasitic females are likely to be young and unmated birds, those that have lost their nests, or already mated females that may lay eggs in

TABLE 2 Estimated Conspecific Clutch Parasitism/Dump-Nesting Rates among Waterfowl[a]

Species	Number of citations[b]	Parasitism frequency (%)
Cavity nesting (17% of all Anatidae)		
North American wood duck	8+	23–95
Common goldeneye	4+	34–38
Common shelduck	3+	27–33
Black-bellied whistling duck	2	to 84
Hooded merganser	2	21–36
Bufflehead	2	5–8
Australian shelduck	2	—
Comb duck	2	—
Cotton pygmy goose	2	—
Gray teal	2	—
Chestnut teal	2	—
Common merganser	2	—
Australian maned duck	1(a)	31
Barrow's goldeneye	1	13
Lesser whistling duck	1	—
New Zealand shelduck	1	—
Ruddy shelduck	1	—
Egyptian goose	1	—
Muscovy duck	1	—
Smew	1	—
Citations subtotal	40+	(approximately 35% of total)
Marsh Nesting (16% of all Anatidae)		
Ruddy duck	4+	19–38
Redhead	4+	17–36
Canvasback	4	to 36(b)
Red-crested pochard	2	17–48
Lesser scaup	2	8–12
Maccoa duck	2	to 14
Common pochard	2	—
Musk duck	2	—
Masked duck	1	38
Pink-eared duck	1	—
Southern pochard	1	—
Ferruginous pochard	1	—
Australasian white-eye	1	—
White-headed duck	1	—
Argentine blue-billed duck	1	—
Australian blue-billed duck	1	—
Fulvous whistling duck	1(b)	—
Citations subtotal	30+	(approximately 25% of total)

TABLE 2 *(continued)*

Species	Number of citations[b]	Parasitism frequency (%)
Ground Nesting (67% of all Anatidae)		
Island-nesting conditions		
Gadwall	5	1–9
Tufted duck	4	10–20
Greater scaup	3	8–11
Mallard	3	1–11
Lesser scaup	2	8–12
Oldsquaw	2	1–3
Red-breasted merganser	1(d)	64
Eurasian wigeon	1	1
Colonial-nesting conditions		
Snow goose	3	0–80
Common eider	3	—
Emperor goose	1	6
Bar-headed goose	1(e)	75
Ross' goose	1(b)	—
Brant	1(b)	—
Noninsular and noncolonial conditions		
Canada goose	4+	1
Red-breasted merganser	3	to 5
Graylag goose	2	3–4
Spur-winged goose	2	—
North American black duck	2	—
Pacific black duck	1	—
Garganey	1	—
Northern shoveler	1	—
Marbled teal	1	—
Black scoter	1	—
King eider	1	—
Harlequin duck	1(b)	—
Citations subtotal	50+	(approximatedly 40% of total)

[a]Waterfowl species list and citation totals as per Rohwer & Freeman (1989) except as otherwise indicated. "Parasitism frequency" refers to percentage of available nests parasitized, not the incidence of parasitically laid eggs. Some species listed as preferential marsh nesters or cavity nesters may also nest on the ground. Likewise, inclusion of the gray and chestnut teal as cavity-nesting species is based on their frequent use of nest-boxes, not natural cavities, in captive breeding facilities. Species are listed by diminishing numbers of citations.
[b]Letters in parentheses indicate citations from sources as follows: (a) Briggs, 1991; (b) Eadie et al., 1988; (c) Sorenson, 1991; (d) Young & Titman, 1988; (e) Weigmann & Lamprecht, 1991.

the nests of others as well as in their own. The lower hatching success of parasitic eggs tends to limit dump nesting to precocial species that have relatively large clutches and to areas where breeding seasons tend to be prolonged, according to Yom-Tov. Yom-Tov, and later Andersson (1984), additionally noted that parasitism is favored under conditions where there is distinct competition for limited nest sites, such as among cavity nesters, or on small islands where breeding populations may be dense but only a few suitable nesting opportunities exist. The

TABLE 3 Estimated Conspecific Clutch Parasitism Rates among Non-waterfowl[a]

Species	Number of nests	Parasitism rate (%)	Reference
Precocial			
American coot (D,A)	417	41.2	Lyon, 1993
Altrical			
European starling (D,H)	180	36.7	Andersson, 1984
European starling (D,H)	241	3–40	Romagnano et al., 1990
Northern masked weaver (C,S)	645	23–35	Jackson, 1992
White-fronted bee-eater (C,H)	164	16	Emlen & Wrege, 1986
South African cliff swallow (C,S)	117	16	Earle, 1986
Eastern bluebird (D,H)	—	15+	Gowaty & Karlin, 1984
Cliff swallow (C,S)	4942	9.9	Brown & Brown, 1989
Tree swallow (D,C)	120	9.2	Lombardo, 1988
House sparrow (D,S)	94	8.5	Kendra et al., 1988
Black-throated weaver (C,S)	154	5.2	Dhindsa, 1983a
Streaked weaver (C,S)	171	2.9	Dhindsa, 1983a
Brewer's blackbird (C,O)	162	3.1	Harmes et al., 1991
Yellow-headed blackbird (C,O)	1227	1.1	Harmes et al., 1991
Yellow warbler (D,O)	1500+	0.7	Sealy et al., 1989
Red-winged blackbird (C,O)	7805	0.4	Harmes et al., 1991

[a]Parasitism rate estimates refer to percentage of affected nests or females, not percentage of eggs laid parasitically. Species are listed by descending estimated parasitism rates. Letters in parentheses refer to colonial (c) versus dispersed (D) nesters, followed by nest types: aquatic (A), cavities or holes (H), open (O), and suspended or spherical nests with lateral or ventral openings (S).

relative ease of finding conspecifics' nests and a lack of territorial defense of the nest site by its owner can also affect the rate of parasitism. However, the genetic relatedness of the females, which through kin selection might theoretically influence the evolution of parasitic tendencies, is unlikely to have any measureable effect on parasitism, judging from the presently available information (Rohwer & Freeman, 1989).

Eadie et al. (1989) concluded that the incidence of intraspecific nest parasitism among waterfowl is positively correlated with low resource availability, specifically nest sites. They also established a weak correlation with general life-history traits; r-type waterfowl species (those generally smaller species that mature early, have large clutch sizes and masses, have short pair-bonds, and demonstrate uniparental care) have higher parasitism rates than do K-type species (those having opposite traits from r-types, such as sea ducks, geese, and swans). Yamauchi (1993) has recently argued that, theoretically, parasitism rates should increase as competition intensities increase among siblings, but this argument, although intuitively convincing, runs counter to the well-documented observation that the rates of intraspecific parasitism are highest among waterfowl and other precocial species, among which food competition among offspring is much lower than it is among altricial species.

A summary of available information on parasitism rates among waterfowl (table 2) indicates that cavity nesters, marsh nesters in emergent vegetation, and island- or colonial-breeding-ground nesters are most frequently involved in parasitism. Cavity-nesting waterfowl engage in dump nesting or brood parasitism at a rate about twice that expected, based on the number of cavity-nesting versus non-cavity-nesting species. Ground-nesting species are involved at a lower than expected rate, with most of the known cases involving either colonial-nesting or island-nesting situations, where crowding and competition for suitable nest sites is likely.

Eadie (1991) has recently suggested that facultative brood parasitism is a primitive trait among North American waterfowl, with the parasitic species tending to have larger clutch sizes, longer egg-laying periods, and longer incubation periods than nonparasites. He also noted that all 7 species of North American cavity-nesting ducks are parasitic, as are 4 of 6 emergent-vegetation nesters, and 8 of 20 upland nesters (mainly island and colonial nesters). Brood parasitism in these species is also positively related to the degree of female philopatry; 10 of 12 species that exhibit strong tendencies to return to natal areas for breeding are frequent brood parasites. Eadie concluded that environmental constraints on nesting (such as nest site limitations) represent one major factor promoting brood parasitism in waterfowl, whereas opportunities for parasitism (potential hosts being readily available and easily located) represent a second major factor.

Among non-waterfowl (table 3), colonial- and hole-nesting species tend to exhibit high rates of intraspecific parasitism, whereas parasitism rates are low in dispersed, territorial species with generalized nest site requirements and open nests, such as the yellow warbler. Of the 14 altricial species listed in table 3, 9 are colonial, 3 are hole or cavity nesters, and only the American coot, yellow warbler, and perhaps the house sparrow lack one or both of these attributes. Estimated rates of intraspecific parasitism (table 3) are subject to much greater risks of error than are those of interspecific parasitism, for obvious reasons associated with the difficulties of identifying parental involvement, and these estimated rates probably should be regarded as minimal in most cases.

Information on the relative hatching and fledging success of eggs associated with egg dumping and/or intraspecific parasitism, as compared to that associated with normal parental (single female or single pair) nesting situations, is summarized in table 4. So far as possible, examples of truly parasitic egg deposition (insertion of eggs into active nests belonging to and actively tended by a single breeding pair or female) have been distinguished in this table from simple dump nesting (the often seemingly random deposition of eggs into a common nest by several females). In every case, the hatching and/or fledging success of parasitic or dumped eggs averages lower than that of eggs in clutches unaffected by intraspecific parasitism or egg dumping, sometimes by a factor as great as six- or sevenfold. In some cases, as in the North American wood duck, the differences in hatching success are not nearly so great. In a few cases the substantially larger clutch sizes typical of dump nests relative to those of normal nests may compensate for the reduced hatching success of the eggs, and may actually result in a greater average number of young hatched per nest (e.g., Grice & Rogers, 1965). Quite possibly, cavity-nesting species such as North American wood ducks can effectively incubate much larger numbers of eggs than their normal clutch sizes without greatly increasing the risk of losing excess eggs due to rolling out of the nest, compared to species that build shallow, cuplike nests. Indeed, cavity-nesting ducks such as the perching ducks and hole-nesting sea ducks tend to have somewhat larger average clutch sizes than do surface nesters such as most *Anas* species, pochards, and ground-nesting sea ducks, perhaps as a reflection of their generally greater effectiveness in incubating large clutches.

TABLE 4 Success of Eggs in Intraspecifically Parasitized Nests and Dump Nests

	Eggs/nests	% Hatched	% Fledged	Reference
Bar-headed goose				
Nonparasitic eggs	84	45	—	Weigmann & Lamprecht, 1991
Parasitic eggs	98	5–6	—	
North American wood duck				
Normal nests	204	67	—	Semel et al., 1988
Parasitized nests	38	53	—	
Eggs in normal nests	4505	47	—	Clawson et al., 1979
Eggs in dump nests	10,620	39.5	—	
Eggs in normal nests	17,700+	71.8[a]	—	7 studies cited in
Eggs in dump nests	37,100+	59.8[a]	—	Semel et al., 1988
Eggs in normal nests	159	87 (8.6/nest)		Grice & Rogers, 1965
Eggs in dump nests	290	81 (14.7/nest)		
Eggs in normal nests	2207	71	—	Richardson & Knapton, 1994
Eggs in dump nests	334	71	—	
Canvasback				
Nonparasitic eggs	752	79	—	Sorenson, 1993
Parasitic eggs	82	29	—	
Redhead				
Nonparasitic eggs	241	79–91	—	Sorenson, 1991
Parasitic eggs	291	35–46	—	
Normal nests	115	18	—	Lokemoen, 1966
Dump nests	18	5.5	—	
Common moorhen				
Nonparasitic eggs	783	44	22	Gibbons, 1986
Parasitic eggs	74	20	23	
American Coot				
Nonparasitic eggs	1701	—	31	Lyon, 1993
Parasitic eggs	128	—	3.6	

[a]Weighted collective means, based on 12,745 hatched eggs in successful normal (single-female) nests and 22,218 hatched eggs in successful dump nests. Sample sizes of other cited studies may refer either to eggs or nests, as indicated.

Eco-geographic Aspects of Brood Parasitism

As others such as Friedmann (1929) have emphasized repeatedly, it seems likely that avian brood parasitism in not a unitary phenomenon. It probably arose independently at various times, in various parts of the world, and in diverse groups of birds. Yet, brood parasitism has been a successful breeding strategy for some groups, and there are now few parts of the world (the Arctic and parts of the Subarctic) that are entirely free of brood parasites (fig. 1).

The current distribution of brood parasites can provide little information on previous dis-

→

FIGURE 1. Species-density map of parasitic cowbirds (CB) and ground-cuckoos (GC) in the New World and of parasitic cuckoos (C), viduine finches (F), and honeyguides (H) in the Old World, by 15° latitude units. Numbers within latilongs represent minimum number of species breeding within each latilong's limits. Approximate minimum tital numbers of brood parasites, by latinudinal zones, are also indicated along map edges for each hemispere.

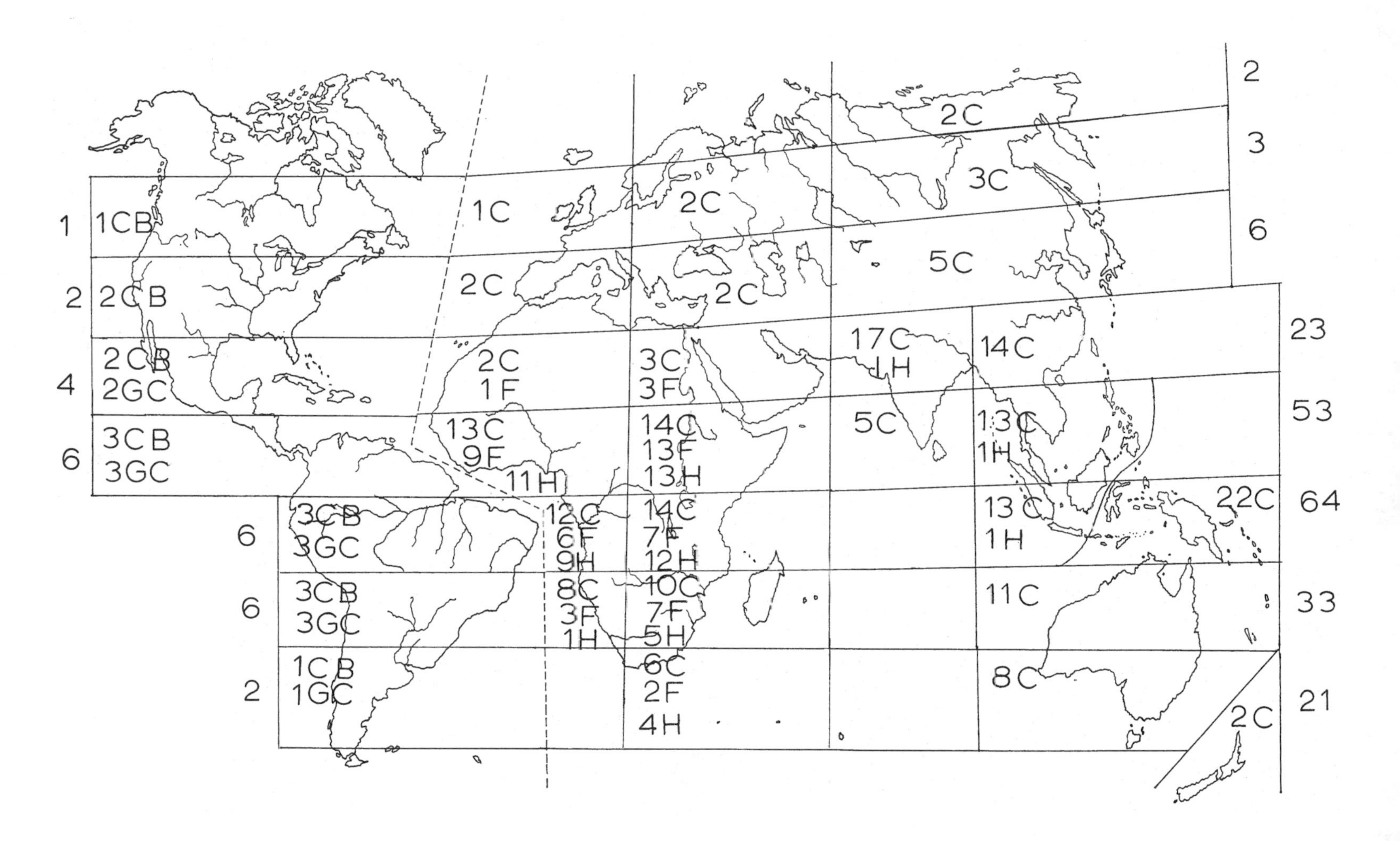
1
1CB
1C
2C
2C
3C
2
6
2
2CB
2C
5C
2C
4
2CB
2GC
2C
1F
3C
3F
17C
1H
14C
23
6
3CB
3GC
13C
9F
11H
14C
13F
13H
5C
13C
1H
53
6
3CB
3GC
12C
6F
9H
14C
7F
12H
13C
1H
22C
64
6
3CB
3GC
8C
3F
1H
10C
7F
5H
11C
33
2
1CB
1GC
6C
2F
4H
8C
2C
21

tribution or the evolution of brood parasitism, but some patterns are apparent. First, there is a gradual increase in the numbers of brood parasites from temperate to equatorial areas, which parallels the general trend of increasing species diversity among birds as a whole. Thus, it cannot be necessarily concluded that brood parasitism is more common in tropical areas, even though breeding seasons are more prolonged there, and therefore one additional predisposing factor favoring the evolution of brood parasitism is present.

As for continental variations in brood parasitism patterns, only in Africa are three major groups of obligate brood parasites present. In some broad subdivisions of that continent (e.g., East Africa's upper Nile and Rift valleys) as many as 40 brood parasites may occur, although the number present in any single locality or habitat would be far smaller. Another focus of brood parasitism occurs in the Indo-Australian area, especially east of Wallace's Line from the Sulawesi to New Guinea, where only a single subfamily (Cuculinae) of brood parasites occurs. In this latter region, more than 20 breeding species and 8 of the 12 cuculine genera (including 3 of the 4 monotypic genera) are found. This general region (Malaysia, Indonesia, and New Guinea) may represent the ancestral home of the cuckoo subfamily Cuculinae; Friedmann (1968) suggested that the genus *Chrysococcyx* originated in the vicinity of southeastern Asia (i.e., in the Indo-Malayan region), from which it spread southeastward to New Guinea, Australia, New Zealand, and their associated Pacific islands, as well as westward to Myanmar, Assam, India, and eventually to Africa. Zoogeographic evidence from other cuculine genera, as well as the high level of endemicity of monotypic genera in the New Guinea region, supports a general pattern of evolutionary radiation of the Cuculinae from southeastern Asia or the East Indies. The seemingly most primitive genera of parasitic cuckoos, *Clamator* and *Oxylophus,* are broadly dispersed throughout both Asia and Africa. Friedmann (1964) was unable to determine whether Africa or Asia represented the most likely source of this stock, but he believed that the pied cuckoo was the most primitive of the crested cuckoos and that it probably spread from Africa to Asia following the separation and drifting of Madagascar from the rest of Africa.

Using this same line of argument, the honeyguides presumably evolved in Africa, although Friedmann (1968) accepted the possibility that they may have spread from an original Asian homeland to Africa, thus accounting for the two surviving species in southern and southeastern Asia. The barbets, perhaps the honeyguide's nearest living relatives and probably their commonest host group, also have their center of geographic distribution in Africa, with that continent supporting at least half of the approximately 80 total barbet species, and 7 of the 13 barbet genera.

Friedmann (1960) agreed with the conclusions of earlier workers, such as Chapin (1931), regarding the zoogeographic origins of the African viduine finches. Chapin had hypothesized an African origin, not only for the African parasitic finches, but also for their estrildine relatives and major hosts and the ploceid (passerid) sparrows generally. Unlike the viduines, however, the early estrildines dispersed from Africa and colonized India, southeast Asia, and even Australia. Friedmann similarly imagined the viduine parasitic finches as representing an offshoot from proto-estrildine stock, and thought both groups had their phyletic roots in even older and more generalized ploceid stock, presumably somewhere in southern Africa.

Geographic Aspects of Host Diversities and Parasitism Pressures

One of the elemental aspects of evolved brood parasitism is the question of the desirability of adaptively parasitizing one or a few species and doing so very efficiently (e.g., narrowly tar-

TABLE 5 Reported Variations in Host Diversities among Brood Parasites[a]

Species	Host Species[b] Major	Minor	Total	References
Waterfowl			Hosts	
Mallard (facultative)	—	—	5	Weller, 1959
Redhead (facultative)	—	—	10	Weller, 1959
Ruddy duck (facultative)	—	—	5	Weller, 1959
Black-headed duck	2	10	12	Weller, 1968
Honeyguides				
Scaly-throated honeyguide	8	4	12	Fry et al., 1988
Greater honeyguide	39	10	49	Fry et al., 1988
Lesser honeyguide	19	11	30	Fry et al., 1988
Thick-billed honeyguide	1	3	4	Fry et al., 1988
Pallid honeyguide	1	2	3	Fry et al., 1988
Cassin's honeyguide	3	1	4	Friedmann, 1955
Green-backed honeyguide	8	2	9	Fry et al., 1988
Wahlberg's honeyguide	3	2	5	Fry et al., 1988
Old World Cuckoos				
Pied cuckoo				
Asia *(O. j pica)*	—	—	36	Friedmann, 1964
Asia *(O. j jacobinus)*	—	—	10	Friedmann, 1964
Africa	4	12+	16+	Fry et al., 1988
Levaillant's cuckoo				
Africa	1	8	9	Fry et al., 1988
Chestnut-winged cuckoo	—	—	23	Friedmann, 1964
Great spotted cuckoo				
Africa	3	16	19	Fry et al., 1988
Europe	1	5	6	Cramp, 1985
Total range	—	—	21	Friedmann, 1964
Thick-billed cuckoo	1	2	3	Fry et al., 1988
Large hawk cuckoo	1	27	28	Baker, 1942
Hodgson's hawk cuckoo				
Japan	4	6	10	Brazil, 1991
India	1	19	20	Baker, 1942
Black cuckoo	3	16	19	Friedmann, 1964, Fry et al., 1988
Indian cuckoo				
India	1	10	11	Baker, 1942
Common cuckoo				
Europe	11	100+	100+	Wylie, 1981, Makatsch, 1976
Japan	16	12	28	Nakamura, 1990
Africa	1	3	4	Fry et al., 1988
African cuckoo	2	—	2	Fry et al., 1988
Oriental cuckoo				
Himalayas	1	15	16	Baker, 1942
Japan	1	13	14	Royama, 1963
Lesser cuckoo				
Asia	1	21	22	Baker, 1942
Japan	3	5	8	Brazil, 1991
Madagascan lesser cuckoo	4	—	4	Fry et al., 1988

(continued)

TABLE 5 *(continued)*

Species	Host Species[b] Major	Minor	Total	References
Pallid cuckoo	32	79	111	Brooker & Brooker, 1989
Brush cuckoo	10	48	58	Brooker & Brooker, 1989b
Fan-tailed cuckoo	17	64	81	Brooker & Brooker, 1989
Plaintive cuckoo				
Burma	1	13	14	Baker, 1942
Black-eared cuckoo	2	20	22	Brooker & Brooker, 1989b
Gould's bronze-cuckoo	—	—	~10	Frith, 1977
Gould's + little bronze cuckoos (Australia)	4	19	23	Brooker & Brooker, 1989b
Little bronze cuckoo				
Asia	1	13	14	Friedmann, 1968
Shining bronze cuckoo				
Australia, New Zealand	2	82	84	Friedmann, 1968
Australia *(plagosis)*	10	79	89	Brooker & Brooker, 1989b
Horsfield's bronze-cuckoo	28	79	97	Brooker & Brooker, 1989b
Asian emerald cuckoo	1	11	12	Friedmann, 1968
Violet cuckoo	1	9	10	Friedmann, 1968
Black-eared cuckoo	1	10	11	Friedmann, 1968
Klass's cuckoo	3	36	39	Fry et al., 1988
African emerald cuckoo	1	17	18	Fry et al., 1988
Dideric cuckoo	2	42	44	Fry et al., 1988
Drongo cuckoo[c]	1	19	20	Baker, 1942
Asian koel	1	12	14	Baker, 1942
Australian koel	6	15	21	Brooker & Brooker, 1989b
Channel-billed cuckoo	5	4	9	Brooker & Brooker, 1989b
Ground Cuckoos				
Striped cuckoo	—	—	20+	(see table 30)
Pheasant cuckoo	—	—	3+	(see table 30)
Pavonine cuckoo	—	—	4+	(see table 30)
Finches				
Village indigobird	1	—	1	Payne, 1973a, 1976b
Jambandu indigobird	1	—	1	(see species account)
Baka indigobird	1	—	1	Payne, 1973a, 1976b
Variable indigobird[d]	2 (?)	—	1	(see species account)
Dusky indigobird	1	—	1	Payne, 1973a, 1976b
Pale-winged indigobird[e]	4 (?)	—	2	(see species account)
Steel-blue whydah	2	2 (?)	3 (?)	Nicolai, 1989
Straw-tailed whydah	1	—	1	Hall & Moreau, 1970
Queen whydah	1	2 (?)	3 (?)	(see table 31)
Pin-tailed whydah	1 (?)	12 (?)	13 (?)	(see table 31)
Northern paradise whydah	1	—	1	Hall & Moreau, 1970
Togo paradise whydah	1	—	1	Hall & Moreau, 1970
Long-tailed paradise whydah	1	—	1	Hall & Moreau, 1970
Eastern paradise whydah	1	—	1	Hall & Moreau, 1970
Broad-tailed paradise whydah	1	—	1	Hall & Moreau, 1970
Parasitic weaver	2 (?)	8 (?)	10 (?)	(see table 31)

TABLE 5 *(continued)*

Species	Host Species[b] Major	Minor	Total	References
Cowbirds				
Screaming cowbird	1	1		Friedmann, 1963
Shiny cowbird	201	53		Friedmann & Kiff, 1985
Bronzed cowbird	77	28		Friedmann & Kiff, 1985
Brown-headed cowbird	220	145		Friedmann & Kiff, 1985
Giant cowbird	7	7		Friedmann, 1963

[a]Numbers do not always correspond to those in table 26, which includes some questionable hosts.
[b]For honeyguides, known hosts and likely hosts are presented and for cowbirds, know hosts and known fosterers are presented rather than major and minor hosts, respectively.
[c]Misidentified by Baker (1942) as banded bay cuckoo eggs (Becking, 1981).
[d]Includes the possibly specifically distinct *codringtoni.*
[e]Includes several forms considered by Payne & Payne (1994) as biologically distinct species.

geting hosts, as do nearly all parasitic finches) versus parasitizing the largest possible number of species, without adapting to, and thus becoming potentially dependent on, the presence and continued survival of some specific host or closely related hosts (broadly targeting hosts, as do by most cowbirds). Clearly, some intermediate strategy may also be selected, and probably most often is, such as targeting specific hosts in some places and times, and adopting a more generalized strategy at other times or places, as is true of many Old World cuckoos and perhaps also the New World ground cuckoos and honeyguides.

Table 5 summarizes the major variations in host diversities reported for representatives of all of the groups of obligate brood parasites and includes a few additional examples of facultative brood parasites among the waterfowl family. Surprisingly, few species are "classic" parasites, in the sense that they are highly adapted to a specific host species. Such a situation seems to exist most clearly in the viduine finches, where a group of closely related and often sympatrically distributed parasites may use their host-specific adaptations (e.g., song mimicry) as behavioral isolating mechanisms, in addition to providing their basic reproductive needs. Male plumage among biologically distinct species of these finches differs little or not at all, and genetic barriers preventing interspecific hybridization also appear to be absent. Thus, reproductive isolation in indigobirds is evidently based largely on differential female attraction to call sites of males, and female mate selection (assortative mating) behavior is probably based on female responses to the host-mimetic songs of males (Payne, 1973a:185).

Some "mistakes" might occasionally be made by females of species that are normally highly host specific, such as viduine finches. Overall, the general pattern that emerges is that many minor hosts are used by most species of brood parasites. This behavior might be potentially advantageous to the parasite as a strategy to seek out new and vulnerable hosts that might supplement or eventually replace the current major hosts. In this view, the idea that host selection and exploitation is a relatively dynamic process, constantly under stress from opposing directions by host and parasite (the "co-evolutionary arms race" concept, discussed in chapter 5), is worth remembering.

An alternative method of examining host specificity is provided in table 6, where the com-

TABLE 6 Host Diversities and Parasitism Pressures among African and Australian Cuckoos[a]

	Host diversity		Parasitism pressure	
Species	Africa	Australia	Africa	Australia
1	3	—	68	74
2	1	1	5	10
3	1	—	—	6
4	—	1	—	—
5	—	1	—	—
6–10	2	3	—	—
11–15	1	—	—	—
16–20	1	1	—	—
21–25	1	1	—	—
>25	—	1	—	—
Total cuckoo spp.	10	9	—	—
Total host spp.	—	—	73	90
Mean host diversity	7.8	12.7	—	—
Mean parasitism pressure	—	—	1.1	1.1

[a]Based on summaries of Rowan (1983) for Africa and of Brooker & Brooker (1989a) for Australia. "Host diversity" refers to the number of cuckoo species (in columns below) parasitizing varied numbers of biological host species (at left); "Parasitism pressure" refers to the number of host species (in columns below) being parasitized by varied numbers of cuckoos (at left).

parative host diversity by parasites, as well as the parasitism pressures on hosts (numbers of parasites affecting a given host species) is presented for the parasitic cuckoos and their hosts in southern Africa and Australia. In both regions, a similar number of parasitic species are present (Australia has a total land area about twice that of southern Africa), and similar degrees of mean host diversities and mean parasite pressures are present.

2

ECO-MORPHOLOGY AND INTERSPECIFIC MIMICRY

One can hardly blame those writers, not professional zoologists, who have considered it impossible that such adaptations [of brood parasites] could be the result of natural selection; but Charles Darwin would rightly have called this a difficulty of the imagination not the reason. Indeed, given that the marvellous adaptations of the brood parasites are a product of natural selection, it is perhaps as hard to concede that this same powerful force is likewise responsible for the dull conventional habits of the monogamous songbirds which raise their own young.

David Lack (1968)

General Relationships between Brood Parasites and Their Hosts

Adult Host-to-Parasite Mass Ratios

One of the significant aspects of host:parasite adaptational adjustments is choosing a host species of an appropriate size. On one hand, the choice of hosts is constrained by the possibility of choosing a host unable to incubate the parasite's eggs effectively or, more probably, unable to provide enough food to support the nestling parasite in its later growth stages, when it may weigh several times more than the host adult. On the other hand, parasites should not select hosts so large that they can easily prevent the parasite from invading their nest space, can readily puncture or remove alien eggs from their nest, or that have young large and strong enough to out-compete parasite chicks for food or other needs. Small hosts have the advantages of perhaps being unable to protect their nest and therefore prevent the parasite from laying its eggs in it or being unable to remove or destroy any alien eggs that might be deposited. A host of nearly the same size as the parasite would seem to be the ideal choice, and yet this is not the usual case, at least among many brood parasites. For example, among the host-intolerant Australian cuckoos, the mean host adult mass ranges from 26% to 74% of the adult parasite's mass, averaging about 58% (table 7). A similar trend in relative adult mass relationships seems to be typical of most of the host-intolerant species of brood parasites, including the common cuckoo (see Table 20), African cuckoos other than the host-tolerant great

TABLE 7 Proportionate Egg Masses of Australian Cuckoos and their Major Hosts[a]

	Parasite		Host	
Cuckoo species	Mean adult mass (g)	Mean egg mass (%)	Mean adult mass (%H)	Mean egg mass (%)
Channel-billed cuckoo	611	19.4 (3.2)	~450 (74)	18.0 (4.0)
Australian koel	225	9.8 (4.4)	94.5 (42)	8.2 (8.7)
Pallid cuckoo	83	3.8 (4.6)	29.9 (36)	3.2 (10.7)
Fan-tailed cuckoo	46	2.6 (5.6)	12.1 (26)	2.0 (16.5)
Brush cuckoo	36	1.8 (4.9)	11.9 (33)	1.7 (14.3)
Black-eared cuckoo	29	2.6 (8.8)	12.3 (42)	2.2 (17.9)
Horsfield's bronze cuckoo	23	1.4 (6.2)	9.9 (44)	1.7 (17.2)
Shining bronze cuckoo	23	1.6 (6.8)	7.9 (34)	1.4 (17.7)
Gould's and little bronze-cuckoos	17	1.8 (11.2)	6.9 (41)	1.5 (21.7)
Collective means	121	4.9 (4.1)	70.5 (58)	4.3 (6.1)

[a]Based on Brooker & Brooker (1989b), and host data shown in tables 13 and 14. Egg masses were calculated from estimated egg volumes, using a correction factor (cc to g) of 1.08 (Johnsgard, 1972). Percentage masses of eggs relative to adults are shown in parentheses, as are percentages of adult mean host mass (unweighted) relative to mean parasite mass (%H). Arranged by diminishing adult mean parasite mass.

spotted (see Table 27), and the apparently host-intolerant New World ground cuckoos (see Table 30). However, no such evident mass relationship exists among the honeyguides (see Table 21), which are also apparently host intolerant.

Among at least some of the host-tolerant brood parasites, whose young are raised with host chicks, this larger parasite–smaller host relationship is not consistently maintained. Thus, in at least some of these species, such as the crested cuckoos, koels, and parasitic cowbirds, commonly exploited hosts may sometimes have adult masses that are as great or even greater than those of the parasite (see tables 9 and 32). In such situations, presumably the parasitic species can only consistently compete with the host if it hatches sooner or begs for food more effectively than the host species chicks. In these species the incubation period of the parasite does seem to be consistently shorter than that of the host, although the nestling period may sometimes be as long or longer.

Among the host-tolerant African parasitic finches, all the host species exhibit adult masses that are less than their parasites; the adult host mass averages about 70% of the parasite's mass (table 8). The incubation periods of host and parasite are similar in these host-specific species, and their nestling periods are perhaps slightly longer.

Proportionate Egg-to-Adult Mass Ratios

Lack (1968) was the first to thoroughly investigate the proportionate relationships between adult mass and egg mass in birds. The trend of smaller species having proportionately heavier eggs than larger ones had been previously noted, and Lack attributed this trend to the fact that smaller chicks have proportionately larger surface areas. Smaller chicks thus lose heat more rapidly than larger ones and hence require larger food reserves at hatching. However, Lack also noted that the parasitic (cuculine) cuckoos have the smallest proportionate egg weights of all families and orders of birds, although the eggs of the nonparasitic cuckoos (Centropodinae and Crotophaginae) in comparison are quite large, and those of Crotophaginae represent the largest proportionate egg mass of any terrestrial and nidicolous bird group. No specific explanation was offered for this extreme difference within the cuckoo family, but Lack pointed out that the egg size (and mass) in most parasitic cuckoos is often similar to that of their hosts. In contrast, he observed that the egg of the parasitic black-headed duck is proportionately large, and the newly hatched duckling is sufficiently precocial as to be able to raise itself with little or no help from its foster parents.

Using currently available information on egg mass (mostly after Schönwetter, 1967–1984) and adult mass (mostly after Dunning, 1993), it is possible to extend Lack's analyses to avian brood parasites. With regard to the cuckoos (fig. 2), it is apparent that two groups of nonparasitic cuckoos do indeed have relatively large proportionate egg masses, with the exception of the roadrunners (genus *Geococcyx*), which exhibit smaller egg masses that fall directly within the ranges typical of brood-parasitic cuckoos. The usual clutch size of roadrunners is not especially large (typically averaging four to six eggs), nor are there any other especially remarkable features of its reproductive biology that set it apart from the other nonparasitic New World cuckoos in terms of potential selection for small egg mass. However, some large clutches (up to 12 eggs) have been reported for the greater roadrunner, suggesting that facultative brood parasitism or egg dumping might occur in this species. However, Ohmart (1973) believed that the greater roadrunner's clutch size is directly related to the individual female's available food supplies during the egg-laying period, and suggested that clutches of as many as 12 eggs might be those of a single female.

Among the parasitic cuckoos, the New World parasitic forms fall within the limits of proportional egg mass typical of the general assemblage of Old World cuckoos (especially *Cuculus* and *Cercococcyx*). Large eggs are characteristic of *Clamator, Oxylophus,* and *Eudynamis,* which are genera that, as Lack has already noted, often parasitize large hosts, and their large eggs might be a result of size mimicry. Additionally, these genera are all host-tolerant parasites, and the young must be able to compete effectively in the nest not only with host chicks,

TABLE 8 Proportionate Egg: Adult Masses of the Parasitic Finches and Their Hosts[a]

Brood parasite		Host species		H:P ratio
Eastern paradise whydah	1.63:22.5 (7.6%)	Green-winged pytilia	1.41:13.9 (10.1%)	0.74
Parasitic weaver	1.59:21.4 (7.4%)	Zitting cisticola	1.07:7.3 (14.6%)	0.34
Broad-tailed paradise whydah	1.64:19.5 (8.4%)	Orange-winged pytilia	1.42:14.5 (9.8%)	0.74
Queen paradise whydah	1.36:15.7 (8.6%)	Common grenadier	1.25:10.8 (11.6%)	0.69
Pin-tailed whydah	1.34:14.4 (9.3%)	Common waxbill	0.87:8.2(10.6%)	0.57
Straw-tailed paradise whydah	1.31:13.6 (9.6%)	Purple grenadier	1.27:13.1 (9.7%)	0.96
Variable indigobird	1.33:13.4 (9.9%)	African firefinch	1.16:10.5 (11%)	0.78
Village indigobird	1.27:13.2 (9.6%)	Red-billed firefinch	0.84:8.7 (9.7%)	0.66
Dusky indigobird	1.30:13.2 (9.7%)	Jameson firefinch	1.04:10.1 (10.3%)	0.75
Mean egg: : adult mass ratio	8.9%		10.7%	
Overall host: parasite adult mass ratio				.69

[a]Species are organized by diminishing adult mass of parasites. Figures following species names sequentially represent mean egg mass (grams), mean adult mass, and the equivalent egg:adult mean mass ratio (shown as a percentage). "H:P ratio" is mean adult mass ratio of host to parasite (shown as a decimal fraction). Mass data mostly from Payne (1977a), with some additions.

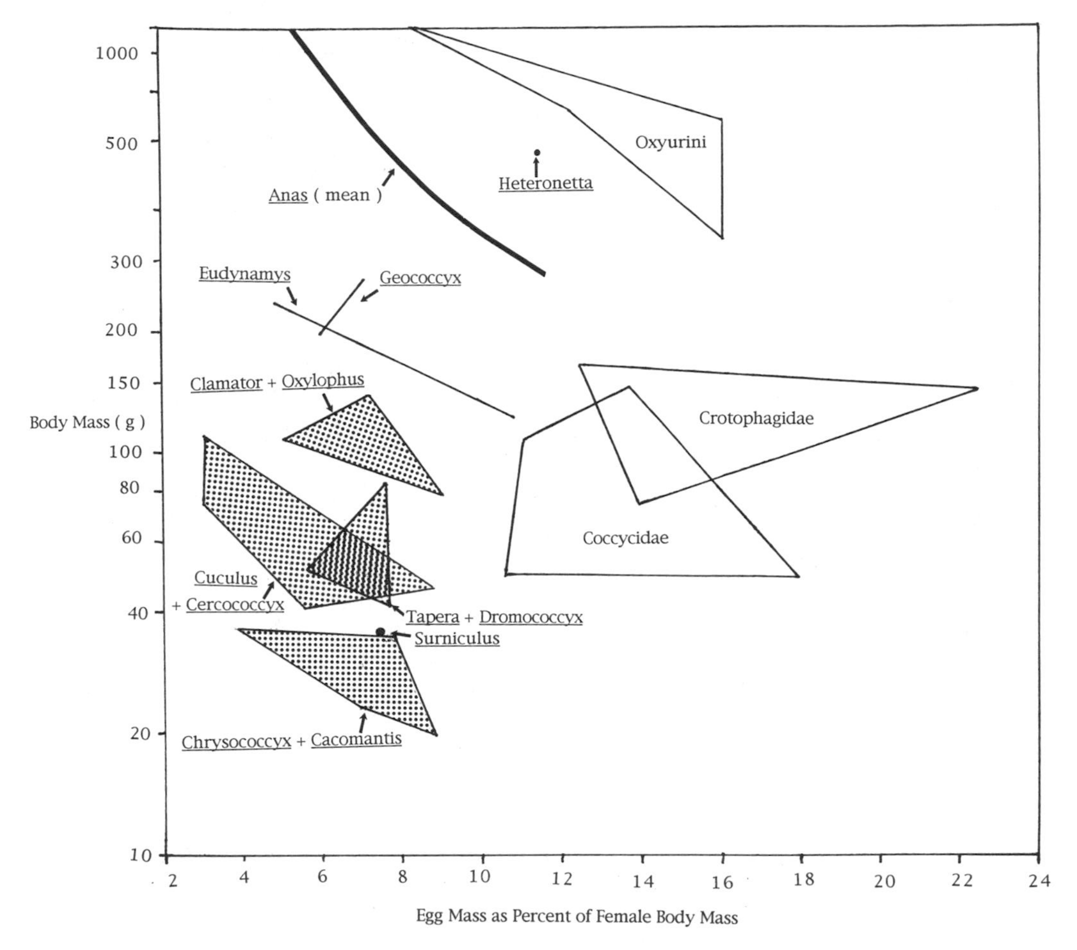

FIGURE 2. Correlation of egg mass and adult female mass in parasitic (shaded) and nonparasitic cuckoos. The relative position (dot) of the black-headed duck, between the curved line representing the mean pattern typical of dabbling ducks (Lack, 1968) and the polygon representing the nonparasitic stiff-tailed ducks, is also shown.

but also with other chicks of their own species, as multiple parasitism is common in these parasites (see table 23). The little-studied channel-billed cuckoo may also be a host-tolerant species (although competing host young rarely survive to fledging); it is much larger in adult mass than any of the other brood-parasitic cuckoos and tends to parasitize quite large hosts. Its eggs approximate those of its largest crow hosts in mass, probably as a reflection of the need for hatching large, strong chicks that can vigorously compete for food during the first few days after hatching.

Figure 2 shows the proportionate egg masses of the black-headed duck relative to the other stifftailed ducks of the waterfowl tribe Oxyurini. The black-headed duck has a somewhat smaller proportionate egg mass than do the other stifftails, but this group as a whole is notable for the large eggs that they lay and for their highly precocial young, which are hatched with a large fat reserve, require little parental attention, and exhibit a well-developed ability to dive for food almost immediately after hatching (Lack, 1968; Johnsgard & Carbonell, 1996). The average proportionate egg-slope characteristic of the typical dabbling ducks (*Anas*), as determined by Lack, is shown as a dotted line; this illustrates the substantial difference between the dabbling ducks and stifftails in terms of parental energy investment in eggs. The black-headed duck is closer to the stifftails than to the dabbling ducks in parental energy investment in eggs, even though it has been suggested that the black-headed duck is a derivative of dabbling duck rather than stifftail ancestral stock.

Lack (1968) did not directly comment on the proportionate egg masses of the icterines, but it is clear from plotting the available data that their proportionate egg masses fall close to the average that Lack drew for passeriform birds generally (fig. 3). The parasitic species of

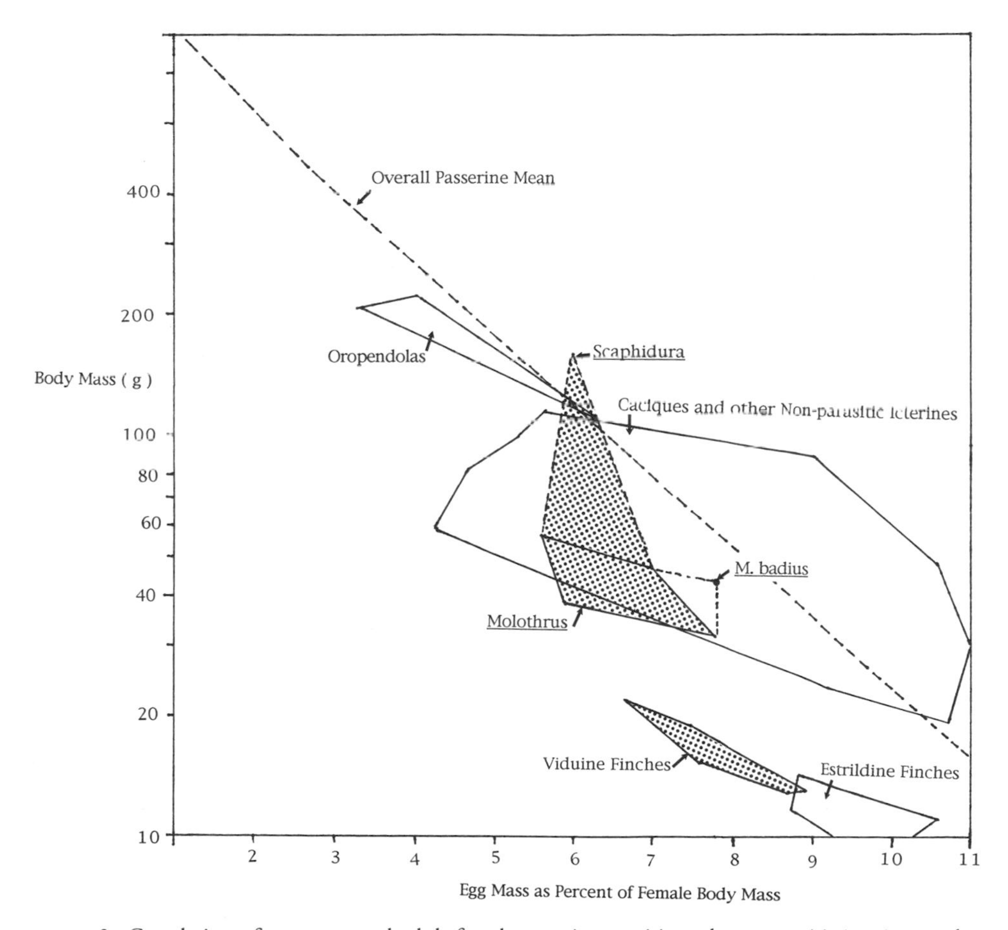

FIGURE 3. Correlation of egg mass and adult female mass in parasitic and nonparasitic icterines and in the viduine finches and their estrildine hosts. Shaded polygons indicate parasitic groups, and the dotted line indicates the mean correlation curve of passerines collectively, according to Lack (1968).

Molothrus fall within the assemblage of caciques and other similarly sized nonparasitic icterines, with somewhat smaller than average proportionate egg masses. However, the single nonparasitic species (the bay-winged cowbird) has a somewhat larger proportionate egg mass, although it is still slightly under the passerline average. The giant cowbird (*Scaphidura*) is both substantially larger in adult mass and has a proportionately larger egg than average for passerines or even for the similar-sized oropendolas that it typically parasitizes, which fall close to the passerine average. Either the cowbirds have not been parasitic long enough to modify their proportional egg masses in ways favorable to brood parasitism, or they have not been required to do so. The only cowbird species that falls slightly outside the polygram formed by the nonparasitic icterines is the brown-headed cowbird, whose somewhat smaller than expected eggs perhaps reflect the fact that it predominantly parasitizes hosts with substantially smaller adult masses than itself (see table 15).

Among the viduine finches and their host-specific estrildines (see table 8 for tabulated data), the proportionate egg masses of both groups fall substantially below the overall passerine average. The two taxa appear to form an essentially unbroken continuum, with the viduine finches averaging larger than their hosts in overall mass, but with the slope of the polygram outlining their proportionate egg masses generally paralleling the pattern typical of passerines. It would appear that, like the cowbirds, the viduine finches have not been forced to modify their proportionate egg mass in ways that specifically reflect their parasitic mode of reproduction.

Incubation and Nestling Periods

In addition to a reasonably close matching of egg sizes and egg color patterns (discussed later), the incubation and nestling periods between a host species and its brood parasite should be similar. Certainly, the incubation period should not be significantly longer than the host's, and the nestling period can be longer only if no surviving nestling young of the host are present and able to leave the nest with their parents before the parasitic chicks achieve their independence.

These similarities between parasite and host need be only general ones. Typically, the parasite's incubation period is a day or two shorter than the host's, even if the parasite's egg is somewhat larger than the host egg (as is often the case). Perhaps this results from the female parasite having carried a ready-to-lay egg in her cloaca for up to a day or possibly even longer, and thus some early embryonic development may have occurred at the time of egg deposition. In any case, host-intolerant brood parasites consistently exhibit shorter incubation periods than do nonparasitic cuckoos of roughly the same adult mass. Similarly, host-tolerant parasitic cuckoos, finches, and cowbirds tend to have incubation periods somewhat shorter than their hosts (table 9). The incubation periods of honeyguides and their hosts are still too poorly documented to conclude whether this same trend applies to them.

Host-tolerant brood parasites seem to have nestling periods that are generally similar to those of their hosts, but these relationships are not consistent (table 9). However, host-intolerant brood parasites often have substantially longer nestling periods than their hosts, probably because these species frequently exploit birds that are much smaller than themselves, and thus the nestling periods for parasitic chicks are likely to be longer than host young simply because of their greater food requirements and the relatively greater foraging efforts required of the much smaller foster parents.

TABLE 9 Adult Mass and Breeding Durations among Brood Parasites, Nonparasitic Relatives, and Hosts[a]

	Adult mass (g)	Breeding period (in days)		
		Incubation	Nestling	Total
Nonparasitic or Facultatively Parasitic Cuckoos				
Dwarf cuckoo	33	13	10+	23+
Black-billed cuckoo	51	10	6–7	16–17
Yellow-billed cuckoo	64	10	6–8	16–18
Groove-billed ani	82	12–13	9	21–22
Smooth-billed ani	105	14	6–7	20–21
White-browed coucal	152	14	18–20	32–34
Senegal coucal	156	18–19	15	33–44
Greater roadrunner	190*	17–18	18	35–36
Host-Intolerant Brood Parasites				
Honeyguides				
Lesser honeyguide	27	12–16	38	50–55
Host: yellow-rumped tinkerbird	12*	~12	~12	~24
Host: cinnamon-chested bee-eater	24*	20 (?)	~25	~45
Scaly-throated honeyguide	48	~18	27–35	45–53
Host: olive woodpecker	41*	15–17	~26	41–43
Greater honeyguide	48	?	~30	—
Host: black-collared barbet	59*	18–19	32–35	51–54
Parasitic cuckoos				
Horsfield's bronze cuckoo	23	12–13	15–18	27–31
Klaas' cuckoo	24	11–14	19–21	30–35
Shining bronze cuckoo	25	13–15	17–23	30–38
Brush cuckoo	31	~13	17–19	30–32
African emerald cuckoo	37*	~13	18–20	31–33
Dideric cuckoo	38	11–12	21	32–33
Fan-tailed cuckoo	50	13–14	16–17	29–31
Striped cuckoo	52	15	18	31
Red-chested cuckoo	72	11–12	20	31–32
Black cuckoo	86*	13–14	20–21	33–35
Thick-billed cuckoo	92*	~13	28–30	41–43
African cuckoo	101	~12	22	~34
Common cuckoo	113	12–13	20–23	32–36
Indian cuckoo	128*	12	21	33
Host-Tolerant Brood Parasites and Hosts				
Parasitic cuckoos				
Pied cuckoo	72	11–12	11–15	22–27
Host: jungle babbler	68*	15	14–16	29–31
Levaillant's cuckoo	122	11	12–17	23–28
Host: arrow-marked babbler	63	12–13	11–16[b]	23–29
Great spotted cuckoo	153	13–15	26–28	39–43
Host: black-billed magpie	178	17–18	22–28	39–46
Asian koel	167*	13–14	21	34–35
Host: house crow	296	16–17	30+	46–47+
Australian koel	205	13–14 (?)	18–28	31–42
Host: magpie lark	89	16	~20	~35

(continued)

TABLE 9 *(continued)*

	Adult mass (g)	Breeding period (in days)		
		Incubation	Nestling	Total
Parasitic finches and hosts				
Pin-tailed whydah	14	~11	20	~31
Host: common waxbill	7.5	11–12	17	28–29
Parasitic weaver	20*	~13	18	~31
Host: black-chested prinia	9	12–13	13–14	25–27
Hosts: various cisticolas	~7	11–14	~14	25–29
Parasitic cowbirds				
Shiny cowbird	35	11–12	13–15	24–27
Brown-headed cowbird	44	10–12	8–10	18–22
Screaming cowbird	48	12–13	12	24–25
Bronzed cowbird	62	12–13	11	23–24
Giant cowbird	161	~12 (?)	(?)	(?)
Various nonparasitic icterines				
Orchard oriole	20	12–15	11–14	23–29
Yellow-hooded blackbird	31	10–11	12	22–23
Bobolink	42	11–13	10–14	21–27
Bay-winged cowbird	44	13	12	25
Red-winged blackbird	52	10–12	10–11	20–23
Brewer's blackbird	62	12–13	13	25–26
Carib grackle	65	12	14	26
Eastern meadowlark	89	13–15	11–12	24–27
Crested oropendola	235	15–19	28–35	43–54
Montezuma oropendola	324	13–18	29–42	42–60

[a]Arranged within categories by increasing mean adult mass. Cuckoo data are partly as summarized by Payne (1977b), with mean adult weights added. Except as indicated (by *), weights are mostly based on Dunning (1993), and pooled means for both sexes are used when they differ. Mean weights in other tables or the text may be based on other sources and thus differ somewhat from these.

[b]Estimation, based on closely related species.

Structural, Plumage, and Acoustic Adaptations

Raptor and Drongo Mimicry

One of the more interesting examples of a structural and behavioral strategy that has been adopted by several cuckoo species but by none of the other brood parasites is the evolution of raptor mimicry. This hypothesis suggests that a raptorlike appearance of a brood parasite might help to intimidate or perhaps decoy a host away from its nest long enough for the parasite to lay its egg (Pycraft, 1910). Such a decoying behavior on the part of the male, while the female surreptitiously approaches the nest and quickly lays her egg, seems to be a regular part of the egg-laying strategy of the crested cuckoos and possibly some other pair-bonding cuckoos.

Kuroda (1966) discussed hawk mimicry in cuckoos at some length and illustrated a threat display by a captive oriental cuckoo that is similar to that of such accipiters as the Eurasian sparrowhawk (fig. 4). He suggested it is more plausible that hawk mimicry serves to attract the host's attention and thus distract than the possibility that the mimicry tends to threaten and intimidate the host species, inasmuch as many small host species fiercely defend their nest against much larger cuckoos. Kuroda hypothesized that the earlier stages of hawk mimicry might serve to alarm and keep host birds away from their nests, but later the more frequently parasitized host species became able to distinguish hawks from cuckoos, making the adaptation primarily of value when acquiring new host species. It has been suggested that the rufous-colored morph that occurs among females of several Eurasian cuckoos is a possible mimic of the common kestrel (Voipio, 1953), whereas the normal gray morph more closely resembles various small accipiters. However, there would seem to be no special advantage to be gained by such plumage dimorphism, and it seems more likely that the gray versus brown plumage morphs reflect differing modes of concealing adaptations under differing habitat conditions.

The flight behavior of many cuckoos, typically characterized by brief alternating periods of flapping and gliding, is similar to that of many falcons and accipiters. This behavioral similarity is enhanced by the wing-barring and general underpart patterning that many cuckoos

FIGURE 4. Adult plumages and defensive postural comparisons of the oriental cuckoo (right) and the Eurasian sparrowhawk (after photos in Kuroda, 1966). Corresponding outer primaries are also shown below.

exhibit while in flight (fig. 5A,B). Some cuckoos such as the hawk-cuckoos, which have squared-off and barred tails rather than elongated and variously spotted tails, show even more convincing hawk mimicry than do the *Cuculus* species (fig. 5C,D). The thick-billed cuckoo is also hawklike in its plumage pattern, and its thickened bill more closely approximates that of a hawk than is the case with most other cuckoos (fig. 5E,F).

Wyllie (1981) suggested that, rather than being directed toward confusing or intimidating hosts, the hawk mimicry of cuckoos may help to protect them against attack by raptors. Thus, the white nape patch common to many cuckoos, especially juveniles, may be a hawk mimicry trait that is also common among many young hawks, and in Wyllie's view might reduce the likelihood of predation by hawks. Although many cuckoos exude highly disagreeable odors when captured, for which neither a biological function nor a physiological origin (perhaps fecal odors) are firmly established, they are evidently fairly edible. Cott and Benson (1970) estimated that the perceived edibility (based on a human taste-panel's low-to-high scale of 3 to 9) of five African cuckoos ranged from 3 to 7.8, and averaged 5.3.

The case of visual mimicry of various drongos by the drongo cuckoo is even more convincing. Both sexes of drongo cuckoos strongly resemble true drongos in their uniformly black and fork-tailed adult plumages (fig. 6), and even to a degree in their vocalizations. There would seem to be little doubt that this is a case of evolved mimicry, rather than chance visual resemblance, and in the past it has been asserted that the drongo cuckoo is primarily a parasite of drongos in India (Ali & Ripley, 1983). However, it is now known that this is not the case, and indeed at least among the Indian population of drongo cuckoos, drongos are not even probable hosts. Instead, this population primarily parasitizes smaller passerines, such as babblers (Becking, 1981). The drongolike adult plumage pattern may serve a similar function as the hawk mimicry of other cuckoos in distracting or intimidating hosts during the egg-laying period, although in such a case it would seem that only males should exhibit drongolike plumages, whereas the female should be as inconspicuously patterned as possible.

Sex-limited mimicry may help explain the presence of plumage polymorphism in the females of several cuckoos, in which a rufous or "hepatic" plumage morph occurs in addition to or intergrading with the gray morph, which is much more malelike (Voipio, 1953). This visual polymorphism, and especially the increased variability of appearance among females, may make it more difficult for host species to recognize these birds as potential threats. Brown plumage may also be a more effective camouflage or provide a more effective mimicry than gray plumage under some environmental conditions; the hepatic morph is reportedly more common in open-country habitats, where small, brownish hawks are also common (Payne, 1967b).

Host-specific Visual Mimicry

Most of the truly convincing examples of visual host mimicry among the avian brood parasites concern egg mimicry and nestling mimicry, and there would seem to be little profit gained by a brood parasite mimicking the adult plumage of its primary host. Such mimicry, if effective, is likely only to stimulate territorial defensive behavior or otherwise aggressive interactions between host and parasite. The all-black plumage of the adult male Asian koel has been attributed to host mimicry of crows and ravens in India (fig. 6D,E). In this scenario, it has been asserted that by resembling the host species, the male is able to distract temporarily

FIGURE 5. Flight profile similarities of the common cuckoo (A) and the Eurasian sparrowhawk (B) (after photos in Wyllie, 1981). General plumage similarities of the common hawk cuckoo (C) with the Eurasian sparrowhawk (D), and comparative head profiles of the thick-billed cuckoo (E) and the little sparrowhawk (F) are also shown.

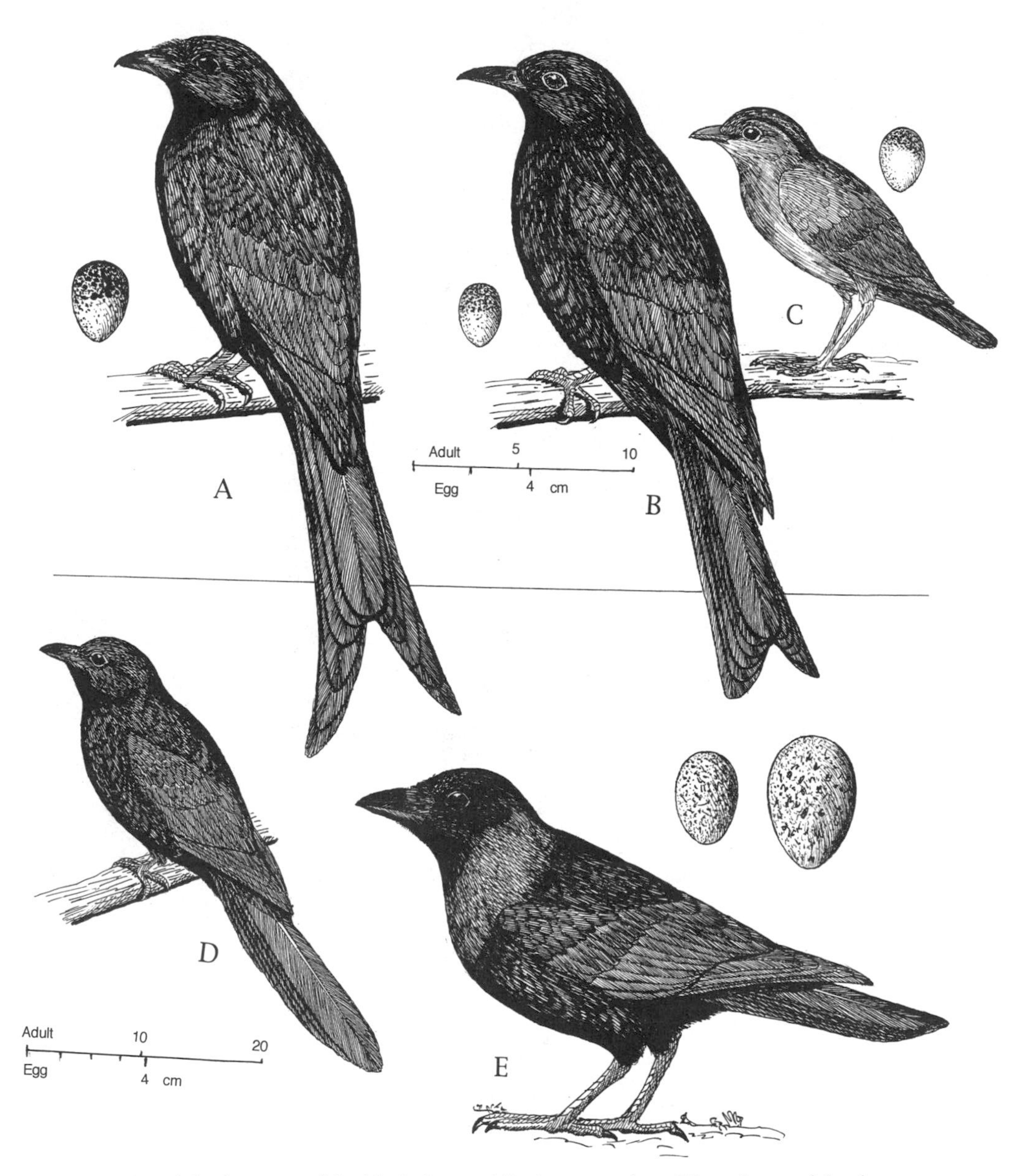

FIGURE 6. Adult plumages of the black drongo (A), drongo-cuckoo (B), and one of the drongo-cuckoo's biological hosts, the Nepal fulvetta (C). Also shown is an adult male Asian koel (D) and an adult house crow host (E). Eggs of these species are also shown.

one or both parents from the nest, while the brown female simultaneously approaches the nest to lay her eggs unobserved (Lamba, 1963, 1975). Lack (1968) doubted the likelihood of this explanation and noted that although adult males of the closely related Australian koel are similarly all black, their host species all consist of various honey-eaters, none of which is similar in plumage pattern to the koel.

Visual mimicry by brood-parasite nestlings of their hosts occurs in several avian groups, most strikingly in the African viduine finches, in which the degree of nestling similarity between the parasite and its species-specific host is little short of unbelievable. Before dealing with this phenomenon, some rather less spectacular cases of apparent nestling mimicry should be discussed.

Jourdain (1925) investigated various cases of cuckoo host mimicry involving nestling plumages and egg pigmentation. He observed that, among the host-tolerant cuckoos that must share the nest with host chicks, it is important for the parasite's head, and especially its crown, to resemble that of the host, since this is the only conspicuous part of the bird in an often crowded nest. He thus pointed out that nestlings (in juvenile plumage) of the great spotted cuckoo have dark crowns similar to those of juvenile magpies, but which are quite different from the gray crown color typical of adults (Fig. 7A,B). Similarly, the nestling juvenile pied cuckoo resembles its large gray babbler host (fig. 7C,D).

The nesting Asian koel has a black head that resembles its house crow host (fig. 7E,F), even though the young bird subsequently molts into a more femalelike brown plumage after leaving the nest (Menon & Shah, 1979). In Australia, where the koel's host species are not crows but rather smaller birds that do not have black heads, koel nestlings have brown, femalelike heads. Additionally, in Australia the nestling koels may eject host chicks from the nest, whereas the Asian koel is evidently host tolerant toward its crow hosts. Perhaps this is because the large crow nestlings would be difficult to eject, and so the ejection habit may have been secondarily lost in Asian koels (Lack, 1968).

No host mimicry occurs among nestlings of the host-intolerant species of cuckoos, such as the two species of *Cuculus* (oriental and banded bay cuckoos; fig. 8A,B) and the presumably host-intolerant drongo cuckoo (fig. 8C).

It is in the viduine finches and their hosts the estrildines that the most remarkable examples of nestling mimicry occur. Friedmann (1960) judged that the remarkable similarities in mouth markings and juvenile plumages between the viduines and estrildines might simply be the result of "community of descent." However, strong evidence of host mimicry has since accumulated (e.g., Nicolai, 1964, 1974; Payne, 1973a, 1982). Nicolai (1974) observed that species-specific host mimicry in the viduine finches includes evolved similarities in (1) size, shape, and color of the hosts' eggs, (2) the interior mouth markings on the host nestlings' palate and tongue, and (3) enlarged, light-reflective and tuberclelike structures ("gape papillae"), or colorful enlargements (mandibular flanges) along the edges and at the base of the mandible in young birds. Mimicry of host juvenile plumage patterns also occurs, as does song mimicry by adult viduine finch males of virtually all of the host species' major vocalizations.

Many examples of remarkable similarities in juvenile plumages between the viduines and their host species are apparent in the later species accounts (see figs. 41–46). One example of similarity in juvenile plumages, as compared with the dissimilarities in adult plumages, is illustrated by the pin-tailed whydah and its common waxbill host (fig. 9). Similarities in their palates, tongues, and mandibular edges are even more remarkable. Nicolai (1974) judged that

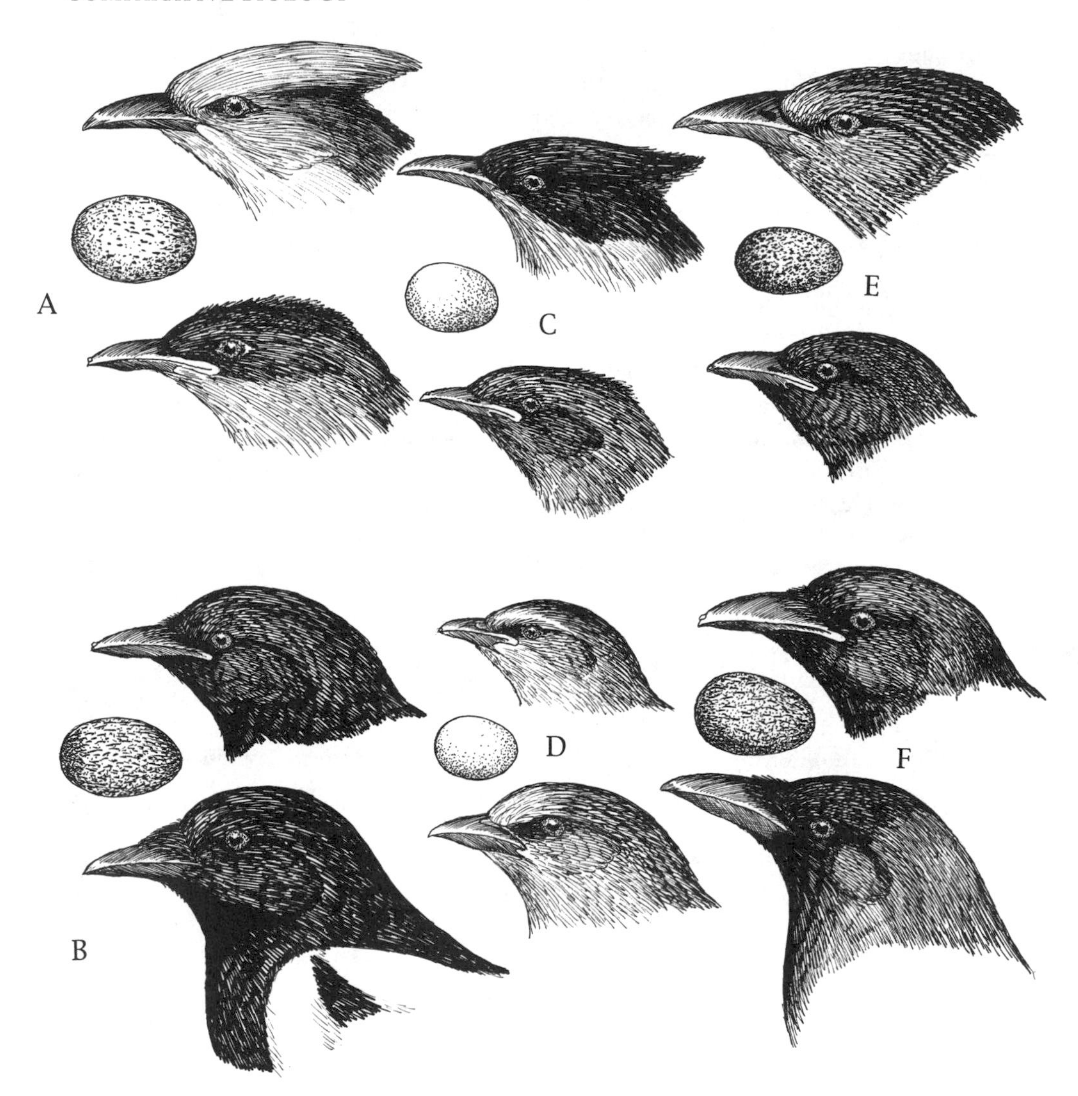

FIGURE 7. Comparison of adults, eggs, and nestling heads of three host-tolerant cuckoos and their hosts: the great spotted cuckoo (A) and its black-billed magpie host (B); the pied cuckoo (C) and its large gray babbler host (D); and the Asian koel (E) and its house crow host (F). Mainly after paintings by N. Gronvøld (in Jourdain, 1925).

no two species of the approximately 125 estrildine finches have identical mouth markings, which usually consist of from three to five black or violet spots arranged on the palate in a semicircular or pentagonlike pattern, supplemented by white, yellow, blue, or violet thickenings or wartlike papillae along the sides of the palate and the edges of the bill. Adult estrildine finches will innately respond to and may feed only those conspecific young exhibition the appropriate species-specific palatal markings, or those parasitic young whose palatal mark-

FIGURE 8. Comparisons of nestling oriental cuckoo (A) and banded bay cuckoo (B), with a Asian koel nestling (C). Individual feathers of A & C are also shown to the right, together with a nestling banded bay cuckoo. Mostly after paintings by P. Barruel (in Becking, 1981).

FIGURE 9. Sketches of a breeding male (A), juvenile (B), and nestling gapes (C) of the pin-tailed whydah, together with comparable views of its common waxbill host (D–F). Also shown are the shared gape patterns of the eastern paradise whydah and its green-winged pytilia host (G), the straw-tailed whydah and its purple grenadier host (H), the steel-blue whydah and its black-cheeked waxbill host (I), and the similar gapes of the village indigobird (J, left) and its red-billed firefinch host (J, right). The middle row shows adult gape patterns of the village (K), variable (L), and pale-winged (M) indigobirds. The bottom row includes nestling gape patterns of two host:parasite pairs, the Jameson's firefinch (N) and dusky indigobird (O), and of the purple grenadier (P) and straw-tailed whydah (Q). Stippled areas indicate light-reflective surfaces. Mainly after Nicolai (1964, 1974) and Payne (1972).

ings are essentially indistinguishable from the host's evolved type. According to Nicolai (1974), nestlings showing even minor deviations from the species' norm are "ruthlessly weeded out by starvation" by their own parents. However, Goodwin (1982) reported several cases of interspecific adoptions of young by captive birds. In some estrildine finches, these palatal patterns persist into adulthood (fig. 9K–M) and have no known function in adults, but in many species they gradually fade and may eventually disappear following the postfledging dependency period, presumably having served their critical functions.

Clearly, host mimicry of mouth patterns must be very precise in this group of parasites if the young are to survive close scrutiny by their host parents and be able to compete effectively for food with one or more host chicks. Indeed, nestling mouth patterns are nearly identical in several host–parasite pairs (fig. 9G–J, N–Q), not only in general patterning, but also in coloration and the presence or absence of reflective tubercles. These highly efficient reflective (but not luminous) structures have complex internal anatomies that allow them to operate as a combination reflecting mirror and refraction–diffraction prism (Friedmann, 1960). As such, they make effective attention-getting devices for stimulating parental feeding in the rather dark environment of an enclosed estrildine nest.

Mimetic Vocalizations and Related Acoustic Adaptations

As in other birds, vocalizations of brood parasites depend upon the syringeal and tracheal anatomy of each species. Even in cuckoos, which exhibit simple tracheal and syringeal structures (fig. 10), too little is known about the role these structures play in determining vocal potentials and constraints to provide much information of value regarding adaptations related to brood parasitism. Lack (1968) characterized the vocalizations of adult parasitic cuckoos as loud, simple, and distinctive; their loudness serving to broadcast the songs over long distances, their simplicity reflecting genetically acquired rather than learned song, and their distinctiveness perhaps facilitating species recognition. Wyllie (1981) noted that nonparasitic as well as parasitic cuckoos have simple and distinctive songs, often with ventriloquial acoustic qualities. It is probable that among these generally secretive and usually well-camouflaged birds, vocalizations are the most effective means of long-distance communication. The advertisement "cuck-oo" song of the common cuckoo may be heard (at least by humans) from distances up to about 1.5 km from the source, and male common cuckoos can accurately locate the direction of a tape-recorded cuckoo advertisement song and respond to it from as far away as about 1 km (Wyllie, 1981).

The singing of many cuckoo species is notable for its persistence and its annoying, repetitive nature; the common name "brainfever bird" applied to several cuckoos not only describes the cadence of the advertising song ("brain fee'-ver") but also implies a degree of insanity in its acoustic characteristics. Among many cuckoos, the pitch of the song phrase rises slightly with each successive repetition (thus, the vernacular name "half-tone" birds), or the intervals between the phrases may shorten, or both, so that the song gets progressively higher in pitch and faster in phrasing until it suddenly stops, only to have a new sequence begin a short time later. Perhaps this apparent frequency scanning provides an acoustically effective way of broadcasting advertising vocalizations over and through a variety of environmental barriers, such as forest vegetation of varying densities. Even the monotonously repetitive male advertising song of the common cuckoo, although simple and distinctive enough to be almost immediately

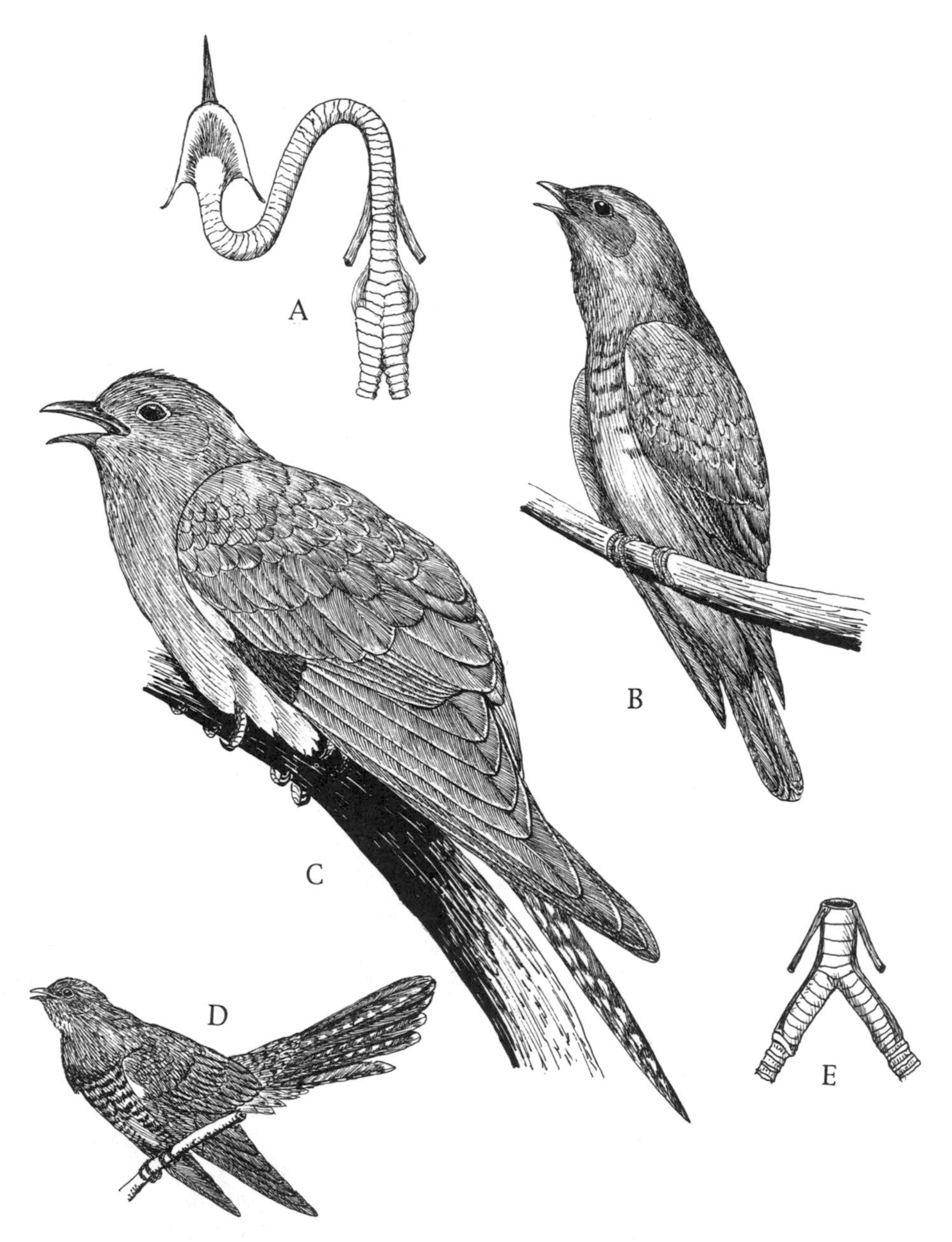

FIGURE 10. Trachea and tracheobronchial syrinx of the black-billed cuckoo (A, from a specimen), as compared with a bronchial syrinx of the lesser ground-cuckoo (E, after Berger, 1960). Also shown are singing postures of the brush cuckoo (B, after a photo in Coates, 1985), pallid cuckoo (C, after a photo in Frith, 1977) and common cuckoo (D, after a sketch in Glutz & Bauer, 1980).

recognized by humans the first time it is heard, is nevertheless sufficiently variable individually so that each resident male can recognize its neighbors by minor acoustic differences. This song is uttered at roughly 1-second intervals in series that usually range from 10 to 20 phrases, but sequences of up to as many as 270 uninterrupted phrases have been heard. In spite of the song's obvious audibility and potential to alarm possible host species, most hosts apparently normally pay no heed to it, although they may respond aggressively to such tape recordings when they are replayed near the nest (Wyllie, 1981).

Cuckoos may sing from either fairly inconspicuous or rather exposed perches. During singing, the bird's throat area is somewhat enlarged by feather ruffling, its wings droop, and, during increasing excitement, its head bobs and its tail lifts (fig. 10). The white feathers at the bend of the wing in several *Cuculus* cuckoos may become more apparent during such wing dropping; this contrasting area perhaps functions in an analogous manner to the white "shoulder spot" that is exposed during display by various species of male grouse that also display under dim-light conditions (Johnsgard, 1973). The long tail is sometimes lifted and partially spread, revealing the white spotting present in many cuckoo species. Cuckoos often begin singing earlier in the morning and continue singing later in the evening than nearly all other diurnal birds. This tendency to sing under darkening conditions no doubt accounts for the common name "rain crow" that has been applied to some cuckoos in various parts of the world, such as North America and Australia.

The vocalizations of the viduine finches provide an even more interesting example of adaptations for brood parasitism and represent the only known case of interspecific "stealing" of species-specific advertising songs among birds (see fig. 46). Payne (1973b, 1976b, 1982, 1990) has extensively investigated this phenomenon and has described several new and biologically distinct species of indigobirds on the basis of such traits as song mimicry, host-parasite gape and palate similarities of nestlings, and host-specific dependency traits that hold even when traditionally used taxonomic features such as male plumage characteristics are inadequate for providing distinction among these taxa. Payne concluded (1973a) that the interspecific vocal mimicry exhibited by male indigobirds is related only indirectly toward acceptance by their hosts and that the mimetic phrases are used as heterosexual signals by males to attract females. The host species' songs, as well as some of their other vocalizations, are evidently learned through imprinting on host parents, and thus the indigobirds' own behavioral reproductive-isolating mechanisms must be learned by each new generation; the males learning the host's specific song types and the females learning to respond to these songs (Payne, 1973b).

Nicolai (1964, 1974) determined by sonographic analysis that the male straw-tailed whydah includes in its song eight separate motifs. Males begin their song using the purple grenadier host's contact call, but then incorporate additional host-species motifs that include a female-attraction call, a "clacking" song-phrase, a chase call, some whisperlike nest calls, and nestling begging calls. Additionally, the whydah uses three different and presumably innate whydah-bird motifs that occur in unpredictable sequences.

Payne (1982) has hypothesized that mouth mimicry by nestlings may have evolved before song mimicry had been perfected in the viduine finches. In support of this view, he observed that in the pin-tailed whydah, mouth mimicry is present (see fig. 9), but species-specific song mimicry is lacking. Several closely related emberizine species are reportedly parasitized by this whydah, in addition to its usual common waxbill host, making song mimicry less likely to be effective than in the more host-specific whydah species.

Not only are the estrildine host's nestling begging vocalizations mimicked by nestling viduine finches, their complex and taxon-specific begging posture and feeding behavior are also mimicked by the parasitic chicks. This remarkable behavior is characterized by the chick crouching while holding its bill wide open. Its neck is turned so that its gaping bill is directed toward the parent's head, and its palatal markings are fully visible to the parent; its head thus is often oriented almost upside down. The tongue may sometimes also be lifted during begging, and the head swung from side to side, but the wing fluttering typically performed by most passerine chicks is usually lacking. In this posture (see fig. 16C), the chick grasps the sides of the parent's bill with its own, while the parent bird tilts its head down and regurgitates seeds with rhythmic pulsating movements of the throat (Morris, 1982).

Nestling vocalizations of brood parasites may be important only to the extent that they must be able to match those of their host well enough to prevent host detection and thus rejection. This is especially true of host-tolerant parasites, where competition with host young for attention and feeding may be great. Mundy (1973) sonographically analyzed the nestling begging calls of the great spotted cuckoo and those of its pied crow host and reported that they are very similar sonographically. He also analyzed the nestling begging call of another host-tolerant species, Levaillant's cuckoo, and found that its begging call is much like the chorus-alarm calls of its arrow-marked babbler host. Mundy summarized evidence that host-tolerant cuckoos may mimic the young of different hosts, in cases where two or more hosts may be exploited. Although one might not expect host-intolerant brood parasites to exhibit host-species mimicry very effectively (since they have little or no opportunity to learn host begging calls before killing or expelling host chicks from the nest), this situation has been reported for both the pallid cuckoo and its white-eared honeyeater host, and for the Horsfield's bronze cuckoo and its superb fairywren host (Courtney, 1967). Courtney has suggested that, if the begging call is not learned from host nestlings during the short time available, it may be acquired from the host's parental feeding call, which may be acoustically similar to the begging call of nestlings. However, as noted later (chapter 4), it must also be considered that some similarities in nestling vocalizations may be the fortuitous result of similar sound-producing structures and relatively limited vocal abilities of nestlings.

Egg and Eggshell Adaptations of Brood Parasites

Egg Shapes and Relative Egg Volumes

It has long been known that the matching of egg sizes (volumes or masses) of most brood parasites and their hosts match rather closely, even though their hosts are often much smaller. Baker (1942) attributed this similarity to the discriminative abilities of hosts, as they eliminated eggs that contrasted too greatly in size with their own. However, no evidence yet exists to substantiate Baker's belief that intraspecific specialization by mimicking egg size (i.e., two or more egg sizes laid by different female gentes adapted to differing hosts that lay markedly different-sized eggs) might have occurred; his support of this idea with regard to the large hawk cuckoo resulted from misidentification and resulting wrongful attribution of some larger-sized (probably common cuckoo) eggs as being those of the large hawk cuckoo (Becking, 1981). Data summarized in tables 10, 11, 12, 17, and 18 illustrate such generalized parasite-to-host

TABLE 10 Egg Traits of the Host Species of Non-*Cuculus* Cuckoos in India[a]

Cuckoo species/host species[b]	Clutch records	Egg volume (%)	Base color	With spots?
Pied cuckoo (egg volume 4.23 cc)[c]				
Lesser necklaced laughingthrush (S,O)	42	156	Blue	No
Jungle babbler (S,O)	32	117	Blue	No
White-throated babbler (S,O)	10	248	Bluc	No
Rufous-necked laughingthrush (S,O)	10	124	Blue	No
Greater necklaced laughingthrush (S,O)	9	218	Blue	No
Common babbler (S,O)	8	66	Blue	No
Chestnut-crowned laughingthrush (S,O)	7	157	Blue	Few
Rusty-fronted barwing (S,O)	4	87	Blue	Yes
Streaked laughingthrush (S,O)	3	100	Blue	No
Red-faced liocichla (S,O)	3	108	Blue	Yes
Rufous-vented laughingthrush (S,O)	3	141	Bluish	No
Spot-breasted laughingthrush (S,O)	3	156	Blue	No
Striated laughingthrush (S,O)	3	219	Blue	No
Slender-billed babbler (S,O)	2	72	Blue	No
Gray-sided laughingthrush (S,O)	2	89	Blue	No
Spotted forktail (G,C)	2	90	Cream	Faint
Brown-capped laughingthrush (S,O)	2	115	White	No
Chestnut-bellied rock-thrush (G,C)	2	128	Cream	Yes
Chinese babax (S,O)	2	132	Blue	No
Mean host egg volume		133	(range 72–219%)	
Chestnut-winged cuckoo (egg volume 7.16 cc)[c]				
Lesser necklaced laughingthrush (S,O)	109	92	Blue	No
Greater necklaced laughingthrush (S,O)	37	128	Blue	No
Striated laughingthrush (S,O)	24	129	Blue	No
Rufous-vented laughingthrush (S,O)	18	84	Blue	No
Blue-winged laughingthrush (S,O)	14	90	Blue	No
Gray-sided laughingthrush (S,O)	10	106	Blue	No
Red-faced liochichla (S,O)	7	64	Blue	Yes
Rufous-necked laughingthrush (S,O)	7	73	Blue	No
Chestnut-crowned laughingthrush (S,O)	7	92	Blue	Few
White-crested laughingthrush (S,O)	7	115	White	No
Rusty-fronted barwing (S,O)	3	51	Blue	Yes
Rufous-chinned laughingthrush (S,O)	2	71	White	No
Mean host egg volume		91	(range 51–129%)	
Gray-bellied cuckoo (egg volume 2.0 cc)				
Zitting cisticola (S,S)	48	50	White	Yes
Common tailorbird (S,O)	43	56	Varied	Yes
Ashy prinia (S,O)	41	59	Red	No
Gray-breasted prinia (S,O)	32	51	Varied	Yes
Striated prinia (S,S)	31	72	Whitish	Yes
Hill prinia (S,S)	25	69	Varied	Yes
Dark-fronted babbler (S,S)	6	94	White	Yes
Dark-necked tailorbird (S,O)	5	51	Varied	Yes
Striped tit-babbler (S,S)	3	67	White	Yes
Plain prinia (S,O)	2	53	Varied	Yes
Mean host egg volume		62	(range 50–94%)	

(continued)

TABLE 10 *(continued)*

Cuckoo species/host species[b]	Clutch records	Egg volume (%)	Base color	With spots?
Rusty-breasted cuckoo (egg volume 1.93 cc)				
Long-tailed shrike (S,O)	2	195	Whitish	Yes
Asian emerald cuckoo (egg volume 1.30 cc)				
Crimson sunbird (S,S)	5	77	White	Yes
Little spider hunter (S,S)	3	125	Pinkish	Yes
Zitting cisticola (S,S)	2	77	White	Yes
Mean host egg volume		93 (range 77–125%)		
Violet cuckoo (egg volume 1.38 cc)				
Little spider hunter (S,S)	5	117	Pinkish	Yes
Drongo cuckoo (egg volume 2.43 cc)				
Nepal fulvetta (S,O)	33	76	Whitish	Yes
Brown bush-warbler (S,O)	5	78	Whitish	Yes
Red-vented bulbul (S,O)	4	106	Pinkish	Yes
Spot-throated babbler (S,S)	3	96	Brownish	Yes
Common tailorbird (S,O)	2	46	Whitish	Yes
Rufous-fronted babbler (S,O)	2	52	White	Yes
Horsfield's babbler (S,O)	2	117	Reddish	Yes
Mean host egg volume		82 (range 46–117%)		
Asian koel (egg volume 8.92 cc)[c]				
House crow (T,O)	142	155	Greenish	Yes
Large-billed crow (T,O)	30	171	Greenish	Yes
Jungle crow (T,O)	18	171	Greenish	Yes
Black-collared starling (T,O)	16	96	Greenish	Yes
Black-billed magpie (S,S)	4	129	Greenish	Yes
Common myna (G,C)	2	85	Blue	No
Blue magpie (S,O)	2	111	Grayish	Yes
Mean host egg volume		131 (range 85–171%)		

[a]Includes all host species for which Baker (1942) possessed at least two clutches parasitized by non-*Cuculus* cuckoos. The drongo cuckoo's hosts were originally misidentified as being those of the banded-bay cuckoo (Becking, 1981). Taxonomic identities of two minor host species (*Sericornis barbara* and *Rhipidura albiscapa*) are questionable and have been excluded.
[b]Nest types in parentheses: G, ground or flat substrate; S, shrub or near-ground site; T, tree level; W, wetlands; C, cavity or crevice; O, open above; S, spherical.
[c]Indicates a host-tolerant species, as opposed to a host-intolerant species.

egg-size relationships for various cuckoos, and similar relationships may be observed among the viduine finches (table 8), brown-headed cowbirds (table 15), and honeyguides (table 21). However, this host–parasite size matching is usually only approximate; the parasite's eggs tend to be not only somewhat larger, but also are usually more rounded. A more nearly spherical egg can store more potential energy into a given volume, and such eggs may also be harder for a host species to pick up, since eggs are usually grasped at their narrower ends when being grasped by the bill (see figs. 14 and 15). Rounded eggs are perhaps also more difficult to pierce with the bill than more elongated and flatter eggs (Hoy & Ottow, 1964).

TABLE 11 Egg Traits of the Common Cuckoo's Indo-Tropical Host Species[a]

Host species[b]	Clutch records	Egg volume (%)	Basic color	With spots?
Zitting cisticola (S,O)	366	31	White	Yes
Richard's pipit (G,O)	282	75	Buff-gray	Yes
Hill prinia (S,O)	187	40	Whitish	Yes
Striated prinia (S,O)	148	43	Greenish	Yes
Pied bushchat (G,O)	120	53	Blue	Yes
Verditer flycatcher (G,C)	113	65	Whitish	Yes
Brown bush-warbler (S,O)	85	58	Whitish	Yes
Long-tailed shrike (S,O)	74	118	Whitish	Yes
Chestnut-bellied rock-thrush (G,O)	69	119	Buffy	Yes
Red-billed leiothrix (S,O)	65	89	Bluish	Yes
Silver-eared mesia (S,O)	45	85	Bluish	Yes
Plumbeous water-redstart (G,C)	42	66	Gray-green	Yes
Rufous-bellied niltava (G,C)	41	82	Buffy	Yes
Spotted forktail (C,O)	32	121	Variable	Yes
Common tailorbird (S,O)	33	35	Whitish	Yes
White-tailed robin (C,O)	23	104	White	Few
Small niltava (C,O)	23	53	Whitish	Yes
Chestnut-bellied rock-thrush (G,O)	22	167	Creamy	Yes
Gray sibia (T,O)	20	117	Grayish	Yes
Tawny-breasted wren-babbler (G,S)	19	57	White	No
Mean host egg volume		78.9 (range 31–167%)		

[a]Includes all host species for which Baker (1942) possessed at least 19 clutches parasitized by *Culculus canorus bakeri.* Egg information from Ali and Ripley (1968).
[b]Nest types in parentheses: G, ground or flat substrate; S, shrub or near-ground site; T, tree level; W, wetlands; C, cavity or crevice; O, open above; S, spherical.

Thickened eggshells are a common adaptation of brood parasites; long ago Rey (1892) provided a simple method (see Glossary) of calculating an index ("Rey's index") to estimate the eggshell's thickness, and thus the egg's susceptibility to cracking, breakage, or puncture. For example, brood-parasitic cowbirds have eggs with substantially thicker and considerably more puncture-resistant shells than is typical of comparably sized passerines, including non-parasitic icterines (Hoy & Ottow, 1964; Blankespoor et al., 1982; Rahn et al., 1988). Such thick shells may serve to reduce the probability of chance egg damage such as cracking, either during egg-laying or subsequent random jostling of the clutch by the host during incubation. However, it is also possible that the primary function of shell reinforcement is to increase resistance to egg destruction or puncture-ejection by host parents after the eggs have been identified as being alien (Spaw & Rohwer, 1987).

Egg Colors, Patterns, and Gentes Evolution

Avian eggs can usually be characterized by two basic color attributes. The first of these is an overall and rather uniform "ground color" (usually whitish, buff, brownish, bluish, or reddish brown) that is incorporated into the shell during the later stages of oviducal passage. Second, many avian eggs have additional, variably darker or contrasting patterns of stipples, spots, streaks,

TABLE 12 Egg Traits of the Indo-Tropical *Cuculus* Cuckoos' Host Species[a]

Cuckoo species/host species[b]	Clutch records	Egg volume (%)	Ground color	With spots?
Large Hawk Cuckoo (egg volume 4.17 cc)				
Streaked spider hunter (S,S)	98	62	Brown	No
Little spider hunter (S,S)	24	34	Pinkish	Yes
Lesser necklaced laughingthrush (S,O)	12	140	Blue	No
Greater necklaced laughingthrush (S,O)	6	195	Blue	No
Blue whistling thrush (G,C)	5	263	Buffy	Faint
Brownish-flanked bush warbler (S,S)	4	34	Brown	No
Small niltava (C,O)	3	87	Buff	Faint
Rufous-fronted babbler (S,O)	2	27	White	Yes
Gray-throated babbler (G,S)	2	42	White	No
Spot-throated babbler (S,S)	2	49	Brownish	Yes
Rusty-cheeked simitar babbler (G,S)	2	125	White	No
Chestnut-crowned laughingthrush (S,O)	2	141	Blue	Few
Scaly thrush (S,O)	2	161	Varied	Yes
Mean host egg volume:		104 (range 27–263%)		
Common Hawk Cuckoo (egg volume 5.32 cc (?))				
Jungle babbler (S,O)	33	93	Blue	No
Large gray babbler (S,O)	7	88	Blue	No
Moustached laughingthrush (S,O)	4	84	Blue	No
Rufous-necked laughingthrush (S,O)	4	99	Blue	No
Rusty-fronted barwing (S,O)	2	69	Blue	Yes
Chestnut-crowned laughingthrush (S,O)	2	125	Blue	Few
Mean host egg volume:		93 (range 69–125%)		
Hodgson's Hawk Cuckoo (egg volume 3.76 cc)				
Small niltava (C,O)	23	109	Buff	Faint
Lesser shortwing (G,S)	14	57	Greenish	Yes
Blue-throated flycatcher (G,C)	7	52	Olive	Yes
Plumbeous water redstart (G,C)	7	57	Grayish	Yes
Indian blue robin (G,O)	3	61	Blue	No
White-browed fantail (S,O)	2	37	Buffy	Yes
Brownish-flanked bush warbler (S,S)	2	42	Brown	No
Little spider hunter (S,S)	2	43	Pinkish	Yes
Streaked spider hunter (S,S)	2	78	Brown	No
Spotted forktail (G,C)	2	101	Cream	Faint
Mean host egg volume:		63.7 (range 42–109%)		
Indian Cuckoo (egg volume 4.62 cc)				
Striated laughingthrush (S,O)	5	200	Blue	No
Indian Gray thrush (S,O)	2	110	Varied	Yes
Mean host egg volume:		155 (range 110–200%)		
Oriental Cuckoo (egg volume 1.87 cc)				
Blyth's leaf-warbler (G,C)	24	65	White	No
Yellow-vented warbler (G,S)	6	56	White	No
Chestnut-crowned warbler (G,S)	6	53	White	No
White-spectacled warbler (G,S)	5	65	White	No
Golden-spectacled warbler (G,S)	3	66	White	No

TABLE 12 *(continued)*

Cuckoo species/host species[b]	Clutch records	Egg volume (%)	Ground color	With spots?
Inornate warbler (G,S)	2	52	White	Yes
Tawny-breasted wren-babbler (G,S)	2	99	White	Yes
Eyebrowed wren-babbler (G,S)	2	115	White	Yes
Mean host egg volume:		71.4 (range 52–115%)		
Asian Lesser Cuckoo (egg volume 2.17 cc)				
Brownish-flanked bush warbler (S,S)	23	73	Brown	No
Blyth's leaf-warbler (G,C)	8	56	White	No
Manchurian bush-warbler (S,O)	5	101	Red	Faint
Yellow-vented warbler (G,S)	4	48	White	No
Western crowned-warbler (G,C)	4	57	White	No
Gray-bellied wren (S,S)	4	68	Pink	Yes
Dark-necked tailorbird (S,O)	3	47	Varied	Yes
Yellow-bellied prinia (S,S)	3	49	Red	No
Scaly-breasted wren-babbler (G,S)	3	89	White	Few
Gray-hooded warbler (G,S)	2	58	White	No
Pygmy wren-babbler (S,S)	2	69	White	No
Mean host egg volume:		65 (range 47–101%)		

[a]Includes all *Cuculus* host species (except for *C. canorus,* which is summarized in the previous table) for which Baker (1942) possessed at least two parasitized clutches. Becking (1981) regarded Baker's identification of the common hawk cuckoo and Indian cuckoo eggs as being unreliable. Species are arranged by diminishing number of clutches and by increasing egg volume for species having the same total number of clutches.

[b]Nest types in parentheses: G, ground or flat substrate; S, shrub or near-ground site; T, tree level; W, wetlands; C, cavity or crevice; O, open above; S, spherical.

blotches, or other more superficial pigments that are laid down on the shell surface near the lower ends of the oviduct or even perhaps deposited while the egg is already in the cloaca, shortly before the egg is deposited. Among the brood parasites, unpatterned, white eggs are typical of the honeyguides, viduine finches, and one of the New World ground cuckoos, the striped cuckoo. These groups parasitize species that lay their eggs in dark places, either in cavities or in constructed but enclosed nests, and which themselves lay white eggs. It is impossible to attribute egg matching to evolved mimicry in these cases, inasmuch as both groups may well have evolved from cavity-nesting ancestors. Similarly, the black-headed duck lays an unmarked, buff-colored egg that differs little from the eggs laid by many dabbling and diving ducks and can scarcely be regarded as host mimetic, although it is sometimes nearly impossible to distinguish the eggs of this species from those of the rosy-billed pochard, which is probably a significant host.

It is in the parasitic cuckoos and icterine brood parasites that examples of egg mimicry can be found. The parasitic cowbirds have generally not been regarded as egg mimics, mainly because of their broad ranges of host usage. However, the screaming cowbird is host specific on the bay-winged cowbird, and the giant cowbird is largely or entirely dependent on the oropendolas and caciques as hosts. In the screaming cowbird, both the ground color and the superficial markings vary considerably, as do those of their host the bay-winged cowbird, and apart from being somewhat more spherical and the markings more vague, their eggs are scarcely distinguishable (Hoy & Ottow, 1964). Whether these similarities are the result of common descent or mimetic matching is debatable.

Likewise, in Panama, the giant cowbird lays eggs that in Smith's (1968) view include some distinctly mimetic types. Smith suggested that mimetic populations of giant cowbirds exist in which not only the eggs of a particular species of oropendula might be mimicked in both size and color, but in which such mimicry might extend to the level of local host populations. Smith listed five types of giant cowbirds, including three different oropendola mimics, a cacique mimic, and a fifth "dumper" type that laid nonmimetic eggs. Later observations (Fleischer & Smith, 1992) have not fully confirmed this view.

Egg mimicry in the Old World cuckoos, and especially the common cuckoo, is a topic of such long history, and with such a large associated literature (e.g., Rey, 1892; Baker, 1913, 1923, 1942; Jourdain, 1925; Southern, 1954; Wyllie, 1981), that it is impossible to do more here than touch on a few highlights. The first to effectively establish the existence of intraspecific host mimicry in the common cuckoo was Baldamus (1853), who concluded that each female lays eggs of a single type, which generally match that of the host in color and pattern. Additional support for host mimicry was almost simultaneously provided by Brehm (1853), who observed this phenomenon in the great spotted cuckoo and its European corvid hosts. Baker's classic studies (1913, 1923, 1942), based on his personal collection of more than 6000 cuckoo eggs and associated host clutches, convinced him that a high level of egg matching has evolved among the cuckoos. This trait was especially apparent in India and surrounding countries, from where Baker's egg collections were more extensive. He believed that in India, where habitats were (at least at that time) less affected by ecological disturbance than in Europe, the evolution of a habitat-based evolution of host-specific female subpopulations ("gentes") was most apparent, whereas in areas such as England, interbreeding on once-distinct populations of host-specific females had resulted in a low degree of egg mimicry.

Baker also believed (1923) that geographic variations in a host species' egg patterns are sometimes matched by that of the parasite, as, for example, the apparent local matching of host egg types (of Indian crows) by the Asian koel. He anticipated the genetic problems of the evolution of local strains of host-specific females producing polymorphic egg types by hypothesizing that egg coloration and markings are controlled by and hereditary in females, and that the males have no influence on the color and pattern of eggs laid by their female offspring. Since in birds it is the female that is the heterogametic sex, the genes regulating egg color must be located on the W chromosome so her egg-pigmentation traits would be directly transmitted to her daughters (Jensen, 1966; Becking & Snow, 1985).

With regard to the local Indian (*telephonus*) race of the common cuckoo that he studied intensively, Baker believed that six recognizable egg types are produced by various strains of females. These egg types range from white eggs with only slight speckling or dull-colored eggs with dense brownish blotching to beautiful blue-colored eggs lacking any markings or with some darker spotting present. Referring to cuckoos collectively, Jourdain (1925) suggested that entirely white eggs, characteristic of most nonparasitic cuckoos, represent the ancestral avian/reptilian type. Unpatterned, uniformly blue- or green-tinted eggs (produced by the bile pigment cyanin) may represent the most primitive type of colored eggs, rather than a highly evolved mimetic type as suggested by Baker. Additions of hemoglobin-derived brown to reddish-brown pigments might follow as later specializations; these pigment patterns might range from fine overall stippling to streaks, smears, spotting, or blotching, the latter often forming a ring or zone around the more rounded end. In Jourdain's view, the most advanced type of egg mimicry is not found in the striking and immaculate blue (or sometimes rust-red) eggs that closely match some of their hosts' equally beautiful eggs, but rather in those eggs precisely matching some of the Old World warbler (*Acrocephalus, Sylvia*) and bunting (*Fringilla, Emberiza*) hosts. These host species have eggs characterized by complex patterns of streaks and spots on a light ground color, which are sometimes almost perfectly matched by the cuckoos. Jourdain agreed with Baker that egg mimicry evolved as a direct result of the selective forces imposed by host discrimination. However, Jourdain suggested that the relative degree of perfection in egg mimicry by brood parasites is not necessarily a reflection of the age of this general process, but rather derives from the intensity of the selective process that has resulted from variable degrees of egg discrimination and subsequent nest desertion or rejection of inadequately mimetic eggs by the host species.

This process of host adaptation might be expected to result in ever more perfect host specialization and perhaps a proliferation of separate gens (see next section), especially in areas where a variety of potential hosts might exist. A possible example of regional host shifting associated with varied competition for hosts, and evidence for alloxenia (occurrence of nonoverlapping host specificity among sympatric brood parasites) has been described in Japan (Haguchi & Sato, 1984). There the oriental cuckoo and lesser cuckoo occur sympatrically in Honshu, where the lesser cuckoo selectively parasitizes the Japanese bush warbler and lays mimetic brown eggs. In that same region, the oriental cuckoo lays whitish eggs and parasitizes various crowned warblers. However, in Hokkaido, which is north of the breeding range of the lesser cuckoo, the oriental cuckoo parasitizes the Japanese bush warbler, and there it too lays mimetic brown eggs.

Host-specific Gentes versus Generic Similarity

Among the brood parasites of the world, egg mimicry was first recognized in the common cuckoo, and it was also in the common cuckoo that variations in egg mimicry of varied hosts by intraspecific brood parasites was first recognized. Newton (1896) described these subpopulations as "gentes" (singular, "gens"). This useful term has no formal taxonomic significance, although the word suggests that a genetic basis for this polymorphism must exist in female host-choice behavior and/or egg mimicry capacity. Clearly, not only is some kind of a genetic basis required for controlling the egg color and pattern generated among individual females, but also an effective mechanism is needed to reduce or eliminate the tendency of females to lay any of their eggs in nests of species having eggs unlike their own. Newton believed that about half of the cuckoo species that had been studied as of that period probably exhibited this trait. Later evidence supporting this idea came primarily from Chance (1922), who provided detailed information on the egg-laying behavior of a few cuckoos over a period of several years. He observed that little variation in the shape, color, or superficial markings occurred among eggs laid by individual females. Additionally, individual females almost invariably laid their eggs in the nests of a single host species. Thus, in one case, over a period of four breeding seasons, one female laid 58 of her total 61 eggs only in meadow pipit nests. Therefore, two of the major criteria that might allow for gentes evolution, individual female specificity in egg characteristics and in host choice, would seem to have been established.

Although a quick perusal of the popular literature of brood-parasitic cuckoos might lead one to believe that a refined degree of host-specific egg mimicry and gentes development is a

TABLE 13 Traits of Major Biological Host Species of the Australian Cuckoos[a]

Cuckoo species/host species	% of ROP	Mass ratio (%)	Egg match	Range match
Pallid Cuckoo (1052/111, 83.1 g)				
Red wattlebird	5.2	120	2	3
Yellow-faced honeyeater	5.0	19	2	2
White-plumed honeyeater	4.4	23	1	2
Yellow-tufted honeyeater	3.9	26	2	1
White-naped honeyeater	3.7	16	2	2
Singing honeyeater	3.3	32	3	3
Willy wagtail	3.1	24	1	3
Rufous whistler	2.5	29	2	3
Brown-headed honeyeater	2.3	14	2	2
White-eared honeyeater	2.2	24	2	2
Bell miner	2.0	36	3	1
White-rumped miner	2.0	69	3	2
Mean host mass		36		
Brush Cuckoo (376/58, 36.2 g)				
Bar-breasted honeyeater	14.6	36	2	2
Gray fantail	12.8	22	3	3
Brown-backed honeyeater	9.8	36	3	1
Scarlet robin	5.8	36	3	1
Leaden flycatcher	5.6	36	3	3
Mean host mass		33		
Fan-Tailed Cuckoo (662/81, 46.4 g)				
White-browed scrubwren	19.9	31	2	2
Brown thornbill	14.9	17	1	2
Yellow-throated scrubwren	5.4	34	0	1
Speckled warbler	3.5	28	1	2
Large-billed scrubwren	3.3	22	3	1
Superb blue fairywren	3.0	22	3	2
Mean host mass		26		
Black-Eared Cuckoo (163/23, 29.1 g)				
Speckled warbler	49.1	44	3	1
Redthroat	23.9	41	3	2
Mean host mass		42.5		
Horsfield/s Bronze Cuckoo (1555/97, 22.7 g)				
Superb blue fairywren	14.6	44	3	1
Splendid fairywren	11.5	40	3	2
Yellow-rumped thornbill	5.3	44	2	2
Blue-and-white fairywren	5.3	35	3	1
Red-capped robin	4.6	40	2	3
Variegated fairywren	4.1	35	3	3
Scarlet robin	3.1	57	1	1
White-fronted chat	2.9	79	2	2
Buff-rumped thornbill	2.9	35	3	1
Brown thornbill	2.7	35	3	1
White-browed scrubwren	2.3	65	0	1

TABLE 13 *(continued)*

Cuckoo species/host species	% of ROP	Mass ratio (%)	Egg match	Range match
Chestnut-rumped thornbill	2.3	26	3	2
Broad-tailed thornbill	1.9	31	3	2
Southern whiteface	1.7	62	2	2
Western thornbill	1.5	31	3	1
Mean host mass		44		
Shining Bronze Cuckoo (909/82, 23.4 g)				
Yellow-rumped thornbill	26.0	43	2	2
Brown thornbill	8.0	34	1	3
Buff-rumped thornbill	5.9	34	1	3
Striated thornbill	4.9	30	1	3
Yellow thornbill	4.6	26	1	3
Western thornbill	4.2	30	1	1
Broad-tailed thornbill	4.1	30	1	1
Superb blue fairywren	3.4	43	1	3
Mean host mass		34		
Gould's and Little Bronze Cuckoos (193/23, 16.9 g)				
Large-billed warbler	52.3	41	1	2
Fairy warbler	7.8	41	1	2
Mean host mass		41		
Australian Koel (196/21, 224.8 g)				
Magpie lark	15.8	42	2	3
Figbird	15.8	52	2	2
Little friarbird	12.7	28	3	3
Noisy friarbird	11.7	41	3	2
Mean host mass		42		
Channel-Billed Cuckoo (138/9, 610.8 g)				
Crows (several species)	51.4	63–101	2	3
Pied currawong	33.3	47	3	2
Mean host mass		~74		
Overall Egg/Range Match	0	1	2	3
Egg similarity (%)	4	25	30	41
Range overlap (%)	—	28.5	43	28.5

[a]Includes all species having at least 20 records of parasitism (ROP) as listed by Brooker & Brooker (1989b). Numbers in parentheses after cuckoo's name are total host records/number of host species, and mean adult mass. "Egg match" categories are 3 = eggs similar in volume, color, and patterning; 2 = eggs similar in two of these traits; 1 = similar in one of these traits; 0 = no similarity in egg traits. "Range match" categories (3 = high, 2 = intermediate, 1 = low) are estimates of breeding range overlap between parasite and host, especially the percentage of the parasite's breeding range that is also occupied by the host species.

common, but not the predominant, breeding strategy employed by this group of parasites, the actual situation is somewhat less impressive. For example, among the major biological hosts of 10 species of parasitic Australian cuckoos (table 13), no similarity in egg traits between the cuckoo parasites and their hosts was apparent in 4% of the cases; 25% were simi-

TABLE 14 Egg Traits of the Australian Cuckoos' Biological Hosts[a]

Cuckoo species	Host species	% Egg match	Mean egg volume (%)	Egg volume, range (%)
Pallid cuckoo	37	22	80.2	40–219
Brush cuckoo	10	60	94.7	70–159
Fan-tailed cuckoo	17	44	74.8	48–172
Black-eared cuckoo	2	100	82	76–88
Horsfield's bronze cuckoo	28	57	115	69–233
Gould's and little bronze cuckoos	4	0	77.7	67–89
Shining bronze cuckoo	10	0	85.3	80–107
Australian koel	6	50	83.5	63–103
Channel-billed cuckoo	9	33	91	77–104
Weighted mean	15	37	79	65–140

[a]Based on data of Brooker & Brooker (1989b). "Percent egg match" refers to percentage of host species having eggs that resemble parasite's in three attributes (volume, color, and patterning), and mean egg volume is that of all hosts relative to parasite.

lar in one of three traits (volume, color, or pattern), 30% were similar in two of these three traits, and 41% were similar in all three traits. The highest level of egg matching occurred in the black-eared cuckoo, whose eggs matched those of both of its two major hosts in all three of these traits (table 14). However, in the shining Gould's and little bronze cuckoos, there was no perfect matching for any of the 10 biological hosts, and the average overall egg-matching score was only 1.1. Similarly, Rey (1892) reported that, of 139 common cuckoo eggs found in red-backed shrike nests, only 12 resembled those of the host. However, Moksnes et al. (1995) reported that, among a sample of 11,870 cuckoo eggs from parasitized nests, blue eggs were found in a higher proportion of parasitized nests of host species laying blue eggs (mainly the redstart), than in host species not laying blue eggs.

Among the parasitic cuckoos of India, evidence of close host parasite volume matching is virtually nil. The volume of the parasite's eggs fall within 10% of the host's egg volume in less than 15% of the 80 host–parasite combinations (tables 10–12). Likewise, in the brown-headed cowbird, for only 12% of the 145 fostering host species does the cowbird's egg volume fall within 10% of the host's (table 15).

The best available discussion of gentes evolution, especially from a modern genetic viewpoint, comes from Southern (1954). He concluded that cuckoo gentes exhibit some characteristics of genetic polymorphs, especially those associated with Batesian mimicry, with as many as four different host-specific gentes occurring in the same general location. He believed

FIGURE 11. Distribution of major host usages by common cuckoos in Europe, plus associated host-parasite egg similarities. Map mainly after Southern (1954), with updating for reed warbler. Egg sketches (after Cramp, 1985) show eggs of cuckoo (left) and corresponding hosts. Host species not identified on the map are the European redstart (open inverted triangles), red-backed shrike (open squares), and sedge warbler (solid diamonds). The lines outlining each host species' major use localities suggest minimum distributions of gentes.

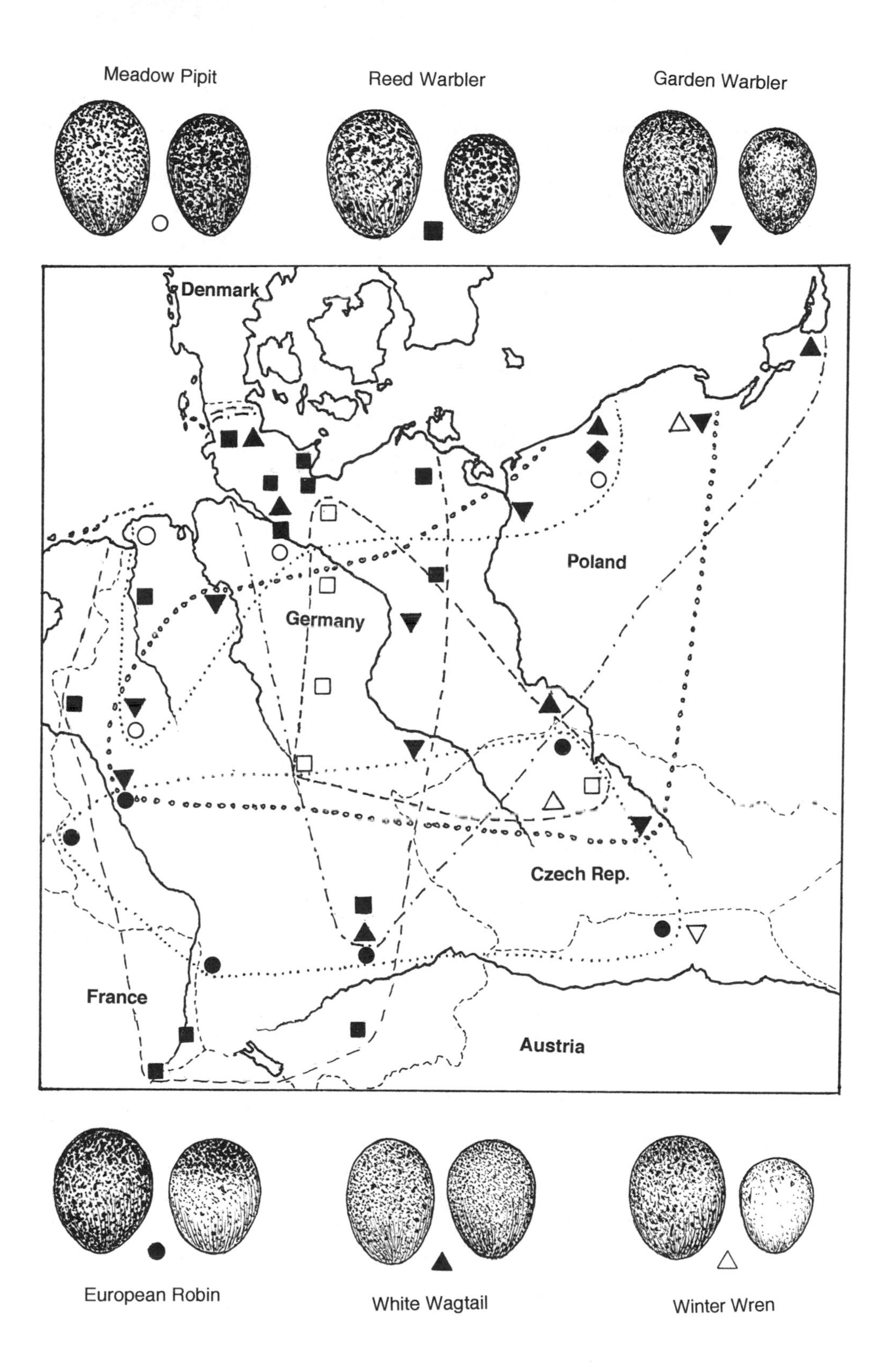

Meadow Pipit
Reed Warbler
Garden Warbler
Denmark
Poland
Germany
Czech Rep.
France
Austria
European Robin
White Wagtail
Winter Wren

that new gentes may arise in restricted areas and gradually spread from such locations through other parts of the species' range. However, some features of gentes are more like those typical of biological races than polymorphs, such as the occurrence of intermediate types of cuckoo eggs in those regions having the least topographic isolation between geographically separated gentes.

Southern also suggested that a female's attachment to its appropriate host probably occurs through early learning while she is still in the nest. However, some egg-laying "mistakes" are probably made, and those females hatching from such nests may become attached to the wrong hosts. Southern believed that cuckoo gentes cannot be maintained by normal genetic polymorphism mechanisms alone. Rather, they must be genetically supported by some degree of microgeographic isolation or habitat differences, thus improving the chances of increased mimicry and reducing the chances of improper laying choices by females within local host populations. Presumably, the degree of egg mimicry is likely to improve gradually with increased duration of such isolation, but the same sorts of selective pressures that might allow gentes to evolve fairly rapidly might also cause them to disappear equally rapidly under conditions of ecological fragmentation or other sources of range alterations among host species.

In attempting to analyze the mimicry and gentes situation in Europe, Southern was forced to describe it as a "desperate tangle," with only generally "muddled" and "often dubious" information available. Southern found evidence of present-day egg mimicry only in regions with large tracts of homogeneous habitat. Two well-adapted European egg mimics were apparent, a pale blue ("redstart type") morph that occurs, for example, in southern Finland, central Sweden, the Czech Republic, Slovakia, and east-central Germany (Saxony). Another egg morph, a densely patterned and freckled type, closely mimics the great reed-warbler in areas such as the marshes of Hungary. Southern noted and mapped several other examples of localized host preference (fig. 11), but the overall modern-day pattern of cuckoo parasitism in western Europe seems to be one of deteriorating, rather than improving, egg mimicry, and correspondingly little contemporary evidence for the occurrence of host-specific gentes.

An alternative reproductive strategy has seemingly been pursued by many brood-parasitic species, such as the brown-headed cowbird. This highly successful brood parasite exhibits no significant regional or local variations in egg volume, color, or patterning; instead, the species has apparently opted for a breeding strategy involving a broadly generalized degree of similarity to the eggs of its many primary hosts, which mainly consist of emberizines, flycatchers, vireos, and New World warblers (see tables 15 and 16). Parasite-to-host egg similarities, at least in ground color and pattern, are fairly close in the case of a few of the cowbird's icterine hosts such as meadowlarks, but these similarities can probably be attributed to closeness of descent rather than the result of evolved egg mimicry. Of its major fostering hosts, only the yellow-breasted chat's eggs are fairly similar in size, color, and pattern to the cowbird's, and of the occasional to infrequent fostering hosts, only the house sparrow, lark sparrow, northern oriole, and horned lark eggs are fairly similar in all three of these qualities. Altogether, only about 5% of the cowbird's hosts have egg characteristics that are fairly well matched in all three qualities by the cowbird.

3

BEHAVIORAL AND REPRODUCTIVE ECOLOGY

> Finally, it may not be a logical deduction, but to my imagination it is far more satisfactory to look at such instincts as the young cuckoo ejecting its foster-brothers, ants making slaves, the larvae of ichneumonidae feeding within the live bodies of caterpillars, not as specially endowed or created instincts, but as small consequences of one general law leading to the advancement of all organic beings—namely, multiply, vary, let the strongest live and the weakest die.
>
> Darwin, *The Origin of Species by Means of Natural Selection*

Behavioral Ecology of Brood Parasites

In his seminal review of the ecology of avian brood parasitism, Payne (1977a) identified host selection and specialization, breeding season synchronization between host and parasite, and several additional reproductive strategies as important ecological aspects of brood parasitism. These additional reproductive strategies include the parasite's mating systems, its population structure, and its demographic characteristics. They also include the parasite's relative egg size, as well as possible host mimicry involving the parasite's eggs or chicks. Last, but of equal importance, are the parasite's fecundity-related adaptations, such as its clutch size, rate of egg laying, seasonal egg production, and egg dispersion patterns. Some of these primarily morphological aspects of brood-parasite strategies, such as egg and chick mimicry, were discussed in chapter 2. Other more strictly behavioral reproductive strategies of brood parasites, such as host-selection behavior, will be dealt with in chapter 4. The remaining strategies of brood parasites that have particular ecological interest are discussed below.

Mating Systems and Breeding Dispersion Patterns

The foremost prediction about mating systems of brood parasites is that polygamy or promiscuity should be the norm, since there is no need for either sex to tend the eggs or young, and it is not only to the male's advantage to disperse his gametes widely, but also for the female to lay her eggs over a broad area holding many host pairs, and thereby perhaps encountering a variety of males able to fertilize her eggs throughout a potentially extended breeding season. If pair-bonds are not formed, there is little purpose for territorial defense of resources by males, and thus singing or other advertising should simply be concerned with female attraction or domination of other sexually competing males.

In line with this expectation, the only brood parasites believed to have fairly strong pair-bonds are the crested cuckoos, in which the participation of the male may be important in luring the generally larger hosts from the nests long enough for the female to deposit her eggs. Otherwise, polygyny or promiscuity appears to be the typical pattern in the parasitic cuckoos. Little specific information is available for any parasitic cuckoos except the common cuckoo, in which polygyny or promiscuity seems to be the most likely breeding system, although this is uncertain. In the common cuckoo, males occupy somewhat overlapping "song ranges" and females maintain similarly partially overlapping "egg ranges." Males may sing within areas of about 30 ha and may travel at least 4 km to forage. Within these singing areas, males are probably able to expel subordinate males and gain preferential sexual access to females. Females may also lay most of their eggs within areas of about 30 ha. However, the ranges of individual males and females do not coincide, as might be expected if monogamy were to prevail (Wyllie, 1981).

In the honeyguides, brief pair-bonds are believed to be formed (Fry et al., 1988). However, in the orange-rumped honeyguide, promiscuous mating has been seen, with as many as 18 females observed mating with a single male (Cronin & Sherman, 1977).

Among the viduine finches, polygyny is the typical mating pattern, with male indigobirds sometimes mating with several different females in the course of a single day. Males sing from call sites that they may solely occupy for several days. These sites are defended by the dominant singing males, although much aggression occurs among males that contest site ownership. Copulations occur only at the call sites and are performed exclusively by dominant males. Call sites are often uniformly spaced at distances that may be a 100 m or less apart, producing a dispersion pattern similar to that of typical avian territoriality, although interspecific as well as intraspecific dispersion patterns are typical of indigobirds (Payne, 1973a).

Among the parasitic cowbirds, only the brown-headed cowbird's mating system has been well studied. Although most evidence supports monogamous pair-bonding (Yokel, 1986, 1987), other field data indicate that a promiscuous mating system exists (Elliott, 1980). The mating system may in fact be a fluid one, with adult sex ratios determining whether males can afford to leave their mate and try to fertilize other females, as may occur when sex ratios approach equality, or whether an excess of males in the population forces males to remain with their mates and guard them more closely from the sexual advances of other males (Teather & Robertson, 1986). Monogamous matings may be possible in areas where host density is high, so that females can occupy a predictable area over a prolonged period, but host-nest scattering may increase female mobility and increase the probability of promiscuous matings (Teather & Robertson, 1986). Monogamous matings might also be favored in some low-density cowbird populations, under which conditions mate-guarding by males might be the most effective mating system (Yokel & Rothstein, 1989). Females are evidently nonterritorial and may travel several kilometers daily between roosting, foraging, and breeding (Thompson, 1993). Similarly, female shiny cowbirds may travel up to 4 km between foraging and breeding areas on a daily basis (Woodsworth, 1993). According to one radiotelemetry study, male brown-headed cowbirds may occupy nonexclusive territories of about 4.5–5.5 ha. These areas typically encompass the comparable but somewhat smaller ranges (of about 2–3.5 ha) occupied by one to several female mates (Teather & Robertson, 1986). There is no good evidence of associated territorial defense of resources by either sex among the parasitic cowbirds. Males probably provide females with nothing more than genes, and females are probably able to make active mate choices on the basis of overall quality (Yokel & Rothstein, 1991).

Breeding Cycle Synchronization Strategies

There is little evidence, pro or con, on the degree to which brood parasites are able to synchronize their own breeding cycles with those of their hosts. In North America, the egg-laying period of the brown-headed cowbird is directly related to latitude and associated lengths of the summer period (see species account for representative durations). These durations must overlap with those of most other locally breeding passerines, including numerous host as well as nonhost species. Of course, this situation of generally synchronized breeding of temperate-zone passerines is advantageous to the cowbird, but it does not require any special physiological adaptations.

In Britain, the laying season of the common cuckoo generally coincides with the peak laying periods of its major hosts. In seasonal laying progression, these include the European robin, hedge accentor, sedge warbler, pied wagtail, and reed warbler, species whose laying periods collectively extend from early May to early July (Lack, 1963). This approximate 12-week breeding period is similar to that of the North American nonparasitic cuckoos (black-billed mostly late May to late July, yellow-billed mid-May to mid-August), although individual females are unlikely to have laying seasons more than 6 or 7 weeks long (Wyllie, 1981).

Using previously collected field data, Southern (1954) analyzed the temporal and ecological aspects of brood parasitism among the 15 Indo-tropical species of brood parasites occurring there. He concluded that competition among these brood parasites may be reduced by interspecific differences in breeding times, with associated altitudinal and/or habitat variables, or by having different host-choice preferences (alloxenia) (fig. 12).

Behavioral Ecology of Host Selection

It is difficult to categorize such host generalists as the brown-headed cowbird as having any specific nest-selection strategies. No other brood parasite has such an extensive list of known victims (more than 200, including unsuitable as well as rejector hosts), nor half as many known biological or fostering hosts. The 145 known fostering hosts (Friedmann & Kiff, 1985) of this species are listed in table 15, together with estimated egg volumes, basic (or ground) color of eggshells, pigmentation patterns of eggs ("spots" here include markings ranging from fine stippling to broad blotching), and estimates of incubation and nestling periods. Four host species (yellow warbler, chipping sparrow, red-eyed vireo, and song sparrow) have at least 1000 records of parasitism, and probably represent about 25% of the total parasitism records. Based on frequencies of occurrence (see table 19 for summary), the most typical features of host-selection behavior are that female cowbirds are likely to select open, cuplike nests (89% of host species) in shrub- or tree-level sites (69% of hosts). The host eggs are likely to be variously patterned (85% of hosts) but are otherwise basically white or whitish (66% of hosts).

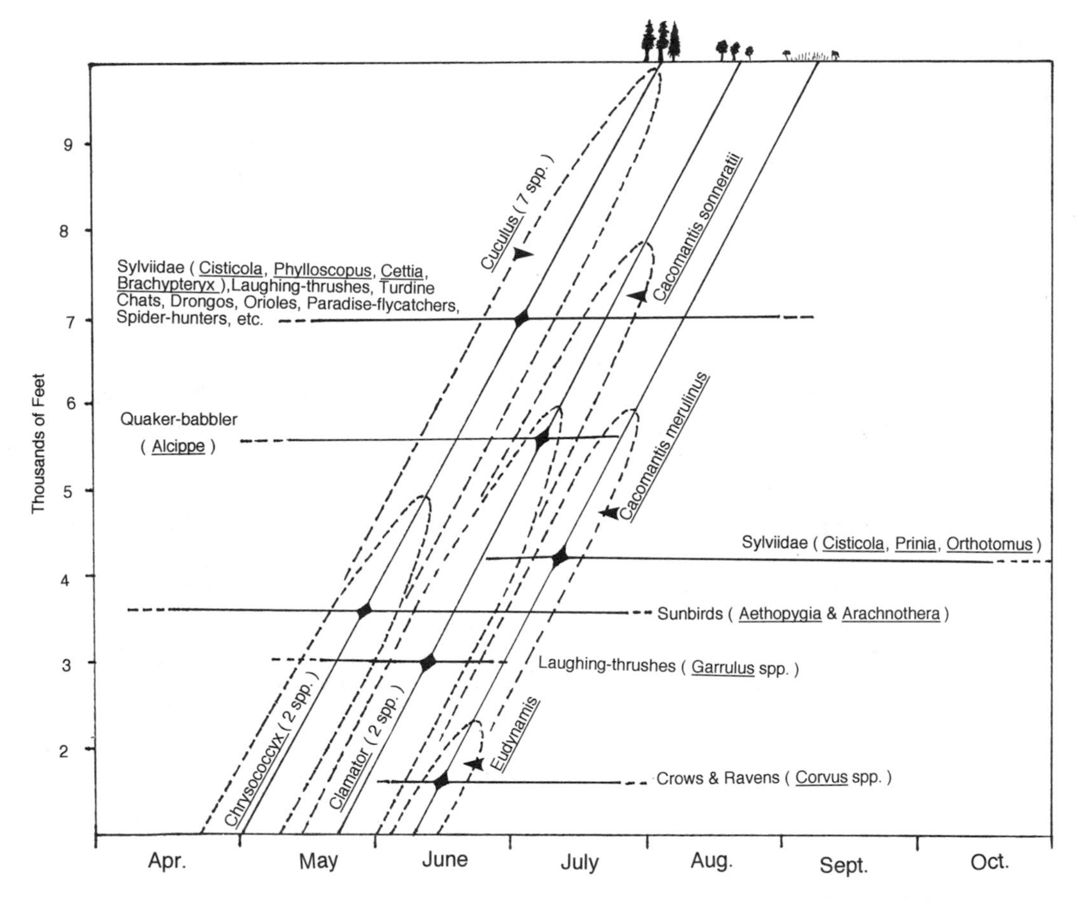

FIGURE 12. Breeding chronologies and ecological distributions of the Indian cuckoos and their major hosts. Horizontal lines represent approximate lengths of hosts' breeding seasons, with enlarged intersections indicating major host-parasite combinations. Diagonal lines represent preferred habitats (moderate to dense forests, light forests or scrub, and open country or villages) and the approximate elevational limits of each cuckoo species or genus. Derived from data of Baker (1942).

The eggs are also likely to be about 30% smaller in volume than the cowbird's, and only occasionally (18% of the host species) are larger. The host species is also likely to have an incubation period of about 13 days (or 1–1.5 days longer than the cowbird) and a fledging period of about 11 days (or about 2.5 days longer than the cowbird). In a similar analysis of host traits, Petit and Petit (1993) concluded that cowbirds primarily parasitize hosts breeding in deciduous forests, open woodlands, shrubby habitats, and, to a lesser degree, grassland and coniferous forest species. Open-cup nesters are preferred, and shrub-level nests are preferred over ground nests, as were smaller hosts (under 40 g) over larger ones. Of the various environmental variables affecting host choice, habitat type was judged to be among the most important, and life-history traits of the host were less important than host habitat type or nest placement characteristics.

TABLE 15 Breeding Traits of the Brown-headed Cowbird's Fostering Host Species[a]

Hosts	Host egg/ nestling traits				
		Eggshell			
	Volume (%)	Base color	With spots?	Incubation (days)	Nestling period (days)
Minor Fostering					
Golden-crowned kinglet (NF,CP)	22	Cream	Yes	14–15	14–19
Ruby-crowned kinglet (NF,CP)	29	White	Yes	14–15	12 (?)
Carolina chickadee (EF,DC)	31	White	Yes	12–14	17
Verdin (SX,SS)	31	Blue	Yes	14	21
Black-capped chickadee (EF,DC)	36	White	Yes	12–14	16
Red-breasted nuthatch (WF,NO)	36	White	Yes	12	14–21
Brown creeper (NF,DC)	36	White	Yes	14–15	14–16
Virginia's warbler (SX,SO)	39	White	Yes	(?)	(?)
Black-throated gray warbler (WF,CO)	39	White	Yes	(?)	(?)
Orange-crowned warbler (WF,GO)	45	White	Yes	12–14	8–10
Townsend's warbler (WF,GO)	48	White	Yes	12 (?)	8–10 (?)
Hermit warbler (WF,CO)	48	White	Yes	(?)	(?)
Yellow-throated warbler (EF,CO)	48	White	Yes	12–13 (?)	(?)
Grace's warbler (SF,CO)	48	White	Yes	(?)	(?)
Palm warbler (NF,SO)	48	White	Yes	12	12
Dusky flycatcher (WF,SO)	48	White	No	12–15	18
Blackpoll warbler (NF,SO)	51	White	Yes	11	10–11
Bay-breasted warbler (NF,SO)	51	White	Yes	12–13	11
Hutton's vireo (WF,DO)	51	White	Few	14–16	14 (?)
Gray flycatcher (WF,SO)	51	White	No	14	16
Tree swallow (NF,DC)	54	White	No	13–16	16–24
Wrentit (WF,SO)	60	Blue	No	15–16	15–16
Philadelphia vireo (NF,DO)	63	White	Few	13–14	13–14
Carolina wren (EF,DC)	72	White	Yes	12–14	12–14
Five-striped sparrow (SX,SO)	76	White	No	12–13	9–10
Seaside sparrow (EL,GO)	90	White	Yes	11–12	9
Olive-sided flycatcher (NF,CO)	94	Cream	Yes	16–17	15–19
Phainopepla (SX,DO)	94	White	Yes	14–15	18–19
Scissor-tailed flycatcher (SF,SO)	108	White	Yes	12–13	14–16
Western kingbird (WF,DO)	112	White	Yes	12–13	16–17
Western tanager (WF,CO)	112	Blue	Yes	13	13–15
Evening grosbeak (NF,CO)	116	Blue	Yes	12–14	13–14
European starling (In,NO)	222	Blue	No	12–15	20–22
Mourning dove (PF,SO) (dubious record)	225	White	No	14–16	13–15
Occasional Fostering					
Lesser goldfinch (WF,DO)	36	Blue	No	12	(?)
Nashville warbler (NF,GO)	36	White	Yes	11–12	11
Northern parula (EF,DP)	39	White	Yes	12–14	(?)
Wilson's warbler (NF,GO)	39	White	Yes	11–13	10–11
Pine siskin (NF,CO)	41	Blue	Yes	13	14–15
Western flycatcher (WF,NO)	48	White	Yes	14–15	14–18
Golden-winged warbler (EF,GO)	48	White	Yes	10–11	9–10
Black-throated blue warbler (EF,SO)	48	White	Yes	12	10
Blackburnian warbler (NW,CO)	48	White	Yes	11–12	(?)

(continued)

TABLE 15 *(continued)*

Hosts	Host egg/ nestling traits: Volume (%)	Eggshell: Base color	Eggshell: With spots?	Incubation (days)	Nestling period (days)
Cerulean warbler (EF,DO)	48	White	Yes	12–13	(?)
Brewer's sparrow (WX,SO)	48	Blue	Yes	13	8–9
Pine warbler (EF,CO)	51	White	Yes	10 (?)	10 (?)
Lincoln's sparrow (NF,GO)	63	Bluish	Yes	13–14	10–12
House finch (WF,DO)	63	Blue	Yes	12–14	14–16
Swainson's warbler (EF,SO)	72	White	No	13–15	12
Savannah sparrow (PG,GO)	72	Bluish	Yes	12	10–14
Orchard oriole (EF,DP)	76	Blue	Yes	12–15	11–14
Hooded oriole (WF,DP)	83	Blue	Yes	13	14
House sparrow (In,NS)	99	White	Yes	11–14	15
Northern oriole (PF,DP)	103	Bluish	Yes	12–14	12–14
Summer tanager (EF,DO)	111	Blue	Yes	12	(?)
Brown towhee (WX,SO)	131	Blue	Few	11	8
Brown thrasher (EF,SO)	157	Bluish	Yes	11–14	9–12
Infrequent Fostering					
Black-tailed gnatcatcher (WX,SO)*	29	Blue	Yes	14	9–15
Lucy's warbler (WX,CO)	31	White	Yes	(?)	(?)
Tenessee warbler (NF,GO)	39	White	Yes	11–12	(?)
Tropical parula (SF,DP)	39	White	Yes	(?)	(?)
Bewick's wren (EF,CO)	45	White	Yes	12–14	14
House wren (PF,CO)	45	White	Yes	13–15	12–18
Vermilion flycatcher (SX,DO)	48	Cream	Yes	14–15	14–16
Magnolia warbler (NW,CO)	48	White	Yes	11–13	8–10
Black-throated green warbler (EF,CO)	48	White	Yes	~12	8–10
Black-and-white warbler (EF,GO)	48	White	Yes	11	8–12
Canada warbler (NF,GO)	48	White	Yes	(?)	(?)
Alder flycatcher (EF,SO)	49	White	Few	12–13	13–14
Black-capped vireo (SF,DO)*	49	White	No	14–17	10–12
Gray vireo (SX,SO)	49	White	Few	13–14	13–14
Golden-cheeked warbler (SF,CO)	49	White	Yes	12	9
Northern waterthrush (NF,GO)	54	White	Yes	12	9–10
Worm-eating warbler (EF,GO)*	56	White	Few	13	10
Western wood-pewee (WF,DO)	60	Cream	Yes	12–13	14–18
Mourning warbler (NF,GO)	60	White	Yes	12–13	7–9
MacGillivray's warbler (WF,SO)	60	White	Yes	13	8–9
Hooded warbler (EF,SO)*	60	White	Yes	12	8–9
LeConte's sparrow (EnG,GO)*	60	White	Yes	11–13	(?)
Rufous-winged sparrow (WX,SO)	63	Bluish	No	(?)	9–10
Dark-eyed junco (NF,GO)	63	White	Yes	11–12	10–13
Barn swallow (PS,NO)	67	White	Yes	14–16	17–24
Rock wren (WX,GC)	72	White	Few	14	14
Swamp sparrow (NL,WO)*	72	Bluish	Yes	12–13	9–10
Purple finch (NF,CO)	76	Blue	Yes	13	14
Louisiana waterthrush (EF,GO)	76	White	Yes	12–14	10
White-throated sparrow (NF,GO)	80	Blue	Yes	11–14	8–9

TABLE 15 *(continued)*

Lark sparrow (WG,GO)*	90	White	Yes	11–13	9–10
White-crowned sparrow (NF,SO)*	90	Blue	Yes	9–15	9–11
Cedar waxwing (PF,DO)	94	Blue	Yes	12–14	16–18
Bobolink (PG,GO)	94	Bluish	Yes	11–13	10–14
Scarlet tanager (EF,DO)	99	Blue	Yes	13–14	15
Fox sparrow (NF,GO)	99	Blue	Yes	12–14	9–11
Horned lark (PG,GO)	99	White	Yes	10–14	9–12
Hermit thrush (NF,SO)	106	Blue	No	12–13	10
Gray catbird (EF,SO)	111	Blue	No	12–13	10–11
Eastern kingbird (EF,DO)	131	White	Yes	12–13	13–14
Northern mockingbird (EF,SO)	136	Blue	Yes	11–14	12–14
American robin (PF,DO)	189	Blue	No	11–14	14–16
Eastern meadowlark (EG,GO)	189	White	Yes	13–15	11–12
Major Fostering					
Blue-gray gnatcatcher (EF,DO)[39]	29	Blue	Yes	15	12–13
Blue-winged warbler (EF,GO)[35]	39	White	Few	10–11	8–10
Prairie warbler (EF,SO)[35]	39	White	Yes	12–14	8–10
American redstart (EF,DO)]200+]	39	White	Yes	12–13	9
American goldfinch (PG,SO)[100+]	39	Blue	No	12–14	11–17
Chestnut-sided warbler (EF,SO)[75+]	41	White	Yes	12–13	10–12
Bell's vireo (EF,SO)[82]	48	White	Few	14	10–12
Yellow warbler (PF,SO)[1300+]	48	White	Yes	11–12	9–12
Yellow-rumped warbler (NF,CO)[60+]	48	White	Yes	12–13	12–14
Comon yellowthroat (PW,SO)[270+]	48	White	Yes	12	9–10
Clay-colored sparrow (EnG,SO)[50+]	48	Blue	Yes	10–11	9–12
Willow flycatcher (PF,SO)[150+]	51	White	Few	13–15	12–15
Chipping sparrow (PF,CO)[1,000+]	51	Blue	Yes	11–14	9–12
Field sparrow (NG,GO)[125]	51	White	Yes	10–11	7–8
Eastern wood-pewee (EF,DO)[60]	60	Cream	Yes	12–13	15–18
Acadian flycatcher (EF,DO)[59]	60	White	Yes	13–14	13–15
Kirtland's warbler (EF,GO)[75+]	60	White	Yes	14–15	12–13
White-eyed vireo (EF,SO)[57]	63	White	Few	12–16	(?)
Warbling vireo (PF,DO)[64]	63	White	Few	12	16
Kentucky warbler (EF,GO)[150+]	63	White	Yes	12–13	8–10
Lazuli bunting (WF,SO)[23]	63	Bluish	No	12	10–15
Indigo bunting (EF,SO)[600+]	63	Bluish	No	12–13	9–13
Painted bunting (SF,SO)[50]	63	White	Yes	11–12	8–9
Grasshopper sparrow (PG,GO)[26]	63	White	Yes	11–12	9
Chestnut-collared longspur (EnG,GO)[22]	63	White	Yes	11–13	9–11
Red-eyed vireo (EF,SO)[1000+]	66	White	Few	11–14	12
Solitary vireo (NF,CO)[60+]	66	White	Few	11–14	14 (?)
Prothonotary warbler (EF,DO)[90+]	67	White	Yes	12–14	10–11
Yellow-throated vireo (EF,DO)[100+]	68	White	Few	14	(?)
Eastern phoebe (EF,NO)[525+]	72	White	No	14–16	15–17
Ovenbird (EF,GO)[280]	76	White	Yes	11–14	8–10
Vesper sparrow (PG,GO)[70]	82	White	Yes	11–13	9–13
Dickcissel (EG,GO)[100+]	90	Blue	No	11–13	7–10
Veery (NF,SO)[150+]	108	Blue	No	10–12	10–12
Yellow-breasted chat (PF,SO)[180+]	108	White	Yes	11	8–11
Blue grosbeak (SF,SO)[30]	108	Blue	No	11	9–13
Song sparrow (PF,GO)[1300+]	108	Bluish	Yes	12–14	9–12
Rufous-sided towhee (PF,GO)[300+]	111	White	Yes	12–13	8–10

(continued)

TABLE 15 *(continued)*

Hosts	Host egg/ nestling traits				
		Eggshell			
	Volume (%)	Base color	With spots?	Incubation (days)	Nestling period (days)
Abert's towhee (WF,SO)[50+]	121	Blue	Yes	(?)	12–13
Northern cardinal (EF,SO)[250+]	136	White	Yes	11–13	9–11
Rose-breasted grosbeak (EF,DO)[43]	136	Blue	Yes	12–14	9–12
Red-winged blackbird (PW,SO)[450+]	136	Blue	Yes	10–12	10–11
Brewer's blackbird (WG,CO)[85]	152	Green	Yes	12–13	13
Wood thrush (EF,DO)[500+]	152	Blue	No	12–14	12–13
Western meadowlark (WG,GO)[160+][b]	207	White	Yes	13–15	12

[a]Host list based on Friedmann & Kiff (1985); parasitism rates judged mostly from earlier summaries, especially Friedmann (1963). Numbers in brackets indicate minimum number of known cases for major hosts, based mostly on Friedmann's summaries but with many additions, including Peck & James (1987). Some poorly documented species (*) may also be major local hosts. Letters in parentheses following each species' name indicates its zoogeographic breeding range relative to the Great Plains: En, endemic to Great Plains; E, east to southeast; N, north to northeast; S, south to southwest; W, west to northwest; P, pandemic; In, introduced, as well as its breeding habitat preferences; F, forest or forest edge; G, grassland; S, substrate dependent; W, wetlands; X, xeric scrub. These categories are mostly as per Johnsgard (1979). Final letters indicate typical nest sites: C, conifers; D, deciduous trees or trees generally; G, on ground or flat substrates; N, in niches or cavities; S, in shrubs or near-ground vegetation; O, open-above nest; P, pendant nest; S, spherical nest. Most breeding data are from Ehrlich et al. (1988) or Harrison (1978); estimated egg volumes are shown as percentages relative to the mean of the brown-headed cowbird (2.9 cc). "Spotted" eggs include those with stipples, streaks, or blotches.

[b]Not included on Friedmann's list of known fostering hosts, but included here on the basis of recent studies (Bowen & Kruse, 1994).

Another way to estimate ecological selectivity of host choice among brown-headed cowbirds is to compare the list of fostering hosts shown in table 15 with an ecological and zoogeographic analysis of all breeding birds of the Great Plains collectively, as shown in table 16. This comparison suggests that cowbirds have greater than random tendencies to parasitize birds nesting in forests, woodlands, and forest edges, are less likely to parasitize wetland species, and parasitize grassland and xeric species at rates close to expected. Somewhat higher than expected rates of parasitism occur among host species with eastern zoogeographic affinities, and lower than expected rates occur among species having pandemic distributions.

The host traits of the common cuckoos (tables 17–19) are in some respects surprisingly similar, considering that the cuckoo is generally regarded as a prime example of a host specialist. More than 125 species have been listed as hosts, including nearly 100 in Europe (Wyllie, 1981). Based on a summary of records of parasitism by fostering hosts in Britain (Glue & Morgan, 1972), four species (hedge accentor, reed warbler, meadow pipit, and European robin) account for 85% of the records of parasitism (table 18), and the first three of these are the most important hosts in the Netherlands, Belgium, and northern France (Wyllie, 1981). In central Europe, major hosts include the garden warbler, meadow pipit, white wagtail, and European robin. In central Russia, the tree pipit, European robin, and white wagtail are major hosts, whereas in southern Russia, the streaked scrub-warbler is a major host, and in the Amur region of Siberia, the thick-billed reed-warbler is one of the primary hosts (Wyllie, 1981). In

Japan, prime hosts include the great reed warbler, bull-headed shrike, and azure-winged magpie (Nakumura, 1990). For Eurasia as a whole, the five most commonly parasitized species are the reed warbler, hedge accentor, azure-winged magpie, great reed warbler, and marsh warbler, which collectively account for about 70% of the available records (table 17).

There is some ecological segregation of host selection in Britain: the reed warbler is selectively parasitized in lowland, freshwater habitats, the head accentor is chosen at most intermediate elevations, especially near human habitations or in woodland habitats, and the meadow pipit is mainly exploited in higher elevations moors, as well as in coastal habitats to some degree. All three of the major hosts exhibit low levels of egg rejection (in 2 of 184 cases), and the estimated fledging success rate for cuckoos is also fairly high in all of these major host species (49% for 383 nesting efforts, regardless of the stage of breeding when the nest was first found). Based on data involving 24 host species from Europe and elsewhere in Eurasia (table 19), the common cuckoo tends to select host species with shrub- or tree-level nests (60% of hosts) that are open and cuplike (92% of hosts). Preferred hosts lay eggs that are variably patterned (96% of hosts), but usually are not white or whitish in ground color (83% of hosts). Their eggs are typically about 75% as large (in volume) as the cuckoo's eggs, have incubation periods of 13–14 days (or about 2 days more than the cuckoo's), and have fledging periods of about 13 days (or about 4 days shorter than the cuckoo's).

Host-preference data for the Indian subcontinent (mainly Assam) are similar but less complete (table 19). In India, more ground- and shrub-level nest sites are parasitized and a greater diversity of nest types is used, including more nests made by species laying bluish or white

TABLE 16 Comparative Affinities of the Brown-Headed Cowbird's Hosts with the Total Great Plains Breeding Avifauna[a]

	Zoogeographic distributional affinities (%)							
Habitat	East	North	West	South	Pandemic	Endemic	Introduced	Total
Cowbird Hosts								
Forests, woodlands, and forest-edge	31	20	13	5	8	—	—	77
Wetlands	1	1	—	—	—	—	—	2
Grasslands	3	1	1	—	4	2	—	11
Xeric scrub	—	—	4	4	—	—	—	8
Other habitats	—	—	—	—	1	—	1	2
Total (145 spp.)	35	22	18	9	13	2	1	
Total Avifauna								
Forests, woodlands, and forest edge	20	11	8	2	10	—	—	51
Wetlands	4	4	5	—	8	1	—	22
Grasslands	1	1	2	—	2	5	—	11
Xeric scrub	—	—	2	2	—	—	—	4
Other habitats	2	1	2	2	3	—	2	12
Total (325 spp.)	27	17	19	6	23	6	2	

[a]Avifaunal affiliations based primarily on Johnsgard (1979); fostering hosts of brown-headed cowbird are listed in Table 15.

TABLE 17 Breeding Traits of the Common Cuckoo's Major Hosts in Eurasia[a]

Host species	Nesting habitat	Nest type	Egg traits: Clutch totals	Volume (%)	Eggshell: Color	Eggshell: Markings	Incubation
Reed warbler*	Marsh	W,O	1080+[b]	53	Greenish	Gray	~12
Hedge accentor*	Brush	S,O	323[c]	68	Blue	None	~13
Azure-winged magpie*	Woods	T,O	252[d]	153	Olive-buff	Brown	17–20
Great reed warbler*	Marsh	W,O	247	90	Greenish	Brown	14
Marsh warbler*	Marsh	W,O	215+[b]	62	Greenish	Gray	~12
Meadow pipit*	Open	G,O	118	66	Variable	Variable	~13
European robin*	Woods	G,O	93	75	White	Red-brown	13–14
Pied wagtail*	Open	G,O	79	76	Whitish	Brownish	~14
Bull-headed shrike*	Open	S,O	67[d]	111	Greenish	Gray-brown	14+
Garden warbler*	Woods	S,O	55	69	Whitish	Green-olive	12
Yellowhammer	Open	G,O	42	90	Variable	Black	13
Red-backed shrike*	Brush	S,O	39	105	Variable	Blackish	14
Common redstart*	Woods	G,C	38	57	Bluish	Few	13
Winter wren*	Woods	G,C	38	43	White	Red-brown	14
Whitethroat*	Brush	S,O	37	57	Greenish	Gray	12
Tree pipit*	Open	G,O	32	79	Variable	Blackish	13–14
Sedge warbler*	Marsh	W,O	27	48	Greenish	Light brown	13
Spotted flycatcher*	Edges	T,O	26	56	Greenish	Red-brown	13
Linnet	Open	S,O	26	50	Bluish	Red-purple	11
Greenfinch	Open	S,O	22	71	Whitish	Red-brown	13
Long-tailed shrike	Open	S,O	20	118	Grayish	Brown	15–16
Brambling*	Edges	T,O	17	65	Variable	Blackish	12
Reed-bunting*	Marsh	W,O	16	67	Olive-buff	Black	14
Blackcap	Woods	S,O	12	67	White	Brown	12

[a]Listed, except as indicated, by descending number of clutches present in Baker's (1942) egg collection. Nest types: G, ground or flat substrate; S, shrub or near-ground site; T, tree level; W, wetlands; C, cavity or crevice; O, open-above; S, spherical. Estimated egg volumes are shown as percentages relative to cuckoo. Species marked with asterisks are known to have reared cuckoos successfully, according to Glue & Morgan (1972) and Willey (1981). Other rarely pasitized species that have also raised cuckoos are the gray wagtail, wood warbler, barn swallow, Eurasian blackbird, and northern shrike.
[b]Overall means of several European studies cited in Schulze-Hagen (1992).
[c]Sample total from Glue & Moran (1972).
[d]Overall means of several Japanese studies cited in Nakamura (1990).

eggs. However, these apparent differences may simply reflect a different array of host species and nest-building strategies in these two regions.

The host data for the pallid cuckoo (tables 11, 14, 17, and 19), which is a fairly broad-spectrum brood parasite known to have at least 37 biological hosts, are also of interest for comparative purposes. This widespread and common Australian species shows considerable similarity to the brown-headed cowbird in its nest-site and nest-type selection tendencies. Of 1052 records of parasitism summarized by Brooker and Brooker (1989a), 18.5% are represented by the four most commonly exploited hosts, which is an indication of its low level of dependency on any particular host species. However, it exhibits greater similarities to the common cuckoo regarding the attributes of the hosts' eggshells. In the pallid cuckoo, as in the common cuckoo and the brown-headed cowbird, there is a strong tendency to select hosts whose eggs are substantially smaller in volume than the parasite's and whose adult body mass is likewise considerably less.

Information on the other Australian parasitic cuckoos (tables 13 and 14) suggests that they exhibit trends present in the pallid cuckoo, especially adult host–parasite mass ratios and relative egg volumes. The same trends can be seen concerning the host–parasite mass ratios and relative egg volumes of the African parasitic cuckoos (table 20). Additionally, the African parasitic cuckoos have incubation periods that range from 1 to 4 days shorter than corresponding host incubation periods.

Host-selection information on the honeyguides is summarized in table 21. This summary suggests that honeyguides exploit a wide variety of cavity-nesting species as hosts. As is typical of cavity nesters, nearly all the usual hosts have white eggs, and, although incubation period data are often limited, it appears that host incubation periods tend to average several days

TABLE 18 Topographic and Ecological Aspects of Common Cuckoo Parasitism Rates in Britain[a]

	% of Records	Total host species	Most frequent host species (%)[b]
Elevation (m)			
0–60	46	17	Reed warbler (35)
60–120	30	13	Hedge accentor (50)
120–180	10	7	Hedge accentor (72)
180–240	5	9	Hedge accentor (46)
240–310	4	5	Meadow pipit (69)
> 310	5	2	Meadow pipit (93)
Habitats			
Farmlands	27	6	Hedge accentor (~80)
Freshwater habitats	20	9	Reed warbler (~75)
Near habitations	18	7	Hedge accentor (~75)
Woodland habitats	16	13	Hedge accentor (~65)
Lowland heaths	8	7	Hedge accentor (~50)
Upland moors	7	8	Meadow pipit (~75)
Coastal habitats	4	6	Meadow pipit (~55)
Primary hosts	**Host nest records**	**Parasitism rate, % (*N*)**	**Fostering host species?**
Meadow pipit (13%)[c]	2659	3.1 (83)	Yes
Reed warbler (14%)	2826	3.0 (85)	Yes
Hedge accentor (53%)	14,788	2.2 (323)	Yes
Pied wagtail (2%)	2125	0.7 (15)	Yes
European robin (5%)	7649	0.4 (31)	Yes
Linnet (3%)	12,400	0.1 (16)	No

[a]Based on records of 613 parasitized nests (Glue & Morgan, 1972). Percentages for habitats were shown graphically and thus are estimated here. Fostering hosts are known to have raised cuckoos to at least 10 days of age. Primary hosts are organized by descending numbers (N) of parasitized nests in the collective sample.

[b]Percentages after host species' names indicate percentage of parasitized nests in the topographic or ecologic subsample indicated.

[c]Percentages after host species' names indicate percentage of all parasitized nests in that species' total nest sample.

longer than parasite incubation periods. The adult masses of these hosts are considerably more variable than those of hosts used by cuckoos or brown-headed cowbirds, and tend to be nearly as large, or even larger, than the honeyguide. This large host size is an unusual trait that is shared with the crested cuckoos, but otherwise seems to be unique among brood parasites. However, it apparently does not pose not a special problem for honeyguides, inasmuch as the

TABLE 19 Comparative Host Traits of the Brown-headed Cowbird and Two Cuckoos[a]

	Brown-headed cowbird	Common cuckoo		Pallid cuckoo
		Eurasia	India	
Total reported host species	145	24	20	32
Usual Host Nest Site (%)				
Ground level	24	21	40	0
Shrub level	29	37.5	40	42
Tree level	40	12.5	5	68
Wetland-habitat sites	2	21	0	0
Niches, cavities, etc.	5	8	15	0
Host Nest Structural Type (%)				
Open, cuplike	89	92	80	100
Other	11	8	20	0
Host Eggshell Ground Color (%)				
White or whitish	66	17	40	21
Blue or bluish	32	8	15	3
Other ground colors	2	75	45	76
Host Eggshell Markings (%)				
Variously patterned eggs	85	96	95	100
Unpatterned eggs	15	4	5	0
Host Egg Volume (relative to parasite) (%)				
0.2–0.39	15	0	10	6
0.4–0.59	29	29	30	18
0.6–0.79	26	46	15	25
0.8–0.99	10	8	15	32
1.0–1.09	5	4	5	0
1.1–1.29	5	8	20	6
1.3–1.99	8	4	5	3
≥2.0	2	0	0	3
Avg. host egg volume (cc)	2.2	2.4	2.6	3.0
Avg. parasite egg volume (inches)	2.9	3.4	3.75	3.7
Parasite : host volume ratio	1:0.76	1:0.71	1:0.69	1:0.8
Mean Host Mass (% of parasite)[b]	20.7 (41%)	26.8 (24%)	—	29.9 (36%)
Host Incubation Period, days (%)				
<12 days	10	4	—	—
12–12.5	41	25	—	—
13–13.5	27	38	—	—
14–14.5	14	25	—	—
15–16.5	8	4	—	—
>17 days	0	4	—	—
Avg. host duration (days)	~13	~13.5		
Avg. parasite duration (days)	11.7	11.6		
Host : parasite duration ratio	1:0.9	1:0.85		

TABLE 19 *(continued)*

	Brown-headed cowbird	Common cuckoo: Eurasia	Common cuckoo: India	Pallid cuckoo
Host Nestling Period, days (%)				
>10 days	24	0	—	—
10–12	32	29	—	—
14.5–16	13	8	—	—
16.5–18	6	0	—	—
>18 days	3	4	—	—
Avg. host duration (days)	~11.2	~12.7		
Avg. parasite duration (days)	8.7	17		
Host : parasite duration ratio	1:0.8	1:1.3		

[a]See tables 11, 15, and 17 for associated host-species lists and sources of host data on cowbird and common cuckoo; pallid cuckoo data mainly from Brooker & Brooker (1989b) and Frith (1977).
[b]Mass data for cowbird is based on 45 major hosts (tablw 15), that of the common cuckoo for 20 hosts (table 18), and the pallid cuckoo mean is for 12 hosts (table 13).

TABLE 20 Breeding Traits of the African Parasitic Cuckoos' Major Hosts[a]

Cuckoo Host (ROP)	% Adult mass	Nest type	% Egg volume	Eggshell color	Incubation (days)
Jacobin Cuckoo (egg 6.2 ml; mean mass 73.5 g; incubation 11–12.5 days)					
Common bulbul (~135)	50	Cup	58	Spotted White	12–14
Cape bulbul (~60)	52	Cup	56	Speckled white	11.5–14
Fiscal shrike (~30)	41	Cup	62	Speckled cream	15–16.5
Sombre greenbul (21)	37	Cup	41	Spotted cream	15–17
African red-eyed bulbul (12)	42	Cup	47	Spotted white	12–13
Overall average	44		53		
Levaillant's Cuckoo (egg 5.43 ml; mean mass 122.5 g; incubation 11–12 days)					
Arrow-marked babbler (30+)	51	Cup	85	Blue-green	(?)
Bare-faced babbler	(?)	Cup	(?)	(?)	(?)
Hartlaub's babbler	(?)	Cup	92	Blue-green	(?)
Brown babbler	55	Cup	61	Pink to blue	(?)
Blackcap babbler	67	Cup	84	Blue	(?)
Overall average	54		80.5		
Great Spotted Cuckoo (egg 10.22 ml; mean mass 140 g; incubation 12–13 days)					
Pied crow (~36)	378	Cup	196	Spotted green	18–19
Pied starling (20)	86	Cavity	68	Blue-green	(?)
Cape rook (18)	498	Cup	180	Spotted pink	18–19
Red-winged glossy starling (10)	(?)	Cavity	88	Blue-green	~16
Hooded crow	246	Cup	149	Spotted greenish	17–19
Fan-tailed raven	532	Cup	235	Spotted bluish	(?)
Black-billed magpie	90	Roof	87	Blotched bluish	17–18
Overall average	305		143		
Thick-Billed Cuckoo (egg 3.6 ml; mean mass c. 104 g; incubation ~13 days)					
Red-billed helmit shrike (11)	41	Cup	98	Spotted green	17+

(continued)

TABLE 20 *(continued)*

Cuckoo Host (ROP)	% Adult mass	Nest type	% Egg volume	Eggshell color	Incubation (days)
Red-Chested Cuckoo (egg 3.73 ml; mean mass 73.4 g; incubation 12–14 days)					
Cape robin chat (108+)	39	Cup	87	Speckled white	13–19
Boulder chat (13)	90	Cavity	121	Spotted blue	
Cape wagtail (13)	28	Cup	68	Speckled yellow	13–14
White-browed robin-chat (12)	57	Cavity	87	Speckled buff	15–17
Ruppell's robin-chat (12)	37	Cup	78	Olive brown	12–13
Overall average	50		88		
Black Cuckoo (egg 4.43 ml; mean mass ~86 g; incubation 13–14 days)					
Tropical boubou (~32)	58	Cup	76	Speckled green	(?)
Crimson-breasted boubou (~22)	57	Cup	84	Spotted greenish	16–17
African golden oriole	(?)	Cup	149	Spotted pink	(?)
Overall average	57.5		103		
Common cuckoo (egg 3.47 ml; mean mass ~125 g; incubation 11–12 days)					
Moussier's redstart	10	Cup	52	Bluish white	(?)
African Cuckoo (egg 3.92 ml; mean mass ~100 g; incubation 11–12 days?)					
Fork-tailed drongo (25+)	46	Cup	109	Spotted pink	16
Klaas's Cuckoo (egg 1.59 ml; mean mass 27.9 g; incubation 13 days)					
Cape crombec (13)	41	Pendant	95	White	~14
Greater double-collared sunbird (13)	44	Pendant	91	Speckled white	15–16
Bronze sunbird (12)	58	Pendant	116	White	14–15
Yellow-bellied eremomela (10)	29	Cup	65	Spoted white	13–14
Cape batis (10)	46	Cup	109	Spotted pink	17
Overall average	44		95		
Emerald Cuckoo (egg 2.83 ml; mean mass 37.5 g; incubation 12–13 days?)					
Sao Tome' weaver (20)	(?)	Roof	93	Blue-green	(?)
Common bulbul (12)	96	Cup	128	Spotted white	12–14
Overall average	96		110		
Dideric Cuckoo (egg 2.52 ml; mean mass 32 g; incubation ~11–13 days)					
Red bishop (245)	51	Pendant	85	Blue-green	11–14
Masked weaver (219)	97	Pendant	81	Variable	(?)
Cape sparrow (118)	72	Roof	89	Spotted white	12–14
Cape weaver (40)	133	Pendant	136	Blue-green	13.5
Red-headed weaver (26)	69	Pendant	87	Blue-green	12–13
Village weaver (25)	125	Pendant	102	Spotted bluish	~12
Spectacled weaver (13)	79	Pendant	97	Speckled whte	13.5
Overall average	89		97		

[a]Host lists compiled from Fry et al. (1988) and other sources; host traits also derived from various sources. Total records of parasitism (ROP) indicated parenthetically for host species with numerous records.

TABLE 21 Breeding Traits of Honeyguide Host Species[a]

Honeyguide species/ host species (ROP)	% Adult mass	Nest type	% Egg volume	Egg color	Incubation period (days)
Scaly-Throated Honeyguide (egg 3.1 ml; adult mass 48 g; incubation 18 days?)					
Yellow-rumped tinkerbird	26	Cavity	45	White	~12
Cardinal woodpecker	66	Cavity	86	White	10–12
Black-collared barbet (15+)	122	Cavity	122	White	~18.5
Gray woodpecker	95	Cavity	123	White	(?)
Whyte's barbet	122	Cavity	127	White	(?)
Nubian woodpecker	129	Cavity	128	White	(?)
Olive woodpecker	85	Cavity	128	White	15–17
Golden-tailed woodpecker	142	Cavity	131	White	~13
Overall average	98		111		
Greater Honeyguide (egg 4.18 ml; adult mass 47.8 g; incubation ?)					
Scarlet-chested sunbird	19	Pendant	41	Speckled white	13–16
Black tit	39	Cavity	45	Speckled white	(?)
Yellow-throated petronia	49	Cavity	45	Spotted brown	(?)
Gray-headed sparrow	50	Cavity	47	Spotted white	(?)
White-throated swallow	48	Cup	50	Speckled white	16
Boehm's bee-eater	35	Cavity	51	White	(?)
Little bee-eater (19)	29	Cavity	57	White	18–20
African pygmy kingfisher	31	Cavity	57	White	18
Abyssinian scimitarbill	62	Cavity	56	Blue	(?)
Banded martin	49	Cavity	61	White	(?)
Rufous-chested swallow	63	Cavity	63	White	16
Scimitarbill	66	Cavity	69	Bluish	(?)
Little green bee-eater	38	Cavity	70	White	(?)
Rufous-breasted wryneck	108	Cavity	72	White	12–15
Pied barbet	67	Cavity	73	White	14–15
Hoopoe (9)	122	Cavity	82	Speckled white	15–16
Swallow-tailed bee-eater	48	Cavity	86	White	(?)
Green wood-hoopoe	147	Cavity	88	Greenish	17–18
White-fronted bee-eater (6)	66	Cavity	97	White	19–21
Knysna woodpecker	(?)	Cavity	90	White	12–19 (?)
Tullberg's woodpecker	109	Cavity	(?)	(?)	(?)
Gray woodpecker	94	Cavity	92	White	(?)
Southern anteater-chat	117	Cavity	93	White	(?)
Nubian woodpecker	134	Cavity	97	White	(?)
Golden-tailed woodpecker	143	Cavity	97	White	~13
Black-collared barbet	123	Cavity	91	White	~18.5
Cinnamon-chested bee-eater	50	Cavity	101	White	20 (?)
Northern anteater chat (50+)	117	Cavity	113	White	(?)
Carmine bee-eater	104	Cavity	128	White	(?)
Red-shouldered glossy starling	209	Cavity	139	Speckled blue	14
Crested bartet	145	Cavity	142	White	13–17
Gray-headed kingfisher	94	Cavity	145	White	(?)
Madagascar bee-eater	91	Cavity	155	White	(?)
Pied starling (12)	219	Cavity	166	Bluish green	(?)
Brown-hooded kingfisher	119	Cavity	201	White	~14
Abyssinian roller	250	Cavity	233	White	(?)
Overall average	93		94		

(continued)

TABLE 21 *(continued)*

Honeyguide species/ host species (ROP)	% Adult mass	Nest type	% Egg volume	Egg color	Incubation period (days)
Lesser Honeyguide (egg 2.97 ml; adult mass 26.5 g; incubation 11–12 days)					
White-throated swallow	86	Cup	69	Speckled white	15–16
Pied barbet	121	Cavity	102	White	14–15
Rufous-breasted wryneck	196	Cavity	102	White	12–15
Red-fronted barbet	113	Cavity	105	White	12+
Anchieta's barbet	178	Cavity	119	White	(?)
Green barbet	186	Cavity	126	White	(?)
Black-collared barbet (35+)	223	Cavity	127	White	~18.5a
Violet-backed starling	396	Cavity	131	Spotted bluish	~12
Whyte's barbet	207	Cavity	136	White	(?)
Golden-tailed woodpecker	258	Cavity	136	White	~13
Chaplin's barbet	(?)	Cavity	136	White	(?)
Cinnamon-chested bee-eater	91	Cavity	142	White	(?)
White-headed barbet	234	Cavity	143	White	15–21
Bennett's woodpecker	279	Cavity	145	White	15–18
Yellow-throated petronia	88	Cavity	164	Spotted brown	(?)
Striped kingfisher	143	Cavity	189	White	(?)
Crested barbet	260	Cavity	199	White	13–17
Pied starling	40	Cavity	236	Bluish green	(?)
Overall average	175		136		
Thick-Billed Honeyguide (egg ?; adult mass ~30 g; incubation ?)					
Gray-throated barbet	19.4	Cavity	3.7 ml	White	(?)
Pallid Honeyguide (egg ?; adult mass ~16; incubation ?)					
Yellow-rumped tinkerbird	12.5	Cavity	1.4 ml	White	~12
Cassin's Honeyguide (egg 1.1 ml; adult mass c. 10; incubation ?)					
Buff-throated apalis	(?)	Cup	(?)	(?)	(?)
Green white-eye	134	Cup	106	Blue-green	10–13
Black-throated wattle-eye	(?)	Cup	160	Spotted white	2
Green-backed Honeyguide (egg 1.1 ml, adult mass 14.4 g; incubation ?)					
Abyssinian white-eye	(?)	Cup	(?)	Blue	(?)
Yellow white-eye	(?)	Cup	105	White	11
Montane white-eye	(?)	Cup	123	White	(?)
Dusky alseonex	(?)	Cavity	154	Speckled White	14–15
Amethyst sunbird	72	Pendant	157	Spotted white	13–18
Black-throated wattle-eye	(?)	Cup	175	Spotted white	17+
African paradise-flycatcher	100	Cup	180	Speckled cream	12–15
Overall average	88		143		
Wahlberg's Honeyguide (egg ?; adult mass 14.2 g; incubation ?)					
Tabora cisticola	63	Roofed	(?)	Variable	(?)
Gray-backed camaroptera	65	Cup	(?)	Speckled white	14–15
Yellow-throated petronia	164	Cavity	(?)	Spotted brown	(?)

[a]Host traits from various sources; number of parasitism records are shown in parentheses after species' name for major hosts. Host species are organized by increasing egg volume so far as data permits.

honeyguide chicks kill the host young soon after they hatch. It is nevertheless interesting that some honeyguides are able to parasitize the nests of various woodpeckers, barbets, or starlings, which have adult masses that may be two or three times greater than their own. This situation suggests that perhaps these cavity-nesting hosts are not so protective of their nest sites as might be expected. Another possibility is that the honeyguides may have evolved effective methods of invading their hosts' nests, either by forcing their way in or by approaching without being recognized as a potential threat and being intercepted before they can enter and lay their eggs. Honeyguides are surprisingly pugnacious and will often threaten or even attack a variety of other bird species, including host species as well as nonhosts.

Reproductive Ecology of Brood Parasites

Egg-laying Rates and Seasonal Egg Production

The egg-laying potential of an individual brown-headed cowbird in nature is still uncertain, but estimates have varied from as few as 11 per season to more than 40 (Payne, 1965, 1976a; Scott & Ankney, 1980, 1983). Under captive conditions, a single 2-year-old female laid an egg every day for 67 consecutive days, and 3 females (out of 24) laid more than 40 eggs each within a single laying season of 89 days. A total of 524 eggs were laid by 24 two-year-old females during this period (Holford & Roby, 1993). It is possible that females of the shiny cowbird in Colombia lay continuously over a 9-month breeding season, interrupting their breeding only for molting and waiting out the dry season (Kattar, 1993).

The seasonal egg production of the common cuckoo is also uncertain. Various estimates have been based on the finding of eggs that, by their color or patterns of maculation, have been attributed to single females. Chance (1940) estimated that a cuckoo may lay up to 25 eggs during a single breeding season (mean of 9 estimates, 12.5), and Wyllie (1981) similarly estimated that a female may lay as many as 15 eggs in a single season (mean of 9 estimates, 7.7). Various other estimates summarized by Wyllie range from 5 to 18 eggs per season (mean of 21 estimates, 12.0). Other estimates have been summarized by Payne (1973b), who judged that a female probably lays 10–20 eggs in a single season. These eggs seem to be laid on an alternate-day basis over variably long periods, producing egg "series" or clutches (mean of 16 such estimates, 6.7 eggs). Intervening periods of 3–10 days (mean of 10 interval estimates, 4.8 days) separate such clusters (Wyllie, 1981). However, these intervening nonlaying intervals may reflect times when eggs were actually deposited by the cuckoo but for various reasons were never found by the observer.

Payne (1973b) also made some estimates of clutch sizes and number of eggs laid during a breeding season for various southern African parasitic cuckoos, based on histological studies of ovaries and oviducts of breeding females. He concluded that eggs are usually laid on alternate days and that most species of parasitic cuckoos probably lay 16–26 eggs per season. Mean clutch size, or number of eggs laid in an unbroken sequence, varied from about 2 to 4 eggs, resulting in the laying of 1.2–2.5 eggs per week. Additionally, the overall egg-laying period for all these species was about 10 weeks, not much different from the estimates for the common cuckoo in Europe.

Payne (1977a) also estimated clutch sizes and numbers of eggs produced per season for 11 species of African parasitic finches, using similar techniques. Clutch sizes averaged about 3.1 for the viduine finches and about 2.9 for the parasitic weaver, with no significant interspecific differences or latitudinal trends apparent within the viduine finches. Generally, from two to three eggs are laid per 10-day period, so the average egg-laying interval must be 3–4 days, which seems rather long. Single female viduines were estimated by Payne to lay 22–26 eggs in a single breeding season. Females also typically lay over a period of about 90 days, although some may remain sexually active for as long as 5 months. Many of the viduine finches, as well as their seed-eating estrildine hosts, breed at about the same time, at the end of the wet season and during the early portion of the dry season, when many grasses and similar seed-bearing plants are producing seeds. As a result, there is no opportunity for temporal reproductive-isolating mechanisms to help reduce risks of hybridization or ameliorate ecological competition in these birds.

Parasitism Rates and Intensities

Probably the most common statistic used for estimating the potential ecological effects of brood parasites is the parasitism rate relative to a particular host. This statistic is usually measured as a percentage of the number of nests of a host species that are parasitized at a particular time and place. Such statistics, although fairly easily obtained, often show enormous regional and temporal variations (table 22). Obvious sources of variation include differences in population

TABLE 22 Representative Estimates of Interspecific Brood Parasitism Rates[a]

Brood parasite/host species	Total nests	% Parasitized	References
Nonobligatory Brood Parasites or Egg-dumpers			
Red-crested pochard			
Mallard	62	31	Amat, 1991
Common pochard			
Red-crested pochard	228	22	Amat, 1993
Redhead			
Canvasback	74	80	Erickson, 1948
Mallard	173	68	Weller, 1959
Canvasback	179	55–66	Sorenson, 1991
Cinnamon teal	56	53	Sorenson, 1991
Mean, 7 island-nesting ducks	178	~37	Lokemoen, 1991
Host-tolerant Obligatory Brood Parasites			
Black-headed duck			
Red-fronted coot	133	55	Weller, 1968
Pied cuckoo			
Jungle babbler	38	71	Gaston, 1976
Common babbler	31	42	Gaston, 1976
Cape bulbul	115	36	Liversidge, 1971
Great spotted cuckoo			
Black-billed magpie	277	63.5	Soler et al., 1994
Pied crow	23	22	Mundy & Cooke, 1977
Black-billed magpie	50	16	Mountfort, 1958
Carrion crow	47	8.5	Soler, 1990
Red-billed chough	162	4.9	Soler, 1990
Eurasian jackdaw	290	2.1	Soler, 1990
Asian koel			
House crow	20	15	Lamba, 1963

TABLE 22 *(continued)*

Brood parasite/host species	Total nests	% Parasitized	References
Village indigobird			
Red-billed firefinch	31	42	Payne, 1977a
Red-billed firefinch	374	36	Morel, 1973
Eastern paradise whydah			
Green-winged pytilia	51	92	Nicolai, 1969
Green-winged pytilia	75?	28	Skead, 1975
Shiny cowbird			
Long-tailed meadowlark	24	96	Gochfield, 1979
Cinereous finch	36	86	Friedmann & Kiff, 1985
Chalk-browed mockingbird	91	84	Salvador, 1984
Chalk-browed mockingbird	65	78	Fraga, 1985
Brown-and-yellow marshbird	74	74	Mermoz & Reboreda, 1994
Yellow-shouldered blackbird	76	74	Post & Wiley, 1977
Rufous-collared sparrow	45	69	Fraga, 197
Rufous-collared sparrow	50	66	King, 1973
Rufous-collared sparrow	83	61	Sick & Ottow, 1958
Rufous-collared sparrow	90	60	Sick, 1993
Common diuca-finch	72	61	Johnson, 1967
Short-tailed field-tyrant	81	42	Friedmann & Kiff, 1985
White-breasted flycatcher	36	42	Friedmann & Kiff, 1985
Chestnut-capped blackbird	213	22.5	Salvador, 1983
Brown-headed cowbird			
Wood thrush (Illinois)	329	90+	Trine, 1993
Hooded warbler	25	80	Dufty, 1994
Kirtland's warbler[b]	(?)	70	DeCapata, 1993
Solitary vireo	78	49	Chace et al., 1993
Western meadowlark	294	47	Bowen & Kruse, 1994
Wood thrush (Midwest)	126	42	Hoover & Brittingham, 1993
House finch	50+	42	Peck & James, 1987
Yellow warbler	109	41	Clark & Robertson, 1981
Purple finch	50+	40	Peck & James, 1987
Red-eyed vireo (Ontario)	50+	38	Peck & James, 1987
Chipping sparrow	50+	32	Peck & James, 1987
Yellow-rumped warbler	50+	31	Peck & James, 1987
Wood thrush (Mid-Atlantic)	381	26.5	Hoover & Brittingham, 1993
Eastern phoebe	391	24	Klaas, 1975
Northern cardinal	70	29	Mengel, 1965
Field sparrow	49	26	Mengel, 1985
Mean, 20 species, Michigan	500	22	Berger, 1951
Prothonotary warbler	172	21	Petit, 1991
Eastern phoebe	494	19	Rothstein, 1975b
Indigo bunting	1721	19.6	Payne, 1992
Mean, 25 species, Kentucky	512	17	Mengel, 1965
Mean, 3 empidonaces, Michigan	142	15	Walkinshaw, 1961
Wood thrush (Northeast USA)	348	15	Hoover & Brittingham, 1993
Willow flycatcher (8 studies)	537	9	McCabe, 1991
Red-winged blackbird	1325	8	Freeman et al., 1990
Bell's vireo	57	7	Brown, 1994
Mean, 86 species, Ontario	44,788	6.7	Peck & James, 1987
American goldfinch	70	3	Berger, 1951

(continued)

TABLE 22 *(continued)*

Honeyguide species/ host species (ROP)	% Adult mass	Nest type	% Egg volume	Egg color	Incubation period (days)
Giant cowbird					
Mean, nondiscriminators		1277	73	Smith, 1968	
Mean, discriminators[c]		1993	28	Smith, 1968	
Host-Intolerant Obligatory Brood Parasites					
Thick-billed cuckoo					
Red-billed helmit-shrike		50	38	Vernon, 1984	
Black cuckoo					
Crimson-breasted boubou		28	36	Jensen & Clinning, 1975	
Common cuckoo					
Azure-winged magpie (Honshu)		146	57.5	Nakamura, 1990	
Great reed warbler					
Hungary		374	50	Molnar, 1950	
Japan (Honshu)		722	18	Nakamura, 1990	
European robin					
France		116	17	Blaise, 1965	
Bull-headed shrike		160	13	Nakamura, 1990	
Reed warbler					
Germany		177	9	Moksnes & Røskaft, 1987	
Mean, 34 European studies		15,461	8.3	Shulze-Hagen, 1992	
England (1972–82)		4101	7.3	Brooke & Davies, 1987	
England (1939–82)		6927	5.5	Glue & Morgan, 1984	
Meadow pipit					
Norway		341	7	Moksnes & Røskaft, 1987	
England		5331	2.7	Glue & Morgan, 1984	
Marsh warbler					
Mean, 18 European studies		2781	6.3	Schulze-Hagen, 1992	
White wagtail		74	3	Moksnes & Røskaft, 1987	
Hedge accentor					
Norway		357	2	Moksnes & Røskaft, 1987	
England (1939–82)		23,352	1.9	Glue & Morgan, 1984	
England (1972–82)		8564	1.5	Brooke & Davies, 1987	
Brush cuckoo					
Brown-backed honeyeater		39	26	Brooker & Brooker, 1989b	
Fan-tailed cuckoo					
Yellow-throated scrubwren		81	2–17	Brooker & Brooker, 1989b	
Shining bronze cuckoo					
Yellow-rumped thornbill		135	26	Brooker & Brooker, 1989a	
Western thornbill		226	8	Brooker & Brooker, 1989a	
Horsfield's bronze cuckoo					
Splendid fairywren		724	20	Brooker & Brooker, 1989a	
Klass's cuckoo					
Dusky sunbird		64	11	Jensen & Clinning, 1975	
Pririt batis		48	8	Jensen & Clinning, 1975	
Dideric cuckoo					
Southern masked-weaver		120	10	Hunter, 1961	
Red bishop		749	10	Jenson & Vernon, 1970	

[a]Adapted in part from Wyllie (1981). Hosts listed by diminishing parasitism rates; brood parasites listed taxonomically.
[b]Before initiation of cowbird-control measures.
[c]Hosts were oropendolas and caciques nesting in sites that were either botfly-free (discriminators) or botfly-infested (nondiscriminators).

densities of hosts and parasites. However, sometimes two similar habitats may have markedly different rates of local parasitism for no apparent reason. Red-winged blackbirds nesting in one wetland site had a parasitism rate 16 times greater (3% vs. 48%) than those nesting in another wetland located only 2 km away (Carello, 1993). Additionally, rates of parasitism may vary seasonally to a marked degree. For example, many late-nesting passerines, or the second or later nesting efforts of various early breeders, probably avoid cowbird parasitism completely, as undoubtedly do those European passerines still nesting after early July. Norris (1947) believed that early brown-headed cowbird eggs are laid in the nests of grassland or open-field hosts, whereas later eggs are deposited mainly in the nests of woodland species.

Several examples of rapid changes in parasitism rates have been documented in recent years. Nakamura (1990) documented several major changes in parasitism rates in Nagano Prefecture, in central Honshu, Japan. Azure-winged magpies expanded into this area in the late 1960s, and by the early 1980s they were being parasitized by common cuckoos at a rate of about 30%. By the late 1980s, the rate was about 80%, which represents a higher parasitism level than currently occurs among the long-standing hosts of common cuckoos in this area.

Less extreme, but still significant, changes in parasitism rates have occurred in Britain during the past 40 years for several important host species. There have been reductions in parasitism rates for the hedge accentor, European robin, and pied wagtail, but the parasitism rate for the reed warbler has more than doubled during that same period. Nevertheless, the degree of mimicry of the reed-warbler's eggs has not noticeably improved during the past half-century (Brooke & Davies, 1987).

Another important measure of the relative ecological effects of brood parasitism is the parasitism intensity, a statistic describing the number of a parasite's eggs in a particular host species' nest. It is believed that individual females of few if any species of brood parasite ever purposefully deposit more than a single egg in a host nest, but the presence of more than one parasitic egg per host nest is not infrequent, especially among some species of brood parasites. As is apparent from table 23, multiple deposition of parasitic eggs is often common in nests of host-tolerant species, such as the crested cuckoos, parasitic cowbirds, and the viduine finches. It might be expected that in such species multiple parasitism is not strongly selected against, inasmuch as two host-tolerant parasite chicks in the same nest might stand approximately the same chance of fledging as would a two-chick brood composed of a single parasite and host.

The presence of more than one parasitic egg in a nest could result from one female laying more than one egg in it or from two or more females independently depositing eggs. If a female, upon visiting a nest and finding that it has already been parasitized, has a reduced tendency to lay an egg in the nest, the probabilities of multiple parasitism are altered from a randomized pattern of egg deposition. Preston (1948) was the first to recognize and test this possibility. Using a Poisson series of fractional probabilities, he tested the hypothesis that any given nest will have a random pattern of egg deposition, regardless of whether any other cowbird eggs are present in the nest. Using data from five field studies, Preston concluded that a female brown-headed cowbird's first egg is not placed in a nest at random, but all the subsequent eggs are. Mayfield (1965) extended Preston's analysis to include his own field data and those of three other field studies. In five of eight examples, Mayfield found that the placement of cowbird eggs, including the first one, closely followed a random distribution, and in eight of nine additional analyzed cases, the placement of eggs subsequent to the first one

TABLE 23 Parasitism Intensities Reported for Various Intraspecific and Interspecific Brood Parasites

	Parasitic eggs present per nest							Total
Species/References	0	1	2	3	4	5	6–13	nests
Intraspecific Parasites								
Bar-headed goose								
Weigmann & Lamprecht, 1991	10	10	4	3	4	3	6	39
Host-Intolerant Honeyguides								
Lesser honeyguide								
Friedmann, 1955	—	14	1	0	0	0	0	15
Host-Intolerant Cuckoos								
Common cuckoo								
Wyllie, 1981	1197	164	6	0	0	0	0	1367
Baker, 1942	—	3530	81	6	1	0	0	3617
Rey, 1892	—	1195	51	0	0	0	0	1246
Total	—	4889	138	6	1	0	0	5034
Pallid cuckoo								
Brooker & Brooker, 1989b	—	832	10	1	0	0	0	843
Brush cuckoo								
Brooker & Brooker, 1989b	—	291	7	1	0	0	0	299
Fan-tailed cuckoo								
Brooker & Brooker, 1989b	—	562	12	0	0	0	0	574
Gould's + little bronze cuckoos								
Brooker & Brooker, 1989b	—	128	16	3	1	0	0	148
Horsfield's bronze cuckoo								
Brooker & Brooker, 1989b	—	985	25	2	0	0	0	1012
Shining bronze cuckoo								
Brooker & Brooker, 1989b	—	802	31	0	0	0	0	833
Black-eared cuckoo								
Brooker & Brooker, 1989b	—	116	0	0	0	0	0	116
Dideric cuckoo								
Friedmann, 1968	—	231	12	2	0	0	0	245
Klaas's cuckoo								
Friedmann, 1968	—	29	3	0	0	0	0	32
Australian koel								
Brooker & Brooker, 1989b	—	120	5	0	0	0	0	125
Total of all species	—	8985	259	21	0	0	0	9265
Expected[a]	—	8995	269	4	0	0	0	9268
Host-Tolerant Cuckoos								
Pied cuckoo								
All hosts								
Baker, 1942	—	84	13	6	2	1	0	106
All hosts								
Friedmann, 1964	—	182	21	9	5	—	3	220
Jungle babbler								
Gaston, 1976	11	13	12	1	1	0	0	38
Common babbler								
Gaston, 1976	19	11	1	0	0	0	0	31

TABLE 23 (*continued*)

Species/References	Parasitic eggs present per nest 0	1	2	3	4	5	6–13	Total nests
Levaillant's cuckoo								
All hosts								
Friedmann, 1964	—	20	2	0	1	0	0	23
Chestnut-winged cuckoo								
All hosts								
Baker, 1942	—	139	18	10	1	2	1	171
Great spotted cuckoo								
All hosts								
Friedmann, 1964	—	82	41	13	13	10	11	172
Four corvid hosts								
Soler, 1990	—	19	10	10	1	1	0	41
South African hosts								
Rowan, 1983	—	21	12	4	10	5	2	64
Asian koel								
Baker, 1942	—	36	33	8	7	4	3	93
Channel-billed cuckoo								
Brooker & Brooker, 1989a	—	33	22	8	1	2	3	69
Total of all species	—	640	185	69	42	25	23	984
Expected[a]	—	504	338	113	25	4	1	985
Expected[b]	—	—	136	126	58	18	4	342
Host-Tolerant Cowbirds								
Brown-headed cowbird								
Prairie warbler								
Nolan, 1978	244	80	12	0	0	0	0	336
Yellow warbler								
Weatherhead, 1989	226	83	9	2	0	0	0	320
Kirtland's warbler								
Mayfield, 1965	62	36	29	9	1	0	0	137
Dickcissel								
Zimmerman, 1983	249	89	85	57	37	15	12	544
Field sparrow								
Mayfield, 1965	482	135	42	5	0	0	0	664
Song sparrow								
Nice, 1937	125	69	26	3	0	0	0	223
Red-winged blackbird								
Orians et al., 1989	2039	156	27	10	4	1	0	2237
Linz & Bolin, 1982	149	66	28	10	0	2	3	258
Weatherhead, 1989	250	93	28	7	1	1	2	382
20 Michigan hosts								
Berger, 1951	388	53	36	15	6	2	2	500
14 Pennsylvania hosts								
Norris, 1947	164	45	21	7	0	0	0	237
86 Ontario hosts								
Peck & James, 1987	—	1635	616	172	56	13	12	2504
Totals, all hosts	—	2540	959	297	105	33	30	3964
Expected[a]	—	2310	1247	336	60	8	1	3962
Expected[b]	—	—	868	434	108	18	2	1432

(continued)

TABLE 23 *(continued)*

Species/References	Parasitic eggs present per nest: 0	1	2	3	4	5	6–13	Total nests
Shiny cowbird								
13 hosts								
Mason, 1986[a]	138	52	19	15	5	4	5	238
Expecteda	107	86	34	9	2	0	0	238
Host-Tolerant Finches								
Village indigobird								
Red-billed firefinch								
Morel, 1973	241	73	36	12	10	1	1	374
Expected[a]	201	124	38	8	1	0	0	374
Payne, 1977a	18	8	5	0	0	0	0	31
Expected[a]	17	10	3	1	0	0	0	31

[a]Based on Poisson distribution, assuming random depositions of the first and all succeeding eggs. Observed numbers differ significantly from expected in all cases except for Morel's (1973) sample.
[b]Based on Poisson distribution, assuming a random egg deposition of the second and succeeding eggs. Observed numbers differ significantly from expected in all cases.

closely approximated a random pattern. A somewhat better statistical fit occurs when data concerning nests with only a single egg are omitted; this procedure avoids problems associated with overlooking nests that may be abandoned by their hosts as soon as the first parasitic egg is deposited and are less likely to be found by human observers (Mayfield, 1965).

Orians et al. (1989) tested the random egg-laying hypothesis using data on the intensity of cowbird parasitism for red-winged blackbird nests and found that it could not be rejected only if parasitized nests were included (excluding the zero-egg category) or when weekly data were separately analyzed. However, when entire breeding-season data were used, the distribution of parasitism intensity was not within the expected range, suggesting that pooling data for entire breeding seasons may produce misleading results.

In spite of the problems associated with pooled data, support for the random-deposition hypothesis can be found in table 23, not only for the brown-headed cowbird but also for several other brood parasites. Similar Poisson-like egg-deposition patterns exist with regard to the shiny cowbird, the host-tolerant cuckoos, and, among the viduine finches, the village indigobird. Intraspecific parasitism, or egg dumping, perhaps follows the same general trend, judging from the limited available data shown in the table. The host-intolerant cuckoos and perhaps the lesser honeyguide clearly exhibit a far lower likelihood of multiple layings, a situation that makes biological sense, especially if at least some of these multiple layings might be the product of a single female.

Breeding Success and Lifetime Productivity

The lifetime reproductive success of any female brood parasite depends in part on the number of eggs that she is able produce per season, as well as the number of breeding seasons in which she participates in breeding. Another important component relates to the hatching and fledging success of the eggs that are laid.

Seasonal egg production rates by various brood parasites were discussed earlier (see Egg-laying Rates and Breeding Chronologies). Much more information about estimated hatching success and fledging success rates of both interspecific and intraspecific brood parasites is now available, especially for such well-studied species as the brown-headed cowbird (Norris, 1947; Young, 1963) and the common cuckoo (Glue & Morgan, 1972; Wyllie, 1981).

Some of this information is summarized in table 24, in which some preference is given to studies involving large sample sizes and to studies in which breeding success rates were estimated from the egg stage until fledging. Such estimates are necessarily fraught with various sources of potential error, owing to such variables as the fact that not all nests are found at the same stages of initiation. For example, nests found at an advanced state of incubation will have artificially high hatching success estimates, and the same applies to fledging success estimates among nests found with nestlings that are already well developed. On the other hand, repeated visits by humans might greatly alter a nest's susceptibility to predation or desertion and reduce breeding success estimates. Thus, as with parasitism rates, there is a seemingly enormous range of estimated breeding success rates, with some of these variations perhaps related to host suitability. Figures for the common cuckoo range from as low as 9% to as high as 76%, averaging 37.5%, those of the great spotted cuckoo 40–56%, averaging 50%, and those of the brown-headed cowbird 5–53%, averaging 20%.

These mean figures are mostly below the overall mean success rate of 45.9% calculated for nearly 22,000 eggs that were collectively associated with 29 studies of open-cup, temperate-zone altricial birds (Nice, 1957). However, some of the common cuckoo studies cited in table 24 also indicate breeding success rates that are similar to, or even substantially above, the overall average calculated by Nice. It is of interest that the brown-headed cowbird shows a considerably lower overall breeding success rate than the common cuckoo (about 20% vs. 37.5%). Presumably, this large difference can be attributed to such factors as a finer degree of perfection in host choice on the part of the cuckoo and to a much lower degree of nestling food competition (none in the cuckoo, compared with varying numbers of additional parasite and host nestlings in the cowbird). Judging from table 23, about 35% of all brown-headed cowbird eggs are placed in nests already containing at least one other cowbird egg, so average levels of food competition among chicks might be expected to be considerably more intense in cowbirds on this basis alone, regardless of additional competition from host nestlings.

Interestingly, the same sort of posthatching nestling competition typical of cowbirds also occurs in nests containing young of the great spotted cuckoo. This species enjoys a high breeding success rate, even by the standards determined by Nice to be typical of nonparasitic, open-cup nesting species. Perhaps the fact that great spotted cuckoos usually parasitize hosts much larger than themselves, which may be better able to feed any nestling cuckoos as well as their own chicks, helps to account for these considerable differences in breeding success between these host-tolerant species.

Brood parasitism not only offers potential benefits for those species that have effectively evolved parasitism strategies, it also offers benefits resulting from costs incurred by host species. These host costs occur in the form of increased requirements for energy expenditures in feeding their own and the parasite's offspring, and may lead to a substantial reduction in host productivity. These costs and the host responses that have evolved to reduce them are discussed in chapter 5.

The number of offspring potentially raised during a female's lifetime is a subject of special

TABLE 24 Estimated Breeding Success Rates of Interspecific Brood Parasites[a]

Parasite species/host species	Total eggs	% Eggs hatched	% Eggs fledged	Reference
Redhead				
Five duck species	650	18	—	Joyner, 1975
Four duck species	57	14	—	Lokemoen, 1966
Mallard & cinnamon teal	(?)[b]	9	—	Weller, 1959
Ruddy duck				
Four duck species	68	31	—	Joyner, 1975
Common cuckoo				
England (various hosts)	—	—	66[c]	Owen, 1933
Germany (various hosts)	189	58	30	Glutz & Bauer, 1980
Reed & sedge warblers	176	65	9	Wyllie, 1981
Total & means	365	60	39.4	
Great spotted cuckoo				
Black-billed magpie	25	59	56	Arias de Reyna & Hidalgo, 1982
Black-billed magpie	31	70	42	Arias de Reyna & Hidalgo, 1982
Black-billed magpie	90	63	51	Soler, 1990
Carrion crow	5	100	40	Soler, 1990
Total and means	151	60	50	
Brown-headed cowbird				
Prothontary warbler	43	65	53	Pettit, 1991
Song sparrow	324	—	52	Smith & Arcese, 1994
Eastern phoebe	169	—	34	Klaas, 1975
Mean, 14 host spp.	108	43	27	Norris, 1949
Mean, 36 host spp	879	38[d]	25	Young, 1963
Yellow warbler	180	—	14	McGeen, 1972
Field sparrow	234	—	12	Gates & Gysel, 1978
Northern cardinal	126	—	8	Scott & Ankney, 1980
Dickcissel	132	—	7	Zimmerman, 1966
Eastern meadowlark	86	—	6	Elliott, 1978
Prairie warbler	102	9	5	Nolan, 1978
Total and means	2383	37	23	
Giant cowbird				
Two host species				
Botfly-free sites	666	84	72	Smith, 1968
Botfly-infested sites	1708	74	40	Smith, 1968

[a]In nearly all these studies, the hatching and fledging success rates are based on all nests found, regardless of their stage of development. Host species sequence organized by diminishing breeding success.
[b]Nest total 229; egg totals unreported.
[c]Percent survival of 213 hatched young, not percentage of eggs laid.
[d]Number of chicks in total sample, egg totals were unreported.
[e]Total of all nests, including those with already hatched young.
[f]Based on 795 eggs of 34 species; no hatching data for two species.

TABLE 25 Variations in Long-Term Fertility Durations and Estimated Longevities of Female Common Cuckoos

Location	Minimum egg-laying years per female								Years studied	Reference
	1	2	3	4	5	6	7	8		
France	30	5	2	1	2	1	2	1	12	Blaise, 1965
England	17	7	1	—	—	—	—	—	3	Wyllie, 1981
Total	47	12	3	1	2	1	2	1		
Expected[a]	—	23.5	12	6	3	1.5	1	0.5		

[a]Chi-square value of 20.66, or 0.08 confidence level. Model assumes an initial population of 47 one-year-old females, with an annual adult survivorship of 52% (cf. Brooke & Davies, 1987), which would produce a mean further life expectancy of 1.42 years (Lack, 1966). The survival data here imply a mean adult female longevity of 1.7 years. If an estimated 1.37 year longevity for Japanese common cuckoos (Nakamura, 1994) is incorporated, an overall mean adult life expectancy of 1.56 years results. Maximum reported longevity for wild birds is 12.9 years.

interest with regard to any brood-parasitic species, but few good data are available concerning this topic. Such information for the common cuckoo comes from two long-term studies by Wyllie (1981) and Blaise (1965), which lasted 3 and 12 years, respectively (table 25). A 12-year observational period is probably long enough to cover the lifetime breeding durations of virtually all wild, common cuckoo females, although maximum lifetimes of 13 years have been documented by banding studies (Glutz & Bauer, 1980). Although the figures in table 25 are necessarily based on rather small samples, they provide a reasonable basis for constructing a hypothetical longevity pattern that can be compared with independent estimates of cuckoo mortality. If this model of cuckoo lifetime reproduction is close to reality, about half of the female common cuckoo breeding population in western Europe consists of birds that are laying for the first time, about a quarter consists of birds breeding in their second season, and the remainder of the population is composed of older and increasingly more experienced birds. This model places a moderate survival value on some potentially important and experience-dependent breeding behaviors, such as learning how to evade host detection effectively and remembering the locations of host nests or specific nest sites from year to year.

Population Demography and Mortality Rates

With the field data currently available concerning annual female egg production and average breeding success on the one hand, and average mortality rates of eggs, nestlings, and older birds on the other, a rudimentary population model can be proposed for the two best-studied brood parasites, the brown-headed cowbird and the common cuckoo (fig. 13).

Given the available data (table 24), one may assume that for every 100 cowbird eggs laid, about 37 will hatch and 20 nestlings will survive to fledging. Survival rates of juveniles between fledging and the end of the first year are not yet known with any confidence, but have been conservatively and indirectly estimated at about 15% (Dyer et al., 1977). Assuming a slightly higher (20%) rate as realistic, 4 birds (of both sexes) out of the 20 fledged nestlings should survive their first year. Both sexes mature in their first year, and adult sex ratios are probably fairly close to unity, given the nearly equal proportions of the sexes that are trapped

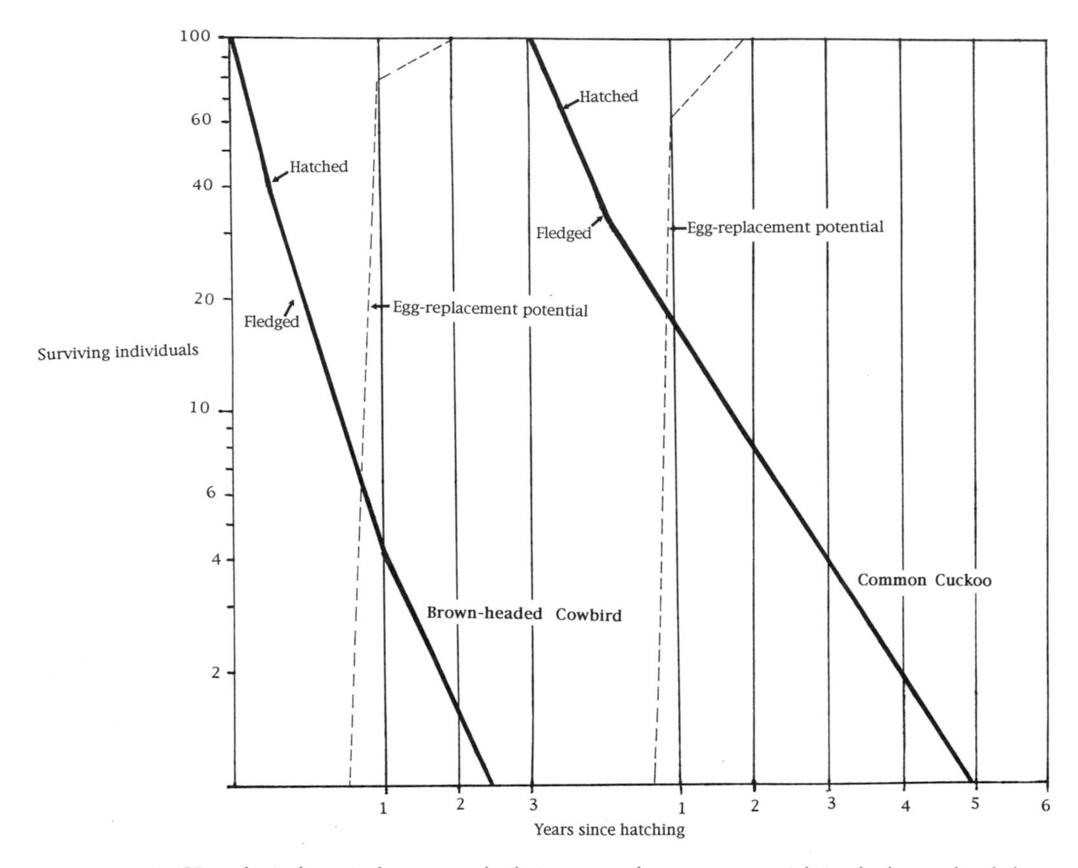

FIGURE 13. Hypothetical survival curves and relative egg-replacement potentials in the brown-headed cowbird (left) and common cuckoo (right). Assumptions for both include first-year survival rates half those of adults, initial reproduction in the first year, and equal sex survival rates. Differing assumptions include species' annual adult survival rates (cowbird 40% cuckoo 50%), annual fecundities (cowbird 40 eggs, cuckoo 9 eggs), hatching success rates (cowbird 40%, cuckoo 60%), and fledging success rates (cowbird 25%, cuckoo 50%).

during large-scale banding operations, so 2 of the surviving first-year birds should be females. These 2 yearling females should lay an average of 40 eggs each, thus regenerating 80 of the 100 original eggs. Adult annual mortality rates among adult females are probably about 60% (Darley, 1971; Fankhauser, 1971; Dyer et al., 1977). As a result, 1.6 birds should survive at the end of the second year, of which the surviving 0.8 female should generate 32 additional eggs, bringing the total egg regeneration to 112, or slightly near the original egg cohort. The few birds surviving to their third year (0.6, including 0.3 females), will generate an additional 9 eggs. However, by the third year there will have been a nearly complete population turnover, based on a 100-egg starting point. This model produces a mortality curve similar to that proposed by Dyer et al. (1977), in which the average age for adults is 1.28–1.72 years.

Comparable data for the common cuckoo offer some interesting differences. Of the 100 original eggs, some 60 should hatch, and about 38 chicks should survive to fledging (table 24). Again, mortality between fledging and the end of the first year is uncertain but is estimated here as 60%, or the same as that suggested for cowbirds. This results in 15 birds surviving through their first year, of which presumably half are females. These 7.5 females can generate, at a projected average rate of 9 eggs per female annually, 67 eggs in their first breeding season. Adult annual mortality rates of the common cuckoo are approximately 50% (table 25), so about 3.75 females should survive to their second year, which can generate another 33 eggs, thus replacing the original egg cohort of 100 eggs within 2 years. Of the females, 1.9 should survive to their third year, and 0.9 to their fourth year, adding another 25 eggs for the population during these 2 years, but a population based on 100 eggs will have virtually turned over by the end of the fifth year. (The somewhat greater longevities projected in the population model presented in table 25 result from a larger initial population, with 47 females, rather than 15 first-year birds of both sexes.)

Both models imply an increasing parasite population because the number of eggs regenerated before the original cohort has been eliminated is about 20 surplus eggs for the cowbirds and about 25 for the cuckoo. The major difference evident in these two suggested population models is that, whereas only about 40% of the female cowbird population consists of birds older than 1 year, in the case of the cuckoo at least half of the breeding population consists of birds 2 years old or older. However, in both models the importance of the breeding success of first-year females is critical to maintaining a viable population.

Few other estimates of annual mortality rates of brood parasites are available, but Payne and Payne (1977) estimated an annual survival rate of 50% for singing males of the village indigobird.

4

BREEDING BEHAVIOR

The cuckoo, as we have said elsewhere, does not make a nest but lays in other birds' nests, mostly in those of the wood pigeons and hypolais and of larks on the ground, and on a tree in the nest of the so-called greenfinch. Now it lays but one egg but does not itself sit on it, but the bird in whose nest it has been laid hatches it and rears it; and (so they say) when the cuckoo chick grows big it throws out her young, and so they are destroyed.

Aristotle, *Historia Animalium,* c. 300 B.C. [*trans.* Balme, 1991]

Mating Systems, Mate Choice, and Egg-laying Behavior

Proximate Controls of Reproduction

In Britain, the common cuckoo typically arrives in mid-April in southwestern and southern regions and in the latter part of April farther north. A large number of dates from various parts of Britain indicate a high degree of consistency in average spring arrival dates (the earliest mean date is April 4, the latest April 26). There is a mean maximum range of 22 days, or 11 days on either side of the mean arrival date, among 16 British locations having migration records ranging in duration from 6 to 189 years (Wyllie, 1981). Males arrive from a few days to as much as 3 weeks before the females and begin singing almost immediately upon arrival. This fairly precise arrival date and the immediate onset of male singing suggests that a reliable environmental timer such as photoperiod is the primary proximate factor controlling the cuckoo's migration, and photoperiod probably is thus at least indirectly responsible for controlling the onset of breeding.

The incidence of male song is highest during the 10 days immediately following their arrival. Singing reaches a secondary peak in mid-May, during the female's prelaying period. Singing then gradually tapers off and terminates by about 80 days later or at the time of departure. Wyllie believed that the incidence of male song might also be related for food availability but was unable to obtain any data relative to this possibility. The females' egg-laying season in Britain lasts about

12 weeks, and "appears to be proximally timed to coincide with the peak laying period of the different host species" (Wyllie, 1982, p. 129). However, individual females are unlikely to lay throughout this entire period, and a maximum egg-laying period of 54 days and a maximum production of 25 eggs have been reported among a sample of 46 individually marked females.

In North America, spring arrival dates of brown-headed cowbirds seem to be more variable than in Britain. Perhaps this variation can be attributed to the highly variable degree of local wintering in the central and southern states from year to year; the annual differences are probably a response to differing degrees of weather severity. Unfortunately, no experimental information exists on the possible proximate controls of migration in cowbirds, but Payne (1967a) determined that males exposed during winter to 17-hour photoperiods developed large gonads, and females developed follicles up to 1.8 mm in diameter, whereas birds of both sexes maintained on short days showed no gonadal enlargement. These results suggest that the increasing photoperiods of spring may control normal gonadal development during that season. Payne's captive female cowbirds did not develop mature ovaries, even though they were provided with potential hosts. However, Jackson and Roby (1992) were able to obtain egg laying among 18 females in a group of 25 yearling captives. They reported that the fecundity of yearling females raised apart from males, but within sight and hearing range of them, did not differ significantly from those females that were housed with males. Thus, direct one-to-one courtship and copulation are not required for stimulating egg laying by females. Among these captive birds, the egg-laying period ranged from 14 to 56 days. By the second week of June, or 4 weeks after the initiation of egg laying in the group, nearly 80% of the total experimental group was laying. Wild females breeding in the same area began their egg laying about 1 month before the captives did, but terminated their laying at about the same time. By the summer solstice, egg production by the captives was well past its peak, suggesting that photoperiod alone is probably not the only proximate timer of egg laying, and since nests containing host eggs were made available through the entire observation period, a reduction in host nests was also not responsible. Likewise, food, vitamins, and a calcium supplement continued to be provided to the captive birds throughout the period, so these factors also can be ruled out as limiting. Rather, it seems likely that some inherent physiological limits on egg laying may well be responsible for the seasonal termination of laying, which may, however, be highly variable individually, such as overall health.

Male Dispersion, Advertisement, and Mating Success

As noted earlier, there are no apparent direct reproductive benefits for either sex of obligate brood parasites to be gained in maintaining long pair-bonds, unless special conditions prevail. Among these include the possibility that a male is needed to help distract host parents from their nests (as in the crested cuckoos and screaming cowbird). Or perhaps monogamy might be advantageous where breeding populations of the parasite are so low or dispersed that it is less beneficial to spend time and energy on searching for new mates than it is to maintain an easily accessible sexual partner and produce as many offspring as that single mating will allow. Monogamous pair-bonding occurs in one of the parasitic cowbirds, the screaming cowbird, where mate guarding and cooperation by both pair members in successfully parasitizing host nests are probable advantages for such a mating system (Mason, 1980).

The form of the cowbird's mating system was classified by Oring (1982) as "male-dominance polygyny: female pursuit." Elliott (1978, 1980) examined the mating systems and dispersion patterns of brown-headed cowbirds in Kansas and observed that the predominant mating system in his study area was a promiscuous one and that long-term pair-bonds were evidently nonexistent. Males occupied what appeared to be overlapping but relatively exclusive home ranges, with no evidence of territorial defense or competitive exclusion behavior. An individually marked female's range overlapped with those of at least two different males. She was observed copulating with different males within the span of a single hour, and in total, nonmonogamous matings occurred at least 5 times among the 25 observed copulations. Additionally, some marked males were observed to court different females on different days, providing additional evidence of nonmonogamous pair-bonding. Most of the male courtship display occurred in a directly competitive or communal context, but some one-to-one courtship also occurred. Yet, in some situations an apparently monogamous bonding occurs; Laskey (1950) found that although no true territoriality was evident in the marked birds that she studied, the dominant male and a single female (also a dominant bird) shared a common "domain," within which both mate guarding and copulation occurred. Males performed the "bowling" or "song-spread" (see fig. 50) display both toward females and to other males. However, it was performed with the greatest intensity toward other males, a situation in which it apparently serves as an intimidation signal.

West et al. (1981) concluded that the potency of the male's advertising song serves as a kind of "bioassay" that may allow female cowbirds to evaluate individual male fitness; males singing the most potent songs obtained the most copulations. Those males placed in visual and auditory isolation, thus having no direct male competitors, sang the most highly potent male songs. Although females to a degree have mate-selection possibilities largely predetermined for them as a result of prior male-to-male competition and established dominance patterns, females are nonetheless able to selectively identify and choose such males for mating and to regulate the location and timing of copulation.

Oring (1982) classified the village indigobird as having a "male-dominance polygyny: intermediate dispersion" mating system. This category is sometimes described as an "exploded lek," and involves several males simultaneously advertising at a localized and often traditionally utilized site. Such exploded lek assemblages have much in common with true leks of some grouse, except that the participating males may be out of sight of, but still in auditory contact with, other competing males (Johnsgard, 1973, 1993). Oring's classification was based on the observations of Payne and Payne (1977), who similarly placed this species among a group of classic lek-displaying grouse, manakins, and shorebirds, in which males form no pair bond and provide no parental care. Mating success among males was nonrandom, with one male (in a population of 14 singing males) obtaining 53% of the observed matings, and three males accounting for 86%. In another local population, a single male obtained 66% of the observed matings. Breeding males defend special display locations ("call sites"), which are used throughout the breeding season and often from year to year by the same male. Call sites are usually dispersed over distances of at least 100 m, and ovulating females regularly visit these sites, where copulation occurs. The females evidently "sample" the males and their sites individually, as a probable basis for selecting a copulation partner.

Male common cuckoos have often been regarded as territorial, owing to their persistent use of regular song posts. However, Wyllie (1981) determined that two or more advertising males might use the same song post in the course of a single breeding season, and that male ranges overlapped with one another over most of the area that he observed. Rather than describing males as territorial, Wyllie instead suggested that they probably have a hierarchically organized social dominance system, within which dominant males can expel subordinate rivals as necessary. However, marked females exhibited somewhat greater tendencies to occupy separate territories in Wyllie's study area, and the birds maintained fairly separate "egg ranges" that overlapped with the "song range" of the males. It is possible that in low-density situations, territoriality is only slightly developed, and male dispersion is at least partly maintained by song alone, whereas under high-density conditions a more exclusive-use behavioral response and associated dispersion pattern is present (Glutz & Bauer, 1980; Cramp, 1985). Population densities evidently vary greatly, with home ranges correspondingly varying with habitat and host density, as well as with the age and social status of the individual (Glutz & Bauer, 1980).

Advertising vocalizations by males from a few regularly used and rather dispersed song posts have been described for various other *Cuculus* and *Chysococcox* species and probably are a common characteristic of many parasitic cuckoos. Transient pair-bonds may be formed in some species such as Klass's cuckoo, in which males occupy isolated territories of about 30 ha, although these territories appear to be relatively impermanent and might be occupied for only a few weeks at a time (Rowan, 1983).

Although earlier observers had conjectured that common cuckoos might have permanent, life-long pair-bonds, the species is now believed to be promiscuous (Cramp, 1985). Wyllie (1981) concluded that "the possibility of true pair-formation seems unlikely." He admitted that his field data were inadequate to make a conclusive determination about the cuckoo's mating system, but he favored the possibility that the species has a promiscuous mating pattern. Male song is the primary component of male courtship display, which is supplemented by visual posturing, display flights (males directly chasing females or the pair soaring in close formation), presentation of plant materials such as leaves or twigs, and perhaps also by courtship feeding. The display flights that involve swift, darting chases of individual females by males were regarded by Wyllie as providing a possible basis for the female's evaluation of a particular male's individual fitness. The frequency of copulation among common cuckoos is still unknown, but it appears be low, judging from the scarcity of available descriptions.

In all of these three representative cases, a pattern of male-dominance polygyny seems to be a basic mating pattern. Male advertisement is achieved by singing from traditional conspicuous or inconspicuous song posts (cuckoo, indigobird), with females attending these posts for direct courtship and mating. Or the males may seek out females and directly compete with one another for their attention (as in the cowbird). By either strategy, the dominant males get preferential access to females. In most brood parasites, males are of comparable size to or have slightly greater adult mass than females. The black-headed duck is the only known exception to this, with adults having a slightly reversed sexual dimorphism that may be related to female egg-laying requirements or other unknown factors. The greatest degrees of sexual dimorphism occur in the giant cowbird, in which the female:male adult mass ratio is 1:1.35; the mean of three other parasitic cowbirds is 1:1.21. Such dimorphism in these species is presumably related to the effects of sexual selection in a polygynous or promiscuous pairing system involving intense male competition for mates and a usually unbalanced adult sex ratio, with females the rarer sex.

Female Home Ranges and Breeding Season Mobility

Nest finding by brood parasites is probably a fairly constant occupation during the egg-laying period and certainly must require a good deal of mobility. Wyllie (1981) tracked the egg-laying ranges of several females over a 4-year period. During one year, three females occupied ranges of about 30 ha, which were centered about 1.2–1.5 km apart. However, one bird tracked by radiotelemetry during the 1979 season was observed moving over a roughly triangular area of about 2 × 5 km, or approximately 5 km^2 in area, although her egg-laying activities were confined to only two fairly small egg-laying sites that were less than 2 km apart. Roosting occurred about 2 km from the nearest egg-laying site. Within her area of primary use, six males were present, and probably three other females used the same egg-laying sites. While in her egg-laying range, the female may spend several hours each day looking for suitable hosts, either by watching them engaged in nest-building without revealing her presence or sometimes visiting a nest site, apparently to check its exact location or stage of development (Wyllie, 1981). Other European studies suggest that egg-laying ranges may be as large as 4–5 km^2, and eggs may be deposited by individual females in nests as close together as 37 m and as far apart as 4 km (Blaise, 1965; Cramp, 1985).

Several home range and breeding-season mobility studies have been performed with brown-headed cowbirds, and some similarities may be seen with the pattern just described for the common cuckoo. Dufty (1982) radio-tracked 4 females and 3 males during one season and 11 birds the following year. Two of these individuals had remained paired and had maintained identical breeding ranges during both of the previous years. In the third year, each of the birds both acquired a new mate and established closely adjoining breeding ("nonfeeding") ranges that partially overlapped with their original ranges of the previous years. However, Dufty judged that these birds formed largely monogamous sexual associations and occupied nonfeeding ranges ranging in area from about 10 to 33 ha (mean 20.4 ha, n = 16). Darley (1968) estimated considerably smaller home ranges, but these estimates were based on visual observations of color-banded birds and therefore may be underestimates. Both studies showed considerable site fidelity in subsequent years, especially among paired birds. Dufty suggested that monogamy in these populations of the northeastern United States and eastern Canada may be related to the birds' fairly small home ranges and greater abundance and diversity of host nests there as compared with the Great Plains region and the corresponding ability of males to guard individual females effectively. Rothstein et al. (1986) described sexual "consortships" between males and females, rather than pair-bonds, and found that such relationships might last 1 month or more, with dominant males (having higher singing rates and a higher incidence of head-up or bill-tilting displays) tending to consort with high-ranking females.

Thompson (1993) judged from telemetry data on 96 individuals radio-tagged in Missouri and Illinois that female cowbirds typically moved about 7 km each day during the breeding season, including a mean of 3.6 km between roosting and breeding sites, 1.2 km between breeding and foraging sites, and 2.6 km between foraging and roosting sites. Similarly, Woodsworth (1993) estimated an average daily movement of 4 km between breeding and foraging sites. Uyehara (1993) found that female brown-headed cowbirds tend to scan a large area while on elevated perches, probably listening and watching for host nesting activities, then fly to specific patches where they actively move about in the vegetation searching for nests.

Nest-searching and Egg-laying Behavior

Egg-laying behavior by brood parasites is difficult to observe; it often occurs under near-dark conditions, when photography is difficult or almost impossible. Norris (1947) reported that brown-headed cowbird females lay their eggs at times correlated with light intensity; on three clear days, the mean time of egg-laying was 18 minutes before sunrise, on an overcast day it was 14 minutes, and on a heavily overcast day it was 3 minutes. Scott (1991) estimated that 9.14 minutes before sunrise was the average egg-laying time for the species, based on 36 records, and noted that several other nonparasitic icterines typically lay an hour or so after sunrise. Thus, this early laying in the brown-headed cowbird appears to be an adaptation for parasitism in this species, and probably also occurs in the shiny cowbird. Burhans (1993) similarly reported that the mean arrival time at eight host nests was 11.4 minutes before sunrise, which averaged slightly sooner than the indigo bunting hosts (10.6 minutes), but later than field sparrow hosts (17.4 minutes). Although the presence or absence of the host species at the nest had little effect in deterring parasitism, the usual presence of the field sparrows at the time of parasitism might have accounted for the high observed rate of nest desertion by these sparrows (seven of nine nests).

Egg-laying behavior in the common cuckoo has attracted a great deal of attention, and its method has been the subject of prolonged debate among ornithologists (e.g., Chance, 1922, 1940; Baker, 1942). Unlike the brown-headed cowbird, the common cuckoo usually lays eggs in late afternoon or near dusk; of 120 observed instances, 35% occurred between 6:00 P.M. and dusk, and 44% between 4:00 and 6:00 P.M., with only two cases of egg laying observed between dawn and noon. Late afternoon laying may allow the female more time for nest searching, and holding a ready-formed egg in the cloaca for an extended period may also allow embryonic development to begin, thereby shortening the period of incubation by the host species (Wyllie, 1982).

When about to parasitize a nest, the female common cuckoo will usually hide within 50 m of the nest, often lying in a horizontal position along a branch in the manner of nightjars. The bird then glides hawklike to the nest, either landing on the nest itself or very nearby. Clinging to the nest, she then picks up a host egg, and settles over the nest in such a way as to bring her cloaca opening above the nest cup (fig. 14A). The egg is then expelled from the protruding cloaca, and the cuckoo leaves the nest site immediately. The host's egg may be swallowed whole or crushed and eaten. Eggs are frequently eaten; rarely even newly hatched chicks may be taken from the nest and eaten by the cuckoo (fig. 15), even when egg-laying by the cuckoo does not directly follow. Such egg robbing may cause considerable nest desertion by the host species and thus stimulate its relaying, thereby extending the cuckoo's available overall laying period (Gärtner, 1981; Cramp, 1985). Rarely, a female may eat as many as three host eggs while at the nest, but she does so before laying her own egg. The time spent at the host nest may be as little as 3 or 4 seconds, and usually requires less than 10 seconds (Wyllie, 1982). In spite of assertions to the contrary, there is no convincing evidence to support the belief that females sometimes lay their eggs on the ground and carry them with their bill to deposit them in a host nest, especially when such nests have such small entrances as to be relatively inaccessible. When the female cannot enter the nest, such as with roofed-over or crevice nests, the female will cling to the entrance and press her somewhat protrudable cloaca against the nest's entrance in order to deposit her egg (Baker, 1942; Cramp, 1985).

Other cuckoos for which fairly detailed information on egg-laying behavior is available include Horsfield's bronze cuckoo. A female of this species has been videotaped entering the roof-over nest of a splendid fairywren, laying an egg while the tips of the wings and tail were still protruding from the nest's entrance, emerging with a host egg in its bill (fig. 14B), and

FIGURE 14. Egg removal by a female common cuckoo while egg laying in a reed warbler nest (A). After a photo by Wyllie (1981). Also shown (B) is a female Horsfield's bronze-cuckoo with a splendid fairywren egg (after a photo in Brooker & Brooker, 1989a) and (C) a brown-headed cowbird with an ovenbird egg (after a photo by Hahn, 1941).

leaving with the egg. In three cases, laying took place within 2 or 3 hours of sunrise, and egg deposition required only about 6 seconds. A similar video recording of a female shining bronze cuckoo indicated that an identical procedure is used, but the bird was in the nest for 18 seconds and laying occurred about 1 hour after sunrise (Brooker & Brooker, 1989a). The female dideric cuckoo may likewise sit motionless and watch the host's nesting colony for 30–40 minutes before silently flying in and going directly to the nest to be parasitized or robbed.

FIGURE 15. Host egg-removal (A) and egg-laying (B) by female common cuckoo at a reed warbler nest, plus egg-ejection behavior (C) by cuckoo nestlings. The inset at left compares a cuckoo's egg (left) with a reed warbler's egg. Mainly after photos in Wyllie (1981).

Her visit may last only about 5 seconds, and the egg that she steals is usually eaten while she is still at the nest (Rowan, 1983). In order for bronze cuckoos to be successful parasites, their eggs must be deposited after egg-laying by the host has begun (earlier cuckoo eggs are likely to buried underneath host eggs), but no later than 4 days after incubation has gotten underway (young of cuckoos hatched from eggs laid later during incubation are unlikely to survive) (Brooker & Brooker, 1989a).

Good observations of egg-laying in the great crested cuckoo are also available. Like the common cuckoo, females of this species may remain hidden and immobile for long periods, watching the movements of the intended hosts and searching specific areas for nests, sometimes for several hours (Cramp, 1985). Unlike most species, males sometimes participate in the egg-laying process by helping to distract the host species, especially if its nest is being closely watched by both parents (Mundy & Cook, 1977; Rowan, 1983). Arias de Reyna et al. (1982) observed that egg-laying nest visits could be completed within 3 seconds from arrival to departure, and in all but one of nine observed cases egg-laying occurred before the host's clutch was complete. These authors also observed that a large percentage of host nests had damaged host eggs present (75% of 28 nests). This damage was evidently a direct result of parasitism, either by pressures produced by the cuckoo's feet while egg laying or, more probably, from the impact caused by dropping the cuckoo egg into the host nest.

Similar egg-laying behavior occurs in Levaillant's and pied cuckoos. In the latter species, the female of a pair may perch near the host's nest for an hour or so before she is joined by the male. They then approach the nest together, and as the male distracts the host pair, the female deposits her egg, often within a period of about 10 seconds. She then flies off to rejoin her mate (Liversidge, 1971). Likewise, in the Levaillant's cuckoo, a joint approach to the nest is typically used. Although the preliminary attempts to approach the nest may be considerably delayed by effective nest-defense behavior on the part of the babbler hosts, the act of laying may require no more than a few seconds (Rowan, 1983).

Little is known of the actual egg-laying behavior of most other brood parasites. The black-headed duck is able to lay an egg within 8 minutes of entering a host nest (Powell, 1979). In the brown-headed cowbird, the required time might be as little as a few seconds (Friedmann, 1929; Hahn, 1941) to 30 seconds (Hahn, 1941), or even as long as about 2 minutes (Howell, 1914). Unlike the common cuckoo, the female cowbird evidently does not regularly remove a host egg at the time of laying, but rather may remove one the previous day, later on the day of laying, or, infrequently, on the following day (Hahn, 1941; Norris, 1947).

Incubation, Hatching, and Behavioral Ontogeny

Incubation Adaptations and Durations

One of the primary characteristics of nearly all brood parasites is that they have incubation periods somewhat shorter than those of their hosts. At least in some species, such as several Old World cuckoos, this characteristic is at least partly related to the fact that egg laying tends to occur on a 48-hour, rather than a 24-hour, cycle, and a fully formed, ready-to-lay egg can be temporarily stored in the female's cloaca for periods approaching 24 hours after its arrival at the base of the oviduct. Embryonic development apparently begins in these unlaid eggs, giving them an important head start on the host species in terms of incubation time needed before hatching (Perrins, 1967). It has also been suggested that the relatively thick shell of the cuckoo's egg might reduce heat loss, both before and during incubation (Wyllie, 1981). Likewise, the more spherical shapes of cuckoo eggs and their generally larger volumes than host eggs would tend to restrain rates of heat loss at times when the host species is off the nest. Rounded eggs are also harder for the host to grasp and puncture than oval eggs and tend to be more resistant to accidental breakage.

Incubation periods among cuckoos, whether parasitic or nonparasitic, are unusually short compared with other families of birds (Lack, 1968). Wyllie (1981) observed that these short durations may help minimize predation during the egg stage, and less convincingly suggested that corresponding earlier hatching might allow the young to develop sufficiently that they can flee the nest if necessary, even before they are fledged. In any case, this short incubation period may have been a significant preadaptation for so many members of the group to evolve brood parasitism as an effective mode of reproduction. The incubation periods of some representative host-tolerant, as well as host-intolerant, parasitic cuckoos consistently range from 10 to 14 days, with little if any direct correlation between incubation period and adult mass. However, incubation periods of nonparasitic cuckoos seem to be correlated more directly with variations in adult mass, and range from 10 days in the smallest species to as long as 19 days in the largest ones (see table 9).

Honeyguide incubation periods have not been studied sufficiently to draw any conclusions regarding the possible adaptive significance of their lengths. Incubation periods of the viduine finches are about the same as, or slightly shorter than, their hosts.

Among at least the smaller parasitic cowbirds, incubation periods and fledging periods are not obviously shorter than those of nonparasitic icterines, including several cowbirds. However, the giant cowbird has a short incubation period relative to its mass and one that is reputedly several days shorter than are those of its oropendola and cacique hosts (Smith, 1968).

Nestling and Postnestling Behavior

No specifically adaptive behavior has been described with respect to the process of hatching in brood parasites; in all species but the highly precocial black-headed duck, the newly hatched chick is blind, entirely or almost entirely naked, and without enough strength to do little more than lift its head occasionally during the first few hours after hatching. In 9 of 10 cases involving hatching of brown-headed cowbird eggs under incubator conditions, the chicks hatched early in the morning (Wetherbee & Wetherbee, 1961). However, Nolan and Thompson (1978) noted that more than 40 artificially incubated cowbird eggs hatched at all times of the day and night. No special advantage is apparent for a consistent or synchronized hatching pattern among brood parasites.

One of the most remarkable aspects of early nestling life in some brood parasites is their disposal of potential competitors in the nest via nest ejection (fig. 15C). This behavior occurs not only in the common cuckoo, where it was first described by Edward Jenner more than two centuries ago (1788), but probably in all other species of *Cuculus,* and in some if not all of the species of *Cacomantis* and *Chrysococcyx* (Friedmann, 1968). Among *Chrysococcyx,* nest ejection has been observed in such African species as the Klass's and dideric cuckoos (Rowan, 1983). In Australia and New Zealand, nest ejection has been observed in Horsfield's and shining bronze cuckoos (Gill, 1983; Brooker & Brooker, 1989a) and in the black-eared cuckoo (Chisholm, 1935). It has been suggested that ejection behavior probably occurs in the Asian and African emerald cuckoos and the violet cuckoo, but observations are lacking (Friedmann, 1968).

Among species of *Cacomantis,* the plaintive cuckoos and the banded bay cuckoos have been described as ejector species (Baker, 1942). Baker also judged that the large hawk cuckoo may perform this behavior, but only on the basis that he had never seen any host nest-mates in company with cuckoo nestlings. Nest-ejection behavior is apparently lacking in some fairly well-studied cuckoo taxa such as the crested cuckoos and the Asian koel, but in all of these forms the host species tend to be as large or larger than the parasite, making the physical possibility of ejection somewhat unlikely.

In his initial description of the ejection behavior of the common cuckoo, Jenner (1788) observed that the back of the nestling cuckoo differs from other newly hatched birds in that its back is very broad and has a considerable medial concavity that disappears after about twelve days. Baker (1942) stated that in his experience, all species of *Cuculus* possess cavities in the back. He also observed that the same condition occurs in very young *Cacomantis* nestlings, but among species of that genus the cavity fills in more quickly than it does in *Cuculus* nestlings, possibly within 5 days after hatching.

Many others since Jenner have described the ejection behavior, but few more carefully. Wyllie (1981, p. 151) described ejection as follows:

> "From about 8–36 hours after hatching the young cuckoo wriggles about in the bottom of the nest until it manoeuvres one of the host's eggs against the side of the nest. Its back has a slight hollow between the scapulars which traps an egg against the nest-wall. The cuckoo's head is held down, almost touching its belly. Then with its feet apart and with muscular thighs, the youngster slowly works the egg up the side of the nest, holding its tiny wings backward to prevent the eggs from rolling off. When it nears the nest rim the wings clasp the top as the legs push up from the side of the nest. Balancing the egg on its back to the top of the nest, the young cuckoo quivers and jerks for a few seconds and hangs there feeling with its wings to make sure the egg has gone over. Then it drops back into the nest-cup."

This process is repeated until the nest is empty of other eggs or chicks.

Several features of this nest-ejection behavior are of special ethological and evolutionary interest. First, it occurs at such an early age that it can only be regarded as an experience independent, or instinctive, response. It must be primarily dependent on tactile cues, since the chick's eyes are still closed and few if any sounds would be available from the host's eggs or newly hatched chicks as cueing devices. Third, and most remarkably, heaving an egg from the nest requires a good deal of careful balancing of a rounded egg on the chick's back or the equally difficult balancing of a wriggling chick. This sophisticated behavior has few if any components that might have existed as logical evolutionary precursors and that might have been available for modification to this new and lethal end. The pipping behavior of a chick might provide the necessary backward-thrusting component, but the preliminary juggling and centering on the back of the host egg or chick requires the presence of complex motor abilities for which any ancestral functions are difficult to imagine.

The ejection response wanes rather rapidly in young cuckoos; the process is usually completed within 2 or 3 days after hatching. In rare cases, it may not be completed until as late as 7 days after hatching. This situation occurred when a shining cuckoo hatched 3 days after a host chick and was unable to evict the warbler host chick until the nestling cuckoo had attained the same weight as the warbler (Gill, 1983).

Just as remarkable as the nest-ejection response of these cuckoos are the "host-assassination" and nest-ejection responses of honeyguides. Friedmann (1955) made a detailed investigation of this behavior and its morphological basis, the presence of mandibular hooks on newly hatched nestlings. Mandibular hooks (see fig. 23) have been observed in only two honeyguide species with certainty, the greater and lesser honeyguides, but probably occur widely among newly hatched honeyguides. These needle-sharp structures, which are derived from sharpened elaborations of the egg teeth, fall off well before the chick is ready to leave its nest. In some cases, this shedding occurs fairly soon after hatching but after the chicks have performed their deadly functions.

With regard to the greater honeyguide, nest ejection of three young crested barbets by a honeyguide only 1 or 2 days old has been observed. The barbets landed below the nest without scratches or bruises, suggesting that they had been pushed out of the nest in a manner similar to that used by cuckoos, rather than stabbed and pulled out by the honeyguide using its sharp bill hooks. Friedmann suggested that these hooks may have been used to grasp the nest wall, thus providing a firm grip during the ejection behavior. Yet, in the lesser honeyguide, there is good reason to believe that a stabbinglike reflex is present in newly hatched chicks only about 2 days old. In chicks of this age, the mandibular hooks are well developed on both the upper and lower mandibles and are situated side by side when the bill is closed. The heel pad is also well developed at this age (see fig. 23). The honeyguide chick will attack any other nestlings, using fierce gripping and biting attacks, and will also attack eggs and attempt to puncture them (Friedmann, 1955).

Later stages of nestling life and the postnestling dependency period of brood parasites are dominated by effectively begging food from their host parents. This begging behavior invariable involves gaping, sometimes with associated neck stretching, wing fluttering, and food-begging calls (fig. 16). As noted earlier (chapter 2), vocal mimicry of food-begging calls of host species are present in some species. Visual begging is usually supplemented by loud vocal begging, especially among older birds. Although brown-headed cowbird chicks utter a faint peeping note from shortly after hatching onward, this food-begging call reaches its maximum development by about 6 days. By then it becomes especially loud and persistent; the parasitic birds continue to call even when the adults have uttered alarm notes that cause the host's young to crouch and become quiet (Friedmann, 1929).

Even more remarkable is the presence of visual host mimicry such as plumage mimicry or gape-pattern mimicry among the nestlings of some brood parasites, as described earlier. The cuckoos in general do not exhibit clear-cut gape mimicry of their host species, although nestlings of many brood-parasitic cuckoos have conspicuous mandibular flanges, and some such as the crested cuckoos have well-developed and brightly colored palatal papillae (see fig. 25). The origins of specialized gape patterns, such as those of the estrildine finches and the viduine mimics, is of evolutionary interest. Goodwin (1982) suggested that gape patterns serve as guides or markers for adults during feeding, especially under the dimly lit conditions of a roofed-over nest. A second and equally probable function is that of species recognition

FIGURE 16. Feeding of a pavonine cuckoo by ochre-faced tody flycatcher host (A) of a shiny cowbird by rufous-collared sparrow host (B, after sketches by P. Barruel, in Sick, 1993), and feeding of a juvenile red-billed firefinch by parent (C, after a photo in Payne, 1973). Nestling gape markings of a parasitic viduine (broad-tailed paradise whydah, D) and of a nonparasitic cuckoo (olive-capped coucal, E) are also shown.

and alerting the parents to avoid feeding alien chicks, including brood parasites. The last possible function, and a possible basis for evolutionary origins when neither of these other explanations suffice, is that bright or conspicuous mouth patterns might help deter predators from eating the chicks. These signals might operate either as a visual "bluff" in those species that are edible or as functional warning signals for vile-tasting species (Swynnerton, 1916). Interestingly, nestlings of several nonparasitic cuckoos such as coucals have strange, sometimes eyelike gape markings (fig. 16). Like many if not most cuckoos, coucal nestings void vile-smelling and viscous feces when disturbed, which probably serves as an antipredator mechanism (Rowan, 1983).

Early Experience and Host Specificity

Weller (1968) noted that one black-headed duck chick hatched early one morning and was gone from the host's nest by the following morning, while in another nest the duckling hatched during the night, and was gone by the time it was no more than 2 days old. One of three chicks that were exposed to human "parents" from shortly after hatching exhibited slight tendencies toward imprinting and associated following behavior. However, the other two exhibited almost no inclination to follow the host "parent," and two ducklings raised by a domestic hen also exhibited limited following behavior. For this species at least, early imprinting by ducklings on a specific host parent is not adaptive.

Teuschl et al. (1994) suggested that there are at least four ways in which host-specific brood parasites such as the common cuckoo can locate their hosts: inherited host preferences, imprinting on their foster parents, imprinting on their natal habitat, or highly developed site fidelity. They regarded habitat imprinting as the most likely mechanism for explaining host specificity in cuckoos and obtained some experimental evidence for this using five cuckoo nestlings that had been raised in different natural or artificial habitats. Brooke and Davies (1991) tested the possibility of host imprinting as a mechanism by studying host responses among seven cuckoos (five females, two males) that had been hatched in reed warbler nests. Two of these were transferred at an early age to European robin nests, while the others remained with the reed warblers. None of the five birds, when observed at 1 and 2 years old, displayed any apparent host preferences.

Brooke and Davies (1991) also suggested that natal philopatry might provide an alternate potential mechanism for gentes establishment, although such a mechanism would likely be effective only in areas of continuous and uniform habitat. In areas of fragmented habitats, the chances of philopatry operating effectively in establishing or maintaining host specificity would probably be greatly reduced. Wyllie (1981) noted that, of 60 common cuckoo nestlings he marked, only 4 (6.7%) returned to the study area in a subsequent year. Nakamura (1994) similarly reported that only 6 of 92 common cuckoo nestlings (6.5%) returned to his study area the following year. Even by assuming a high degree of first-year mortality, neither of these studies support a high degree of natal philopatry in the common cuckoo.

5

HOST RETALIATION

The Coevolutionary Arms Race

"Now here, you see, it takes all the running you can do, to keep in the same place. If you want to get somewhere else, you must run twice as fast as that."

Lewis Carroll, *Through the Looking Glass*

As the Red Queen remarked to Alice, sometimes it is necessary to compete as strongly as possible just to remain in the same position relative to others. So it is with brood parasites and their hosts; just as rapidly as a parasite is able to exploit and deleteriously affect the reproductive potential of a host species, the host actively undertakes protective strategies that tend to avoid or at least ameliorate the parasite's effects. The result is a coevolutionary arms race of varying degrees of intensity and speed (Dawkins & Krebs, 1979), with no endpoint or clearcut ultimate benefit resulting to either species.

The Costs to Hosts of Being Parasitized

Among many human financial undertakings, strict accounting measures allow a cost–benefit analysis that can be objectively tallied, and the "bottom line" of overall costs relative to benefits can thus be calculated. In contrast, the biological costs of brood parasitism are much more difficult to calculate because they are difficult to measure and unpredictable in nature. Briefly, some of the more likely costs, which obviously may not apply in all cases of brood parasitism, might be outlined as follows:

A. Probable Overall Costs to Host Species
   1. Increased vigilance-behavior and antiparasite responses required for nest protection
   2. Increased foraging-energy costs associated with feeding parasite's offspring

3. Increased energy costs for egg replacement caused by egg stealing and nest desertion following parasitism
4. Undesirably prolonged breeding season caused by parasite-induced nest desertion and resultant needs for renesting
5. Increased probability of rejection of species' own eggs, as egg-rejection defensive behaviors are evolving.

B. Direct Effects on Host Species' Recruitment Potential
1. Reduced hatching success associated with parasite's egg destruction or egg stealing
2. Reduced fledging success resulting from parasite outcompeting, smothering, or killing host chicks
3. Decreased vitality of surviving host chicks caused by parasite chick dominating most parental feedings or by aggressive behavior of parasite chick toward nest-mates
4. Increased susceptibility of entire brood to nest predation due to conspicuous begging behavior of parasite.

Of all of these potential costs, only the first two direct reproductive costs can be measured with any degree of accuracy. Some such attempts are tabulated in table 26 for five host-tolerant brood parasites. This summary is similar to one made for six brood parasites by Payne (1977b). However, his estimated host costs were based on total egg-to-fledging survival rates, and his resultant estimates differ somewhat from these. Thus, the estimated effects on recruitment rate reductions for seven hosts of the brown-headed cowbird ranged from 5 to 42.5% in Payne's analysis, whereas in table 26 the estimates for this species range from 10 to 53%. Likewise, Payne estimated of the mean recruitment reduction in babbler hosts caused by the pied cuckoo as 14%, whereas the same data resulted in a 31% estimated reduction by the method used in table 26. In any case, the effects of brood parasitism in terms of directly reducing host fledging success can clearly be significant in reducing overall host recruitment rates, possibly by as much as 50% in some cases.

Estimates on fledging success for host-intolerant species such as the common cuckoo cannot be made using these techniques, since almost invariably the fledging success of host species is nil in all parasitized nests that hatch successfully, regardless of whether the parasite chick itself successfully fledges. Additionally, egg predation by the brood parasite may seriously affect host hatching success, even if the parasite's own egg never hatches. Schulze-Hagen (1992) provided some data correlating incidence of common cuckoo parasitism on reed warblers with their hatching success. Whereas unparasitized reed warbler nests had (in three studies) a mean hatching success of nearly 80%, hatching success diminished (mostly through losses by egg predation) in a straight-line manner by about half, or an estimated mean value of less than 40%, under conditions of a 20% incidence of nest parasitism. Additional parasite-related nestling losses would, of course, occur after hatching, through nest-ejection behavior by the nestling cuckoo. These statistics suggest that the overall impact of host-intolerant parasites such as common cuckoos is potentially quite devastating on the recruitment potentials of susceptible host species, especially among those hosts having parasitism rates of 20% or more. Among 34 studies of reed warbler parasitism in western and central Europe, the mean parasitism rate was 8.3%, with a maximum observed 63% rate (Schulze-Hagen, 1992). Other representative parasitism rates for common cuckoos as well as other host-intolerant cuckoos such as the bronze cuckoos are shown in table 22, and these species generally appear to have lower parasitism rates than those found in such host-tolerant species as the cowbirds.

TABLE 26 Productivity Costs to Hosts from Interspecific Brood Parasitism

Parasite/host	Fledging Success[a]		Rate of parasitism (%)	Cost of parasitism (%)	Reference
	UPN	PN			
Redhead					
Canvasback[b]	7.3	6.0 (−18%)	80	−14	Erickson, 1948
Pied cuckoo					
Jungle babbler	2.5	1.1 (−54%)	71	−39	Caston, 1976
Common babbler	2.6	1.1 (−56%)	42	−23	Gaston, 1976
Unweighted means			−55	−31	
Great spotted cuckoo					
Black-billed magpie	3.29	1.6 (−51%)	43	−22	Soler, 1990
Carrion crow	3.31	2.5 (−25%)	40	−10	Soler, 1990
Unweighted means			−38	−16	
Brown-headed cowbird					
Red-eyed vireo	2.92	0.79 (−73%)	72	−53	Southern, 1954
Kirtland's warbler	1.28	0.28 (−78%)	~63	~−50	Walkinshaw, 1983
Bell's vireo	1.40	0.29 (−79%)	58	−47	Goldwasser et al., 1980
Indigo bunting	1.5	0.40 (−73%)	56	−41	Berger, 1951
Solitary vireo	2.35	0.50 (−79%)	49	−38	Chace, 1993
Lark sparrow	1.78	0.60 (−66%)	45	−30	Newman, 1970
Yellow warbler	2.28	0.88 (−62%)	30	−19	Weatherhead, 1989
Eastern phoebe	4.4	0.32 (−93%)	19	−18	Rothstein, 1975a
Acadian flycatcher	1.68	0.38 (−77%)	24	−18	Walkinshaw, 1961
Song sparow	3.41	2.03 (−38%)	44	−17	Nice, 1937
Field sparrow	2.38	0.60 (−75%)	18	−14	Berger, 1951
Yellow warbler	1.60	1.2 (−27%)	41	−13	Clark & Robertson, 1979
Song sparrow	1.34	.64 (−52%)	25	−13	Smith & Arcese, 1994
Various hosts	2.94	2.05 (−31%)	30	−10	Norris, 1947
Unweighted means			−64	−26	
Shiny cowbird					
Rufous-collared sparrow	.69	.29 (−52%)	61	−32	Sick & Ottow, 1958
Rufous-collared sparrow	1.0	.47 (−47%)	66	−31	King, 1973
Yellow-shouldered blackbird	.75	.39 (−48%)	74	−35	Post & Wiley, 1976
Yellow-shouldered blackbird	1.68	.38 (−77%)	24	−18	Post & Wiley, 1977
Unweighted means			−56	−29	

[a]PN, parasitized nests, UPN, unparasitized nests. "Fledging success" represents mean number of host young fledged per active nest. Negative numbers in parentheses indicate percentage reduction of host young produced in parasitized relative to unparasitized nests. Species are organized by diminishing estimated costs to hosts. Cowbird data partly after May and Robinson (1985).
[b]Canvasback means are for brood size at hatching.

Parasite Recognition, Nest Concealment, and Nest Defense

It is impossible to distinguish the nest-concealment behavior of host species that has developed to avoid nest predation from that which has developed to avoid brood parasitism, but it is probably true that a higher degree of concealment is required for effective protection against brood parasitism. Many observers (see references in Payne, 1977b) have commented on the patience with which a female brood parasite will watch host species in the process of their nest building and egg laying, assiduously search for nests in various microhabitats, and obtain cues for locating nests from the alarm reactions of host species. By such strategies, brood parasites often manage to locate and parasitize well-hidden nests. Although lateral, ventral, or even purselike entrances are sometimes present in host nests and probably serve partly as protection against nest predators, such entrances are ineffective defenses against intrusion by many cuckoo species (Baker, 1942).

Host responses to the presence of brood parasites in the vicinity of their nests vary considerably. Edwards et al. (1950) reported on the aggressive responses of several passerine species toward mounted European cuckoos. Smith and Hosking (1955) determined that willow warblers will furiously attack a mounted specimen of a common cuckoo (even when only the cuckoo's head is provided) when it is placed near their nest. They also noted that host species exploited frequently by common cuckoos react much more strongly to the visual stimuli provided by cuckoo mounts than do those species that are rarely parasitized. Wyllie (1981) determined that various British hosts such as the reed warbler also aggressively respond to tape recordings of cuckoo songs when these are played back near their nests. When such songs are used in conjunction with a mounted cuckoo specimen, the birds may "lose all fear of man" in their attempts to attack the cuckoo specimen.

Moksnes et al. (1991b) tested responses of cuckoo mounts against three commonly parasitized species in Norway, the hedge accentor, European redstart, and meadow pipit, but obtained strong responses only from the meadow pipit. However, other important frequently used as well as several rarely used hosts responded more strongly to the mount, whereas most of the unsuitable hosts exhibited little or no aggression. The strongest responses from the pipits occurred early in the breeding period. Moksnes and Røskaft (1989) also observed that the meadow pipits exhibited considerably stronger aggression toward cuckoo mounts when both parents were at the nest than when only one was present.

Robertson and Norman (1976, 1977) have similarly investigated the defense mechanisms of various North American hosts against brown-headed cowbirds. They used cowbird models placed near the nest. Both male and female models in various positions and postures were tested, including some in a head-down "invitation to preening" posture (see fig. 49) that has been reputed to reduce aggression levels among hosts. The authors generally found that "acceptor" host species displayed levels of aggression that were proportional to the rates at which they were being parasitized. Most host species reacted less strongly to cowbird models exhibiting the head-down posture than to other postures, as might be expected if this posture truly functions as a hostility-reducing signal. However, this diminished antagonistic response was not found in the red-winged blackbird, whose aggressive song-spread display has some postural similarities to the head-down display of cowbirds. On several occasions live cowbirds were also seen in the study area near host nesting sites, and these cowbirds were actively chased away by northern orioles as well as by red-winged blackbirds. Robertson and Norman rejected a hypothesis relative to possible alternative functions of these hostile interactions, namely, that host aggression may be exploited by cowbirds as a means of helping them locate host nests. Instead they concluded that aggression toward cowbirds is simply an adaptive behavioral nest-defense mechanism of the host species.

In another recent study, Mark and Stutchbury (1994) presented mounted female cowbird specimens, as well as mounts of a nonparasitic species (veery) as a control, to 25 incubating hooded warblers. Tape recordings of cowbirds or of veerys were used as supplementary stimuli. The female warblers were more aggressive to the cowbird mounts than to veery mounts, and this discrimination ability extended to yearling warbler females. The authors suggested that some experience-independent (innate) basis for cowbird recognition may be present in controlling the elicitation of aggressive behavior among these naive birds.

Briskie et al. (1992) observed that greater levels of nest defense toward mounted cowbirds occurred among nesting female yellow warblers in an area of sympatry than in an allopatric population, even though females from both areas exhibited similar levels of egg-rejection behavior. Although it has been logically speculated (e.g., Soler & Møller, 1990) that the incidence of such egg-rejection behavior by hosts is directly related to the duration of sympatry between host and parasite, Zuniga and Redondo (1992) were unable to establish such a correlation in the case of the great spotted cuckoo and its black-billed magpie host.

A similar study to that of Mark and Stutchbury was done by Folkers and Lowther (1985), using mounts of brown-headed cowbirds and song sparrows that were placed near active nests of red-winged blackbirds and yellow warblers. Both host species showed more aggression to the cowbird model than to the song sparrow. Additionally, the yellow warblers responded more aggressively early in their breeding season, although this pattern was only apparent among male red-winged blackbirds and those females at nonparasitized nests. The authors suggested that the differences they observed in responses to cowbirds and song sparrow models, and the differing responses of red-winged blackbird, were associated with prior exposure to the cowbirds and thus reflected a learned response.

Briskie and Sealy (1989) tested the seasonal response to the threat of cowbird parasitism using the responses of nesting least flycatchers to a female brown-headed cowbird mount and, as a control, a mounted fox sparrow. As expected, aggressive displays by the flycatchers toward the cowbird were stronger than those toward the sparrow. Additionally, the rates of flycatcher threat responses (tail-spreading displays) to the cowbird model were highest during the egg-laying period. It is during this period that the host's costs of being parasitized are greatest, inasmuch as cowbird eggs laid after the host's incubation behavior is underway are less likely to hatch and pose a significant threat to the host. However, other agonistic responses (general defensive behavior and alarm vocalizations) remained at a similar level throughout the nesting period, perhaps because cowbirds sometimes operate as predators on eggs and occasionally even on small nestlings.

Recogniton and Rejection of Alien Eggs

Adaptively responding to the presence of a brood-parasite's egg requires the host to distinguish it from its own eggs. Should the egg be laid before its own clutch has begun, such recognition is simple, but most brood parasites wait until egg laying by the host has begun before depositing their own egg. Additionally, most parasites are likely to remove a host egg about the time that they lay their own, so egg counting by the host is generally not a reliable means of detecting alien eggs.

Egg mimicry has already been discussed (chapter 2), but the discriminative abilities and responses of hosts to parasitic eggs having varying degrees of similarity to their own need attention here. Recognition of an alien egg in a clutch may be achieved by one of two methods. The first simply involves removing any egg that differs in some aspect of appearance from the others. This egg-discrimination strategy has been termed "rejection via discordancy" (Rothstein, 1982). This capacity might be widespread in birds, as it would be advantageous in the recognition of cracked, broken, or otherwise abnormal eggs laid by the host species. The other, called "true egg recognition" by Rothstein (1982), involves the actual recognition and selective preferential treatment of the species' own eggs, even when they might be in the minority (Rothstein, 1975a). These recognition abilities sometimes even extend to intraspecific recognition of self-laid eggs, at least in species laying eggs with highly variable surface patterns. Such host polymorphism in egg patterning might be especially advantageous in helping to reduce the probabilities of effective parasitic egg mimicry (Victoria, 1972).

The available responses for a host, on recognizing the presence of a parasite's egg in its nest, include (1) accepting the egg and incubating it together with its own clutch (see "adaptive tolerance" in the following section), (2) burying the egg at the bottom of the nest and adding a new nest lining, (3) destroying or ejecting the egg from its nest, or (4) deserting the nest and perhaps constructing an entirely new one if time permits. For some host species, whose bills are too small or who are otherwise unable to pierce or grasp and eject the egg from their nest, these choices may be reduced to egg acceptance, egg burial, or nest desertion.

Among these various choices, the least costly is that of egg rejection, simply disposing of the alien egg by burying it in the nest lining or by ejecting it. Both options require the recognition of the egg as alien and thus run the risk of mistaking one's own egg for an alien egg and thereby reducing productivity. Marchetti (1992) has shown that inornate warblers rejected one of their own eggs at about 10% of the 157 nests where artificial eggs had been experimentally introduced, and in several cases some of their own eggs were accidentally broken while rejection of the larger artificial egg was being attempted.

Egg burial is probably most common among species that are unable, because their bills are too small, to grasp the parasitic egg or not strong and sharp enough to pierce it. However, many birds seem unable to selectively bury only a single egg, and instead cover over not only the parasitically laid egg but also any of their own they may have already laid. This is a common phenomenon among yellow warblers, where stacked nests several layers deep, each with a cowbird egg and one or more of their own eggs, are not uncommon. Bent (1953) noted that two-story nests are common, and nests with three, four, five, and even six-stories have been reported, typically with a cowbird egg and sometimes also a warbler egg in each layer. Similarly, reed warbler nests have been found with common cuckoo eggs present at various depths (Wyllie, 1981).

Nest abandonment is another common strategy for coping with alien eggs. Baker (1942) reported that nest-desertion behavior occurred more frequently in unusual hosts than in regularly used hosts among Indian races of the common cuckoo, and Wyllie (1981) obtained additional support for this idea for European hosts. Reported nest desertion rates have been as low as zero for the hedge accentor to as high as 100% for the wood warbler, chiffchaff, and whitethroat, according to Wyllie.

Moksnes et al. (1991a) summarized acceptance/rejection data for nonmimetic eggs introduced into nests of four of the most commonly exploited hosts of the common cuckoo in northern Europe (meadow pipit, white wagtail, hedge accentor, and European redstart). They also summarized acceptance rates for 5 frequently used hosts (tree pipit, yellow wagtail, garden warbler, spotted flycatcher, and Lapland longspur), 9 rarely exploited hosts, and 15 unsuitable hosts. The overall acceptance rate for the most common hosts was 41 of 47 nests (median 86%), for the frequently used hosts, 9 of 20 nests (median 33%), for the rare hosts, 21 of 113 nests (median 10%), and for the unsuitable hosts, 151 of 187 nests (median 100%). The median acceptance rate for the rare hosts was higher (but not significantly so) from that of the most common or the most frequently used hosts. However, the acceptance rate for the unsuitable hosts was significantly higher than for that of the rare hosts. Species that exhibited high rejection rates were also likely to exhibit high rates of aggression toward the cuckoo mount, and low aggression rates were typical of rarely parasitized species.

Moksnes et al. (1991a) introduced white plastic cuckoo egg models in the nests of the meadow pipits, rather than using mimetically colored models, and found that the rejection rate for white dummy eggs was only slightly higher than the rejection rates for well-painted dummy eggs (8% vs. 5%). However, they suggested that this situation may reflect an evolutionary stage in which the cuckoo population has evolved a mimetic egg morph even when a proportion of that population will accept nonmimetic eggs. They also noted that the rejection rate among meadow pipits for nonmimetic eggs is significantly higher in Britain (48% of 58 nests). Surprisingly, the same was found to be true in Iceland, where the rejection rate for nonmimetic eggs of various pattern types was also higher (19% of 27 nests), even though the cuckoo does not currently breed there (Davies & Brooke, 1989a). Additionally, Davies and Brooke found (1989b) no differences in the distinctiveness of egg markings between those species that have interacted with cuckoos and those that have not, and observed no differences in the patterning present in the eggs of British (parasitized) and Icelandic (unparasitized) populations of meadow pipits and white wagtails. They also found no evidence of an evolution of unique host egg patterns in defensive response to cuckoo parasitism.

Soler et al. (1994) estimated that the egg-ejection rate of great spotted cuckoo eggs by magpies in southern Spain increased at a mean rate of 0.5 per year during the 1982–1992 period and that rates of host ejection of mimetic model eggs introduced experimentally into magpie nests increased at a mean rate of 4.7% per year. They postulated that this trend toward increased rejection behavior by the host species might be the result of an evolutionary (genetic) change or a conditional (learned) response.

Rothstein (1982) tested the behavior of the American robin and the gray catbird, two species that are rejectors of cowbird eggs, toward the introduction of egg models that varied in ground color, maculation (patterning), and size. Using 10 model types at 137 robin nests, Rothstein found that robins usually did not reject eggs that differed from their own eggs in one of these respects, but usually did reject eggs that differed in two of the three categories. Small egg size (robin eggs are larger than cowbird eggs) was the most effective of these rejection-stimulating parameters in producing an early response, although ground color and maculation had a greater valence than size in determining whether eventual host rejection occurred. Using the same approach at 37 catbird nests, Rothstein found that ground color rather than maculation stimulated rejection behavior (cowbird and catbird eggs are nearly the same size, so possible size effects were not tested).

Davies and Brooke (1989a) experimentally parasitized 711 nests of 24 British passerines with model cuckoo eggs and found that rejection behavior occurred at 225 nests. The majority these rejections involved egg ejection (51%), and in most of the remaining cases (44%) the nest was deserted. In a few cases the hosts built a new nest lining, burying any eggs already present, or the birds persistently pecked at the model egg until the investigators removed it. Three of the four most commonly exploited British hosts (reed warbler, meadow pipit, and pied wagtail) exhibited strong tendencies toward discrimination against nonmimetic eggs, but the hedge accentor did not. However, some rarely exploited but seemingly acceptable species discriminated at least as strongly as did the most strongly rejecting major hosts, whereas unsuitable host species showed little if any rejection behavior. Davies and Brooke thus concluded that egg discrimination by suitable hosts has generally not evolved in response to cuckoo parasitism. However, they observed that two European species (European redstart and great reed warbler) that endure extremely high rates of local parasitism (43.5% and 50%, respectively) are also exposed to the highest degree of cuckoo egg mimicry. Not surprisingly, they also noted that smaller-billed but suitable host species tended to reject eggs by means of nest desertion, whereas those with larger bills tended to reject model cuckoo eggs by employing egg-ejection behavior. Similar relative bill-length to egg-width relationships have also been observed among American host species of the brown-headed cowbird by Rothstein (1975b), who judged that birds having a bill-length:egg-width ratio of less than 1:0.7 may not be able to remove cowbird eggs by direct grasp-ejection methods, but might nevertheless be able to handle and remove the egg by spearing it (puncture ejection). Rothstein suggested that the most critical factors influencing a host's egg-rejection behavior are its relative nest concealment and its bill size; poorly concealed nests may have resulted in high rates of parasitism, and those species with long bills may have been able to evolve rejection behavior more readily than other smaller-billed ones.

Since egg-rejection behaviors are presumably evolved responses, one might expect all members of a host species to be uniform in their egg-discrimination abilities, but many species exhibit intermediate rates of rejection. This has often been explained by assuming that such behavior represents the effects of an evolutionary lag on the part of the host (Rohwer & Spaw, 1988), whose rejection behavior has not yet fully caught up with the selective pressures being placed on it by the parasite. However, as with differential age-related aggressive responses toward mounts of brown-headed cowbirds, age-related variables in egg rejection behavior have been detected among common cuckoo hosts, as well as evidence for a host learning characteristics of its own eggs (Lotem et al., 1992, 1995). The evolution of interclutch variability in host egg appearance may also represent an important coevolutionary response to parasitism (Ølen et al., 1995).

Davies and Brooke (1989b) speculated that the usual chronology of coevolutionary events occurring during a parasite–host interactive history may begin with the host exhibiting no tendency to reject eggs unlike its own. Soon after parasitism begins, egg discrimination is favored, with the rate of acquisition of egg discrimination depending on the rate of parasitism. Possibly only a few hundred generations of selection on the host may be required for it to acquire egg-rejection abilities under parasitism rates of 20% or more.

Selection for laying mimetic eggs by the parasite will increasingly occur as egg-rejection behavior by the host evolves, but with egg mimicry by the parasite evolving more rapidly than the host's capacity for evolving egg-rejection behavior (Dawkins & Krebs, 1979). Intermediate stages may thus occur in which the parasite population lays a mimetic egg morph, but not all individual hosts reject eggs unlike their own.

If parasitism rates are excessive, the host population may be driven to extinction, which would also cause such host-dependent parasite populations to disappear. However, as egg-rejection abilities improve in one species, alternate hosts may be chosen by the cuckoo, leading to the evolution of local parasite gentes. As many as 10 gentes have evolved among common cuckoos in Europe (Moksnes & Røskaft, in press). If eventually free of parasitism, an egg-rejecting host might slowly revert back to becoming an acceptor of unlike eggs. A possible example of this has occurred with a New World population of the African village weaver, which was introduced before 1797 to Hispaniola. Unlike its ancestral population, this introduced population has lost most of its rejection tendencies (Cruz & Wiley, 1989).

Rothstein (1990) argued that, because there are about 50 species of parasitic cuckoos but only 5 parasitic cowbirds, it is likely that cuckoos have been brood parasites for a much longer period, and coevolutionary arms races have gone on for a much longer time. In line with this idea, he determined that most potentially suitable cuckoo hosts exhibit rejection behavior toward nonmimetic eggs, whereas most potentially suitable cowbird hosts do not. Additionally, cowbirds never exhibit egg mimicry, whereas the cuckoo does, which might help reduce the rate of evolution of rejection behavior. Furthermore, cuckoo parasitism is a fairly rare event (rarely reaching a 30% parasitism rate), whereas in many cowbird hosts the rate of parasitism exceeds 70%, and multiple cowbird parasitism is fairly common, so overall selection pressures favoring the evolution of rejection behavior might be generally weaker among cuckoo hosts than among those of cowbirds.

Posthatching Discrimination of Parasitic Young

It is interesting that, although many species of hosts have evolved effective mechanisms of recognizing and rejecting nonmimetic eggs, there is no direct evidence of host species modifying their own egg coloration as a means of retaliation against a parasite's egg mimicry, nor have most host species managed to evolve a means of rejecting young parasites after they have hatched. Evidently, the hosts' innate parental tendencies to brood any hatched chicks overwhelm any discrimination capacities that might be present. Brooke and Davies (1987) have argued that the risk of the hosts rejecting their own offspring in trying to discriminate against parasitic nestlings has prevented effective antiparasite behaviors from evolving, even among those species that may have evolved excellent egg-discrimination and egg-rejection abilities. Indeed, there are many descriptions of host parents standing by helplessly as their own young are being ejected from the nest by a baby cuckoo. Apparently the only hosts that may require well-developed nestling mimicry by their parasites are the viduine finches, which reportedly will not feed those nestlings that do not have the complex mouth patterns of their own species (Nicolai, 1974; Payne, 1982). Some theoretical explanations for the general absence of nestling discrimination have been advanced, such as the possibility that adult cuckoo predation on nestlings may have evolved to stop nestling discrimination by hosts (Zahavi, 1979), but such explanations have some logical and internal inconsistencies (Guilford & Read, 1990).

It is possible that the similar, if not identical, appearances of bay-winged and screaming cowbird nestlings also represent an example of evolved nestling mimicry by the latter species, but it could also be argued that the similarities between these two congeneric species are instead simply the product of common descent. Lack (1968) has similarly argued that the striped back pattern of juvenile parasitic weavers is not a case of evolved plumage mimicry of nestling cisticolas, but rather both species simply have evolved similar cryptic plumages as a general antipredation device. However, the apparent juvenile plumage mimicry of the great spotted and pied cuckoos with their hosts and the Asian koel with its corvid hosts (Jourdain, 1925; Lack, 1968) and of various African viduines with their estrildine hosts (Nicolai, 1974) suggest that host discrimination abilities might have carried selection for effective mimicry forward into this stage of reproduction. All of these examples of nestling mimicry involve host-tolerant brood parasites, the only group of parasites for which such mimicry is potentially a serious problem.

The tendency for many cuckoos to have well-developed mandibular flanges and brightly colored palates that perhaps operate as super-normal releasers, effectively countering any incipient host discrimination tendencies, may be an important factor influencing host discriminative abilities. Cuckoo mimicry of host species' begging calls may also be relevant here. However, McLean and Waas (1987) suggested that in addition to actual evolved mimicry, the similarities of host and parasitic nestling vocalizations might result from behavior-matching pressures associated with common environmental factors, chance acoustic convergence (owing, for example, to common structural characteristics associated with sound production), or common developmental experiences and opportunities for learning. Regardless of the explanation for such similarities, they might well be important devices for neutralizing the effects of any potential host-discrimination behavior.

Alternate End-games: Adaptive Tolerance and Mutualism

For those species having a low rate of brood parasitism, it is perhaps less costly to accept occasional brood parasitism than to incur the potentially severe costs of rejection (Davies & Brooke, 1988, 1989b; Moksnes et al., 1991). Such species are likely to represent the most ideal hosts available to the parasite, since they have not yet been exposed to a sufficient degree of selection to evolve adequate antiparasite responses to make the costs of parasitizing them unrewarding.

Even for species with a fairly high incidence of brood parasitism, it may be less costly to endure such rates and associated reductions in productivity than to abandon their nests and start over. In a study of the effects of brown-headed cowbird parasitism on the hole-nesting prothonotary warbler, Petit (1991) observed a 21% parasitism rate among 172 warbler nests in her study area, and an estimated 25% reduction in nesting success associated with such parasitism, mainly as a result of egg removal by cowbirds and reduced warbler hatching success. She determined that the incidence of nest desertion was contingent upon the availability of nest sites, which were generally in short supply. Thus, females whose males were defending three or more nest sites within their territory were more likely to abandon their nests and start over elsewhere than were females of those pairs having fewer available sites within their territories. Since the majority of prothonotary warblers in this population are double-brooded, and since productivity of the first brood was reduced only by 25%, accepting the cowbird eggs and attendant reduced productivity on the first breeding effort while still having time to undertake a second brood has a higher reward than starting over and perhaps producing only a single brood.

In the yellow warbler, there are varied kinds of host responses to cowbird parasitism. According to Clark and Robertson (1981), the most frequent response (48% of observed cases) is egg burial, which is most common when parasitism occurs early in the warbler's laying cycle. Nest desertion represents an alternative rejection response (24% of observed cases), but might cause substantial delays in renesting and reduce the potential for renesting. Egg acceptance occurred in 29% of the cases, and such nests produced a mean of 0.53 warbler offspring, as compared with 0.8 offspring for unparasitized nests, representing a 33% reduction in warbler productivity. As with the prothonotary warbler, such a cost may be the most adaptive choice, especially if there is little or no chance of completing a successful renesting.

Given enough time, it is possible that coevolved adjustments between the parasite and its host or hosts will reduce the costs of brood parasitism for the host species to a minimum, thus facilitating the survival of both. A few studies have suggested that a mutualistic relationship may even exist under certain conditions between host and parasite. Morel (1973) reported that hatching success of the host red-billed firefinch was higher in those nests parasitized by village indigobirds than in unparasitized nests, apparently because of improved parental tending behavior in nests with greater numbers of eggs. Overall, the numbers of firefinches reared per unparasitized nest averaged slightly higher than in parasitized nests (2.6 vs. 2.1), but these small differences are of little if any significance, and overall breeding success rates were virtually identical for both nest categories.

In another intriguing study, Smith (1968) compared the breeding behavior and nesting success of the chestnut-headed and other oropendolas and the yellow-rumped cacique under varied nesting conditions and degrees of brood parasitism by the giant cowbird. These large, colonial-nesting icterines evidently prefer to nest in sites closely situated to bee or wasp colonies, whose self-protective behavior also serves to shield the birds from parasitic botflies as well as from disturbance by vertebrate predators. However, dependence on these insect colonies incurs some built-in costs or at least presents some undesirable uncertainties to the birds, such as the possibility that the bees or wasps may desert their sites without apparent reason. The bees and wasps also tend to build their nests late in the birds' breeding seasons, thus forcing the birds to delay their own nesting season if they are to take full advantage of the antibotfly protection provided by the insects.

Those caciques and oropendolas that nested close to and thus under the passive protection of the bees and wasps were found by Smith to discriminate against intrusion by giant cowbirds and their eggs. The female cowbirds laying in such colonies typically laid mimetic eggs, which were surreptitiously deposited at the usual rate of one per nest, and usually only in nests containing a single host egg. However, the oropendolas and caciques nesting in colonies that were not protected by bees or wasps were classified by Smith as "nondiscriminators," and he observed that female cowbirds were allowed to lay their often nonmimetic eggs in the nests of such colonies without interference. The breeding success rates of these

TABLE 27 Effects of Giant Cowbird Parasitism on Host Productivity[a]

	Discriminator colonies		Nondiscriminator colonies	
	Total nests	Hosts fledged per nest	Total nests	Hosts fledged per nest
Unparasitized	1526 (74%)	0.32	562 (45%)	0.18
One cowbird chick	438 (21%)	0.25	401 (32%)	0.52
Cost or benefit		−0.06		+0.34
Two Cowbird chicks	99 (5%)	0.20	295 (23%)	0.43
Cost or benefit		−0.12		+0.25

[a]Based on data of Smith (1968), representing a 4-year data sample. Costs and benefits estimated relative to productivity for unparasitized nests.

two categories of hosts were surprisingly different (table 27), since the nondiscriminator colonies having one or even two cowbird nestlings present produced a larger number of fledged offspring per nest than did the unparasitized nests in either the discriminator or nondiscriminator categories. Smith attributed this remarkable effect to the fact that nestling cowbirds not only self-preen, thereby removing the larvae of botflies from their own bodies, but they also preen all other nestlings, including host nestlings. Since the major cause of icterine nestling mortality is probably the result of botfly parasitism, it is more advantageous for the icterine hosts to raise one or even two cowbirds along with their own chicks than to risk the effects of botfly parasitism. The presence of wasps or bees evidently serves to keep botflies away from the icterine colonies associated with them, and thus no benefits derive to such birds for hosting giant cowbirds when nestling under these special protective conditions.

Part II

THE AVIAN BROOD PARASITES

Whitehead host feeding a fledgling long-tailed koel. After a photo by A. Wilkinson (in Oliver, 1950).

6

WATERFOWL

Family Anatidae

The only known obligate brood parasite among the waterfowl, and indeed among all species of birds having precocial young, is the South American black-headed duck. Now generally accepted to be a member of the stiff-tailed duck assemblage (Johnsgard and Carbonell, 1996), the black-headed duck is fairly widespread but relatively uncommon throughout much of temperate South America, especially Argentina. In this same general region, there are two other species of stiff-tailed ducks, as well as several other marsh-breeding ducks, coots, and other marshland birds that represent potential host species.

Although it has been known to be one of the numerous species of waterfowl parasitizing the nests of other waterfowl species, it was not until recently that biologists became convinced that the black-headed duck is an obligate, rather than simply a facultative, brood parasite. Weller (1968, p. 203) concluded that "parasitism appears to be the sole means of reproduction, as no nests or brood care is known for the species." Weller further judged that, of all the brood-parasitic birds, the black-headed duck "appears to be the least damaging to its host, and in that sense is the 'most perfected' of the brood parasites." In addition to being the only species of Anatidae that is an obligate brood parasite, the black-headed duck is also the only waterfowl species in which the adult female has a somewhat larger mean body mass (averaging about 10% larger) than the male. Weller believed that this reversed sexual dimorphism is not clearly attributable to the species' evolution of brood parasitism, although selection for large body mass in females might well be a correlate of selection for the female's ability to produce a large number of eggs.

Weller (1968) attributed part of the black-headed duck's success as a brood parasite to the fact that it exploits a large number of host species, including at least five fostering hosts (two coots, rosy-billed pochard, white-faced ibis, and brown-hooded gulls). Survival of the young, in the nests of hosts with different diets from that of the black-headed duck, is facilitated because of the considerable precocity of the ducklings, which are able to fend for themselves and forage independently only a few days after hatching.

BLACK-HEADED DUCK

(Heteronetta atricapilla)

Other Vernacular Names: None in general English use.

Distribution of Species (see map 1): Resident in central Chile and in central to southern Argentina, possibly breeding north to Uruguay.

Measurements (mm)

Wing, males 157.5–178 (avg. 168), females 154–182 (avg. 48.7). Tail, males 44–57 (avg. 48.7), females 44–59 (avg. 52.1) (Weller, 1967).

Egg, avg. 58.05 × 43.23 (range 55.1–62.6 × 40.5–43.3) (Weller, 1968). Shape index 1.35 (= oval).*

Masses (g)

Adult males 434–580 (avg. 512.6, n = 11), adult females 470–630 (avg. 565.2, n = 13) (Weller, 1967a). Estimated egg weight 55.5. Egg:adult female mass ratio 9.8%.

Identification

In the field: Black-headed ducks resemble surface-feeding (*Anas*) ducks more than stiff-tailed ducks and usually forage while swimming at the surface or by tipping-up rather than diving. The black head of the male, with a bluish bill that has bright pink to red at the base, is distinctive. Females resemble females of various teal species, especially in their head patterning, but like males they have very short tails and their wings lack iridescence (Fig. 17). Immature individuals resemble adult females. Vocalizations are few and are largely limited to repeated soft "pik" sounds produced by the displaying male while jerking its inflated throat. The birds take off readily, unlike other stiff-tailed ducks, and fly swiftly, with rapid wingbeats in a manner similar to surface-feeding ducks.

In the hand: In contrast to true stiff-tailed ducks and in common with surface-feeding ducks, this teal-sized species has a short tail about the

MAP 1. Range of black-headed duck.

*Egg shape definitions: length:breadth ratios 1.05–1.2:1 = subspherical; 1.21–1.35:1 = broad oval (or broad elliptical); 1.36–1.5:1 = oval (or elliptical); and >1.5:1 = long oval (or long elliptical).

FIGURE 17. Comparison of adult females, ducklings, and eggs of the black-headed duck (A) and two of its hosts, the rosybill (B) and the red-gartered coot (D).

same length as the bill. Its blue to grayish bill is narrower (<20 mm) and longer than those of true stifftails (*Oxyura*), whose bills become flatter and broader toward the tip. Its hind toes also lack the strong lobing typical of the true stiff-tailed ducks.

Habitats

This species is associated with freshwater to somewhat alkaline marshy swamps, especially those with well-developed stands of emergent vegetation such as bulrushes (*Scirpus* spp.) and small-leaved floating vegetation such as duckweeds. Lakes and deep roadside ditches are also used outside the breeding season, and black-headed ducks are sometimes seen on flooded fields and reservoirs.

Host Species

According to Weller (1968), about 14 species of birds have been parasitized by the black-headed duck. These are the white-faced ibis, black-crowned night heron, roseate spoonbill, southern screamer, coscoroba swan, limpkin, spotted rail, maguari stork, chimango, brown-headed gull, red-fronted coot, red-gartered coot, and rosy-billed pochard. Weller's data suggested that the primary hosts are the two coot species and the rosybill, and only these species were observed to be successful fostering hosts (i.e., they actually hatched ducklings).

Egg Characteristics

Eggs of this species are whitish to buffy in color, oval in shape, and finely pitted in surface texture. They have a more granular appearance when candled, as compared with the similar eggs of the rosy-billed pochard. The eggs of black-headed ducks and rosy-billed pochards are virtually identical in measurements, but those of the pochard are slightly longer and tend to be more elliptical in shape. However, the eggs of *Heteronetta* are rather varied in both shape and color and may not always be separable from those of the rosy-billed pochard (Weller, 1968).

Breeding Season

In the northern parts of their range, black-headed ducks breed during the fall (March–June), when late-summer flooding allows nesting by various marsh birds. In the area around Buenos Aires, spring breeding is typical, with courtship starting in September, egg-laying the second half of that month, peaking in October, and terminating by mid-December. This corresponds well with the breeding period of the rosy-billed pochard, but less well with the two coot species, whose breeding season peaks during the latter half of September and early October (Weller, 1968).

Breeding Biology

Nest selection and egg laying. Powell (1979) provided the only first-hand description of egg-laying behavior so far available. A female black-headed duck was observed to lay an egg within an 8-minute visit to a rosy-billed pochard nest at the Wildfowl Trust. During the next 9 days, it laid a total of seven eggs, three in the pochard nest, two in a moorhen nest, and two in a cinnamon teal nest. Weller (1968) list 13 hosts, which include many species that are probably unsuitable as fostering hosts, such as the maguari stork and the chimango caracara, both of which are potential predators of ducklings. Among 82 nests in one area, Weller found that 60 nests (73%) held a single parasitic egg, 17 (21%) had 2, 3 (4%) had 4, and 2 nests had 5 and 8 eggs. Nests containing three to five host eggs were usually chosen for parasitism, but many were laid after incubation was well underway; sometimes the host eggs were nearly ready to hatch.

Incubation and hatching. Weller (1968) estimated the incubation period as lasting 24–25 days, or 3–4 days shorter than that of the rosy-billed pochard, which lays an egg of almost exactly the same size. The periods of the two usual coot hosts range from 24.5–25 (last-laid egg) to 28–29 days (first-laid egg).

Nestling period. The precocial condition of the ducklings allows them to feed themselves almost immediately, and they may leave the host's nest within 1 or 2 days of hatching, perhaps at most returning at night for brooding. The fledging period for birds raised in captivity at the Wildfowl Trust has been approximately 10 weeks, a relatively long period for such a small duck (Johnsgard and Carbonell, 1996).

Population Dynamics

Parasitism rate. Weller (1968) found a 54% incidence of parasitism for 114 nests of red-fronted coots in one study area, and a 58% incidence for 19 nests in another area. A higher rate (83%) was found in a sample of only five rosy-billed pochard nests.

Hatching and fledging success. In one of Weller's (1968) study areas, the hatching success of 76 parasitic eggs was only 18% in nests of the most important host species, the red-fronted coot. In another area the hatching success of 14 eggs was 64% in red-fronted coot nests.

Host–parasite relations. Coots evidently bury the eggs of black-headed ducks fairly frequently; Weller (1968) observed red-fronted coots incubating their own eggs that were placed above a layer of black-headed duck eggs. No such evidence for rejection was found for rosy-billed pochard nests, where the similarity in the eggs of the two species makes such behavior unlikely. Weller found no evidence that female black-headed ducks remove, damage, or destroy host eggs in the process of laying their own. He reported that at five nests of red-fronted coots, at one rosy-billed pochard nest, and at one white-faced ibis nest, the foster parents brooded the black-headed duck ducklings as if they were their own.

7

HONEYGUIDES

Family Indicatoridae

The honeyguides are a predominately African family, with a center of species diversity in tropical West Africa and in the headwater region of the Congo River (Fig. 18). Snow (1968) stated that all of the four honeyguide genera are associated with evergreen forest in Africa and that the evergreen forest species appear to be morphologically primitive. In his view the family may have evolved as evergreen forest forms, which subsequently became adapted to more arid habitats. Friedmann (1955) stated that, of the 11 species he recognized, 6 are to adapted forests, 3 are associated with open habitats, and 2 occur in both habitat types. Although several species are not yet proven to be brood parasites, there is no evidence to support the possibility that they are not.

Friedmann (1955) reviewed the morphological evidence relative to the phyletic relationships of the honeyguides and judged that their nearest affinities are with the barbets (Capitonidae), toucans (Ramphastidae), puffbirds (Bucconidae), and woodpeckers (Picidae), although he was uncertain as to the degree of phyletic closeness that they share with each of these groups. Sibley and Monroe (1990) placed the families Picidae and Indicatoridae within infraorder Picides, thus separating them from the barbets and toucans of the same order Piciformes, and even farther from the puffbirds, which were placed in a separate order (Galbuliformes).

Regarding the evolution of the brood parasitism in the honeyguides, Freidmann (1955) judged it to be an older trait than the splitting of the lines into the present-day generic components, inasmuch as at least the two most divergent genera (*Indicator* and *Prodotiscus*) are

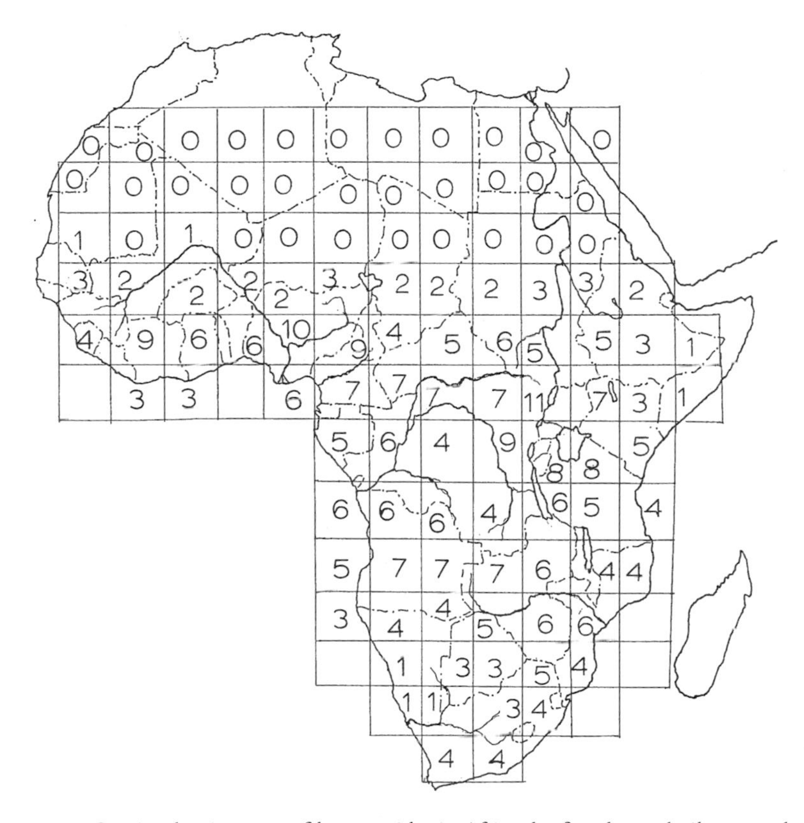

FIGURE 18. Species-density map of honeyguides in Africa, by five-degree latilong quadrants.

known to include parasitic species. He believed that the honeyguides exhibit some "losses" associated with parasitism, such as a reduction or loss in pair-bonding behavior. Additionally, several specializations, or "gains" in specializations associated with parasitism include the development of bill hooks in newly hatched chicks of several species and an apparent matching or near-matching of incubation periods between parasite and host. Traits of uncertain significance relative to honeyguide adaptations for parasitism include their relatively long period of nestling development and the fact that their eggs are sometimes are considerably smaller than those of their hosts. Payne (1989) has discussed the adaptive significance of relative egg size in honeyguides.

SPOTTED HONEYGUIDE

(Indicator maculatus)

Other Vernacular Names: None in general English use.

Distribution of Species (see map 2): Sub-Saharan Africa from Gambia east to Sudan, southwestern Uganda, and east-central Zaire.

Subspecies

I. m. maculatus: Gambia to Nigeria.

I. m. stictothorax: Cameroon to Zaire and Uganda.

Measurements (mm)

7–7.5" (18–19 cm)

Wings, males 99–104 (avg. 102), females

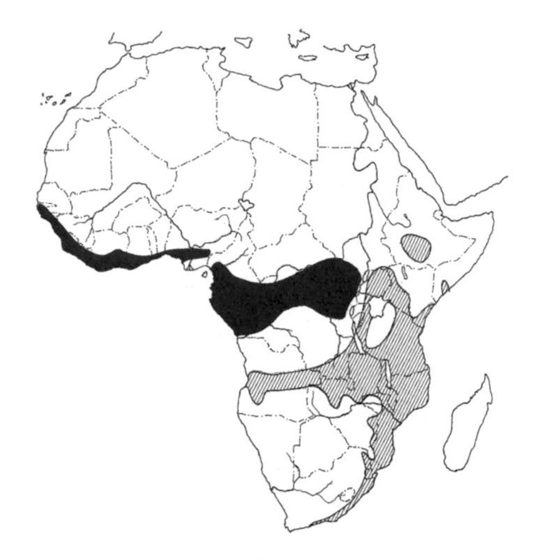

MAP 2. Ranges of scaly-throated (hatched) and spotted honeyguides (filled).

97–100 (avg. 98.6). Tail, males 62–69 (avg. 66.2), females 56–63 (avg. 60.8) (Fry et al., 1988).

Egg, *maculatus* 23.4 × 18; *stictothorax* 21.5 × 17.5 (Fry et al., 1988). Shape index 1.23.

Masses (g)

No body weights available. Estimated egg weight of *maculatus* 3.9, of *stictothorax* 3.4.

Identification

In the field: This medium-sized honeyguide is unique in having a breast and underpart coloration distinctly spotted with rounded, olive-yellow markings (fig. 10). The throat is streaked with black, and the outer tail feathers are mostly white. Immature individuals are streaked, rather than spotted in these areas. The calls include cheeping 'woe-woe-woe' notes, and a purring or trilling 'brrrrrrr' with erected throat and body feathers.

In the hand: Among African honeyguides, this species is unique in having a combination of no white on the rump, a throat that is spotted with yellow, and yellowish underparts. The sexes are alike, but females have slightly smaller measurements. Immature individuals resemble adults on the upperparts but are generally greener and more yellowish, with dark fuscous markings on the bases of the crown feathers. The immature individuals are darker below and are more obviously spotted and streaked with buffy white markings. The rectrices are more pointed than in adults.

Habitats

This species occupies forest and gallery forest, from sea level to about 2100 m.

Host Species

No host species yet proven, but the buff-spotted woodpecker, the gray-throated barbet, and perhaps other barbets of this genus are suspected hosts (Fry et al., 1988).

Egg Characteristics

The egg is white, but no other details have been reported.

Breeding Season

Little information exists. Breeding records exist for June (Ivory Coast) and January (Rio Muni) (Fry et al., 1988).

Breeding Biology

No information.

Population Dynamics

No information.

SCALY-THROATED HONEYGUIDE

(Indicator variegatus)

Other Vernacular Names: Variegated honeyguide.

Distribution of Species (see map 2): Sub-Saharan Africa from Sudan and Somalia south to Angola and northeastern South Africa.

Measurements (mm)

7.5″ (18–19 cm)

Wing, males 102–117 (avg. 109), females 100–116 (avg. 106). Tail, males 63–81 (avg. 69.9), females 62–75 (avg. 66.8) (Fry et al., 1988).

Egg, avg. 21.3 × 16.9 (range 20–22.8 × 16.5–18) (Fry et al., 1988). Shape index 1.26.

Masses (g)

Males 40–56 (avg. 48.5, n = 24), females 36.5–55 (avg. 47.8, n = 15) (Fry et al., 1988). Estimated egg weight 3.1. Egg:adult mass ratio 6.5%.

Identification

In the field: This species resembles the similar-sized spotted honeyguide in having a streaked or freckled throat, but its underparts are vaguely streaked or spotted with olive, rather than distinctly spotted with yellow (fig. 10). Immature individuals are more spotted with black on the breast and underparts. The usual primary song or call is a trill that ranges from a croaking noise to a an insectlike sound to a vibrant whistle, usually rising in pitch at the end, and repeated at intervals of 1 or 2 minutes. Several other vocalizations have been described.

In the hand: Separated from other African honeyguides by the lack of white on the rump and by the yellow to olive breast and throat spotting, which in this species is limited to the breast, whereas the abdomen becomes whitish (not yellowish as in the spotted honeyguide) and lacks spots or streaks. Females are smaller than males; markings on the chin, throat and breast are paler and sparse. The lores and crown are more washed with olive, and the iris is grayish brown rather than dark brown. Immature individuals resemble females but are sometimes more yellowish, with the spotting on the throat and breast darker, and the rectrices narrower and more pointed at the tips.

Habitats

This species occupies woodlands, forests, thickets, wooded grasslands, and gallery forests, but avoids dense forests. From near sea level to about 3000 m.

Host Species

Eight known host species are listed in table 28. Six additional likely hosts, all barbets and woodpeckers, are listed by Fry et al. (1988).

Egg Characteristics

The eggs are glossy white, oval, and are not distinguishable from those of the lesser honeyguide.

Breeding Season

Laying data suggest spring breeding in southeastern and southern Africa: Malawi (September, October), Zimbabwe (September to January), Zambia (October), Natal (October), and Transvaal (December). May and August laying records exist for Kenya, with gonadal breeding data for January, May, and June, and nestlings reported in June and July. In Tanzania, juveniles have been seen during 5 months between February and November, and in Uganda during February and from April to August (Fry et al., 1988).

Breeding Biology

Nest selection, egg laying. Little information exists. Dowsett-Lemaire (1983) suggested that the female may lay her eggs by standing at the tiny entrance of a tinkerbird's nest with her tail fanned, evidently dropping the eggs through the hole without actually entering the nest.

Incubation and hatching. The incubation period has been estimated at 18 days (Dowsett-Lemaire, 1983).

Nestling period. The fledging period has been estimated as 27–35 days (Dowsett-Lemaire, 1983).

Population Dynamics

Parasitism rate. No information.

Hatching and fledging success. No information.

Host–parasite relations. It is believed that the newly hatched young kill their hosts' chicks and destroy any remaining eggs, using their well-developed bill hooks, as in other better-studied honeyguides (Dowsett-Lemaire, 1983).

GREATER HONEYGUIDE

(Indicator indicator)

Other Vernacular Names: Black-throated honeyguide.

Distribution of Species (see map 3): Sub-Saharan Africa from Gambia to Somalia, south to Namibia and South Africa.

Measurements (mm)

8″ (20 cm)

TABLE 28 Reported Host Species of African Honeyguides[a]

Nonpasserine host species	Passerine host species
Scaly-Throated Honeyguide	
Black-collared barbet	
Whyte's barbet	
Yellow-rumped tinkerbird	
Nubian woodpecker	
Golden-tailed woodpecker	
Olive woodpecker	
Gray woodpecker	
Cardinal woodpecker	
Greater Honeyguide	
African pygmy kingfisher	Greater striped martin
African gray kingfisher	White-throated swallow
Gray-headed kingfisher	Rufous-chested swallow
Brown-hooded kingfisher	Banded martin
Cinnamon-chested bee-eater	Pied starling
Swallow-tailed bee-eater	Red-shouldered glossy starling
White-fronted bee-eater	Black tit
Carmine bee-eater	Scarlet-chested sunbird
Little bee-eater	Southern anteater chat
Little green bee-eater	Northern anteater chat
Madagascar bee-eater	Yellow-throated petronia
Boehm's bee-eater	Gray-headed sparrow
Abyssinian roller	
Abyssinian scimitarbill	
Scimitarbill	
Green wood-hoppoe	
Hoopoe	
Pied barbet	
Black-collared barbet	
Crested barbet	
Rufous-breasted wryneck	
Nubian woodpecker	
Golden-tailed woodpecker	
Knysna woodpecker	
Tullberg's woodpecker	
Gray woodpecker	
Lesser Honeyguide	
Striped kingfisher	White-throated swallow
Little bee-eater	Pied starling
Cinnamon-chested bee-eater	Violet-backed starling
Whyte's barbet	Yellow-throated petronia
Anchieta's barbet	
Green barbet	
Pied barbet	
Red-fronted barbet	
Chaplin's barbet	
White-headed barbet	

TABLE 28 *(continued)*

Nonpasserine host species	Passerine host species
Black-collared barbet	
Crested barbet	
Rufous-breasted wryneck	
Bennett's woodpecker	
Golden-tailed woodpecker	
Thick-Billed Honeyguide	
Gray-throated barbet	
Pallid Honeyguide	
Yellow-rumped tinkerbird	
Cassin's Honeyguide	
	Black-throated wattle-eye
	Buff-throated apalis
	Green white-eye
Green-Backed Honeyguide	
	Black-throated wattle-eye
	Dusky alseonax (flycatcher)
	African paradise-flycatcher
	Abyssinian white-eye
	Montane white-eye
	Yellow white-eye
	Amethyst sunbird
Wahlberg's Honeyguide	
	Yellow-throated petronia
	Gray-backed cameroptera
	Tabora cisticola

[a]Host list based mainly on Fry et al. (1988).

Wing, males 102–117 (avg. 111), females 97–114 (avg. 106). Tail, males 66–82 (avg. 73.6), females 60–78 (avg. 68) (Fry et al., 1988).

Egg, range 22.5–26 × 16.8–20 (Fry et al., 1988). Shape ratio ~ 1:1.3 (= broad oval).

Masses (g)

Males 41–52 (avg. 48.9, $n = 18$), females 40–52 (avg. 46.8, $n = 14$). Weight of fresh egg 5.2 (Fry et al., 1988.). Egg:adult mass ratio 10.9%.

Identification

In the field: This is the largest of the honeyguides and the most colorful (fig. 19). Adult males have black throats outlined above by a pale gray ear-patch and below by a nearly white breast and underpart coloration. Males also have a small patch of golden yellow feathers on the anterior (lesser) wing coverts (probably visible only in flight). Females are uniformly grayish brown above and have white underparts similar to those of the male. Immature males generally resemble females,

MAP 3. Range of greater honeyguide.

but have a bright yellow wash on the breast and upper flanks. The male's song is a long series of two-note phrases, "burr-wit" (also variously described as "whit-purr," "vic-tor" and "sweet beer") that each last for about 0.5 second and are regularly repeated from song posts. Males also display in a circular and undulating "rustling" flight, producing audible noises ("bvooommm") during the swooping phase, noises evidently produced by wing or tail feather vibrations. After such winnowing flights, they may perch and begin to sing. Several other vocalizations occur, especially during guiding behavior.

In the hand: Easily distinguished from all other honeyguides by its mostly white rump and upper tail coverts (the feathers are edged with white, with wide, brown shaft streaks). Adult males have a distinctive black throat and a pinkish white bill; females have a white throat and a brownish horn bill. Juveniles and immature individuals are mostly olive-brown on the upperparts and crown, with yellow-green feather edging, and with a yellow case to the feathers of the chin, throat, and breast. The rump of juveniles is clear white in the center; this white rump area extends to the upper tail coverts, which in adults is streaked with olive-brown. The rectrices have narrower brown terminal banding than adults. The yellowish green colors of the crown and the yellow throat disappear about the time that the golden wing marking of adults appear, and black feathers soon begin to appear on the throat of males. Adult plumage takes about 8 months to attain. The last adult trait to appear is the pink color of the male's bill (Fry et al., 1988).

Habitats

This species favors grasslands with scattered trees, including acacia savanna, open gallery woodlands, and forest edges, but avoids entering dense forests. It occurs from near sea level to about 3000 m, but usually is found below 2000 m.

Host Species

Thirty-eight known host species are listed in table 28. Of 161 host records, 8 bee-eater species accounted for 47 (Friedmann, 1955).

Egg Characteristics

The eggs are white, rather glossy, thick-shelled, and broad oval in shape, with little difference in the shell curvature characteristics of the two poles. They average larger than those of other *Indicator* species, but have overlapping measurements.

Breeding Season

In southern Africa breeding occurs during spring, from September to January, with a peak in Natal in November and December. Similarly, egg data from Malawi indicate a September–October breeding period, in Zimbabwe from September to November, and in Zambia during July and again in September–October. East African laying records are mostly from July to December. Gonadal data from Kenya suggest peaks in breeding activity occurring during May–July and September–February. Various evidence for breeding in West Africa suggests a January to July breeding season (Fry et al., 1988).

Breeding Biology

Nest selection, egg laying. Both hole-nesting and deep-nest species have been exploited by this honeyguide. Although most of the species lay white eggs, some do not. In the process of egg laying, the host's eggs are often broken by the visiting honeyguide, apparently with its bill or possibly with its claws. As many as 12 peck marks have been found on a single hoopoe's eggs, suggesting that this behavior is purposeful rather than accidental. The number of eggs laid by a single female is unknown, as is the egg-laying interval, but in 46 cases of parasitism, only a single honeyguide egg was present in the nest. As many as three honeyguide eggs have been seen in a single nest, but in at least one case the variations in their dimensions suggested that they had been laid by different females (Friedmann, 1955).

Incubation and hatching. No information exists on the incubation period.

Nestling period. The paired bill hooks with which the chick is hatched and which fit crossbill-like beside one another when the bill is closed persist for about 14 days, during which time any nestmates are stabbed and killed, probably during the first week after hatching. There is also some evidence that the honeyguide may at times evict the host young from the nest, possibly using the bill hooks to grasp the nest wall, in order to push out any nest companion (Friedmann, 1955). There are two records of two honeyguide chicks present in the same nest. The nestling period lasts about 30 days, and the honeyguide fledgling becomes independent of its foster parent after another 7–10 days (Fry et al., 1988).

Population Dynamics

Parasitism rate. No detailed information. Friedmann (1955) reported that all of four little bee-eater nests were parasitized, and a correspondent informed him that he had never found a hoopoe nest that was not parasitized.

Hatching and fledging success. No specific information.

Host–parasite relations. No specific information, but there seems to be no data suggesting that any host young are raised successfully in the presence of honeyguide chicks.

MALAYSIAN HONEYGUIDE

(Indicator archipelagicus)

Other Vernacular Names: None in general English use.

Distribution of Species (see map 4): Malay Peninsula, Sumatra, and Borneo.

Measurements (mm)

6.5″ (17 cm)

Wing, males 91.5–101.6, females 86.4–89. Tail, males 65.5–71, females 56.5–66 (Friedmann, 1955). Wing, both sexes 79–100. Tail, both sexes 50–73 (Medway and Wells, 1976).

Egg, no information.

Masses (g)

Avg. of 2 males, 38.5 (Dunning, 1993).

Identification

In the field: The small lemon-yellow patch of anterior shoulder feathers is diagnostic for males, but this feature may be difficult to observe in nonflying birds. Otherwise the adults are generally

MAP 4. Range of Malaysian honeyguide.

olive-brown, with short tails and a heavy, rather blunt bill. Immature individuals resemble females but are more streaked below. The call is said to resemble a cat's "meow," often repeated, and with a terminal rattle or churr that rises in pitch like a toy airplane, as in "miaw, miaw, krrwuu."

In the hand: Distinguished as a honeyguide by its zygodactyl feet and short bill. Females have smaller measurements than males (see above) and lack the yellow shoulder feathers. Juvenile males have only a trace of the yellow shoulder feathers, and the upperparts, throat, and breast are somewhat more washed with yellow.

Habitats

This species occupies lowland and mid-level forests up to more than 2000 m in mainland Malaysia and to about 1000 m in the Greater Sundas.

Host Species

No host species are yet know.

Egg Characteristics

No information.

Breeding Season

No information.

Breeding Biology

No information.

Population Dynamics

No information.

MAP 5. Ranges of lesser (filled) and cone-billed (hatched) honeyguides.

LESSER HONEYGUIDE

(Indicator minor)

Other Vernacular Names: None in general English use.

Distribution of Species (see map 5): Sub-Saharan Africa from Senegambia east to Somalia and south to Namibia and South Africa.

Subspecies

I. m. minor: South Africa to southern Botswana.

I. m. damarensis: Southern Angola to Namibia.

I. m. senegalensis: Senegambia to northern Cameroon, Chad, and Sudan.

I. m. riggenbachi: Central Cameroon to Uganda and Zaire.

I. m. diadematus: Ethiopia to Sudan.

i. m. teitensis: Uganda and Sudan to Angola and Mozambique.

Measurements (mm)

6″ (15 cm)

Wing (all subspecies), males 82–98, females 77–98. Tail (*I. m. teitensis*), males 50–57 (avg. 56.7), females 47–63 (avg. 53.1) (Fry et al., 1988).

Egg, range 20.3–22.5 × 15.5–17.5 (avg. 21.4 × 16.5) (Fry et al., 1988). Shape index 1.3.

Masses (g)

Males 23–36.5 (avg. 27.5, $n = 21$), females 22–35 (avg. 25.6, $n = 18$) (Fry et al.., 1988). Estimated egg weight 2.9. Egg:adult mass ratio 10.9%.

Identification

In the field: This is a small, widespread honeyguide that has a gray head, a blackish malar stripe, and a white line extending from the lores forward along the bill (fig. 10). The underparts are uniformly light gray. The bird resembles a pale version of the thick-billed honeyguide and similarly has a distinctly stout bill. The mostly white outer tail feathers are conspicuous in flight, as in other honeyguides. Immature individuals lack the loral mark and dark malar streak but have a streaked throat and grayer underparts. The usual call is a monotonous, continuously repeated "pew" note, and the song is a series of 10–30 such notes, uttered at the rate of about 2 notes per second, with singing periods lasting several hours. There are also several other vocalizations such as a trilled call, squeaking notes, and an in-flight clapping noise that sounds like an irregularly firing motor.

In the hand: This species differs from the other honeyguides in having nearly uniform grayish white underparts, extending from the chin to the under-tail coverts. It is larger (wing more than 80 mm) than the smallest species of *Indicator* and differs from the closely related (previously considered conspecific) thick-billed honeyguide mainly in a lighter tone of gray below. Females differ from males in being somewhat smaller and have a less definite white line from the culmen to the lores, and the black malar streak less distinct. Immature individuals are even less well defined in these markings, there is some streaking on the throat, and they are generally grayer, although the breast is more buffy. The rectrices of immature individuals are also narrower, more pointed, and their dark terminal markings are narrower than in adults (Friedmann, 1955; Fry et al., 1988).

Habitats

This honeyguide occupies brushy habitats in dry areas, including savannas with scattered trees, forest edges, gallery forests, and wooded gardens. It occurs from sea level to about 3000 m but is usually found below 1800 m.

Host Species

Nineteen known host species are listed in tables 12 and 13. Several additional probable or possible hosts are listed by Fry et al. (1988). They noted that 35 of 50 host records from Zaire and 23 of 33 from Zimbabwe involved barbets; 20 of the latter were black-collared barbets. Friedmann (1955) also judged that the black-collared barbet (fig. 20) is the most frequent host species of the lesser honeyguide.

Egg Characteristics

The eggs are glossy white and are broad oval. The black-collared barbet's eggs average 24.3 × 17.5 mm, which represents a host:parasite volume ratio of 1.27:1.

Breeding Season

In South Africa, this species breeds from September to January (Ginn et al., 1989), mainly between October and December. The same span applies to Zimbabwe, and in Zambia and Malawi a September to November spread of breeding occurs. In East Africa the breeding occurs at various times, such as from April to August (central Kenya, eastern Tanzania), May to June (coastal Kenya and coastal Somalia), and January to February (southwestern Kenya, northeastern Tanzania) (Fry et al., 1988).

Breeding Biology

Nest selection, egg laying. Of the numerous hosts listed by Friedmann, only one (the white-throated swallow) builds a cuplike nest; all the others nest in cavities or in fully enclosed nests.

Incubation and hatching. The incubation period was originally estimated at 16.5 days by Skead (1951), but later judged to be no more than 12 days, and possibly as short as 11 days (Friedmann, 1955).

Nestling period. The nestling period lasted 38 days in two cases, or somewhat longer than the 32–35 days typical of its black-collared barbet host. The barbet nest-mates are killed within a week of the honeyguide's hatching by stabbing and biting using the specialized mandible tips (fig. 21). The heel pads of young nestlings are also unusually well developed; presumably this adaptation is useful for gripping the substrate when engaged in biting or fighting. In one observed case the honeyguide's biting response was well developed by the second day, and a single barbet brood-mate was attacked from that day until it died by the fifth day (Friedmann, 1955).

FIGURE 19. Profile sketches of 10 honeyguides: adult male and (inset) female of the greater honeyguide (A), plus adults (sexes alike) of the scaly-throated (C), thick-billed (D), lesser (E), least (F), pallid (G), Cassin's (H), green-backed (I), and Wahlburg's (J) honeyguides. A ventral view of the lyre-tailed honeyguide's tail is also shown at larger scale (B).

FIGURE 20. Eggs of lesser honeyguide (A) and black-collared barbet host (B). Developmental stages of lesser honeyguide are also shown: bill (C) and foot (D) of young chick, naked (E) and feathered (F) nestlings, and heads of feathered nestling (G), fledgling (G), and adult (H). Heads of three host species are also illustrated: striped kingfisher (I), black-collared barbet (J) and cinnamon-chested bee-eater (K). Mostly after photos by G. Ranger (in Friedmann, 1955).

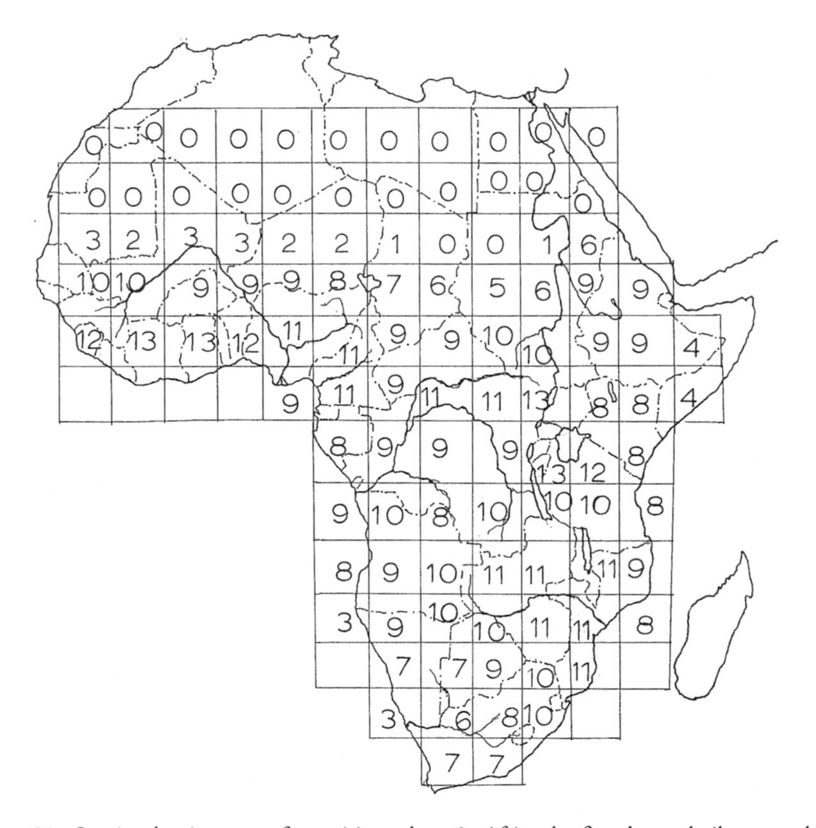

FIGURE 21. Species-density map of parasitic cuckoos in Africa, by five-degree latilong quadrants.

Population Dynamics

Parasitism rate. No information.

Hatching and fledging success. No information.

Host–parasite relations. Friedmann (1955) presents several accounts of black-collared barbets trying to prevent lesser honeyguides from entering their nests, and other species have also been observed attempting to drive honeyguides away from the vicinity of their nests.

THICK-BILLED HONEYGUIDE

(Indicator conirostris)

Other Vernacular Names: None in general English use.

Distribution of Species (see map 5): Sub-Saharan Africa from Liberia east to Uganda and western Kenya, and south to Zaire.

Subspecies

I. c. conirostris: Nigeria to Kenya, Zaire and Angola

I. c. ussheri: Liberia to Ghana

Measurements (mm)

6.5″ (17 cm)

Wing, males 87–95 (avg. 90.6), females 81–95 (avg. 86.2). Tail, males 55–63 (avg. 57.3), females 48–59 (avg. 53.3) (Fry et al., 1988).

Egg, no information.

Masses (g)

Males 24.5, 33.5, females 30.7, 32.5 (Fry et al., 1988). Mean of 8 of both sexes, 31.2 (Dunning, 1993).

Identification

In the field: Closely resembling the last species, this honeyguide is darker below, and the white of the lores and base of bill is reduced or absent, whereas the malar streak is better defined (fig. 19). Immature individuals are streaked below and lack contrasting markings on the lores. The song is a "wheew-wheet, wheet, wheet . . ." that is almost indistinguishable from that of the lesser honeyguide, but it may be slightly faster. Other calls, such as trills and squeaks, are also similar.

In the hand: Closely related to the lesser honeyguide and separated from it in the hand by the distinctions mentioned above. Females differ from males only in being slightly smaller and having a less evident malar stripe. Immature individuals are greener and darker than adults, with unmarked lores, grayer eyes, and rectrices that are narrower, more pointed, and have narrower dark terminal markings.

Habitats

This species inhabits dense forest growth, gallery forest, and heavy second growth, from near sea level to about 2300 m.

Host Species

The only known host is the gray-throated barbet (Fry et al., 1988), but the naked-faced barbet is a highly probable host, as are the two other *Gymnobucco* species.

Egg Characteristics

The eggs remain undescribed.

Breeding Season

In Cameroon there are October and February laying records; there is a December record for Zaire, and a September record for Liberia. Other breeding data, mainly based on gonadal activity, indicate breeding during February in Ghana, during November in Liberia, from October to December in Central African Republic, during December in Kenya, and during February in Angola (Fry et al., 1988).

Breeding Biology

No information.

Population Dynamics

No information.

WILLCOCK'S HONEYGUIDE

(Indicator willcocksi)

Other Vernacular Names: None in general English use.

Distribution of Species (see map 6): Sub-Saharan Africa from Guinea-Bissau east to Zaire and southwestern Uganda.

Subspecies

I. w. willcocksi: Nigeria and Ghana to Zaire and Uganda

I. w. hutsoni: Nigeria and Cameroon to southwestern Sudan

I. w. ansorgei: Guinea-Bissau

Measurements (mm)

5″ (13 cm)

Wing, males 71–79 (avg. 76.2), females 65–73 (avg. 69.3). Tail, males 48–52 (avg. 49.3), females 39–44 (avg. 41.3) (Fry et al., 1988).

Egg, no information.

Masses (g)

Males 19.5, 20.5, females 13–17.7 (avg. 16, $n = 7$) (Fry et al., 1988).

MAP 6. Ranges of Willcock's (filled) and dwarf (arrows) honeyguides.

Identification

In the field: This small honeyguide is olive-green, with a short, stubby bill and no malar stripe or loral mark. It is probably hard to distinguish from the sympatric dwarf honeyguide in eastern Zaire; the slightly larger least honeyguide has definite loral and malar markings. The song consists of a long series of mechanical sounding notes that last about 0.4 second each and terminate in a snapping sound.

In the hand: Adults of this species can be separated from the least honeyguide by their lack of a black malar streak or white on the forehead and upper lores. It is also more greenish on the crown, usually with somewhat blackish streaks. Females are smaller than males. Immature birds have softer and more pointed tips to the rectrices, and their rectrix tips have relatively narrow, dark fringes.

Habitats

This species favors forest edges and gallery forests, but extends to dense woods among grasslands and forest edges. It occurs from sea level to the lower zone of montane woodlands at about 1500 m.

Host Species

No host species are known.

Egg Characteristics

The eggs are undescribed.

Breeding Season

Gonadal cycle information suggests that breeding occurs during August–September in Liberia, during January and May in Liberia, during February in Ghana, and during February, April, June, and September in Zaire (Fry et al., 1988).

Breeding Biology

No information.

Population Dynamics

No information.

LEAST HONEYGUIDE

(Indicator exilis)

Other Vernacular Names: Western least honeyguide.

Distribution of Species (see map 7): Sub-Saharan Africa from Senegal to western Kenya, and south to Zaire and Angola.

Subspecies

I. e. exilis: Senegambia to Angola, Zambia and Zaire.

I. e. poensis: Bioko Island (Gulf of Guinea).

I. e. pachyrhynchus: Sudan and Uganda to Kenya, Zaire, Burundi, and Tanzania.

Measurements (mm)

5″ (13 cm)

Wing (all subspecies); males 70–85, females 65–72. Tail, males 42–51, females 41–47 (Fry et al., 1988).

Egg, 17.2–18 × 13–13.3 (avg. 17.6 × 13.15, $n = 2$) (Fry et al., 1988). Shape index 1.34 (= broad oval).

Masses (g)

Males of *exilis* 18–19.5, females 14.5–21. Males of *pachyrhynchus* 16–23 (avg. 19.8, $n = 48$), females 15–22 (avg. 17.6, $n = 33$) (Fry et al., 1988). Estimated egg weight 1.6. Egg:adult mass ratio 8.5%

MAP 7. Ranges of least (filled) and pallid (shaded) honeyguides.

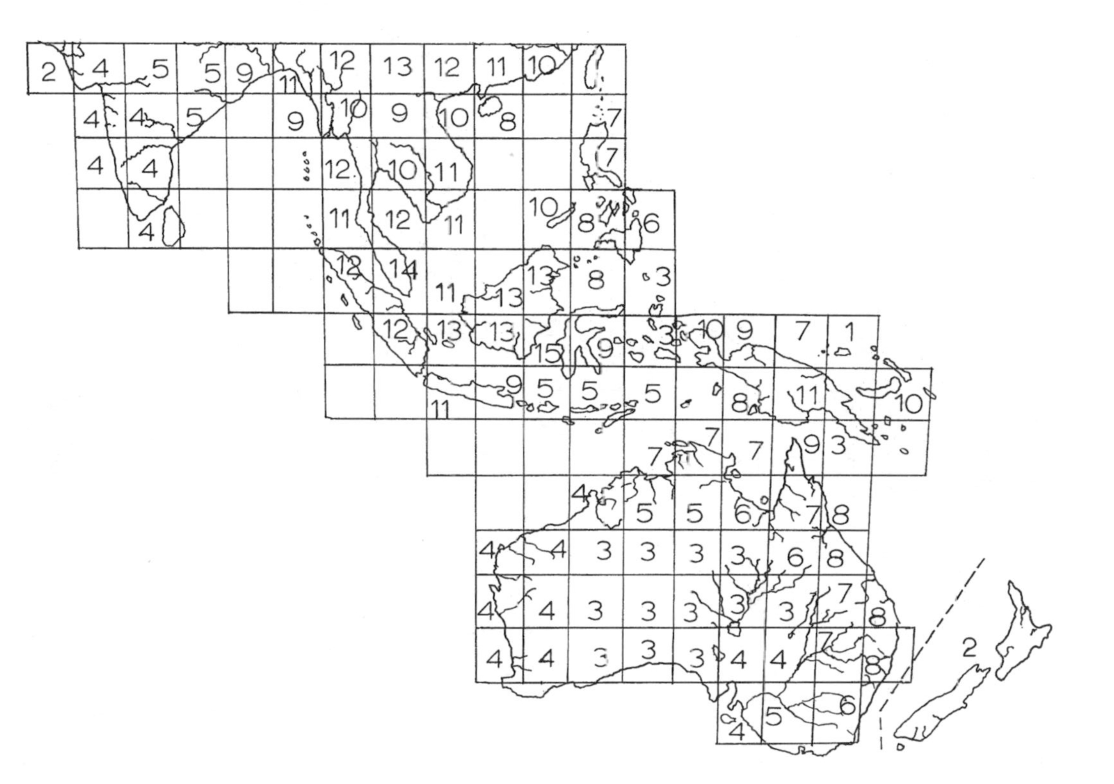

FIGURE 22. Species-density map of parasitic cuckoos in southeastern Asia and adjoining Australasian region, by 5° latilong quadrants.

Identification

In the field: This species has the most distinctive malar stripe and white lores of the small honeyguides; it also has definite, dark flank streaking. Immature individuals have gray, not black, malar stripes, blackish lores, and grayer flank streaks. The song is a high-pitched "pew-pew-whew-whew-whew. . . ," and a long trill or rattling "kwiew" call is also frequently uttered.

In the hand: As noted above, this species has a blackish malar stripe and a narrow white line extending from the front of the eye nearly to the nostril. The crown is deep gray, with an olive wash, and the crown feathers lack distinct streaking but may have darker feather centers. Females have distinctly smaller measurements than males. Immature individuals have soft-tipped rectrices.

Habitats

This species occupies forest edge, gallery forest, secondary forest, and scattered trees from sea level to about 2400 m.

Host Species

No host species are known, but the yellow-rumped tinkerbird is a highly probable host (Fry et al., 1988).

Egg Characteristics

Two oviducal eggs were white and broad oval in shape. No other information is available.

Breeding Season

Breeding records are few. In Nigeria, breeding records exist for November and in Cameroon for February and May. January and June laying records exist in Zaire (indirect evidence suggest egg laying occurs from January to September), and in eastern or southern Africa (Kenya, Zambia) the limited data suggests that breeding perhaps occurs between May and October (Fry et al., 1988).

Breeding Biology

No information.

Population Dynamics

No information

DWARF HONEYGUIDE

(Indicator pumilio)

Other Vernacular Names: None in general English use.

Distribution of Species (see map 6): Sub-Saharan Africa in eastern Zaire, extreme southwestern Uganda, and adjacent Rwanda and Burundi.

Measurements (mm)

5″ (13 cm)

Wing, males 69–75 (avg. 72.9), females 64–68 (avg. 65.7). Tail, males 46–51 (avg. 48.9), females 40–45 (avg. 42.3) (Fry et al., 1988).

Egg, no information.

Masses (g)

Males 13.5, 15, females 12–13.5 (Fry et al., 1988).

Identification

In the field: This tiny, stubby-billed species has a very restricted range that is sympatric with the nearly identical Willcock's honeyguide. A faint loral spot present in the dwarf but absent in the Willock's may help to identify it. The vocalizations include a "tuutwi" call, but little is known of its behavior.

In the hand: Very similar to the Willcock's honeyguide, lacking a malar stripe and having a greenish crown. However, it has a smaller bill (culmen 8–9 in males, 7.3–8 in females vs. 9–9.2 in males and 8.5–9.5 in females of the Willcock's) and a definite whitish loral patch or streak. Females are distinctly smaller than males. Immature individuals have narrow, black borders on the tips of their white outer rectrices.

Habits

This honeyguide is limited to moist forests at levels of 1500–2400 m, which is generally above the elevation typical of Willcock's honeyguide.

Host Species

No host species are yet known.

Egg Characteristics

No information.

Breeding Season

Gonadal data indicate January and April breeding in Zaire, with juveniles seen from June to October (Fry et al., 1988).

Breeding Biology

No information.

Population Dynamics

No information.

PALLID HONEYGUIDE

(Indicator meliphilus)

Other Vernacular Names: Eastern least honeyguide.

Distribution of Species (see map 7): Sub-Saharan Africa from Uganda and Kenya southwest to central Angola and south to northern Mozambique.

Measurements (mm)

Wings, males 67–83 (avg. 75.1), females 67–77 (avg. 73.3). Tail, males 42–54 (avg. 47.9), females 42–49 (avg. 45.9) (Fry et al., 1988). Egg, no information.

Masses (g)

Males 15–17 (avg. 16, $n = 5$), females 11, 12.5 (Fry et al., 1988).

Identification

In the field: This is another small, stubby-billed honeyguide lacking most fieldmarks, but it has distinctly pale gray underparts and a variably conspicuous whitish loral streak, setting it apart from the other small *Indicator* species (fig. 19). Immature individuals are darker than adults, lack the loral patch, and are grayer below. The song is a "pwee, pa-wee, pa-wee-wet, pa-wee-witp," with the individual phrases repeated up to 23 times at the rate of about 1 per second. Males also "winnow" by flying in a circle, producing snipelike sounds that may be generated by vibrations of the tail or wing feathers.

In the hand: Distinction of this species from the three similar previous ones is possible by its pale gray underparts. There is no definite blackish malar streak, but a whitish loral streak is present. The crown feathers are greenish, with some dusky streaking. There may also be some streaking on the otherwise whitish throat. Females are slightly smaller than males, but with overlapping measurements. Immature individuals are darker than adults, and the tail has more white, with narrower brown terminal markings and more pointed tips (Chapin, 1962; Fry et al., 1988).

Habitats

This honeyguide occurs in miombo (*Brachystegia*) and acacia woodlands, forest edges, and gallery forests, ranging in elevation from sea level to 2000 m.

Host Species

The yellow-rumped tinkerbird is the only known host species (Fry et al., 1988).

Egg Characteristics

No egg descriptions exist.

Breeding Season

Little information exists, but gonadal data suggest breeding during February in Tanzania, and breeding behavior has been recorded during January and February in Kenya (Fry et al., 1988).

Breeding Biology

No information.

Population Dynamics

No information.

YELLOW-RUMPED HONEYGUIDE

(Indicator xanthonotus)

Other Vernacular Names: Orange-rumped honeyguide.

Distribution of Species (see map 8): Asia from Afghanistan east to Bhutan and northern Burma (Myanmar).

Subspecies

I. x. xanthonotus: Nepal, Sikkim, Bhutan, and northern Burma

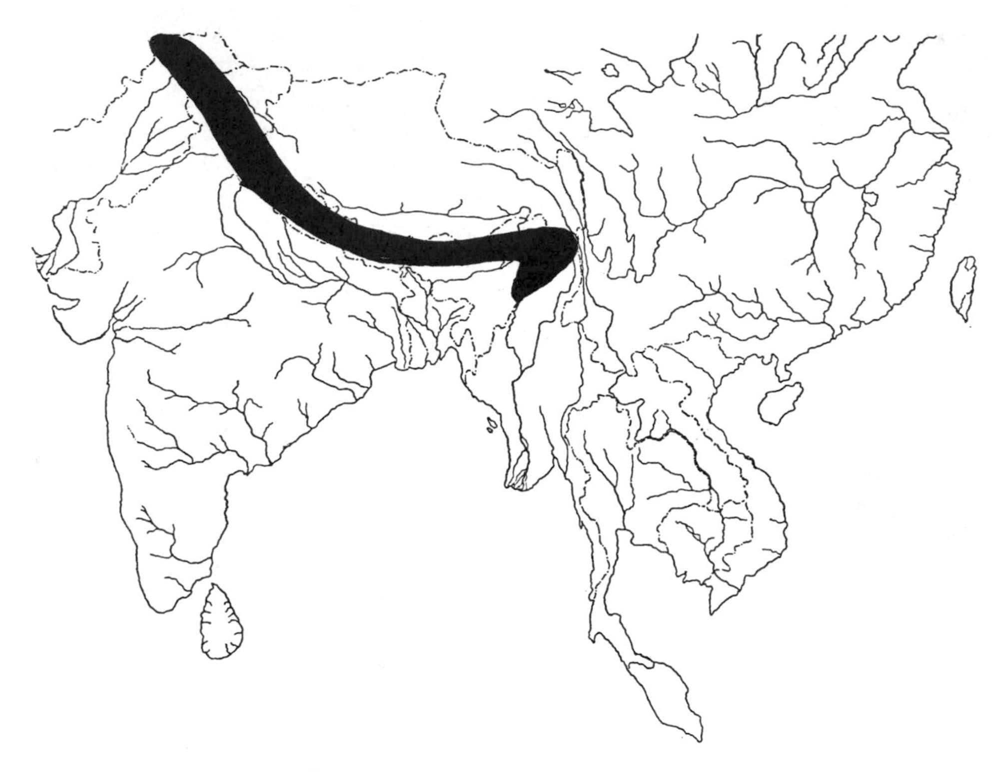

MAP 8. Range of yellow-rumped honeyguide.

I. x. radcliffi: Western Himalayas to eastern Assam

I. x. fulvus: Naga Hills of Assam

Measurements (mm)

6″ (15–16 cm)

Wing, males 92–96, females 82–92. Tail, males 56–61, female 55 (Friedmann, 1955).

Egg, no information.

Masses (g)

Male 29 (Friedmann, 1955), female 26 (Ali & Ripley, 1983). Five males averaged 30.9, 16 females averaged 26.3 (Cronin & Sherman, 1977). Both sexes avg. 28 (Dunning, 1993).

Identification

In the field: This is the only honeyguide in the Indian region (no known overlap occurs with the Malaysian species), and the only small bird species in the region with both a bright sulfur-yellow forehead and an orange-yellow rump. Vocalizations are still undescribed in detail.

In the hand: Easily recognized by the orange forehead, lores, and rump. The bill is unusually small and finchlike. Females have less extensive yellow on the head and throat, and their rump coloration is more yellow then orange. Immature plumages are still undescribed.

Habitats

This species occupies broadleaved and coniferous forests, from about 1500–3500 m elevation, especially in sites where cliffs and rock outcrops support bees colonies. The birds have been observed in such habitats as broadleaved, wet tropical forests, dry deciduous forests, and pine–oak forests.

Host Species

No host species are known.

Egg Characteristics

No egg descriptions exist.

Breeding Season

Apparent courtship behavior has been observed in Bhutan during mid-May and June and copulation observed in mid-May, suggesting a spring breeding period (Hussain & Ali, 1984).

Breeding Biology

No information.

Population Dynamics

No information.

LYRE-TAILED HONEYGUIDE

(Melichneutes robustus)

Other Vernacular Names: None in general English use.

Distribution of Species (see map 9): Sub-Saharan Africa from Ivory Coast east to western Uganda and south to northwestern Angola.

Measurements (mm)

7.5″ (19 cm)

Wing, males 93–99 (avg. 96), females 90–96 (avg. 93.6). Tail, males 55–61 (avg. 58.4), females 53–61 (avg. 55.8) (Fry et al., 1988).

Egg, no information.

Masses (g)

Males 52.3–61.5, females 46.9–57 (Fry et al., 1988).

MAP 9. Range of lyre-tailed honeyguide.

Identification

In the field: The lyre-shaped tail of this species is highly distinctive; the longer central rectrices are black, and the shorter, outer ones are white, as are the under-tail coverts (fig. 19). Otherwise the birds are mostly olive-green above and pale greenish yellow below. Immature individuals also have lyre-shaped tails but are much darker than the adults. Males display in the air, producing "winnowing" or "tooting" sounds (presumably produced by the specialized tail feathers), described as "wow-wow-wow . . . ," that accelerate and increase in loudness as the bird descends. These notes consist of a pulsed series of up to 30 sound elements that perhaps represent individual wingbeats variably influencing the rate of tail vibration, as occurs during aerial display in many snipes.

In the hand: Easily recognized in any plumage by the distinctive tail feathers. In females the longest under-tail coverts are somewhat shorter than in males (the feathers not extending beyond the angle made by the adjoining curved pairs of rectrices), the tail is shorter, the iris is more brownish (less orange to reddish), and the bare skin around the eye is olive-gray, not pinkish brown. Immature individuals not only have darker upperparts than adults but also have rather sooty abdomens; their under-tail coverts are paler and greener. The bill and feet of immature individuals are blackish, not brown, and the base of the bill is pale. The curvature of the longest tail feathers is only slightly developed in juvenile birds.

Habitats

This honeyguide occupies undisturbed lowland and lower montane forests, including forest edges and fairly open forests, ranging from sea level to 2000 m.

Host Species

No host species are yet known, but barbets of the genus *Gymnobucco* are suspected hosts (Fry et al., 1988).

Egg Characteristics

No egg descriptions exist.

Breeding Season

In Zaire, enlarged ovaries or nestlings have been reported in April and August, and breeding has been reported from March to September. An August–September breeding period has been reported in Angola (Fry et al., 1988).

Breeding Biology

No information.

Population Dynamics

No information.

YELLOW-FOOTED HONEYGUIDE

(Melignomon eisentrauti)

Other Vernacular Names: Eisentraut's honeyguide.
Distribution of Species (see map 10): Endemic to Liberia and Cameroon; possibly also breeds in Sierra Leone, Ghana, and Ivory Coast.

MAP 10. Ranges of Zenker's (filled) and yellow-footed (arrows) honeyguides.

Measurements (mm)

6″ (15 cm)

Wing, males 79–86 (avg. 83.2), females 76–86 (avg. 79.5). Tail, males 47–52 (avg. 49.8), females 46–50 (avg. 47.6) (Fry et al., 1988).

Egg, no information.

Masses (g)

Males 21–29 (avg. 25.5, $n = 6$), females 18–25 (avg. 23.1, $n = 7$) (Fry et al., 1988).

Identification

In the field: This is a rather small, nondescript sharp-billed species of honeyguide that lacks malar and loral markings, is unstreaked and unspotted below, and lacks white on the upper-tail coverts. Like Zenker's honeyguide, it has olive-yellow to pale yellow legs and feet and has a yellow gape. Young birds are generally paler and more yellowish throughout than adults. The species' vocalizations are unknown.

In the hand: Difficult to separate from the Zenker's honeyguide, as noted above. As in that species, the sexes are alike, but females average slightly smaller. Immature individuals have paler and more yellowish green upperpads. Their rectrices are more pointed, whiter, and have narrower dark tips than in adults. The orbital skin of immature individuals is yellowish (rather than dull green), and a subadult had a mostly orange-yellow (rather than a yellowish brown or olive-yellow) bill.

Habitats

This species favors evergreen forest and secondary forests on lower montane slopes.

Host Species

No host species are yet known.

Egg Characteristics

No information on this species' eggs exists.

Breeding Season

Breeding in Liberia may occur from November to June (Fry et al., 1988).

Breeding Biology

No information.

Population Dynamics
No information.

ZENKER'S HONEYGUIDE

(Melignomon zenkeri)

Other Vernacular Names: None in general English use.
Distribution of Species (see map 10): Sub-Saharan Africa, in Cameroon, Gabon, Zaire, and southwestern Uganda.
Measurements (mm)
6″ (15 cm)
Wing, males 78.5–86.5 (avg. 82.6), females 71–78.5 (avg. 75). Tail, males 51–56.5 (avg. 53.8), females 47–54 (avg. 51) (Fry et al., 1988).
Egg, no information.
Masses (g)
Males 24, 25 (Fry et al., 1988).

Identification

In the field: This species is sympatric with the yellow-footed honeyguide and is virtually identical to it. It is somewhat darker in color, and it has more olive tinted underparts and considerably darker upperparts. The vocalizations of both species are still undescribed.

In the hand: The sexes are alike as adults. Immature individuals are more yellowish to live-green and pale grayish (less yellowish) on the underparts, and the tail is whiter, probably with narrower and more pointed rectrices.

Habitats

This honeyguide occupies dense lowland evergreen forests and sometimes higher areas up to about 1500 m.

Host Species

No host species are yet known.

Egg Characteristics

No information is available.

Breeding Season

Breeding in Cameroon and eastern Zaire may occur from January to March, and in the latter region breeding may also occur during the dry season of July–August (Fry et al., 1988).

Breeding Biology
No information.

Population Dynamics
No information.

CASSIN'S HONEYGUIDE

(Prodotiscus insignis)

Other Vernacular Names: Green-backed honeyguide, slender-billed honeyguide.
Distribution of Species (see map 11): Sub-Saharan Africa from Liberia and Sudan south to northern Angola and east to eastern Zaire and western Kenya.
Subspecies
P. i. insignis: Nigeria to Angola and the Rift Valley lake district of eastern Africa.
P. i. flaviodorsalis: Sierra Leone to Nigeria.
Measurements (mm)
5″ (13 cm)
Wing, males 62–68, females 61–72. Tail, males

MAP 11. Ranges of Cassin's (filled) and green-backed (shaded) honeyguides.

42–47 (avg. 44.4), females 41–50 (avg. 45.5) (Fry et al., 1988).

Egg, avg. 15 × 12 (sample size unreported) (Fry et al., 1988). Shape index 1.25 (= broad oval).

Masses (g)

Males 8.9–11 (avg. 9.8, *n* = 5), females 9.1–11.4 (Fry et al., 1988). Estimated egg weight 1.1. Egg:adult mass ratio ~11%.

Identification

In the field: This is a tiny, sharp-billed honeyguide that, in common with the other *Prodotiscus* species, has a white, but often invisible, patch of erectile feathers extending from the edge of the rump to the flanks (fig. 19). This species differs from Wahlberg's honeyguide in having entirely white outer tail feathers, and from the green-backed honeyguide in having paler underparts. The vocalizations include a chattering note and a "whi-hihi" or "ski-a" call. Spreading the white outer tail feathers during an undulating flight is a typical display.

In the hand: The primary markings distinguishing this species are noted above. The sexes are alike as adults, with generally dark olive-brown plumage above and below, the white outer tail feathers and white rump patch the only contrasting markings. There is a faint eye-ring of dark gray, and the feet and toes are also mostly grayish. Juveniles resemble adults but the feathers are duskier above and have a less pronounced green wash and feather margins.

Habitats

This honeyguide occupies primary forest, gallery forests, forest edge, and second-growth woodlands, from sea level to about 2200 m elevation.

Host Species

The known hosts are flycatchers, including the black-throated wattle-eye (which has spotted, greenish white eggs), and warblers of the genus *Apalis* (whose eggs are rather varied in color and spotting) such as the buff-throated apalis. White-eyes such as the green white-eye (which has pale blue or bluish green eggs) are almost certainly hosts (Friedmann, 1955; Fry et al., 1988).

Egg Characteristics

Nothing of significance is known of the eggs beyond their linear measurements. An oviducal egg was white.

Breeding Season

In the northeastern Uganda, western Kenya region, egg laying may occur in April. Ovulating females have been collected in March and July in Kenya and Uganda, respectively. Such females have also been reported during September in Angola and during November in Cameroon (Fry et al., 1988).

Breeding Biology

No information.

Population Dynamics

No information.

GREEN-BACKED HONEYGUIDE

(Prodotiscus zambesiae)

Other Vernacular Names: Slender-billed honeyguide, Zambesi honeyguide.

Distribution of Species (see map 11): Sub-Saharan Africa from Ethiopia south to south-central Angola and Mozambique.

Subspecies

P. z. zambesiae: Angola to Tanzania and Mozambique

P. z. ellenbecki: Ethiopia, Kenya, and northeastern Tanzania.

Measurements (mm)

15″ (13 cm)

P. z. zambesiae. Wing, males 72–77 (avg. 74.2), females 71–75 (avg. 72.9). Tail, males 45–52 (avg., 47.4), females 45–51 (avg., 47.4) (Fry et al., 1988).

P. z. ellenbecki. Wing, male 66–72 (avg. 69.9), females 67–71 (avg. 69.7) (Fry et al., 1988)

Egg, one oviduct egg 15 × 12 (Fry et al., 1988). Shape index 1.25 (= broad oval).

Masses (g)

Male 16.5, females 12, 12.3 (Fry et al., 1988). Estimated egg weight 1.1. Egg:adult mass ratio 8.1%.

Identification

In the field: This species is probably not sympatric with the similar Cassin's honeyguide, and differs from it mainly in having paler underparts (fig. 19). Like the other birds in the genus, the bill is short and narrow, and there is an erectile patch of feathers above the rear flanks. Few vocalizations are known; they include a harsh stuttering chatter and a "skeee-aaa" flight call that might serve as a courtship signal.

In the hand: This species was earlier regarded as a subspecies of *insignis,* and differs from it mainly in having the sides of the face pale gray, and the abdomen and under-tail coverts white, rather than dingy olive-brown. Females are probably not distinguishable from males, but immature birds are both paler and grayer and tend to be more yellow above and more buffy below. In immature birds, outer three pairs of rectrices are entirely white, thus wholly lacking brownish tips (Fry et al., 1988). These feathers probably also are narrower, softer, and more pointed in young birds than in adults. Nestlings have a dark brown skin and a grayish bill with a yellow base and a bright yellow-orange gape.

Habitats

This species inhabits miombo (*Brachystegia*) and other similar woodland habitats, including forest edges, clearings, and gallery forests, from sea level to about 2100 m.

Host Species

Seven known host species are listed in table 28. Of these, the white-eyes are probably the primary hosts (Fry et al., 1988). As many white-eyes have pale blue or pale greenish rather than white eggs, the egg color of hosts may be significant in recognizing parasitic eggs in a clutch.

Egg Characteristics

One oviducal egg was described as white and about 15 × 12 mm (Friedmann, 1960).

Breeding Season

Breeding records exist for December, February, and April-July in Kenya and northeastern Tanzania, and for August–October in Zimbabwe. Evidently breeding often occurs during and after rains (Fry et al., 1988). In southern Africa breeding probably occurs from September to December, during the austral spring, judging from display activity (Ginn et al., 1989).

Breeding Biology

Nest selection, egg laying. The primary hosts consist of white-eyes, and all of the known hosts are open-cup nesters.

Incubation and hatching. No information.

Nestling period. No information.

Population Dynamics

No information.

WAHLBERG'S HONEYGUIDE

(Prodotiscus regulus)

Other Vernacular Names: Brown-backed honeyguide, sharp-billed honeyguide.

Distribution of Species (see map 12): Sub-Saharan

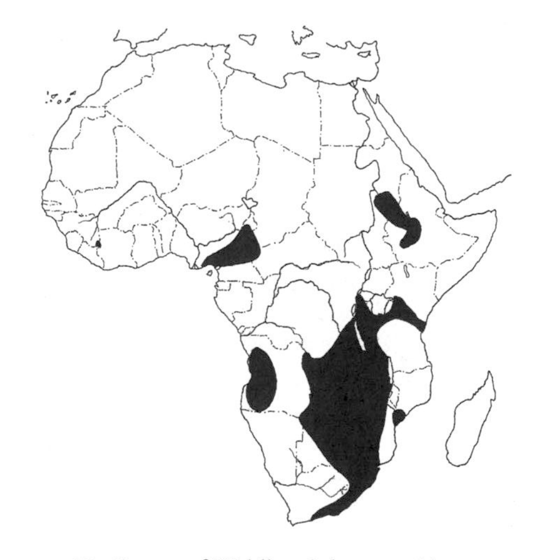

MAP 12. Range of Wahlberg's honeyguide.

Africa from Ivory Coast and Kenya south to South Africa.

Subspecies

P. r. regulus: Sudan to Angola and South Africa.

P. r. camerunensis: Guinea to Central African Republic.

Measurements (mm)

5″ (13 cm)

P. r. regulus. Wing, males 74–81.5 (avg., 77.5), females 72–80 (avg., 75.7). Tail, males 47–55 (avg. 50.5), females 46–51 (avg. 48.5). (Fry et al., 1988).

P. r. camerunensis. Wing, males 79–84.5, female 83. Tail, males 54–59, female 56 (Fry et al., 1988).

Egg, no mensural information. Eggs perhaps of this species measured about 18 × 15 mm and were white (Mackworth-Praed & Grant, 1962).

Masses (g)

Males 13–15 (avg. 14.1, $n = 4$), females 13.4–15.5 (avg. 14.4, $n = 4$) (Fry et al., 1988).

Identification

In the field: Like the two previous species, this one has a small, pointed bill, and an erectile patch of white feathers on either side of the rump, but few other definite fieldmarks (fig. 19). It is probably locally sympatric with both of these two species. Its calls include a song that is a buzzy trill lasting up to about 2 seconds, and a metallic-sounding "zwick" is produced during a "dipping" (presumably undulating) flight display.

In the hand: Compared with the other species of *Prodotiscus,* adults of this species have browner backs, dark-tipped outer rectrices, more buffy underparts, and lack pale eye-rings. The interior of the mouth is flesh or pale flesh in color. The sexes are alike as adults. Immature individuals differ from adults in having their outer three pairs of rectrices white-tipped (narrow, brown edging may be present on the second and third pairs); their back feathers are more brownish than those of the Cassin's and green-backed honeyguides (Fry et al., 1988).

Habitats

This species occupies miombo (*Brachystegia*), acacia, and other woodlands, as well as more arid scrub and savanna habitats, plus plantations, orchards, and gardens, ranging in elevation from sea level to about 2000 m.

Host Species

Three known host species are listed in table 28. The scarlet-chested sunbird and tinkling camaroptera are also probable hosts (Fry et al., 1988).

Egg Characteristics

No specific information is available on the eggs of this species. The yellow-throated petronia has cream-colored eggs that are heavily marbled and mottled and measure about 18 × 14.5 mm. The tabora cisticola has white eggs with a few reddish spots and measure about 14.5 × 11.5 mm. The gray-backed cameroptera has eggs that vary from blue to white and from plain-colored to spotted, averaging about 16 × 12 mm (Mackworth-Praed & Grant, 1973).

Breeding Season

Females with oviducal eggs have been reported in September (Nigeria) and October (Natal), and in Zambia there are December and March breeding records (Fry et al., 1988). Display in southern Africa occurs from September to December (Ginn et al., 1989).

Breeding Biology

No information.

Population Dynamics

No information.

8

OLD WORLD CUCKOOS

Family Cuculidae

The typical Old World cuckoos comprise, in the classification of Sibley and Monroe (1990), a group of nearly 80 species that include 12 genera and more than 50 species that are almost exclusively brood parasites. The group has its primary area of species diversity in southeastern Asia and the East Indies (centering on Borneo), with a secondary area of high species diversity in tropical West Africa and the upper drainages of the Congo Basin (figs. 22 and 23). The family, at least as it was recognized by Sibley and Monroe, also includes an additional group comprising four genera and 25 species (malkohas, Old World ground cuckoos, and couas) that rear their own offspring. Other more traditional and more widely accepted classifications of the Cuculidae (e.g., that used by Fry, et al., 1988, and by Cramp, 1985) recognize a much larger cuculid family, comprising seven subfamilies, of which the Cuculinae represent all the brood parasites of the Old World.

Regardless of their broad taxonomic classification, all cuckoos, whether parasitic and nonparasitic, have zygodactylous feet (first and fourth toe reversed), and their ribs nearly always lack uncinate processes. The vocal organs or syringes of some cuckoo genera are highly unusual among birds in being entirely bronchial in location, as in such New World cuckoos as *Dromococcyx*. However, they more often are of the usual passerine tracheobronchial type (as in *Tapera* and all the Old World parasitic cuckoos so far studied), or occasionally may be intermediate in position (Beddard, 1885; Berger, 1960). Nearly all the Old World parasitic cuckoos may be distinguished by their small, round nostrils with raised

Brown shrike feeding a juvenile Indian cuckoo. After a photo by I. Neufeldt (1966).

edges (exceptions occur only among *Oxylophus, Clamator,* and *Pachycoccyx*), whereas the nonparasitic cuckoos (and the parasitic species of New World ground cuckoos) invariably have narrow nostrils lacking raised edges and are partially covered by a membranous operculum.

Of all the members of the parasitic Old World cuckoo group, the crested cuckoos of the genera *Clamator* and *Oxylophus* are most generally accepted as being the most "primitive." Friedmann (1960) suggested that the present-day malkohas of Asia, or perhaps the African yellowbill, most closely approximate the ancestral type from which *Clamator* and *Oxylophus* may have evolved, and that the pied cuckoo most closely represents the most primitive of the existing stocks of crested cuckoos. Morphological evidence supporting this position includes the fact that the nostrils are linear and flat rather than rounded with raised edges, there are only 13 cervical vertebrae (as in the maklohas, and in contrast to the other cuckoos) (Friedmann, 1960). Additionally, in at least *Clamator,* the outer flight feathers (primaries 6–10) exhibit a complex molting pattern, in which these primaries are molted almost simultaneously in nonadjacent pairs, typically in the sequence 6–9, followed by 7–10, and finally 8–5. This unique pattern differs from the equally complex pattern typical of *Cuculus* and nearly all other Old World parasitic cuckoos, but both patterns are clearly derived from the molt pattern typical of the nonparasitic cuckoos (Stresemann & Stresemann, 1961, 1969).

Considering these morphological evidences of relationships, it would seem that parasitism arose at least twice, and perhaps several times, from nonparasitic ancestors among the Old World cuckoos. Berger (1960, p. 80) commented that "one must discount either myological data or breeding behavior in deciding the relationships among the cuckoos," and noted furthermore that if myological data are to be accepted as a sole criterion of phylogeny, then it must be assumed that "parasitism has developed as many as four times in this one family (which seems highly unlikely) or that the parasitic habit . . . developed in the primitive cuckoos . . . while still in the Old World ancestral home of the family." Although it may be that parasitism did not evolve as many as four times in the Old World cuckoos, multiple origins of the behavior seem no less unlikely than do multiple evolutionary origins of, for example, similar adult plumage patterns or nest-structure types.

Old World Cuckoos

PIED CUCKOO

(Oxylophus jacobinus)

Other Vernacular Names: Black cuckoo, black-and-white cuckoo, black-crested cuckoo, crested cuckoo, gray-breasted cuckoo, Jacobin cuckoo.

Distribution of Species (see map 13): Disjunctive, with one population breeding in sub-Saharan Africa; another breeds over most of the Indian subcontinent of Asia. Part of this latter population migrates to eastern Africa over winter, but wintering also occurs in southern India and Sri Lanka (Ceylon).

Subspecies

O. j. serratus: Southern Africa from Zambia south.

O. j. pica: Senegambia to Red Sea, south to Zambia and Malawi.

O. j. jacobinus: Asia, from Pakistan, India and Sri Lanka to Burma (Myanmar).

Measurements (mm)

13″ (33 cm)

O. j. jacobinus. Wing, both sexes 136–144; tail, both sexes 147–163 (Ali & Ripley, 1983).

O. j. pica. Wing, males 148–157 (avg. 153), females 151–164 (avg. 156); tail, males 170–186 (avg. 176), females 171–195 (avg. 181) (Cramp, 1985). Mean wing:tail ratio 1:1.16.

O. j. serratus. Wing, males 148–165 (avg. 156), females 148–160 (avg. 153); tail: males 170–190 (avg. 182), females 170–194 (avg. 178) (Fry et al., 1988). Mean wing:tail ratio 1:1.16.

Egg *(O. j. pica).* Avg. 24.1 × 19.8 (range 22.1–26 × 18.1–22 (Schönwetter,

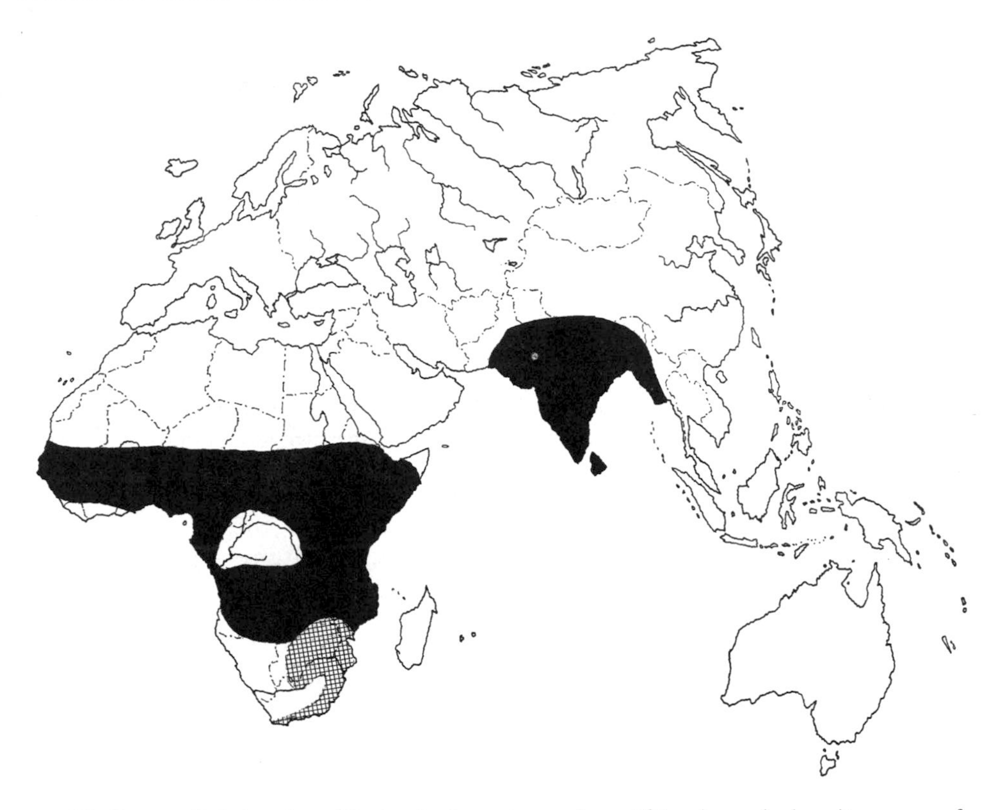

MAP 13. Range of pied cuckoo. The hatched area in southern Africa shows the breeding range of the race *serratus.* Filled areas show the breeding range of *picta* plus the nonbreeding range of all races.

1967–84). Shape index 1.22 (= broad oval). Rey's index = 1.08.

Masses (g)

O. j. jacobinus, 13 unsexed, 50–73 (Ali & Ripley, 1983).

O. j. pica, males 66–83, females 80, 84 (Cramp, 1985).

O. j. serratus, 12 males 58–81, avg. 71; 10 females 52–94, avg. 76 (Fry et al., 1988).

Estimated egg weight 5.1 (Schönwetter, 1967–84). Egg:adult mass ratio ~7%.

Identification

In the field: Recognizable by the well-developed crest and the generally black-and-white/gray overall plumage (fig. 23). Young birds have less well-developed crests and duller upperparts, with the white areas smaller or tinted with fulvous tones. In the African range of *serratus,* the dark-morph birds appear almost entirely black, but in addition to being crested they have white wing bars that distinguish them (even when at rest) from the black cuckoo. Dark-morph birds are more common in coastal areas, and the lighter morphs greatly predominant in interior regions. The most common male advertisement call is a fluty, wild-sounding, and somewhat metallic "peeu" or "plie-ue" note that is uttered at the rate of about one per second and lasts for 4–8 seconds. This series often grades into or alternates with a rapid chatter

FIGURE 23. Profile sketches of the crested cuckoos: juvenile (A) and adult (B) of chestnut-winged cuckoo, a dark morph (C) and typical-plumage adult (D) of Levaillant's cuckoo, an adult (E) and juvenile (F) of great spotted cuckoo, and a typical-plumage (G) and dark morph (H) adult of pied cuckoo. Their eggs are shown at an enlarged scale.

lasting another 5 seconds. In courtship situations the usual call becomes a "kru-kru-kru-kleeuuu." Liversidge (1971) recognized a total of eight calls typical of the African population. The race breeding on the Indian subcontinent has a similar repeated "peearr" or "piu" male advertisement note that may be repeated 5–12 times and may also be interspersed with a 3-syllable "piu-piparr," the second and third notes uttered quickly and merging acoustically.

In the hand: This crested cuckoo has a wing length that is no more than 165 mm (vs. a minimum of 170 mm in Levaillant's cuckoo), underparts that are usually white or gray (rarely entirely black, in the melanistic morph), and a throat and foreneck that are heavily streaked with black. Like all other crested cuckoos (and the thick-billed cuckoo), the nostrils are narrow oval to slitlike in shape, and the primaries are not much longer than the longest secondaries. The all-black plumage variant of this species may be separated from that of the Levaillant's cuckoo by its shorter wing length. Immature individuals are duller black above (and also below, in the melanistic morph), with poorly developed crests and have pale yellowish (not brown) eyes. They already show the white wing patch characteristic of adults, and typical or light-morph individuals also have white tips on their tail feathers. The two color morphs are distinct on acquiring their juvenal plumages, being either dark brown or buffy below. Nestlings have pink to orange-pink skin initially, which darkens soon to purplish brown. They have scarlet to orange mouth linings, and a yellow bill, or at least yellowish edges to the bill (Rowan, 1983; Fry et al., 1988).

Habitats

This species is primarily found in fairly open woodlands, including dry thornveldt savannas, especially those near thickets used by bulbul hosts, and is absent from closed forests and cleared forests. It also occurs in low-growing coastal scrub in South Africa. It has been observed to 3660 m elevation in Nepal, but usually is found from 300 to 1500 m.

Host Species

Nineteen host species reported by Baker (1942) for India are summarized in table 10. Of these approximately 140 records, laughingthrushes and babblers make up the majority of host records. Fry et al. (1988) listed 17 host species from Africa (table 29). At least in southern Africa, these are primarily bulbuls, especially the common (about 135 records), Cape (60 records), sombre (21 records), and African red-eyed (12 records). The fiscal shrike is also an important host in southern Africa, with about 30 host records. Rowan (1983) listed 6 confirmed biological hosts (those seen with nestling or fledgling cuckoos) and 14 additional alleged hosts for southern Africa. She listed 33 records of nestlings or fledglings for the common ("black-eyed") bulbul, 27 for the Cape, and 5 each for the sombre bulbul and fiscal shrike. Fourteen species were listed as unconfirmed hosts (those seen only with eggs in their nests). Babblers are also likely hosts in eastern and northeastern Africa, but in contrast to India, there are no proven records of parasitism by babblers in Africa. The few host records available for eastern and western Africa are for common bulbuls.

Egg Characteristics

Like other parasitic cuckoos, the eggs of this species are relatively thick and resistant to breakage, with a shell weight averaging 0.47–0.6 g (Schönwetter, 1967–84). This cuckoo's eggs are perhaps indistinguishable from those of the common hawk cuckoo (Ali & Ripley, 1983). Although somewhat heavier than the chestnut-winged cuckoo, the pied cuckoo's eggs are smaller, and it correspondingly parasitizes smaller host species. The eggs of the nominate Indian population are sky-blue to pale blue, of varied color intensity, and are rounded-oval in shape. They are relatively rounded, lack gloss, and tend to be larger and more rounded than those of their usual babbler hosts. African females of the race *pica* also mainly lay blue to bluish green eggs similar in color to those of various babblers (a group not yet proven to be parasitized in Africa), but the endemic southern

TABLE 29 Reported Hosts of African Parasitic Cuckoos[a]

Pied Cuckoo	
Fiscal shrike (M)	Sombre greenbul (M)
Common bulbul (M)	African red-eyed bulbul (M)
Levaillant's Cuckoo	
Chestnut-bellied starling	Hartlaub's babbler (M)
Arrow-marked babbler (M)	Brown babbler (M)
Bare-faced babbler (M)	Blackcap babbler (M)
Great Spotted Cuckoo	
Pied starling (M)	Ruppell's glossy starling
Hildebrandt's starling	Indian myna
White-crowned starling	Pied crow (M)
Red-winged starling	Hooded crow (M)
Pale-winged starling	Cape rook (M)
Splendid starling	Common raven
Greater blue-eared starling	Fan-tailed raven (M)
Burchell's starling	Brown-necked raven
Long-tailed starling	Black-billed magpie (M)
Red-winged glossy starling	
Thick-Billed Cuckoo	
Red-billed helmit shrike	
Red-Chested Cuckoo	
Cape wagtail	Stonechat
Pied wagtail	Starred robin
Long-tailed wagtail	Swinnerton's robin
Cape robin chat (M)	Boulder chat (M)
Chorister robin-chat	Mocking chat
Natal robin-chat	Familiar chat
White-browed robin chat (M)	Kurrichane thrush
Ruppell's robin chat (M)	Olive thrush
White-throated robin chat	Cape rock thrush
Eastern bearded scrub robin	African paradise flycatcher
White-browed scrub robin	Busky alseonax
Black Cuckoo	
Tropical boubou (M)	African golden oriole (M)
Crimson-breasted boubou	
Common Cuckoo	
Woodchat shrike	Dartford warbler
Moussier's redstart (M)	Tristam's warbler
African Cuckoo	
Yellow-billed shrike	Fork-tailed drongo (M)
Klaas's Cuckoo	
Common bulbul	African paradise flycatcher
Stonechat	Collared sunbird
Cape crombec (M)	Pygmy sunbird
Green crombec	Amethyst sunbird

(continued)

TABLE 29 *(continued)*

Yellow-bellied eromomela (M)	Scarlet-chested sunbird
Bar-throated apalis	Bronze sunbird (M)
Green-backed camaroptera	Beautiful sunbird
Greater swamp-warbler	Malachite sunbird
African reed-warbler	Mariqua sunbird
Piping cisticola	Greater double-collared sunbird (M)
Singing cisticola	White-breasted sunbird
Red-faced cisticola	Variable sunbird
Chattering cisticola	Dusky sunbird (M)
Gray tit-flycatcher	Mouse-colored sunbird
Dusky alseonax	Green-headed sunbird
Pale flycatcher	Copper sunbird
Cape batis (M)	Yellow white-eye
Pririt batis (M)	Yellow-eyed canary
Chinspot batis	Cabanis' bunting
Black-throated wattle-eye	
African Emerald Cuckoo	
Southern puffback	Malachite sunbird
Common bulbul (M)	Red-chested sunbird
Yellow-whiskered greenbul	Olive-bellied sunbird
Starred robin	Olive sunbird
Brown illadopsis	Scarlet-chested sunbird
Bleating bush warbler	Amethyst sunbird
Black-throated wattle-eye	Newton's sunbird
Brown-throated wattle-eye	Baglafecht weaver
African paradise flycatcher	Sao Tome' weaver (M)
Dideric Cuckoo	
Masked weaver (M)	Black-necked weaver (M)
Spectacled weaver (M)	Northern masked weaver
Cape weaver (M)	Southern brown-throated weaver
Bocage's weaver	Southern rufous sparrow
African golden-weaver (M)	Gray-headed sparrow
Village weaver (M)	Chestnut sparrow
Vieillot's black weaver (M)	Cape sparrow (M)
Black-headed weaver	Red-headed weaver (M)
Holub's golden weaver	Crested malimbe
Lesser masked weaver	White-browed sparrow-weaver
Baglafecht weaver	Black-capped social-weaver
Chestnut weaver	Red bishop (M)
Speke's weaver	Black-winged bishop
Slender-billed weaver	Yellow bishop
Heuglin's masked weaver (M?)	Red-collared widowbird
Golden-backed weaver	Yellow-spotted petronia

[a]Based mostly on summaries by Fry et al. (1988) and Friedmann (1960). Due to confusion between their eggs and young, the host lists for emerald and Klass' cuckoos include some highly probable but unproven hosts. Major host species are indicated by (M); host list is organized taxonomically.

African race *serratus* lays white eggs and mainly parasitizes bulbuls and shrikes, whose speckled or variously colored eggs are readily distinguishable visually from those of the cuckoo.

Breeding Season

Breeding in the Indian region occurs nearly throughout the year. However, in peninsular India nesting is mainly concentrated from June to September, and from February to May in Sri Lanka (Ali & Ripley, 1983). In southern Africa breeding is mostly concentrated during spring and summer, the egg records extending from October to February or March in South Africa, from December to April in Namibia, from October to March in Zimbabwe, and from October to April in Malawi. Angolan and Zambian records are for November to February. At the northern (transequatorial) edge of the range in Mali and Nigeria, breeding occurs during spring and early summer, from May to July, and from March to June in Somalia. East African breeding in the vicinity of the equator is quite variable, with March–April and June–July breeding in northern Uganda and western Kenya, as well as March and May breeding in eastern Kenya and northeastern Tanzania (Fry et al., 1988).

Breeding Biology

Nest selection, egg laying. As with the other crested cuckoos, both sexes participate in and cooperate in nest selection and parasitism behavior. Initial inspection of a potential nest may occur during afternoon hours, although most egg laying occurs shortly after sunrise. At dawn, the female approaches and remains perched close to the host's nest for a time before joining the male and approaching the nest together. At this time the male begins to make himself conspicuous by calling loudly from the tops of bushes, while the female remains well hidden. Finally, the male perches directly above the chosen nest, awaiting attack by the owners. As this happens, the female quickly moves to the nest and lays an egg, sometimes in as little as 5 seconds (Gaston, 1976) or perhaps within no more than 10 seconds (Liversidge, 1971). Neither Gaston nor Liversidge found any evidence that a host's egg is removed at this time, although other observers have reported the disappearance of one host egg or the presence of a broken host egg below the nest of a newly parasitized clutch (Friedmann, 1964; Rowan, 1983). There may thus be individual variation in this aspect of parasitic behavior. Eggs are probably laid by individual female cuckoos at about 48-hour intervals (Liversidge, 1970; Payne, 1973b). It is possible that a female cuckoo may thus lay about 2.5 eggs per week over a 10-week breeding season, for a total of about 25 eggs per season (Payne, 1973b). Of 31 eggs found by Liversidge, 3 were laid a day in advance of the host's first egg, 16 were laid during the host's egg-laying period, 10 were laid during the host's first day of incubation, and 2 were laid within 4 days of the initiation of incubation by the host. The 2-day shorter incubation period of the cuckoo than that of the bulbul host should have allowed most of these eggs to hatch. Similarly, Gaston (1976) found that 52.5% of 38 cuckoo eggs found in central India were laid during the host's (babblers) laying period and that eggs laid 8 days after the initiation of the host's clutch did not survive to fledging.

Incubation and hatching. The incubation periods of six eggs incubated by host bulbuls ranged from slightly fewer than 11 days to a maximum of 12.5 days (Liversidge, 1970). It is likely that, by the time the egg is laid, it has been in the female's oviduct for about a day, so a significant amount of embryonic development may have already occurred when the egg is laid (Payne, 1973b). A single egg is normally laid in each host nest (75% of the records from southern Africa), but sometimes two (15%) or three (4%) cuckoo eggs are present, with a maximum report from this region of seven eggs (Rowan, 1983). Baker (1942) reported that, in India, a single egg was present in 84 of 106 nests (79%), 2 in 13 nests, 3 in 6 nests, 4 in 2 nests, and 6 in 1 nest. More recently in India, Gaston (1976) found that among 39 parasitized babbler nests (jungle and common), 61% of the nests had single cuckoo eggs present. Most of the remaining cases (3%) involved two cuckoo eggs, with a maximum of four eggs being found. Gaston believed that many of the cases of multiple laying involved the same female, but also judged that this occurred only when no suitable unparasitized nests were available.

Nestling period. In one case, the fledging period was determined to be 16 days under natural conditions in Africa (Liversidge, 1971), and in another case it was judged to be 17 days (Skead, 1962), although fledging occurred later. Dependence on the host for food may last 25–30 days. Although occasionally one or more host chicks survive long enough to fledge successfully, this is evidently rather unusual. Liversidge (1971) found a case of this in 1 of 10 parasitized nests of the Cape bulbul, but more often the host's young simply disappear or are found dead below the nest, probably from starvation or being tramped by the young cuckoo (Rowan, 1983).

Population Dynamics

Parasitism rate. In southern Africa, the overall incidence of parasitism of bulbul nests has been estimated (Payne & Payne, 1967) as 11.8% (729 nests of the garden bulbul), 12.7% (104 nests of the sombre bulbul), and 16.3% (263 nests of the Cape bulbul), but with higher rates of about 30% in the eastern Cape region. A similar 4-year overall rate of about 36% (41 of 115 nests) was reported by Liversidge (1971) for the Cape bulbul, but with substantial annual variations (12–72%). Gaston (1976) reported parasitism rates for three Indian babblers and ranging from 28.6% (large gray babbler) to 71% (jungle babbler), the three species collectively averaging 53% (83 total nests).

Hatching and fledging success. Data obtained by Liversidge (recalculated by Rowan, 1983) indicate that over a 4-year study period, about 20% of 50 cuckoo eggs hatched and survived to fledging (range of annual egg-to-fledging success rates: 0–55%). Gaston (1976) suggested that each breeding pair of cuckoos in his study area produced about six fledged young per breeding season, although this was based on a crude estimate of the local cuckoo population.

Host–parasite relations. Using Liversidge's field data for his 4-year study, Rowan (1983) calculated that the nesting success of Cape bulbuls was reduced from 33% to 24% (a 2% reduction in success) when parasitized. Assuming an approximate 36% parasitism rate, the overall reduction in local bulbul productivity as a result of parasitism might be approximately 10%. In Gaston's (1976) 3-year study, there was an egg-to-fledging reduction of jungle babbler productivity from 4.2 eggs to 2.5 fledged young in unparasitized nests (a 40.5% egg-to-fledging loss), and a corresponding reduction of 3.6 to 1.14 in parasitized nests (a 68.4% egg-to-fledging loss). Thus, 32% of the eggs resulted in fledged young in parasitized nests, versus 60% in parasitized nests, representing a cost of parasitism of about 38%. Similarly, there was a 23% egg-to-fledging loss in unparasitized common babbler nests, as compared with a 67% loss in parasitized ones. This represents a cost of parasitism of about 42%. Taking the frequency of parasitism into account, the effective reduction in the local babbler population's productivity of about 39% for the jungle babbler and 42% for the common babbler. Additional potential but unmeasured effects of parasitism include the possibly increased rate of predation on parasitized nests owing to the conspicuous calling of young cuckoos and the effects of the possibly subnormal physical condition of those babbler young that survived to fledging in spite of intense food competition with the cuckoos.

LEVAILLANT'S CUCKOO

(Oxylophus levaillanti)

Other Vernacular Names: African striped cuckoo, striped crested cuckoo, stripe-breasted cuckoo, striped cuckoo.

Distribution of Species (see map 14): Africa from Senegal and Somalia south to Namibia and South Africa.

Measurements (mm)

16″ (41 cm)

Wing, males 170–189 (avg. 180), females, 171–189 (avg. 178). Tail, males 215–238

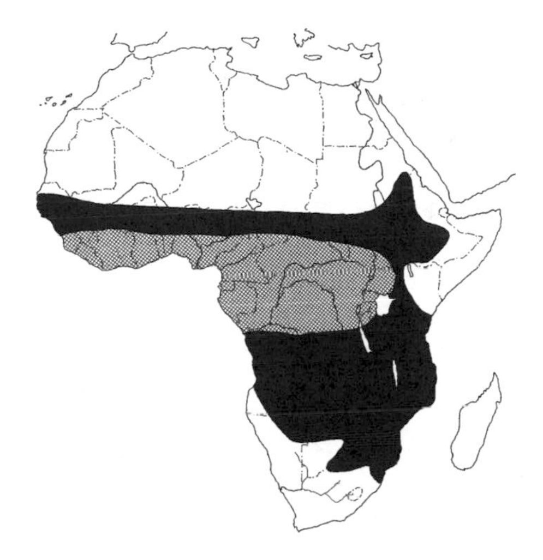

MAP 14. Breeding (inked) and nonbreeding (hatched) ranges of Levaillant's cuckoo.

(avg. 226), females 210–242 (avg. 223) (Fry et al., 1988). Mean wing:tail ratio 1:1.25.

Egg, avg. 25.5 × 20.1 (range 23.7–27.5 × 19.5–21) (Schönwetter, 1967–84). Shape index 1.27 (= broad oval). Rey's index 1.02.

Masses (g)

Males 106–140 (avg. 123, $n = 5$), females 102–141 (avg. 122, $n = 8$) (Fry et al., 1988). Estimated egg weight 5.6 (Schönwetter, 1967–84). Egg:adult mass ratio 4.6%.

Identification

In the field: Similar to but slightly larger than the pied cuckoo, with much streaking on the chin and throat (fig. 23). Like the pied, there is a rare black or melanistic plumage morph that has white only on the flight feathers and sometimes on the tips of the tail feathers. The usual call is a fluty, repeated "piu" note that is uttered at the rate of about once per second for about 20 seconds, alternating with short bursts of harsh chattering.

In the hand: This is a crested cuckoo with a wing length of at least 170 mm and blackish upperparts but usually white (sometimes entirely black) underparts, and a chin and upper breast that is streaked with black or, in the melanistic morph, is entirely black. Females cannot be easily distinguished externally from males, but juveniles (until about 5–6 weeks old) have a short crest and are brown above, and their tail feathers are also brown, with rusty-colored rather than white tips. Nestlings are initially naked, with dark pink skin, a black upper mandible, an orange-red gape (but no mandibular flanges, as are prominent in babblers), and thick yellow eyelids. The skin color darkens with age, as in other cuckoos, becoming blackish by about 5 days after hatching.

Habitats

This species is associated with fairly dense woodlands, streamside shrubs, riparian thicket, and similarly heavy cover. It sometimes occurs in gardens.

Host Species

A list of six known host species is provided in table 29. Hosts are primarily babblers; the arrow-marked babbler is the nearly exclusive host in southern Africa, although the bare-faced babbler and pied babbler are perhaps sometimes also parasitized. In Zambia the Hartlaub's babbler is a known host, and in West Africa the brown and black-capped babblers are exploited (Rowan, 1983; Fry et al., 1988).

Egg Characteristics

The eggs of this species are oval, slightly more rounded than those of most hosts, and have little gloss. The eggshell averages 0.5 g and is 0.16 mm thick (Schönwetter, 1967–84). Like the previous species, the Levaillant's also primarily lays turquoise-blue eggs, which match those of many of its babbler hosts, but tend to be slightly broader, with the surface more pitted and slightly more glossy than arrow-marked babbler eggs. However, in Nigeria bright to pale pink eggs are reportedly laid, matching those of the locally parasitized brown babbler (Fry et al., 1988).

Breeding Season

Egg laying over occurs during the rains, often late in the rainy season. In Transvall there are egg records from November to May, in Zimbabwe from October to June, in Zambia from November to April, and in Malawi for October and March–April. In Senegambia breeding occurs from August to December, in Mali from June to October, and in Nigeria from April to September or October. In East Africa a few egg records are for March to May (Fry et al., 1988).

Breeding Biology

Nest selection, egg laying. As in other crested cuckoos, Levaillant's operate in pairs to facilitate their egg laying. The pair approaches the nest together, with the male being much more conspicuous and serving to attract the attention of the babblers, which typically attack it in a collective effort. While the male thus tries to lead the babblers away, the female makes her way to the nest, and, upon reaching the nest, may lay her egg in only a few seconds. The female then rejoins her mate, and the two make a joint retreat, often with the babblers in strong pursuit. Frequently a babbler egg may be found damaged in the nest or broken and on the ground below, suggesting that at least on occasion the female may attempt to destroy or remove a host egg while laying one of her own (Steyn, 1973; Steyn & Howells, 1975; Rowan, 1983).

Incubation and hatching. The incubation period is probably 11–12 days, with some oviducal incubation occurring before egg laying (Rowan, 1983; Jones, 1985).

Nestling period. The nestling period is evidently very short, only 9–10 days (Jones, 1985). From the fifth day after hatching, the young cuckoo excretes a foul-smelling, dark brown fluid when disturbed, and it also utters food-soliciting calls that are identical to those of young babblers. Nestling babblers are not killed or evicted from the nest by the cuckoo, but food competition among the nestlings is apparently intense. The young cuckoo remains with its foster parents for several more weeks, begging food from them (Rowan, 1983).

Population Dynamics

Parasitism rate. Payne & Payne (1967) estimated that 7.8% of 217 arrow-marked babbler nests in southern Africa were parasitized.

Hatching and fledging success. No information exists on cuckoo success rates. Among six parasitized nests of babblers, one or more host chicks were reared successfully in the presence of a cuckoo, but in a seventh parasitized nest, the babbler chicks seemed unlikely to survive (Rowan, 1983).

Host–parasite relations. Babblers are always aware of the presence of cuckoo eggs in their nests, and there are several cases of arrow-marked babblers abandoning their nests immediately after they have been parasitized, even if at that time their own clutches have been completed and incubation is underway (Steyn, 1973). In parasitized nests, the young babbler chicks often huddle close to the cuckoo, sometimes even sitting on it, and all the nest members are fed by the tending adults (Rowan, 1983).

CHESTNUT-WINGED CUCKOO

(Clamator coromandus)

Other Vernacular Names: Red-winged cuckoo, red-winged crested cuckoo.

Distribution of Species (see map 15): India east to China and Indochina; wintering south to Sumatra, Sulawesi, and the Philippines.

Measurements (mm)

Wing, both sexes 157–166; tail, both sexes 231–245 (Ali & Ripley, 1983). Wing, both sexes 148–167; tail, both sexes 212–240 (Medway & Wells, 1976). Wing, both sexes 152–172; tail, both sexes 231–245 (Delacour & Jabouille, 1931). Mean wing:tail ratio ~1:1.5.

Egg, avg. 27 × 23 (range 25.4–29.9 × 20.3–24.4) (Schönwetter, 1967–84). Shape index 1.17 (= subspherical). Rey's index 1.08 (Becking, 1981).

Masses (g)

Unsexed birds 61–75 (avg. 70, $n = 6$) (Ali & Ripley, 1983). Average of 9 unsexed birds, 78.9 (Becking, 1981). Estimated egg weight 7.85 (Schönwetter, 1967–84), range 7.4–8.0 (Becking, 1981). Egg:adult mass ratio ~10.5%.

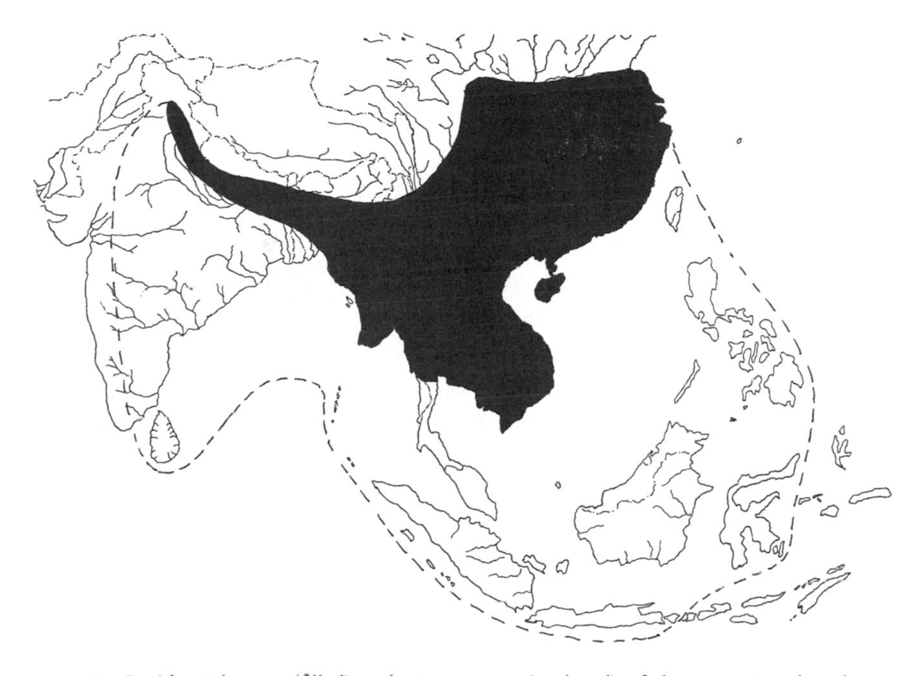

MAP 15. Residential range (filled) and winter range (enclosed) of chestnut-winged cuckoo.

Identification

In the field: This is a very large, well-crested cuckoo with chestnut-brown wings (fig. 23). It is generally blackish above and white to buffy below, with a white collar extending around the base of the neck, and white tips on the long tail. Young birds are brownish above and are less fully crested, with many of their feathers white-tipped. The call consists of a loud, harsh screeching "chee-ke-kek" or "creech-creech-creech," as well as a hoarse whistle. Soft "too-too" notes may also be uttered.

In the hand: This is the only crested cuckoo with chestnut-rufous wings (some noncrested *Centropus* cuckoos are of about the same size and have similar rufous wings, but lack slitlike nostrils). Additional diagnostic features are the white nape collar, the rusty tinge to the chin and upper breast, and the smoky brown thighs. Females cannot be separated easily from males by their plumage traits, but immature individuals have a poorly developed crest, are mostly brown (rather than glossy black) above, have rufous feather edgings, have an orange gape, and the basal half of the lower mandible is also orange.

Habitats

This cuckoo is associated with lowland and low montane evergreen and moist-deciduous forests, including foothills forests, teak forests, scrub-and-bush jungle, and sometimes gardens. It ranges up to about 700 m in India, from 250 to 350 m (rarely to 1400 m) in Nepal, and from the foothills to about 1800 m in Myanmar.

Host Species

A list of 12 host species is provided in table 10, based on eggs in Baker's (1942) collection. Nearly all of these host species are laughingthrushes; the lesser necklaced laughingthrush alone is responsible for about 45% of the total 245 host records.

Egg Characteristics

This species lays unmarked blue eggs that range considerably in shape from broad elliptical to nearly spherical (shape index average = 1.17, or subspherical). They have an average shell weight of 0.61 and an average thickness of 0.16 mm (Schönwetter, 1967–84). The eggs lack gloss and have a fine surface texture (Becking, 1981).

Breeding Season

Breeding in the Indian subcontinent mainly occurs during May and June, but extremes are from April to August (Ali & Ripley, 1983). In Myanmar its laughingthrush hosts breed mainly from March to May, with second broods extending into August (Smythies, 1953).

Breeding Biology

Nest selection, egg laying. Little information available. Up to four eggs have been found in a single nest, and usually at least two cuckoo eggs are present. It is believed that removal of host eggs may also occur (Ali & Ripley, 1983).

Incubation and hatching. No information.

Nestling period. No detailed information, but the nestling period is probably similar to that of better-studied species. A crest is developed in young birds by 5 weeks, and adultlike plumage is attained by 3 months (Ali & Ripley, 1983).

Population Dynamics

No information.

GREAT SPOTTED CUCKOO

(Clamator glandarius)

Other Vernacular Names: None in general English use.

Distribution of Species (see map 16): Southwestern Palearctic east to Turkey; also Africa from Senegal to Somalia and south to South Africa. Mediterranean populations winter in Africa.

Measurements (mm)

14–16″ (35–40 cm)

Wing, males 187–204 (avg. 197), females 183–197 (avg. 191). Tail, males 192–220 (avg. 206), females 183–205 (avg. 193) (Fry et al., 1988). Mean wing:tail ratio 1:1.03–1.05.

Egg, avg. 31.8 × 24 (range 28.4–35.6 × 21.5–26.7) (Schönwetter, 1967–84). Shape index 1.32 (= broad oval). Rey's index 0.94.

Masses (g)

Breeding males 153–193 (avg. 169, n = 6), female 138. Avg. of 10 breeding females 130 (Cramp, 1985). Estimated egg weight 9.85 (Schönwetter, 1967–84). Actual egg mass (32 eggs), 9.52 (Soler, 1990). Egg:adult female mass ratio 7.6%.

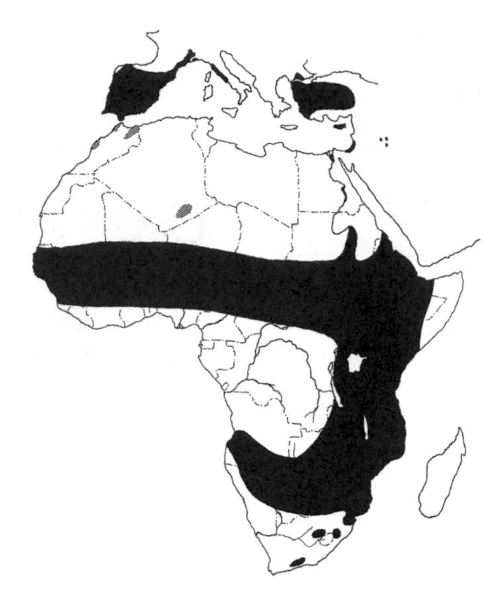

MAP 16. Breeding (filled) and winter or migratory ranges (hatched) of great spotted cuckoo.

Identification

In the field: The large size, bushy crest, white-spotted, dark grayish dorsal plumage, and long, white-tipped tail are distinctive (fig. 23). Both young and adults are white to buffy below (becoming more tawny on the throat); young birds have blackish crowns rather than grayish crests and are generally more brownish to bronze-colored above, especially on the primaries, with lighter spotting at the feather tips. A wide variety of calls have been described. One of the commonest vocalizations is a rapid and accelerating sequence of "kow," "keyer," or "kirrow" notes that sometimes ends in a loud trumpeting series of "euak" notes and probably corresponds to the male's advertisement song. Another common vocalization is a series of quarrelsome "gi" or "kreee" notes resembling a woodpecker's call or the call of a tern. The female has a bubbling "burroo-burroo" call that is mainly uttered during the breeding season. Young birds may learn host-specific food-begging calls and other host-related calls.

In the hand: Easily recognized by the combination of a crested head and the extensive dorsal spotting. The wings are relatively more pointed than in *Oxylophus,* but the nostrils are similarly slitlike, rather than rounded with raised edges as in most cuckoos. Females cannot be readily distinguished from adult males by plumage traits, but juveniles have creamy rather than white dorsal spots and have mostly chestnut primaries. Immature individuals retain chestnut brown at the bases of the primaries and have duller wing-covert spotting for their first year (Cramp, 1985). Nestlings are initially naked, with yellow to pinkish skin and pinkish to orange-red or orange-yellow mouth linings. The palate also has two conspicuous reddish papillae that may serve as an important visual stimulus for feeding behavior by hosts (see fig. 25G). The commissural junction is yellowish white, and the bill is laterally edged with white, but there are no conspicuous mandibular flanges as typically occur in host corvids.

Habitats

This species occurs primarily in dry savannalike habitats, especially acacia savannas in Africa, and is absent from thick woodlands. In western Africa it is especially associated with palms, open thornbush, and fig trees, and in southern Africa it is associated with open woodlands. In Europe it is found in heathlands having oaks, brambles, junipers, and various other low trees or shrubs. It also occurs in cultivated and other human-modified habitats, including olive groves, parklands, and suburban areas.

Host Species

A list of 19 host species known from Africa is provided in table 29. At least in southern Africa, the pied crow (about 36 records), Cape rook (18 records), and pied starling (20 records) are major host species. In northwestern Africa black-billed magpies are the primary hosts, and in northeastern Africa the hooded crow is a major host (Fry et al., 1988). In the western Palearctic the host list is shorter; the black-billed magpie is the primary host in Iberia, with the hooded crow, azure-winged magpie, and Eurasian jay secondary hosts. There is a single curious record involving the parasitism of a common kestrel that was using an abandoned magpie nest (Friedmann, 1948; Cramp, 1985).

Egg Characteristics

Eggs of this species are broad ovals, with a somewhat glossy surface, and range from pale blue to greenish blue, with spots or streaks of reddish brown and lilac that are generally similar to those of its primary corvid hosts. Those laid by sub-Saharan African birds are somewhat bluer and less spotted than are the eggs of European populations (Fry et al., 1988). Hosts in southern Africa include starlings, whose eggs are often similarly bluish, as well as the Cape rook, which uniquely has salmon-pink eggs. The eggshell averages 0.81 g in weight and 0.17 mm in thickness (Schönwetter, 1967–84); the egg dimensions are considerably smaller than those of crows, but larger than starling eggs.

Breeding Season

Nesting occurs from April to June in southern Europe, with a peak during May (Cramp, 1985). In northwestern Africa breeding similarly occurs from March to June, and in Egypt from January to June. South of the Sahara it generally breeds from April to July or August in West Africa, and April to June in Somalia. In East Africa the dates are highly variable, but generally range from February to March or April in western areas (Uganda, western Kenya), from June to December in eastern Kenya and northeastern Tanzania, and at various times throughout the year in western and southern Tanzania. In southeastern Africa (Malawi, Zimbabwe) the breeding dates generally run from about August to February, and in South Africa from October to January (Fry et al., 1988).

Breeding Biology

Nest selection, egg laying. In contrast to most parasitic cuckoos, pair-bonding occurs in this and the other crested cuckoo species. The two pair members often remain in close vocal and visual contact, and the pair-bonds apparently last through at least one breeding season, which in one case (Frisch, 1969) lasted 57 days. Soler (1990) suggested that an advantage of monogamy and cooperative egg-laying behavior in this genus lies in the fact that it parasitizes species larger than itself. The members of the pair are almost constantly together during the breeding season, and (in common with many other cuckoos), the male typically but not invariably feeds his mate a prey morsel just before copulation, which both may grasp during treading (Fig. 24A). Visits by a female to a potential nest may occur only minutes after copulation has occurred, and during such visits stabbing-like movements with the bill may

FIGURE 24. Copulation behavior (A) of the great spotted cuckoo, plus a comparison of eggs, hatchling heads, and nestlings (two of host, one parasite) of the black-billed magpie (B) and the cuckoo (C). A 13-day-old cuckoo is also shown (D). After sketches and photos in Glutz & Bauer (1980), and in von Frisch & von Frisch (1966).

also occur, and may be directed at host eggs. Just before laying, the female may call to attract her mate. The male typically approaches the nest directly, often calling as it flies from perch to perch, while the female surreptitiously approaches the nest. Egg laying may require as little as 3 seconds (Frisch, 1973; Arias de Reyna & Hidalgo, 1982), and it is likely that during egg laying the female simply stands on the rim of the nest and lets its egg fall into the cup from above. Neither host eggs nor other cuckoo eggs are removed during the laying act (Soler, 1990). However, one or more host (or occasionally other cuckoo) eggs may be damaged, evidently not by direct pecking, but rather as a result of the newly laid cuckoo egg dropping on and damaging host eggs as it is laid. Rarely, the host's entire clutch may thus be damaged (Montfort & Ferguson-Lees, 1961); the thicker eggshells may help limit damage to cuckoo eggs. Damage to the host eggs is especially valuable for improving chick survival in this group of cuckoos because the host chicks are not ejected or killed, and, as they are larger than the cuckoo chicks, they might soon be able to outcompete them for attention and food.

In most cases (in 8 of 9 in one study, and in 10 of 16 in another) the cuckoo's egg is laid before the host's clutch is complete, but at times the egg may be laid even after incubation is underway. Eggs are laid at 24-hour intervals, and up to 16 may be laid during a 38-day laying period (Frisch, 1969). Three laying cycles, or "clutches," usually totalling no more than 18 eggs, may be laid during a breeding season in Spain, with 5–8 days between laying cycles (Arias de Reyna & Hidalgo, 1982). South African records of 64 parasitized starling and crow nests (Rowan, 1983) indicate that more than one-third (38%) had a single cuckoo egg and 22% had two eggs, and as many as 13 cuckoo eggs were recorded in a single nest. Apparently females may lay more than one egg in a nest, and several females may also parasitize the same nest. Soler (1990) reported that among 39 nests, a single female had laid in 33 nests, 5 nests had eggs of 2 females, and 1 nest had been parasitized by 3 females. Of 41 parasitized nests (involving 4 host species), 19 nests (45%) had one cuckoo egg present, 10 had 2, 10 had 3, and there were 4 and 5 eggs in single cases.

Incubation and hatching. Incubation lasts an average of 12.8 days (range 11–15) (Arias de Reyna & Hidalgo, 1982), which is about 6 days shorter than the incubation periods of the cuckoo's usual corvid hosts, but is similar to that of starlings.

Nestling period. Developmental stages of this species are illustrated in figs. 24 and 25. The conspicuous, palatal papillae of the nestling cuckoos are notable, as is the vocal mimicry of their pied crow hosts (but not proven for other host species) by the cuckoo nestlings (Mundy, 1973). In one Nigerian field study, two cuckoo chicks fledged at 22 and 26 days (Mundy & Cook, 1977). Other studies suggest that the chicks may leave the nest at ages ranging from 16 to 21 days, although fledging usually occurs somewhat later, at 20–26 days, averaging about 24 days (Valverde, 1971; Cramp, 1985). Soler et al. (1994) reported that the cuckoo's postfledging dependency period averages 33.2 days (range 25–59), as compared with only 20 days for the black-billed magpie. Of 38 cuckoo chicks equipped with radiotransmitters just before fledging, 24 (63%) survived to independence and 14 died, the deaths either by predation or starvation (Soler et al., 1994). All of four similarly tracked magpie fledglings that had fledged in parasitized nests died of starvation or were killed by predators within 12 days of fledging.

Population Dynamics

Parasitism rate. An estimated 12.7% of 196 nests of the pied crow, 10.0% of 159 Cape rook nests, and 5.3% of 189 nests of the hole-nesting pied starling were parasitized in South Africa (Payne & Payne, 1967). In a Spanish study, Soler (1990) found that parasitism rates ranged from as low as 2.1% for 290 jackdaw nests to as high as 43.5% for 69 black-billed magpie nests. Soler et al. (1994) later reported even higher magpie parasitism rates of 58.6% of 111 nests in 1991 and 66.9% of 166 nests in 1992.

Hatching and fledging success. In a Nigerian study, 25 crow eggs and 9 cuckoo eggs were pres-

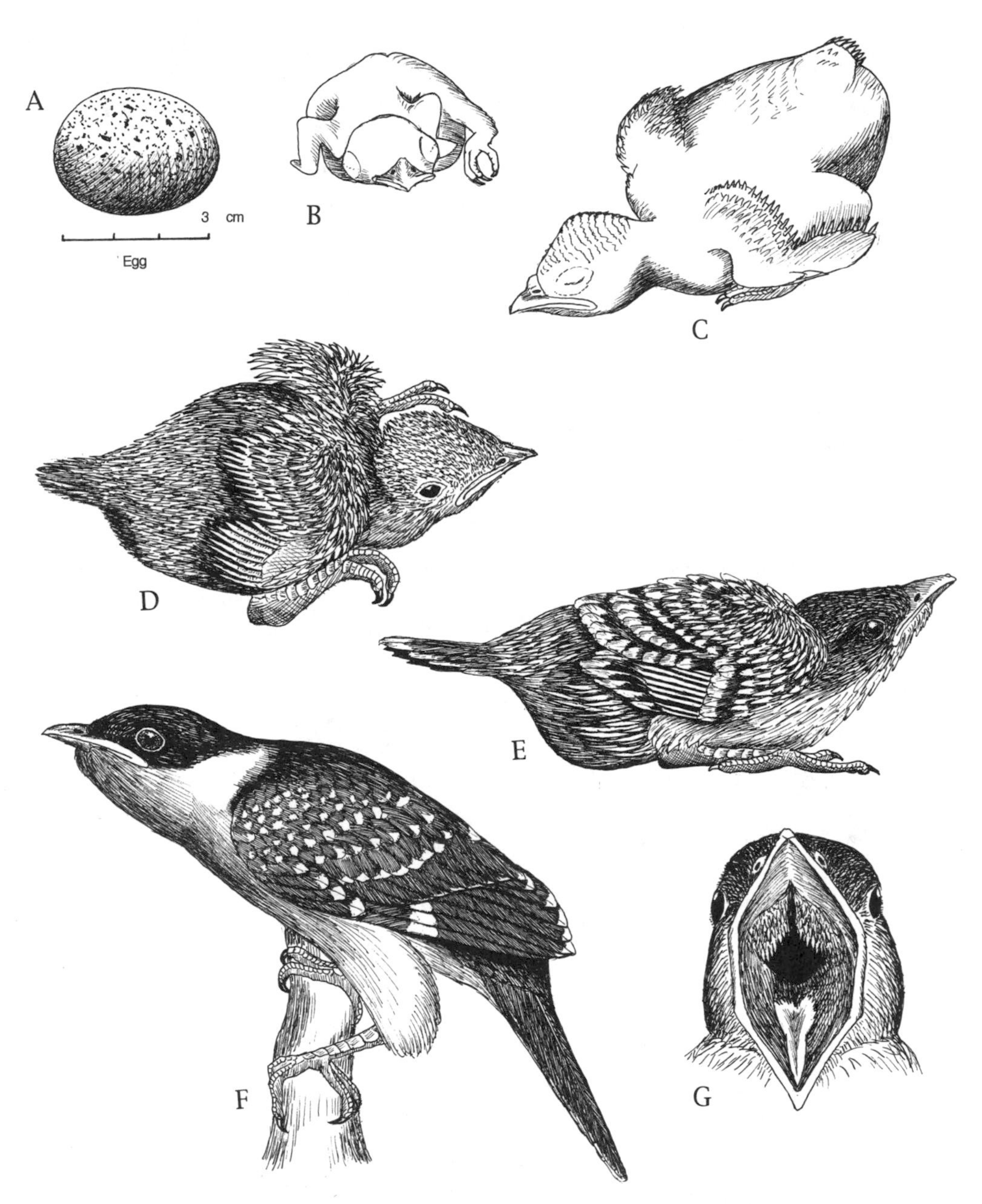

FIGURE 25 Ontogeny of the great spotted cuckoo, including its egg (A) and chicks at 2 (B), 6 (C), 9 (D), 12 (E), and 18 (F) days. Gaping by a 16-day-old juvenile (G) is also shown. After photos in Valverde (1971).

ent in 5 nests. Of these, two cuckoos and two crows evidently survived to fledging (Mundy & Cook, 1977). Soler (1990) determined that first- and second-laid eggs were more successful than third-laid eggs and that eggs laid early in the egg-laying period were more successful than those laid later during the egg-laying period or after the clutch had been completed. When more than one female cuckoo parasitized a nest, only those eggs of the first female to lay were successful. Among 26 parasitized magpie nests, 63% of 57 cuckoo eggs hatched, and 83% of the 35 hatched chicks fledged. In four parasitized carrion crow nests, only two of five hatched young fledged perhaps because the larger size of the crow's young gave them a better competitive advantage. In another 2-year study, 56% of 25 eggs laid the first year of observation resulted in fledged young, and 42% of 31 eggs laid during the second year of observation did so (Arias de Reyna & Hidalgo, 1982).

Host–parasite relations. Studies by Soler (1990) of four host species in Spain indicated that only one host (black-billed magpie) rejected eggs; the others (carrion crow, Eurasian jackdaw, red-billed chough) all accepted the eggs, although their eggs are distinctly larger than the cuckoo's. The cuckoo's eggs are similar to those of magpies, both in color patterning and in dimensions; magpies have been found to reject all nonmimetic eggs (Arias de Reyna & Hidalgo, 1982; Soler, 1990). Recent observations by Soler et al. (1995) in Spain have suggested that a "Maffia-like" interaction may exist, in that among 26 magpie nests in which cuckoo eggs were experimentally removed, some or all of the magpie young were lost by egg destruction or by being pecked to death, presumably by the cuckoos, whereas in 28 nests containing a cuckoo chick only 3 nests were destroyed, probably by crows.

THICK-BILLED CUCKOO

(Pachycoccyx audeberti)

Other Vernacular Names: None in general English use.

Distribution of Species (see map 17): Sub-Saharan Africa from Sierra Leone and Kenya south to South Africa. Previously occurred in Madagascar, but now apparently extirpated.

MAP 17. Residential African range and historic Madagascan range (shaded) of thick-billed cuckoo.

Subspecies

P. a. audeberti: Madagascar (no records since 1992)

P. a. validus: Angola and South Africa to Kenya

P. a. brazzae: Zaire to Guinea

Measurements (mm)

13.5″ (34 cm)

Wing, males 214–240 (avg. 225), females 218–236 (avg. 224). Tail, males 166–205 (avg. 183), females 163–198 (avg. 180) (Fry et al., 1988). Mean wing:tail ratio 1:0.8.

Egg, (of *validus*): avg. 19 × 14.5 (range 18–20 × 14–15) (Schönwetter, 1967–84). Shape index 1.31 (= broad oval).

Masses (g)

Male 92, females 100–120 (avg. 115, $n = 4$) (Fry et al., 1988). Estimated egg weight 2.2 (Schönwetter, 1967–84). Egg:adult mass ratio 1.9%.

Identification

In the field: This cuckoo is quite distinctive as an adult; it is strongly two-toned, being dark above and whitish below, with no crest (see fig. 27). Immature individuals are much paler, the head being spotted with black and white, and the flight feathers and upper-wing coverts strongly tipped with white. The species' vocalizations include a loud, penetrating two- or three-syllable whistle, interpreted as "were-wick" or "whe-yes-yes," as well as loud chattering notes. This call is often uttered during a buoyant flight display, which is typically slow, with undulating or erratic movements, and similar to that of the crested cuckoos. There is also an "undulating" call that begins slowly and rises to a crescendo before fading away, "kloo, kooo, kla, kla, kla, kloo, kloo, kloo." A high-pitched two-syllable squeak is apparently used by young birds to mimic the food-begging plea of young helmet shrikes and also might be used during adult cuckoo courtship.

In the hand: In common with the crested cuckoos, but in contrast to *Cuculus* species, the nostrils of this species are elongated and slitlike. The outermost pair of rectrices is noticeably shorter than the other pairs (which are of similar length), and these feathers are only slightly longer than the longest under-tail coverts. The bill is unusually stout and swollen (culmen length ~3 cm; bill ~1.5 cm wide and 1.0 cm deep at base), and the upper mandible is distinctly decurved toward the tip, producing a somewhat hawklike bill profile. The eye-ring is bright yellow in adults. The sexes are alike in plumage as adults, but immature individuals can be distinguished by the extensive amounts of white on their heads, the white feather edgings on their gray (not brown) upperparts, and their broader white rectrix tips. In juveniles the entire bill is black (rather than black-tipped with a greenish or yellowish base), and the eye-ring is dull yellowish rather than bright yellow. Newly hatched nestlings differ from their helmet shrike hosts in having orange rather than mauve-colored skin, an orange or orange-yellow (not pink) gape color, and pale yellow (not pinkish) feet. The egg tooth is prominent, and the slitlike nostrils and zygodactyl feet also provide easy identification (Benson & Irwin, 1972; Rowan, 1983; Fry et al., 1988).

Habitats

This species is associated with moist, open woodlands, especially miombo (*Brachystegia*) woodlands, gallery forests, and forest edges. It is mostly found in lowland habitats, from near sea level in coastal locations to rarely as high as about 1000 m in interior regions.

Host Species

The only known host of this species is the red-billed helmet shrike (Rowan, 1983). Circumstantial evidence exists for parasitism of the chestnut-fronted helmet shrike in East Africa, and the chestnut-bellied helmet shrike is probably parasitized in West Africa (Fry et al., 1988).

Egg Characteristics

The eggs are pale creamy green to blue-green, with blotches of gray, brown, and lilac, especially at the more rounded end (Fry et al., 1988). They are similar to those of the red-billed helmet shrike host, which are pale greenish, spotted and blotched with gray, brown, and purple, and average about 23 × 18 mm (Vernon, 1984).

Breeding Season

There are few records of laying, but in Zimbabwe and South Africa breeding may extend from September to early in the calendar year (mainly September to November), and there are September through November records for Kenya, Tanzania and Zambia. There is also a March breeding record for Cameroon (Fry et al., 1988).

Breeding Biology

Nest selection, egg laying. In contrast to the crested cuckoos, there is no evidence that the male participates in the egg-laying activities. The female approaches the nest alone, in one case by a series of short and undulating flights, when it closely resembled a small accipiter. The bird then flew directly to the already incubating helmet shrike, forcing the latter off its nest. The cuckoo then landed, removed an egg with its bill, presumably laid an egg, and quickly flew away with the host in close pursuit. On the next day the female reappeared and flew directly to the nest, again displacing the incubating bird, and again removing a host egg after mantling the nest briefly. These events occurred after the hosts had been incubating their clutch of three eggs for 3 days, so the cuckoo eggs were laid at least 4 days into the host's incubation. In three of four parasitized nests, a single cuckoo egg was present, and the fourth contained two (Vernon, 1984).

Incubation and hatching. The incubation period is probably no more than 13 days, as compared with 17 days for the helmit shrike host. In three of four studied nests, the cuckoo chick hatched 1 or 2 days before the host chicks. Within no more than 5 days after hatching, the chick evicts other chicks or eggs from the nest, in the same general manner as occurs in *Cuculus* and many other parasitic cuckoos (Vernon, 1984).

Nestling period. The nestling period lasts 28–30 days, with the chick's gape darkening to blackish, except for two large and contrasting orange palatal spots, and the tongue similarly a conspicuous orange color. After fledging, the chick may continue to follow its helmet shrike host parents for several weeks, and in one case was fed by them until it was at least 56 days old.

Population Dynamics

Parasitism rate. Vernon (1984) observed a minimum parasitism rate of 35% in a sample of 51 helmet shrike breeding records. When only nests containing nestlings were considered in the sample, the estimated incidence of parasitism increased to 55%.

Hatching and fledging success. Vernon (1984) estimated the cuckoo's overall breeding (egg-to-fledging) success rate as about 20% (63% hatching success in 29 nests, and 31% fledging success among 16 cuckoo chicks). He also judged that the host helmet shrikes achieved a breeding success rate of only 14%, which represents about half of the mean breeding success rate estimated for other insectivorous passerine species nesting in that area. Assuming an approximate 50% incidence of parasitism and a nearly complete lack of host production in parasitized nests, a 50% reduction in host fecundity seems possible.

Host–parasite relations. As noted above, the helmet shrike population evidently suffers a considerable reduction in fecundity as a result of cuckoo parasitism; in one case Vernon (1984) reported that the hosts did not rear any of their own chicks over a 5-year period in spite of the species' tendency for two breeding efforts per season. Among seven host nests having chicks present, six were parasitized, and in the seventh nest the young helmet-shrike nestlings were forcibly evicted from their nest by a cuckoo. Occasionally helmet shrikes will also abandon or destroy their own nests after they have been parasitized, especially if parasitic eggs are laid before incubation has begun.

SULAWESI HAWK CUCKOO

(Cuculus crassirostris)

Other Vernacular Names: Celebes hawk cuckoo.
Distribution of Species (see map 18): Sulawesi.
Measurements (mm)

13" (33 cm)

Wing, adults 206–216 (White & Bruce, 1986). Tail, unsexed adult 155 (U.S. National Museum specimen). Wing:tail ratio ~ 1:0.75.

Masses (g)

No information on body or egg weights.

Identification

In the field: This is the only hawk cuckoo breeding on Sulawesi, so it should be possible to assign any large cuckoo with a strongly black-barred breast and somewhat banded tail to this species. The head is mostly gray, contrasting with the rufous back and the heavily barred breast, and, unlike the other hawk cuckoos, there are no buffy wingbars on the flight feathers. The song is similar to that of the common cuckoo, but with the second syllable lower, and the first, higher note often doubled.

In the hand: This species has a rich rufous to medium brown back, which contrasts with a gray crown. The flight feathers are not barred with buff, and the underparts are white, with black throat

MAP 18. Ranges of moustached (filled) and Sulawesi (shaded) hawk cuckoos.

spotting and heavy black barring on the breast and abdomen. A first-year male had a blackish crown, with much of the breast and abdomen pure white; a recently fledged nestling had rufous wings, while the head, nick, and underparts were cream colored and lacked dark markings (White & Bruce, 1986).

Habitats

This species is reported to occur in wooded and forested hills and mountains of Sulawesi, between 500 and 1400 m elevation. No other information is available on this species, but it is probably similar to the other hawk cuckoos in its woodland habitat preferences.

Host Species

No host species have yet been reported.

Egg Characteristics

No information.

Breeding Season

No information.

Breeding Biology

No information.

Population Dynamics

No information.

LARGE HAWK CUCKOO

(Cuculus sparveroides)

Other Vernacular Names: None in general English use.

Distribution of Species (see map 19): From Pakistan east to southern China, south to Burma, Thailand, and Indochina, plus Sumatra (scarce) and Borneo (common). Winters south to Java (where rare), Sulawesi, and the Philippines.

Subspecies

C. s. sparverioides: Himalayan foothills to southeast Asia, the Philippines, and Sulawesi.

C. s. bocki: Malaysia, Sumatra, and Borneo.

Measurements (mm)

15–16″ (38–40 cm)

C. s. bocki. Wing, both sexes 185–193 (Medway & Wells, 1976).

C. s. sparveroides. Wing, both sexes 213–226; tail;: both sexes 175–220 (Ali & Ripley, 1983). Wing, both sexes 201–245; tail, both sexes 197–228 (Delacour & Jabouille, 1931). Mean wing:tail ratio ~ 1:0.9.

Egg (of *sparveroides*), avg. 27.2 × 18.8 (range 25–29.7 × 17.4–21.1 (Schönwetter, 1967–84). Avg. of 70, 26.6 × 18.6 (Becking, 1981). Shape index 1.43–1.45 (= oval). Rey's index 1.59 (Becking, 1981).

Masses (g)

Males 116, 131 (Ali & Ripley, 1983). Estimated egg weight 5.05 (Schönwetter, 1967–84). 4.7–5.4 (Becking, 1981). Egg:adult mass ratio ~ 4%.

Identification

In the field: The large size, relatively short, rounded wings, generally hawklike appearance, and a white-tipped and rather rounded tail with a subterminal darker bar serve to identify this species (fig. 26). It is the largest of the hawk cuckoos and has a barred belly but a rufous-streaked throat and chest. Like all hawk cuckoos, the tail is banded with several brown bars, the flight feathers are strongly barred, and the eyes are surrounded by bright yellow eye-rings. Adults are less strongly barred below and above than are immature individuals and are brownish gray, rather than dark brown, above. The call is a loud, repeated "brain FE-ver" or "pi-PEE-ha." These notes tend to increase in speed and pitch until they reach a frantic climax. The call is reportedly not so shrill or loud as that of the common hawk cuckoo. The Malayan race *bocki* also utters a repeated "ha-ha"

MAP 19. Breeding (filled) and wintering (enclosed area) ranges of large hawk cuckoo.

FIGURE 26. Profile sketches of the common (A), large (B), Hodgson's (C), and moustached (D) hawk cuckoo adults, plus a juvenile of the common hawk cuckoo (E). Their eggs and diagrams of typical song phrases are also shown.

call, with each successive phrase higher in pitch until a crescendo occurs, and the notes then drop off.

In the hand: This is the largest of the hawk cuckoos, with a minimum wing length of 185 mm (Malayan race) or 201 mm (nominate race) and a tail that is usually longer than 190 mm. Adults have a gray crown and nape, the dorsal color becoming more brownish gray on the back, and the hawklike tail has several brown to blackish bands. The throat is mottled with gray, white, and rufous, and the breast is more uniformly rufous. The underwing surface is closely barred in a hawklike manner with white and brownish gray. The sexes are alike, but immature individuals have light rufous bars and edgings on the upperparts, and the underparts are heavily streaked and spotted with blackish brown. In both adults and young, the eyelids are yellow, and the iris color varies from yellow (adults) to dull grayish brown (juveniles). The interior of the mouth is bright yellow in very young birds, as are the feet and toes (Deignan, 1945).

Habitats

This species is associated with open woods on hillsides at elevations of about 1800–2900 m in Nepal, about 900–2700 m in India, and to about 1800 m in Myanmar. It ranges higher in summer than do the Indian and Himalayan cuckoos and breeds in oak as well as in coniferous (fir or pine) forests.

Host Species

Host species in India with two or more records of parasitism, according to Baker (1942), are listed in table 12. Ten additional species (two laughingthrushes, two babblers, two yuhinas, two thrushes, a bulbul, and a shrike) were listed as having single records of parasitism in Baker's egg collection. A few additional species (two laughing thrushes, a quaker babbler, a barwing, and a *Ficedula* flycatcher) were mentioned as reported hosts, but parasitized clutches were not represented in his collection. The streaked spider hunter is the species with the largest number of records associated with Baker's "brown-type" cuckoo eggs, and the lesser necklaced laughingthrush had the largest number of clutches associated with his "blue-type" eggs. When Becking (1981) later analyzed Baker's egg collection, he concluded that only the "brown-type" eggs were those of the large hawk cuckoo; the blue ones were probably misidentified common cuckoo eggs. The majority (68%) of the hawk cuckoo eggs thus identified by Becking were among streaked spider hunter clutches; a few were associated with the smaller but more common little spider hunter. The Nepal short wing is probably also a significant host.

Egg Characteristics

As noted above, Becking (1981) considered only the "brown-type" eggs in Baker's (1941) collection as certainly belonging to this species. Eggs of this type are uniformly olive-brown, with occasional darker olive-brown speckling, especially near the more rounded end. Their eggshell thickness averages 1.07 mm, and their mass averages 0.31 g (Becking, 1981).

Breeding Season

The breeding season of the streaked spider hunter, this species' primary host in India, is from March to July. The little spider hunter, another important host, mainly breeds from May to August in Assam, but from December to August in southwestern India (Ali & Ripley, 1983). In Myanmar singing is mostly heard from early February to the end of June (Smythies, 1953).

Breeding Biology

Nest selection, egg laying. The streaked spider hunter, which builds an open and pendulous cuplike nest, has been reportedly parasitized by several species of cuckoos. In contrast, the small spiker hunter builds a more tunnel-like pendulous nest with a lateral entrance. Such a nest must be very difficult for the much larger cuckoo to parasitize, and the small spider hunter accounted for only 2% of the brown egg morph. Both host species attach their nests to the underside of a banana leaf or similar large-leaved plant. The average size of the streaked spider hunter's brown eggs are 22.7 × 15.9 mm, and the pinkish eggs of the small spider hunter average 18.4 × 13.1 mm, both of which are well below the egg measurements typical of the hawk cuckoo (minimum 27 × 17.4 mm). The Nepal short wing accounted for 17% of the brown eggs in the Baker collection; this species is a forest-dwelling ground nester that builds an oval ball-like nest with an upper lateral entrance and lays brown eggs averaging 19.5 × 14.6 mm. Both the streaked spider hunter and the Nepal short wing have eggshells that average less than half the mass of the hawk cuckoo's (<0.15 g vs. > 0.3 g) (Baker, 1943).

Incubation and hatching. No information on the incubation periods of either major host species or the cuckoo is available.

Nestling period. No information.

Population Dynamics

No information.

COMMON HAWK CUCKOO

(Cuculus varius)

Other Vernacular Names: Brainfever bird, Ceylon hawk cuckoo (*ciceliae*), Indian hawk-cuckoo (*varius*).

Distribution of Species (see map 20): India, Nepal, Sri Lanka, and Myanmar (Burma).

Subspecies

C. v. varius: India, Nepal, and (rarely) Myanmar.

C. v. ciceliae: Sri Lanka.

Measurements (mm)

13–13.5″ (33–34 cm)

Wing, males 193–213, females 192–207. Tail, males 157–188, females 156–180 (Ali & Ripley, 1983). Mean wing:tail ratio ~1:0.77.

Egg avg. 27.1 × 20.4 (range 26–28.9 × 19.7–21.4) (Schönwetter, 1967–84). Egg measurements given by Baker (1942) are unreliable, but two oviducal eggs averaged 24 × 20.3 (Becking, 1981). Shape index 1.18 (= subspherical). Rey's index (avg. of two oviducal eggs) 1.16 (Becking, 1981).

Masses (g)

One female 104 (Ali & Ripley, 1983). Avg. of three unsexed birds 104, range 100–108 (Dunning, 1993). Estimated egg weight 6.1 (Schönwetter, 1967–84). Egg:adult mass ratio 5.8%.

Identification

In the field: This species is slightly smaller than the large hawk cuckoo but still quite large, with the usual hawk cuckoo pattern of a banded, rounded, and pale-tipped tail, strong barring on the rather short wings, and a bright yellow eye-ring (fig. 26). The dark brown throat and breast markings are less distinct, but a rufous breast-band is more evident, and the upperparts of adults are more uniformly grayish, rather than brownish. Immature individuals are heavily spotted with dark brown on their white underparts, barred brownish above, and the tail is strongly banded and tipped with pale rufous. The usual advertising song is much like that of the large hawk cuckoo but perhaps more shrill. It consists of a series of "brain FE-ver" calls that occur in sequences of four to six increasingly frantic phrases that reach a crescendo and then suddenly stop, only to begin again 1 or 2 minutes later. Calling is especially evident on dark, cloudy days and on moonlit nights, as is typical of most cuckoos.

In the hand: The strongly banded tail identifies this as a hawk cuckoo. This species has a tail that ranges from 150 to 190 mm, placing it roughly intermediate in length between the large hawk cuckoo and Hodgson's. Additionally, the chin and throat of this species is uniform gray (not black as in the Hodgson's hawk cuckoo, or distinctly mottled as in the large hawk cuckoo), and the abdomen and flanks are barred with brown (rather than unbarred, as in Hodgson's). The sexes are alike as adults. Immature individuals have a tail pattern that is barred with rufous and black rather than whitish and black, and additionally they are distinctly streaked, rather than barred on the flanks and underparts. The nestling is undescribed.

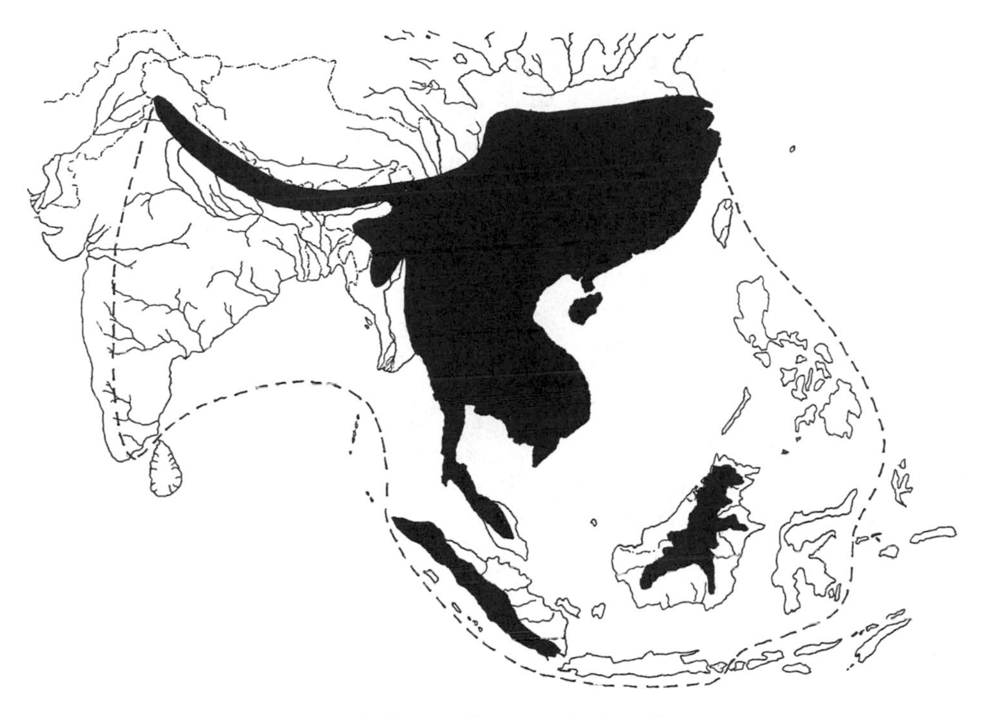

MAP 20. Range of common hawk cuckoo

Habitats

This cuckoo is associated with sparsely wooded to densely wooded woodlands at lower altitudes than those used by *sparveroides*, usually no higher than 1000 m in India, and from 120 to 1000 m (rarely to 1370) elevation in Nepal. It also frequently uses gardens, groves, and other open habitats associated with human activities. In Sri Lanka it is evidently a highland-adapted, rather than a lowland, species.

Host Species

A list of six reported host species, representing those with two or more records of parasitism in Baker's egg collection, is provided in table 12. Baker listed an additional nine species represented by single parasitized clutches in his collection. These consisted of various babblers, laughing thrushes, the rufous-bellied niltava, the Asian fairy bluebird, and the Asian paradise flycatcher. Becking (1981) has questioned the species identification of all these eggs and regarded them as a hodge-podge of blue eggs of the common cuckoo and crested cuckoos. Young common hawk cuckoos have been seen being tended by jungle babblers.

Egg Characteristics

As Becking (1981) concluded, the eggs of this species are still not adequately known. Two seemingly unquestionable (oviducal) eggs in Baker's collection were nearly round (22.1 × 20.2 mm and 26 × 20.4 mm; shape index 1.08 and 1.27). They weighed 0.41 and 0.44 g, and thus were very thick-shelled (Rey's index 1.03 and 1.29), resembling in shape and thickness those of crested cuckoos. These eggs have since been lost, and their colors were unreported, although all the other eggs attributed by Baker to this species were turquoise blue. The eggs of the jungle babbler and various other related babblers (the presumed important host species) are also blue, which may increase the probability that this cuckoo also has blue eggs. Becking speculated that the olive-green eggs that have been attributed to Hodgson's hawk cuckoo might belong to this species, and the turquoise blue eggs attributed by Baker to this species in turn might be those of the Hodgson's hawk cuckoo.

Breeding Season

No specific, reliable information, but it presumably breeds during spring, as do its presumptive hosts. Ali and Riley (1983) indicated that in the Indian region breeding occurs from January until at least April. Few records for this species exist for Myanmar, where calling has been reported in April.

Breeding Biology

Little reliable information exists on the egg-laying phase of breeding. Fledgling cuckoos have been repeatedly seen in the company of jungle babblers and other species of babblers, so there seems little doubt that babblers represent the cuckoo's primary hosts. The nests of the jungle babbler are cuplike and are placed in easily accessible sites among bushes or trees. The incubation periods of *Turdoides* babblers are in the range of 14–17 days, and their eggs range from pale blue to dark blue.

Population Dynamics

No information.

MOUSTACHED HAWK CUCKOO

(Cuculus vagans)

Other Vernacular Names: Dwarf hawk cuckoo, lesser hawk cuckoo.

Distribution of Species (see map 19): Malay Peninsula, Sumatra, Java, and Borneo.

Measurements (mm)

10–12″ (25–30 cm)

Wing, both sexes 135–148 (avg. 141, $n = 4$). Tail, both sexes 119–138 (avg. 128, $n = 4$) (specimens in American Museum of Natural History and U.S. National Museum). Wing:tail ratio ~1:0.9.

Egg, no information

Masses

No information

Identification

In the field: Similar to the large hawk cuckoo, but smaller. The whitish underparts have extensive, dark shaft streaks (but not horizontal barring) and conspicuous dark malar or moustachial stripes (fig. 26). Like most other hawk cuckoos, there is a conspicuously brown-banded and buffy- or white-tipped tail, buffy barring on the flight feathers, and a bright yellow eye-ring. The back is uniformly brown, with the nape and crown more grayish in adults; younger birds are probably more extensively barred with brown dorsally. The usual male vocalization is a disyllabic "kang-koh," with each note inflected downward slightly, and the phrases are uttered about 2 seconds apart. There is also a mellow "peu-peu" phrase that is uttered repeatedly with gradually ascending pitch and speed until it reaches a frantic trill, and then suddenly stops.

In the hand: The in-hand traits for identifying this rare and essentially unstudied cuckoo are the same as the fieldmarks noted above, especially the conspicuous malar stripe. This species is probably distinguishable from other hawk cuckoos on the basis of its smaller wing and tail measurements, but it possibly overlaps with Hodgson's hawk cuckoo in tail length. Adults of Hodgson's hawk cuckoo and the large hawk cuckoo reportedly have pale yellow and orange eyes, respectively, rather than the brown iris color of this species. Like the other hawk cuckoos, there is a prominent yellow eye-ring, and barred tail feathers. The sexes are evidently externally identical as adults. Detailed descriptions of juvenile or immature plumage are unavailable, and the nestling is undescribed.

Habitats

This cuckoo is found in middle and low levels of tropical evergreen forests and second growth, from sea level up to about 800 m.

Host Species

No information.

Egg Characteristics

No information.

Breeding Season
No information.

Breeding Biology
No information.

Population Dynamics
No information.

HODGSON'S HAWK CUCKOO
(Cuculus fugax)

Other Vernacular Names: Fugitive hawk cuckoo.
Distribution of Species (see map 21): India to Siberia and Japan; south to Malay Peninsula and Indochina; also breeds on Sumatra, Borneo, and the Philippines. Winters to Sulawesi (Celebes) and Buru Island.
Subspecies
C. f. fugax: Malaysia, Sumatra, Java (rare), and Borneo.
C. f. hyperythrus: Siberia, China, Japan, and Indochina.
C. f. nisicolor: Northeastern India to Malaysia; Sumatra.
C. f. pectoralis: Philippines.
Measurements (mm)
11–12″ (28–30 cm)
C. f. nisicolor. Wing, both sexes 178–182; tail; both sexes 141–144 (Ali & Ripley, 1983). Wing, both sexes 178–200; tail, both sexes 141–144 (Delacour & Jabouille, 1931). Mean wing:tail ratio ~ 1:0.8.
Egg, avg of *hyperythrus* 27.2 × 19.6 (range 26.2–28.2 × 19.3–20), of *nisicolor* 23.5 × 15.7 (range 21.8–24.6 × 15.2–16.3), of *fugax* 22 × 16 (Schönwetter, 1967–84). Shape index 1.37–1.50 (= oval). Becking (1981) reported 14 eggs of *nisicolor* from India as averaging 23.8 × 15.8, with a shape index of 1.5 (oval) and a Rey's index of 1.98. Five eggs from Japan averaged 27.4 × 19.8, with a shape index of 1.79 (long oval), and a Rey's index of 1.49.
Masses (g)
Males (of *pectoralis*) 73.9–86 (avg. 78.2, *n* = 9); females 72.8–89.2 (avg. 83.0, *n* = 3) (museum specimens). Estimated egg weight (*nisicolor* and *fugax*) 3.1 (Schönwetter, 1967–84); *hyperthyrus* 5.1 (Balatski, 1994). Egg:adult mass ratio ~ 3.7%.

Identification

In the field: Like other hawk cuckoos, this species is notable for its pale-tipped and strongly brown-banded tail and similarly strongly barred flight feathers (fig. 26). The tip of the tail is pale rufous, rather than buffy white as in other hawk cuckoos. Adults are dark brownish gray above, with a bright yellow eye-ring and with mostly rusty-brown tones or streaking on the breast, without the dark brown streaking or barring typical of the large and common hawk cuckoos. Immature individuals are entirely brown on the upperparts. Like other hawk cuckoos, the usual song is a loud and repeated "pee-pee" or "gee-whiz" that is uttered initially at a rather slow rate but that becomes increasingly rapid and soon reaches a frantic trilled or slurred peak, only to stop abruptly and soon begin again. A similar "pee-weet" or "gee-whiz" call may be uttered for up to about 20 times at consistent 1-second intervals. Another reported vocalization is a staccato, stuttering screech that ascends the scale and comes halfway down again as it speeds up and abruptly stops.

In the hand: This is the smallest of the mainland hawk cuckoos, and it has the shortest tail of any (maximum 144 mm vs. at least 150 mm in the common hawk cuckoo). It also has the most uniformly rufous underparts, with little or no dark barring or spotting present on the flanks. Both adults and immature individuals have tails that are barred with black and brownish gray and are tipped with rufous. The tail's terminal dark bar is broad, but the adjoining one is narrower than the more anterior ones. Immature individuals have breasts that are mostly white rather than rufous, and they are distinctively barred or streaked with blackish on the underparts. The nestling is undescribed.

Habitats

This cuckoo is associated with the understory of fairly dense hillside forests, mostly occurring

MAP 21. Breeding (filled) and wintering ranges (enclosed area) of Hodgson's hawk cuckoo.

from 650 to 1800 m elevation in India and Nepal, and using deciduous, semideciduous, and evergreen woodlands. It also has been found in bamboo thickets and tree plantations. In Myanmar the species frequents dense and evergreen forests. Mixed deciduous woods or open evergreen forests at elevations of less than 300 m seem to be the preferred habitats in Thailand, where the species is quite rare. In Japan it occupies broadleaf or mixed montane forests, at elevations up to about 1500 m and sometimes to 2300 m. In the Philippines it extends in dense forests up to elevations of about 2300 m, but in Borneo and Sumatra it reportedly only reaches elevations of about 1400 m.

Host Species

A list of 10 host species represented by two more clutches in Baker's (1942) egg collection is provided in table 12. Baker listed another 10 host species represented by single parasitism records; these mostly consisted of flycatchers and babblers. The small niltava is evidently a prime host, its 23 host records represent about 30% of the total parasitized clutches in Baker's collection. Flycatchers and short wings are probably important hosts in Myanmar (Smythies, 1953). In Japan the host list is fairly long, consisting of at least 11 species, but the main hosts reportedly include the red-flanked bluetail, the Japanese robin, the Siberian blue robin, and the blue-and-white flycatcher (Brazil, 1991). In far-eastern Soviet Asia the hosts include these latter two species and the Mugimaki flycatcher (Balatski, 1994).

Egg Characteristics

According to Becking (1981), two unquestionable (oviducal) eggs of this species from India were oval (22.6 × 16.3, shape index 1.39), with an olive-brown ground color and an indistinct darker ring of brown near the more rounded end. Others that were collected by Baker and that in Becking's judgment probably also belong to this species are generally light to medium brownish olive, with darker brown specks. They are nearly as long as those of the large hawk cuckoo (23.8 mm vs. 27 mm) but are much narrower in width (average width 15.8 mm vs. 18.8 mm for the large hawk cuckoo: average shape index 1.5, or nearly long oval). A collection of 14 eggs from the Baker collection that were identified by Becking as belonging to this species were olive-green, with the same width and shape index averages just mentioned, suggesting that some egg color polymorphism might exist. The apparent primary Indian host, the small niltava, lays white to yellowish eggs averaging about 18 × 14 mm, with darker blotches or freckles, which is not a good match in color or in size. Interestingly, some eggs from Japan that have been attributed to this species are pale blue and are somewhat smaller (see egg measurement data above), but in Becking's view their identity needs additional confirmation. A reputed hawk cuckoo egg from Borneo that was found in a black-and-red broadbill nest was bluish white and measured 30.5 × 20.3 mm (shape index 1.5). Another cuckoo egg from Borneo that was found in a gray-headed canary-flycatcher nest and attributed to this species was ivory-yellow with darker specks and measured 22.1 × 16.4 mm (shape index 1.35, Rey's index 2.03). Size and color differences between these two eggs make it unlikely that both belonged to the same cuckoo species, and they also don't closely match the hawk cuckoo eggs reported from India or Japan (Becking, 1983).

Breeding Season

No specific information is available, but the species' major host in India (the small niltava) breeds from April to July. In Japan this cuckoo breeds from mid-May to mid-July (Brazil, 1991), during the peak period of small passerine breeding activity, and females with enlarged ovaries have been collected during April and May in the Philippines (Dickinson et al., 1991).

Breeding Biology

Nest selection, egg laying. No information exists. The small niltava builds a well-concealed nest, with a cuplike opening, in crevices along stream banks. The reported major Japanese hosts are ground-nesting or cavity-nesting species that typically breed in forests with dense and often damp undergrowth.

Incubation and hatching. No detailed information exists. There are no records of fledgling cuckoos of this species in host nests or being fed by foster hosts in India (Becking, 1981). The major hosts in Japan are reportedly the red-flanked bluetail, Japanese robin, Siberian blue robin, and blue-and-white flycatcher (Royama, 1963). These species have incubation periods of generally 13–15 days, and at least some of them lay bluish eggs, the reputed egg color of the Japanese population of hawk cuckoos.

Nestling period. No information.

Population Dynamics

No information.

RED-CHESTED CUCKOO

(Cuculus solitarius)

Other Vernacular Names: Red-throated dusky cuckoo.

Distribution of Species (see map 22): Sub-Saharan Africa from Senegal and Somalia south to South Africa.

Subspecies

C. s. solitarius: Mainland Africa.

C. s. magnirostris: Bioko Island (Gulf of Guinea).

Measurements (mm)

12″ (30 cm)

Wing, males 168–196 (avg. 177), females 166–190 (avg. 176). Tail, males 137–160 (avg. 148), females 138–158 (avg. 148) (Fry et al., 1988). Wing:tail ratio 1:0.8.

Egg, avg. 22.4 × 16.4 (range 22–26 × 16–19) (Schönwetter, 1967–84). Shape index 1.37 (= oval). Rey's index 1.93.

Masses (g)

Males 68–90 (avg. 75.3, n = 15), females 67–74 (avg. 71.6, n = 5) (Fry et al., 1988). Estimated egg weight 3.22 (Schönwetter, 1967–84). Egg:adult female mass ratio 4.5%.

Identification

In the field: The rufous chest and barred underparts of this species are conspicuous fieldmarks (fig. 27), but both traits also occur in the forest-dwelling race of the black cuckoo. Additionally, the birds are rather uniformly blackish above, with no conspicuous white tail or wing markings except for narrow whitish tips on the rectrices. Immature individuals are dark brown above, have a dark brown rather than rufous breast band, and a contrasting white nape patch. The distinctive vocalization of the male is a loud and resonant "ee-eye-ow," or "whip, whip, whee-oo," that descends in pitch, with the last note most strongly accented and slurred downward slightly. The Afrikaans vernacular name, *piet my vrou,* also describes the song well, as do "quid pro quo" and "whip-poor-will." These phrases last about 1–1.5 seconds, with slightly shorter intervening intervals. Adults of one or both sexes also sometimes utter rapid series of "kwik" notes. Males and females sometimes duet in this manner; the female's voice is higher pitched. Another call is very excited series of "hahehehehehehe" notes that sound like exaggerated panting, with the emphasis on the first "ha" note.

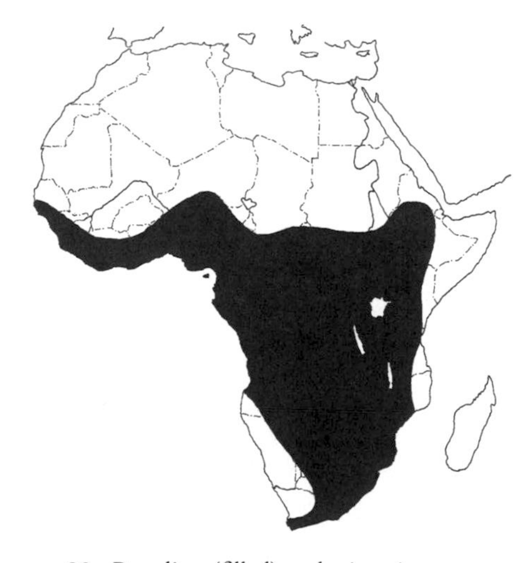

MAP 22. Breeding (filled) and wintering ranges (enclosed area) of red-chested cuckoo.

In the hand: This medium-sized cuckoo is distinctive in having a rufous upper breast contrasting with an otherwise generally gray head and upperpart coloration, a rather squared-off to slightly rounded tail that is tipped as well as slightly spotted and barred with white. The eye-ring and feet of adults are bright yellow, and the gape or mouth lining is orange. Females are more barred below than are adult males, and the breast color of females is less cinnamon-colored and more distinctly barred. Immature individuals are rather uniform brown above and more heavily barred with brown below than are adults; the cinnamon of the breast is replaced with dark brown, there is a white nape-patch, and the yellow eye-ring is less bright. Immature individuals resemble those of the African cuckoo but are darker throughout, with blackish heads and throats and black bills. Nestlings have a bluish black to purplish brown skin color and a deep-yellow gape, which darkens during the first 2

FIGURE 27. Profile sketches of four African cuckoos: juvenile (A) and adult (B) of red-chested cuckoo, adults of *gabonensis* (C), and nominate race (D) of the black cuckoo, adult of barred long-tailed cuckoo (E), and adult (F) and juvenile (G) of thick-billed cuckoo. Typical egg morphs are also shown.

days to become a rich orange-red. The beak is initially dark horn, with a prominent egg tooth. The upper mandible gradually becomes black with increasing age, and the lower one becomes tipped with black. The feet also change within 15 days from a dark flesh to bright chrome yellow. A yellow eye-ring is apparent by the third week of life (Reed, 1969; Rowan, 1983).

Habitats

This species is associated with heavily wooded savannas, forest edges, and leafy thickets, at elevations up to about 3000 m.

Host Species

Twenty-two known host species are listed in table 20, based on a summary provided by Fry et al. (1988). These hosts mostly consist of thrushes and robin chats, of which the Cape robin chat is certainly the most frequent host in South Africa, with more than 90 records. It is also an important host in southeastern and East Africa. The boulder chat and Cape wagtail each have 13 host records, and the Ruppell's robin chat and white-browed robin chat ("Hueglin's robin") each account for 12 records. Rowan (1983) listed 15 biological hosts (those with which nestlings or fledglings have been seen) for southern Africa, of which the Cape robin chat (55 records) is easily the most important, followed by the Cape wagtail (6 records) and white-throated robin chat (4 records).

Egg Characteristics

The eggs of this species are broad oval, with shiny surfaces and polymorphic coloration. The commonest egg morph is a glossy chocolate to olive-brown in ground color, without darker markings, and is an apparent (but poor) mimic of Cape robin chat eggs, which are variably colored (cream, pink, greenish blue, turquoise), but never brown. The size match of this cuckoo's eggs and those of the Cape robin chat is nevertheless very close (the robin chat's eggs average 23.2 × 17.9 mm). Some other robin chats that are known hosts do lay brownish eggs much like those of the cuckoo, such as the Natal, white-browed, and chorister robin chats. This cuckoo also occasionally lays bluish, olive-green to pale greenish eggs, with reddish brown speckling at the larger end, which may mimic those of the boulder chat. Similar pale green to olive-green or bluish eggs with pinkish brown freckles are also laid and closely mimic the eggs of the bearded scrub robin, and olive-green eggs without any markings have also been recorded (Rowan, 1983; Fry et al., 1988). The mean eggshell weights (two races) are 0.28–0.3 g, and the average shell thickness is 0.1–0.11 mm (Schönwetter, 1968–84).

Breeding Season

In South Africa the egg season extends from October to January, with a November (Cape, Transvaal) or December (Natal) peak. Zimbabwe and Zambia records are for October to December or January, and Malawi records are for October and January. West African records (Senegambia to Cameroon) are for March to August, and Ethiopian breedings probably extend from April to July. East African records are, as usual, well scattered, but generally extend from January to July in Uganda and western Tanzania and from March to July in northeastern Tanzania and across much of Kenya (Fry et al., 1988).

Breeding Biology

Nest selection, egg laying. Most of the known host species have rather easily accessible and open-cup nests that should pose no problems in egg-laying for the cuckoo, although the nest of the starred robin is so small and enclosed that perhaps the cuckoo must eject its egg into it by pressing its cloaca against the nest's opening (Rowan, 1983). Reed (1969) observed egg-laying activities associated with seven of the open-cup nests of Cape robin chats. In three nests, and perhaps as many as five, the female cuckoo removed a host egg when laying hers, but in two others this was not the case. In one case the cuckoo laid its egg before the host female began laying, but in five others the egg was laid during the host's egg-laying period or shortly afterward.

Incubation and hatching. The incubation period lasts about 12–14 days, as compared with a 15-day incubation period for the Cape robin chat (Fry et al., 1988). The young cuckoo evicts any other nestlings for eggs beginning about the second day after hatching; this eviction tendency lasts until the fourth day of posthatching life (Rowan, 1983).

Nestling period. The nestling period has been generally estimated as ranging from 17.5 days to about 20 days (Reed, 1969). Thereafter the young bird remains with its foster parents from another 25–32 days, and sometimes for even longer periods, some self-foraging may even begin as soon as 12 days after fledging (Reed, 1969; Rowan, 1983).

Population Dynamics

Parasitism rate. Payne and Payne (1967) estimated that the rate of parasitism for Cape robin chats in southern Africa ranged from 6.3% in Zimbabwe to 22.3% in the Transvaal, with an overall rate of 4.5% for 689 nests throughout southern Africa. Parasitism rates were 1.3% for 229 nests of the Karoo scrub robin, and 2.8% for 144 nests of the red-backed scrub robin. In a Natal study (Oatley, 1970), 13 of 84 (16%) Cape robin chat nests initiated during the cuckoo's laying period were parasitized. The overall parasitism rate was lower, because many of the total 115 host nests were already being incubated before the cuckoo began laying.

Hatching and fledging success. No detailed information exists. Two of the 13 nests of Cape robin chats found by Oatley (1970) were deserted; in one of these the cuckoo had laid after the robin chat had begun incubating its own clutch. Three such cases of late parasitism were reported by Oatley.

Host–parasite relations. Given the rather high incidence of parasitism and the fact that the young cuckoo evicts all of the host nestlings, the reproductive impact of this cuckoo on Cape robin chats is likely to be serious, but specific estimates are unavailable.

BLACK CUCKOO
(Cuculus clamosus)

Other Vernacular Names: Gaboon cuckoo (*gabonensis*,) noisy cuckoo.

Distribution of Species (see map 23): Sub-Saharan Africa from Gambia and Somalia south to South Africa.

Subspecies

C. c. clamosus: Ethiopia and Somalia to Tanzania.

C. c. gabonensis: Liberia and Nigeria to Kenya.

Measurements (mm)

12″ (30 cm)

Wing, males 166–187, females 167–183. Tail, males 138–156, females 140–154 (Fry et al., 1988). Wing:tail ratio ~1:0.8.

Egg: One *clamosus* egg 23.4 × 18; avg. of *gabonensis* 24.1 × 18 (range 23.5–24.7 × 17–19.1) (Schönwetter, 1967–84). Shape index 1.3–1.33 (= broad oval). Rey's index 1.68.

Masses (g)

Males (of *clamosus*) 78–94 (avg. 85, $n = 9$), females 79–92 (avg. 87.4, $n = 6$). Males (of *gabonensis*) 81.5 and 89, females 78 and 87

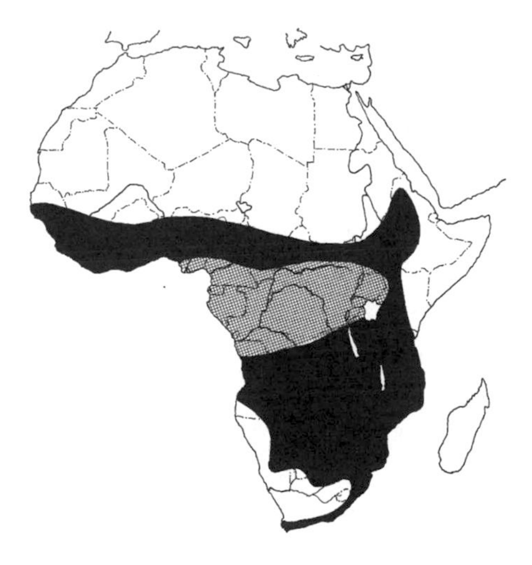

MAP 23. Breeding (filled) and nonbreeding or spare breeding ranges (hatched) of black cuckoo.

(Fry et al., 1988). Estimated egg weight 3.22 (Schönwetter, 1967–84). Egg:adult female mass ratio ~3.7%.

Identification

In the field: Like most cuckoos, this species is more easily recognized by its song than by its appearance. Adults are blackish above and heavily barred with blackish below, with brownish tones on the breast and faint tail-banding in adults. The similarly blackish pied cuckoo is crested and has a white wing patch. The forest race *gabonensis* is whitish on the under-tail coverts and lower flanks and belly, thus closely resembling the red-chested cuckoo, but has a reddish brown rather than grayish chin. Its songs differ from that of the red-chested cuckoo, typically consisting of three notes that last nearly 2 seconds, are irregularly spaced and rise in pitch, with the third note sometimes repeated. The phrase "ten past . . . FIVE," with the "five" upslurred and louder, provides an approximation of the song, as do "I'm so SAD" and "no more RAIN." The song at times may have only two or as many as four notes, all on roughly the same pitch, and the usual interval between phrases is about 2 seconds. A bubbling or "whirling" trill of 20–30 "ho" or "yow" syllables that gradually rise and then diminish in volume has also been reported. This may correspond to a similarly described "hurry-hurry" call of paired syllables that are repeated about 10 times, these notes at first increasing in loudness and then dying away. Calling occurs throughout the day and often extends late into the evening or sometimes even lasts throughout the night.

In the hand: This species has similar measurements to the red-chested cuckoo and generally resembles a darker version of this species. It is blackish above and similarly black below or, in *gabonenis,* densely barred with black and buff or black and rufous below. Adults have a much less conspicuous eye-ring than do red-chested cuckoos and have less white on the tail feathers, where there is some narrow, white barring and whitish tips to the rectrices. Females are similar to males but may show ventral barring below; immature individuals are almost entirely black, with no white tips on the rectrices, and the flight feathers are vermiculated rather than barred on their inner webs (Fry et al., 1988). Nestlings are pale brownish pink at hatching, but the skin darkens to purplish black within a day or so. The inside of the mouth is pink (rather than orange to scarlet, as in the similar red-chested cuckoo), the feet are pale fleshy-colored, and the rounded nostrils are prominent (Jensen & Clining, 1974).

Habitats

This species is associated with lowland rainforest (*gabonensis*) and also drier habitats such as mesic savannas and arid thornbush or dry riparian woodlands (*ciamosus*).

Host Species

There are only three known hosts of the black cuckoo, as listed in table 20. Of these, the boubous ("bush shrikes") of the genus *Laniarus* are the primary hosts, at least in southern Africa, with about 52 records for the 2 species listed. The African golden oriole is an important host in miombe woodland from Zimbabwe northward. Probable additional hosts include the white helmet shrike, the black flycatcher, and the southern puffback. Rowan (1983) listed only three boubous as authenticated biological hosts in southern Africa. These include the crimson-breasted boubou with 15 records, the southern (regarded by Fry et al. as conspecific with the tropical) with 14 records, and the tropical with 3 records. Several other alleged hosts were also listed.

Egg Characteristics

The eggs of this species are nearly elliptical, slightly glossy, and range from white to creamy or greenish, with reddish brown and lilac spots and freckles, especially near the rounder end. They tend to be slightly larger and blunter than those of their usual hosts. For example, the eggs of the crimson-breasted boubou average about 24 × 17.7 mm, but are nearly identical in both ground color and patterning (Jensen & Clining, 1974). The cuckoo's mean shell weight is 0.3 g, and its mean shell thickness is 0.11 mm (Schönwetter, 1968–84).

Breeding Season

South African records are for November–January, and in Namibia for November–March (mostly February). Zambia and Zimbabwe records are for September–January, and Angolan records are for October–November. West African records are few, including a September record for Cameroon, and some East African records (from Tanzania) are for March and April (Fry et al., 1988).

Breeding Biology

Nest selection, egg laying. Egg-laying behavior has not yet been observed, although a female cuckoo was once seen flying into a bush and coming out again a moment later carrying an egg in its beak. On investigation, the nest was found to have a single egg. On the next day a second, somewhat different-appearing egg was fund in the nest, which was presumed to be that of the cuckoo. These eggs that are removed by the cuckoo are apparently eaten (Rowan, 1983).

Incubation and hatching. The incubation period lasts 13–14 days, as compared with 16–17 days for the host boubous. In at least one case, eviction of the host's young occurred between 16 and 30 hours after the cuckoo's hatching, and in a second case during the second to third day after hatching (Jensen & Clining, 1974).

Nestling period. The nestling period is 20–21 days, and the young bird remains dependent on its foster parents for at least 19 days, and probably as long as 3 to 4 weeks (Jensen & Clining, 1974).

Population Dynamics

Parasitism rate. Payne and Payne (1967) estimated a nest parasitism rate of 2.1% for the tropical and southern boubous in southern Africa and a similar 2.6% rate for the crimson-breasted boubou in Zimbabwe. Jensen & Clining (1974) reported a much higher (36%) parasitism rate for 39 crimson-breasted boubou nests in Namibia over a 4-year period of study. In this area the cuckoo evidently concentrates on a single host species.

Hatching and fledging success. No information.

Host–parasite relations. No specific information exists, but the low average levels of parasitism in southern Africa suggest that the primary boubou hosts may not be seriously impacted in their reproductive efficiency by the presence of cuckoos.

INDIAN CUCKOO

(Cuculus micropterus)

Other Vernacular Names: Short-winged cuckoo.

Distribution of Species (see map 24): Asia from Pakistan east to China and Siberia and south to Sumatra, Borneo (Kalimantan), and Java.

Subspecies

C. m. micropterus: Breeds south to Burma, parts of Thailand, and Indochina; migratory.

C. m. concretus: Resident of the Malay Peninsula and southeast to Java and Borneo.

Measurements (mm)

12–15″ (30–33 cm)

C. m. micropterus, wing, both sexes 185.5–207; tail; both sexes 142–161 (Ali & Ripley, 1983). Wing (both sexes) 186–226; tail 144–170 (Medway & Wells, 1976). Wing:tail ratio ~1:0.77.

C. m. concretus, wing, both sexes 158–185 (Medway & Wells, 1976).

Egg, avg. of *micropterus* 24 × 18.2 (range 22.8–26 × 17–20) (Schönwetter, 1967–84). Avg. of *concretus* 23.6 × 17.7 (Becking, 1981). Shape index 1.3–1.46 (= broad oval to oval) (Becking, 1981). Rey's index 1.99 (Becking, 1981).

Masses (g)

Males 112–129 (avg. 119, $n = 6$); one female 119 (Neufeldt, 1966). One unsexed bird, 128 (Ali & Ripley, 1983). Two unsexed birds 114.6 and 121.8 (Becking, 1981).

MAP 24. Breeding (filled) and wintering (enclosed area) ranges of Indian cuckoo.

Estimated egg weight 4.2 (Schönwetter, 1967–84); 4.2–4.6 (Becking, 1981).
Egg:adult mass ratio 3.5%.

Identification

In the field: This is a medium-sized, mostly grayish cuckoo, with a blackish subterminal tail band, setting it apart from both the small cuckoo and the common cuckoo, which have more uniformly gray to blackish tails. The eye-ring is also duller and more grayish in the Indian cuckoo than in these other two species. Otherwise the three species are quite similar as adults, with barred black and white flanks and more uniformly gray back, head, and breast coloration. Sexes are similar in this species, but females have a paler gray throat, and the breast area has brownish or rufous tinge. Immature individuals are strongly barred with white or rufous-white on the head and neck, and the body feathers are broadly tipped with these colors. The tail is more heavily barred and more rufous than in adults. The species can be recognized by the male's distinctive song, a loud, clearly defined four-note whistle, sounding like "crossword puzzle" (also variously described as "What's your trouble?," "blanda mabok," "orange pekoe"), lasting about 1 second, with the final note lower in pitch than the first three. Becking (1988) provided a sonogram of this vocalization. These songs are repeated numerous times, each song sequence usually separated by 1 or 2 seconds, so that about 20–25 songs per minute are common. The female's courtship vocalization is an interrupted warble, similar to the female common cuckoo's "chuckle," but with some higher tones. It is commonly uttered on the day of copulation (Neufeldt, 1966).

In the hand: Distinguished from the other typical Asian *Cuculus* cuckoos by the presence of a broad, subterminal black band on the otherwise mostly gray tail. The species is almost exactly the same size as the common cuckoo, but its wings are substantially shorter (under 200 mm). Females differ from adult males in having a browner throat and breast. Juveniles differ from adults, as well as from young of other Asian *Cuculus* cuckoos, in being extensively spotted and barred with white and rufous white; their rectrices are strongly banded near the tip as in adults, but the tail is otherwise rufous rather than gray and is strongly barred throughout. Juveniles also have dark brown irises, salmon-orange gapes, and yellow eye-rings. Day-old nestlings have yellowish pink body skin, an orange-red gape and tongue, with the commissural junctions and tip of the tongue yellow. The skin soon darkens dorsally to deep gray with a violet shade by the fourth or fifth day, while the ventral area remains yellowish.

Habitats

This cuckoo is associated with fairly open subtropical to temperate wooded habitats up to 2300 m or even sometimes to 3700 m in the Himalayas, but it is probably more common at lower elevations. Subtropical forests such as oak and pine woodlands, at elevations of about 1500–2500 m, seem to be favored in northern India. In Nepal the species mainly occurs from 300 to 2100 m, but it has been recorded during summer to 3700 m. In Thailand it occurs in evergreen broadleaf as well as in pine forests. Very dense forests are apparently avoided; groves and sparse or stunted forests appear to be favored habitats (Neufeldt, 1966).

Host Species

Baker (1942) listed only two host species (striated laughingthrush, Indian gray thrush) with at least two parasitism records involving the Indian cuckoo, based on his collection of parasitized clutches (see table 12). However, many of these records (and perhaps all) probably involved eggs of the common cuckoo (Becking, 1981). Baker listed six additional putative host species represented by single clutches, including the Asian paradise flycatcher, the white-browed fantail, the golden bush robin, the Indian blue robin, the common stonechat, and Blyth's leaf warbler, but at least some of these cases also represent misidentified eggs of the common cuckoo. Baker also described an egg type that is specifically adapted to drongos, including the black drongo. Becking (1981) confirmed that drongos are indeed the primary hosts, including the black drongo and ashy drongo in India, and the racket-tailed drongo in Java, but doubted that such reputed hosts as paradise flycatchers, orioles, or spider hunters are parasitized by this species. In the Amur region of eastern Siberia, the brown shrike is the primary host, and the azure-winged magpie is another presumptive host, since this is a commonly parasitized species south of the Amur River in northeastern China (Neufeldt, 1966).

Egg Characteristics

According to Becking (1981), the blue eggs attributed by Baker to this species were misidentified, and the only certain eggs from India are those associated with drongo nests. Eggs of this cuckoo are broad-oval in shape, with pink to whitish pink ground color, spots or blotches of violet to carmine, and more grayish underlying markings, matching drongo eggs very closely. Those from India average somewhat larger (26.2 × 17.9 mm), and those from Java somewhat smaller (23.6 × 17.7 mm), than eggs from the Amur region of Siberia (25.2 × 19.5 mm). The eggs in that region are close mimics of the brown shrike, but are slightly smaller. They also are fairly similar in color to the eggs of the azure-winged magpie, various drongos, and even the streaked spider hunter.

Breeding Season

In northern and peninsular India most calling occurs from mid-March to early August, and egg-laying occurs between March and June. In Sri Lanka calling by males is loudest between March and May (Ali & Ripley, 1983). In the vicinity of Bejing, China, laying occurs in June, and likewise in the Amur region of eastern Siberia fresh eggs were recorded by Neufeldt (1966) throughout most of June.

Breeding Biology

Nest selection, egg laying. Like the crested cuckoos, it is probable that to at least some degree this species pairs during the breeding season. The male openly and conspicuously attempts to divert the attention of the nest owners, as the female makes her way to the nest. In one nest a cuckoo egg was deposited among a full clutch of six brown shrike eggs, the cuckoo evidently in that case did not remove a host egg. In two other cases only three to five host eggs were present, but these might have represented still uncompleted rather than depleted clutches (Neufeldt, 1966).

Incubation and hatching. The incubation period is about 12 days, as compared with 14 days for the brown shrike. By the second day after hatching, the ejection reaction is apparent in the cuckoo chick, but ejection behavior did not occur until the third and fourth days in the nests observed by Neufeldt.

Nestling period. Neufeldt (1966) observed that by day 18 after hatching, the young cuckoo was perching on the edge of the nest, and when 21 days old it left the nest and perched on a stump. It was still not capable of active flight at that age. However, it was flying by 30–40 days, and by 45 days of age it had acquired its complete juvenal plumage.

Population Dynamics

Parasitism rate. Little information is available. Neufeldt (1966) described three parasitized nests of brown shrikes, from among a total of 50 shrike nests that she located.

Hatching and fledging success. No information.

Host–parasite relations. Neufeldt (1966) noted that the female brown shrikes fed their cuckoo chicks less willingly than did the males and sometimes did not feed them at all. The young cuckoos soon learned to recognize the males and responded only to their calls.

COMMON CUCKOO

(Cuculus canorus)

Other Vernacular Names: Asiatic cuckoo (*subtelephonus*), Chinese cuckoo (*fallax*), Corsican cuckoo (*kleinschmidti*), Eurasian cuckoo (*canorus*), gray cuckoo, Iberian cuckoo (*bangsi*), Japanese cuckoo (*telephonus*), Khasi Hills cuckoo (*bakeri*).

Distribution of Species (see map 25): Breeds throughout most of Palearctic, breeding from Europe east to Kamchatka and south to northern Africa, Pakistan, Burma, China, and Japan. Winters south to sub-Saharan Africa and tropical Asia, at least occasionally reaching New Guinea.

Subspecies

C. c. canorus: Europe and western Siberia.
C. c. telephonus: Northeast Asia and Japan.
C. c. bangsi: Iberia, adjacent northern Africa.
C. c. kleinschmidti: Corsica, Sardinia.
C. c. subtelephonus: Transcaspia to western Chinese Turkestan.
C. c. johanseni: Central Asia.
C. c. fallax: Central and southern China.
C. c. bakeri: Northwestern China, Burma, and Indochina.

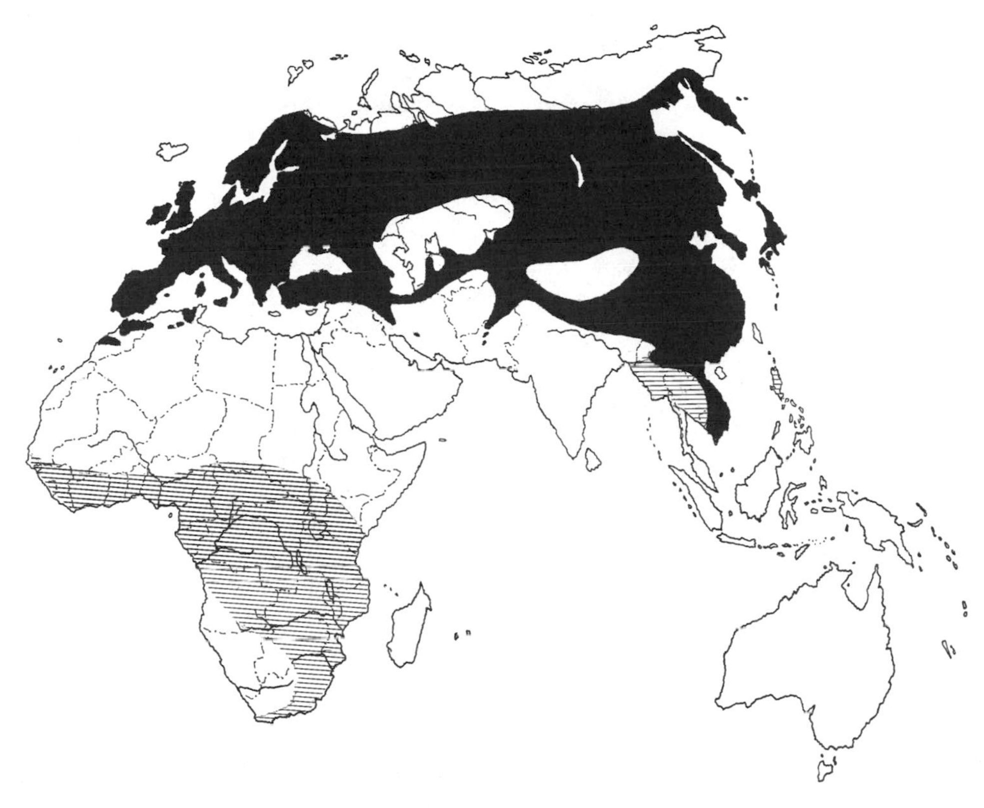

MAP 25. Breeding (filled) and wintering (hatched) ranges of common cuckoo.

Measurements (mm)

13″ (32–33 cm)

C. c. bakeri, wing, both sexes 225–235; tail, both sexes 158–180 (Delacour & Jabouille, 1931).

C. c. canorus, wing, males 215–232 (avg. 223), females 200–224 (avg. 215); tail, both sexes 165–184 (avg. 172) (Fry et al., 1988). Wing:tail ratio 1:0.75–0.8.

C. c. telephonus, wing, males 204–220 (avg. 216), females 184–216 (avg. 205) (Fry et al., 1988).

Egg, avg. of European *canorus* 22.3 × 16.5; British *canorus* 23.05 × 17.23; *bangsi* 21.6 × 16.3; *telephonus* 23.6 × 18; *bakeri* 23.1 × 17.6; *subtelephonus* 23 × 16.7. Overall range of species 19.7–26.9 × 14.7–19.8 (Schönwetter, 1967–84). Shape index 1.31–1.37 (broad oval to oval). Rey's index: *canorus* 1.6, *telephonus* 1.63, *bangsi* 1.68, *subtelephonus* 1.82, *bakeri* 1.84.

Masses (g)

C. canorus, Southern Africa, males 103–125 (avg. 115), females 134, 142. Avg. of 84 British males 117, of 12 females 106 (Fry et al., 1988). British males have monthly means of 114–133, and females 106–112 between April and July (Cramp, 1985).

C. c. subtelephonus, spring to fall weights, males 81–128, females 81–94. Males from China 71–127 (avg. 96, $n = 35$), females 70–138 (avg. 97, $n = 14$) (Cramp, 1985).

Estimated egg weight, *bangsi* 3.05, European *canorus* 3.22, *subtelephonus* 3.55, British

canorus 3.6, *bakeri* 3.95, *telephonus* 4.1 (Schönwetter, 1967–84).
Egg: adult female mass ratio ~ 3.0% (*canorus*) to 3.6% (*subtelephonus*).

Identification

In the field: Over its European range, this is the only bird with the familiar "cuc-koo" song, which is uttered only by breeding males at a regular rate of about once per second. This distinctive song (like that of a cuckoo clock) has the second syllable lower in pitch. Females of this species, as well as those of the Indian and Oriental cuckoos, all utter essentially similar "water-bubbling" notes, a hinnylike call of about 15 descending notes lasting nearly 3 seconds. Males produce harsh chuckling notes (the probable counterpart of female bubbling), and several other calls including hissing, growling, and mewing sounds have been described.

In Africa it is difficult to visually distinguish wintering birds from the resident African cuckoo (note the amount of yellow on the upper mandible). In *canorus* the mandible is mostly black terminally, with yellow only basally apparent, especially around the nares (fig. 28, 29). In Asia this species is equally difficult to distinguish from the Oriental cuckoo; the grayish barring on the carpal feathers at the wrist helps identify the Eurasian species—in the Oriental species this area is entirely white. Adult females sometimes occur as a "hepatic" plumage morph, in which the body plumage is heavily barred with rufous and dark brown, and degrees of brown hues may occur. Normal gray-morph adult females have only a rufous tinge on the breast, and sometimes even this is lacking. Hepatic morphs may perhaps rarely occur in postjuvenile males (Voipio, 1953), although this needs further study. Immature individuals resemble hepatic-morph females but have conspicuous white nape patches.

In the hand: In-hand identification of this species poses no problem except in Africa and perhaps in southeastern Asia. In the latter area it may be distinguished from the similar-sized Indian cuckoo by its uniformly blackish gray tail (not a gray tail with a black subterminal band), and sometimes (but not always) can be distinguished from the Oriental cuckoo by the latter's white carpal marking at the bend of the wing and several other minor traits mentioned in that species' account. In Africa it is distinguishable from the African cuckoo by the bill color and bill shape and by the amount of white spotting or barring on the outer tail feathers (see following species account). Females of the usual gray (eumelanin-based) plumage morph are usually but not always somewhat more barred and buffy to rufous-tinted on the upper breast than are adult males. There is also a relatively rare rufous or hepatic plumage morph among adult females, in which the overall dorsal plumage is generally strongly barred with rufous pheomelanins. Voipio (1953) has suggested that the occurrence of this rufous morph plumage phenotype depends on a single recessive gene that is typically expressed only in females, but occasionally is also apparent in young males. Immature individuals of both sexes include a rufous- and chestnut-rich hepatic morph (mostly present in females), as well as a (probably intergrading) morph that averages more brownish and less rufous in overall hue, and a third even more grayish extreme.

Newly hatched young are naked, with flesh-colored (at hatching) to blackish (after 3 days) skin and pinkish feet and a grayish brown to brown iris. They have a pale orange (at hatching) to orange-red (after a few days) gape, with yellow (initially) to orange (after a few days) mandibular flanges at the commissural junction. As juvenal feathering develops, there are white fringes present at their tips, and a rather persistent white nape patch (and often a mid-crown patch) develops. Feathers begin to emerge by the fourth day, and the eyes start to open on the fifth day. By 9 days old, the bird is well feathered, and by 2 weeks it appears fully feathered (see fig. 30). Juveniles average darker on their upperparts than adults, have no blue-gray on the breast, and the rufous barring on the rectrices and remiges is more broken. Juveniles vary greatly in plumage

FIGURE 28. Profile sketches of five cuckoo species in the *canorus* species-group: an immature (A) and adult (B) of African cuckoo, an adult male (C) and female (D) of oriental cuckoo, and adults of the Madagascar cuckoo (E), lesser cuckoo (F), and Indian cuckoo (G). Morph variations among their eggs and diagrams of typical song phrases are also shown.

FIGURE 29. Ontogeny of the Indian cuckoo, including its egg (A) and that of its brown shrike host (B), plus a 2-day-old cuckoo (C). A 3-week-old cuckoo with a brown shrike is also shown (D). After photos by Neufeldt (1966).

FIGURE 30. Arrival, egg-stealing, and egg-laying behavior of the common cuckoo at a reed warbler nest. After photos by Wyllie (1981).

color, from grayish to nearly as rufous as rufous-morph adult females, but rarely are the rump and tail coverts so uniformly rufous as in rufous-morph females. The iris color of juveniles is more brownish or grayish, the eye-ring and base of mandible a paler yellow, and the anterior parts of the mandible a duller brown than in adults (Cramp, 1985).

Habitats

This species is extremely diverse in its habitat usage, avoiding only deserts and arctic tundra in the western Palearctic. Forest edges, forest steppes, heaths, open woodlands, wetlands with emergent vegetation, and various human-modified habitats are all used. At least in Britain, lowland elevations (those that support the largest number of important host species) are favored over higher ones (those at least 300 m elevation), as are habitats providing song posts, look-out sites, hiding places, and the presence of suitable hosts (see table 18). However, in central Honshu, Japan, the species is widely distributed over elevations from less than 200 m to nearly 2000 m, with the highest estimated densities (6.7 birds/km^2) occurring at about 1200–1400 m (Nakamura, 1990).

Host Species

A list of 20 major hosts of the common cuckoo and their corresponding breeding traits from the Indian region is presented in table 11, and 4 African hosts are listed in table 20. Twenty-six major European and Japanese hosts and their breeding traits are listed in table 17, partly on the basis of Baker's (1942) egg collection and partly on the basis of more recent literature. Wyllie (1981) has similarly classified 56 host species in Europe as to their relative host-frequency status (frequent, occasional, rare). At least 22 of these are known to have served as fostering or "biological" hosts (i.e., those that have reared cuckoos successfully). An additional 27 species were identified by Wyllie as probably having been victimized on occasion but insignificant as potential hosts. Thirteen other reputed host species were discounted by Wyllie as representing erroneous records.

Lack (1963) provided a the first summary of cuckoo hosts in England using available nest records through 1962. Glue and Morgan (1972) identified the 26 most commonly exploited host species (including 17 biological hosts) in Britain according to habitat and elevational characteristics, based on an analysis of 613 nest-record cards through 1971. They concluded that at low altitudes, the reed warbler is the cuckoo's most important host (accounting for 35% of the parasitism records at this elevation and 14% of the total records). At intermediate altitudes (60–240 m), the hedge accentor becomes most important, especially on farmlands, woodlands, and around human habitations. Numerically it also is the most important host in the overall sample, representing 49% of the total parasitism records. The meadow pipit is the cuckoo's primary host at higher altitudes. This zone represented only 9% of the total records, but the pipit was virtually the only host on heather moors at elevations above 310 m and was responsible for 14% of the total records.

More recent changes (to 1982) in host usage in Britain, at least of the six principal host species, were evaluated by Glue and Morgan (1984) and by Brooke and Davies (1987). The latter authors documented some recent increases in the overall rate of reed warbler exploitation. However, they also reported that significant declines have occurred in rates of parasitism among several other host species (hedge accentor, European robin, pied wagtail), but they did not detect any new host species. Host specificity and associated egg mimicry may be the result of traits associated with innate predispositions affecting host-selection tendencies or might result from nestling imprinting on foster parents, natal areas, or specific habitat characteristics of the natal environment. Some recent observations on nestlings raised in natural and artificial environments favor the last of these possibilities (Teuschl et al., 1994).

Nakamura (1990) provided a list of 28 known Japanese hosts, including 12 major hosts. He estimated parasitism rates for 10 of these species in central Honshu, Japan, where 20 host species have been documented. By comparison, in northern Japan (Hokkaido) a total of 14 hosts have been documented, and in southern Japan (Kyushu) only 5 host species have been recorded. Nakamura also documented the acquisition of one major new host species (the azure-winged magpie) within the past 50 years. Indeed, in central Honshu the primary hosts now appear to be the azure-winged magpie and the great reed warbler.

Egg Characteristics

See chapter 2 for a description of egg polymorphism and the evolution of host-specific gentes and associated egg-mimicry in this species. Eggs of this species vary only slightly in measurements and are broadly oval in shape. They are, however, highly variable in color. The ground color varies from white or buffy to varying degrees of blue; the blue egg morphs are generally unspotted, whereas the white to buffy morphs are variously speckled or spotted with gray to brown. A museum study of about 12,000 European clutches indicated that about 15 different egg morphs have evolved there, which individually resemble the eggs of the most frequently used hosts (Røskraft & Moksnes, 1994). The eggshell averages 0.21–0.26 g and is 0.09–0.1 mm thick (Schönwetter, 1968–84).

Breeding Season

In Europe and temperate Asia the common cuckoo breeds during spring and summer, with May and June being the peak period for egg records. In Japan the breeding season extends from late May to July, rarely extending into August, and peaking in June. In India the breeding season is primarily from April or May to June or July, and in northwestern Africa the relatively few egg records are for April and May.

Breeding Biology

Nest selection, egg laying. An extended discussion of host selection was presented earlier in chapters 3 and 4. Egg-laying is timed to coincide with the peak periods of breeding in host species, and at least in Britain the egg-laying period lasts about 12 weeks (Wyllie, 1981). Individual females probably lay over a 6- to 7-week period (mean of 9.2 eggs laid over a period of 27.7 days, with a maximum of 25 eggs per female and a maximum 54-day laying period, $n = 46$ cases). From 2 to 23 eggs are laid in uninterrupted clusters or series, these series sometimes termed "clutches."

Within-clutch eggs are produced at approximate 2-day intervals, which are separated by intervening between-clutch intervals of about 4 days (mean 4.8 days, $n = 10$). Three marked males were found by Wyllie (1981) to sing and court females as yearlings, and there is a record of a yearling female carrying an oviducal egg, so at least some birds must breed in their first year. Indeed, Nakamura (1994) reported that almost half of the breeding birds in a Japanese study were 1 year olds, and radio-tracking revealed that each female parasitized a single host species. As noted earlier, females are able to land on a nest, take a host egg from it, and deposit their own egg (fig. 31), all in the matter of a few seconds.

Incubation and hatching. Wyllie (1981) reported that the mean laying-to-hatching interval of nine eggs was 12.4 days (range 11.5–13.5). It is likely that incubation begins within the oviduct of the female cuckoo, which probably represents an approximate additional 24-hour incubation period. Wyllie suggested that the cuckoo's thick shell may influence the rate of heat loss by cuckoo eggs as compared with their hosts, as would their typically somewhat larger volumes.

Nestling period. The ejection reflex develops early in this species and is best exhibited between 8 and 36 hours after hatching. This ejection behavior may occur while the foster parent is brooding, and these host birds make no effort to help or retrieve their own endangered offspring. When, as rarely occurs, two cuckoos hatch in the same nest, the first to hatch will usually eject the other. In the rare instances where a host young survives the cuckoo's ejection period, it is usually smothered to death by the rapidly developing cuckoo. Nestling cuckoos respond to disturbance with feather-ruffling and gaping (fig. 32); on being handled they void vile-smelling fluid feces. After about 17 days (range 13–20) the chick is likely to leave the nest, although it cannot fly until at least 21–22

FIGURE 31. Egg (A) and ontogeny of the common cuckoo, including a 2-day-old chick (B), resting poses and gape pattern of 7-day-old chicks (C), and juveniles at 12 (D), 16 (E), 21 (F), 29 (G), and 37 (H) days (after Heinroth & Heinroth, 1928–32). Also shown are the wing covert markings of the oriental (I) and common cuckoos (J).

FIGURE 32. Nestling common cuckoo social behavior, including threat-gape (A), begging (B, C), and receiving food from a host parent (D). After sketches in Glutz and Bauer (1980) and (D) a photo by E. Hosking (in Reade & Hosking, 1967).

days. There is another 2- to 3-week period of dependency on the adults (average age at independence from reed warbler foster parents, 33 days), which at times may be extended to as long as 45 days (Wyllie, 1981).

Population Dynamics

Parasitism rate. Some information on parasitism rates is summarized in tables 8 and 14, and in the "Host Species" section above. As already noted, substantial changes in parasitism rates have recently occurred among various species in Honshu, Japan (Nakamura, 1990). This includes a great increase in parasitism rates of the azure-winged magpie in Nagano Prefecture (from none in the early 1960s to 79.6% in 1988), and less marked increases in rates for the great reed warbler (from none to 19.5%) and bull-headed shrike (from none to 13.1%). With the vastly increased use of the azure-winged magpie as a host (the recent parasitism rates being the highest reported for any species of brood parasite), the cuckoo has moved its breeding zone downward altitudinally to encompass the entire breeding zone of the magpie. Correspondingly, the local parasitism rates of the Siberian meadow bunting have declined during this same period, and this species is now only rarely parasitized in that area.

Less marked but significant changes in parasitism rates have also occurred in Britain over a similar time span, judging from the analysis of Brooke & Davies (1987) of 1061 records of parasitism occurring between 1939 and 1982. Five species exhibited significant changes in parasitism rates during that period, with the rates of the reed warbler more than doubling (from about 2.7% before 1962 to about 7.3% for 1972–82). Correspondingly, those of four other host species (hedge accentor, meadow pipit, pied wagtail, and European robin) all declined during this period. For example, the incidence of hedge accentor parasitism declined from about 2.7% during 1939–61 to only about 1.5% in 1972–82, and the meadow pipit from about 2.5% to 2.2% during these same periods. The incidence of minor host usage also declined over the same overall time span.

Hatching and fledging success. Some of the best comparative data on hatching and fledging success come from Britain, where Brooke & Davies (1987) have summarized data on the six principal hosts (those having at least 10 records of parasitism located during the cuckoo's egg stage). The most frequent host, the reed warbler (with 442 records of egg parasitism), resulted in a relatively high 31.9% breeding (egg-to-fledging) success rate for the cuckoo. The authors suggested that the high breeding success rate of the cuckoo with reed warblers may have accounted for the recent increase in parasitism rates of reed warblers, although there was no evidence of improved egg mimicry of reed warbler eggs occurring throughout this 50-year period. The hedge accentor, with 281 egg parasitism records, provided a 26.5% success rate for cuckoos, and the meadow pipit, with 52 records, produced a 23.1% success rate. The collective breeding success rate for cuckoos among 833 nest records was 29.2%, which is not dissimilar to breeding success rates estimated for various nonparasitic insectivorous passerines (e.g., Payne, 1977b). If, as estimated, a female cuckoo produces an average of 8.21 eggs per breeding season, the annual production of fledged young under these circumstances should be 2.4. Judging from the unpublished banding data of D. C. Seel (cited by Brooke & Davies, 1987), there may be an approximate 28% survival rate of birds between fledging and the end of their first year, followed by an approximate 52% annual survival rate for adults thereafter. This would mean that only about 1% of the cuckoos surviving to breed during their first year of adulthood would still be alive 8 years later (see fig. 13). Nakamura (1994) reported that only 6.5% of 92 birds that were banded as nestlings returned the following year, and the subsequent mean life spans were 2.14 and 1.37 years for males and females, respectively. These life spans would suggest annual adult survival rates of about 60% and 50% for the respective sexes.

Wyllie (1981) found that only 22% of the 74 hatched cuckoos whose histories he was able to follow survived to fledging. A fledging rate of 51% was reported for West Germany by Glutz & Bauer (1980). Payne (1973b) estimated 62% laying-to-hatching success, and a 27% laying-to-fledging success rate. Perhaps the highest fledging rates so far reported for the species is 66% of the cuckoos hatched by various hosts in a local English population (Owen, 1933). Glue and Morgan (1972) provided crude reproductive success rate estimates for active, parasitized nests (regardless of the breeding stage at the time the nest was found, thus including already hatched chicks but excluding abandoned nests or rejected eggs) as 76% in 62 nests of the meadow pipit, 59% in 64 reed warbler nests, and 48% in 257 hedge accentor nests. Predation was responsible for most cuckoo losses, both of eggs and young. A few other representative hatching and fledging success rates as reported from various host species and regions were summarized earlier (table 24).

Host–parasite relations. Because hosts essentially never raise any of their own young when successfully parasitized by common cuckoos, the effective level of lost fecundity is roughly equal to the incidence of nest parasitism (see table 22). At least in Britain those species that are currently used as principal hosts are the ones that tend to be nondiscriminating as to egg characteristics. Species now used only rarely as hosts are more discriminating in their tolerance for foreign eggs, suggesting that they too might have been more important hosts in the past, but their increased egg-rejection rates have thus forced the cuckoo to turn to new, less discriminating hosts (Brooke & Davies, 1987). Davies (1992) concluded that the hedge accentor may be such a recent host, inasmuch as this species exhibits no apparent egg discrimination of cuckoo eggs, and the cuckoo subpopulation adapted to parasitizing hedge accentors does not lay mimetic eggs.

AFRICAN CUCKOO
(Cuculus gularis)

Other Vernacular Names: Gray cuckoo, yellow-billed cuckoo.

Distribution of Species (see map 26): Sub-Saharan Africa from Senegal and Somalia south to South Africa.

Measurements (mm)

13″ (33 cm)

Wing, males 205–223 (avg. 213), females 202–218 (avg. 210). Tail, males 152–168 (avg. 160), females 146–166 (avg. 155) (Fry et al., 1988). Wing:tail ratio 1:0.74.

Egg, avg. 23 × 16.7 (range 23.3–23.6 × 16.5–16.9) (Schönwetter, 1967–84). Shape index 1.37 (= oval). Rey's index 1.82.

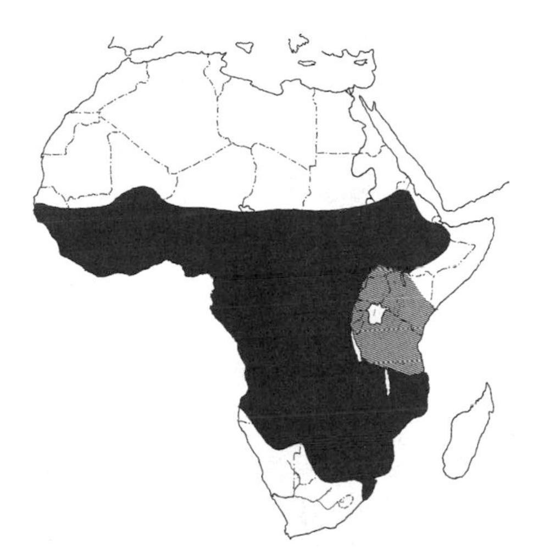

MAP 26. Breeding (filled) and nonbreeding (hatched) ranges of African cuckoo.

Masses (g)

Males 95–113 (avg. 104, $n = 6$), females 96 and 99 (Fry et al., 1988). Estimated egg weight 3.3 (Schönwetter, 1967–84). Egg:adult female mass ratio ~3.4%.

Identification

In the field: In its African range, this species is likely to be confused with wintering individuals of the common cuckoo, but unlike the common cuckoo, yellow color on the upper mandible extends well beyond the nostrils. The male's typical breeding song is an "oo-oo" that has the second syllable very slightly higher pitched (not lower, as in *canorus*). The song is reported to be more monotonous than is the common cuckoo's. It is uttered at an average rate of about 20 songs per minute and in bursts from 3 to 4 song phrases to as many as 50 or more. Females lack a hepatic plumage morph, and immature individuals are more grayish than young common cuckoos, although both have white nape patches. The outer rectrices are barred, rather than spotted, with white in this species. Immature birds also have white-fringed upperparts and more white on their rectrices than do young common cuckoos; in both species the bills of young birds are mostly blackish, but the color grades into somewhat lighter (more yellowish) tones toward the base.

In the hand: Compared with the common cuckoo, the native African species has a heavier and broader bill (average 8 mm wide at the base vs. 7.25 mm in the common cuckoo) with an orange-yellow base. The outer rectrices tend to be barred (with five to six transverse white bands), rather than spotted, and the flank barring tends to be finer than in the common cuckoo. Females are similar to males but may have shadowy barring extending forward to the gray breast area and often have a buff or tawny wash on the throat and upper breast. Immature birds are grayer than young common cuckoos (no brown or hepatic morphs are known), the throat is often paler, and the white spotting on the outer rectrices is larger. The white edges of the upperpart feathers form a distinct white nape patch. The birds are also lighter and grayer overall than young red-chested cuckoos. Young nestlings have dark purple to blackish skins, a bright orange gape, and yellow feet. Just before fledging, the legs and feet are yellow and the gape still bright orange, the iris is black, and the brownish bill is becoming yellowish toward the base (Tarboton, 1975; Fry et al., 1988).

Habitats

This cuckoo inhabits savanna or similar open woodlands, especially acacia savannas, but avoids both open plains and dense forests.

Host Species

Only two species are known hosts of this African endemic: the fork-tailed drongo, with more than 25 records, and the yellow-billed shrike, for which the evidence is much poorer and perhaps is an accidental host (Fry et al., 1988). Rowan (1983) listed only the drongo as an authenticated host.

Egg Characteristics

Eggs of this species are blunt oval in shape and range in color from white or cream to pinkish buff in ground color with mauve and brown spots and blotches, especially near the more rounded end (Fry et al., 1988). Sometimes they are indistinguishable from those of the fork-tailed drongo, but tend to be slightly smaller. The drongo's eggs are highly variable, ranging from white to pinkish or salmon in ground color, with brown or reddish brown freckles, spots or blotches. Tarboton (1986) judged that these cuckoos tend to lay their eggs in nests having eggs closely matching their own in pattern and ground color. The eggshell averages 0.21 g in mass and 0.09 mm in thickness (Schönwetter, 1967–84).

Breeding Season

In West Africa (Ghana to Cameroon and Zaire) egg records are for January–April, and in southern Africa are mostly for the latter half of the year. These records include September–December (Zambia, Zimbabwe), October–December (South Africa), December (Namibia), and December and February (Angola) (Fry et al., 1988).

Breeding Biology

Nest selection, egg laying. Tarboton (1986) stated that in his study area the drongos began nesting in September and continued until December, with a peak in October. He observed parasitism in late October and early November. Six eggs were laid over a 12-day period, quite possibly by the same female. All these eggs were laid within an area of about 100 ha, and all the eggs were nearly identical in appearance. In contrast to the situation in the European cuckoo, it has been reported that the male may participate in egg laying by allowing the host to mob him, while the female silently approaches the nest to lay her egg (Pitman, 1957). Tarbarton (1986) found some cuckoo feathers under the nests of drongos, suggesting that nest defense behavior might be strong. He observed that the cuckoos invariably laid their eggs in uncompleted clutches of drongos, so an early hatching of the cuckoo would be assured.

Incubation and hatching. The incubation period is believed to be between 11 and 17 days, most probably about 12 days as in the European cuckoo. The comparable period for the host drongo is 16 days (Tarboton, 1975; Fry et al., 1988).

Nestling period. The nestling period of one bird was 22 days (Tarboton, 1975), but fledglings remain dependent on their host parents for some time thereafter (Fry et al., 1988).

Population Dynamics

Parasitism rate. Payne & Payne (1967) estimated a 1.3% rate of parasitism of fork-tailed drongo nests in southern and central Africa. Tarboton (1986) located 27 nests of this species and found that 6 of them were parasitized (22% parasitism rate). At a Zambian location, an estimated 8% of the drongo nests were parasitized (Fry et al., 1988). The breeding synchrony between the drongo and cuckoo is not perfect in Transvaal, so that early nesting efforts by drongos escape cuckoo parasitism (Tarboton, 1975).

Hatching and fledging success. Tarboton (1986) reported that of six cuckoo eggs he was able to follow, four were rejected by the host drongos and two were lost to predation, together with the host's eggs. At another nearby location, one of seven drongo nests was parasitized, and it produced a fledged cuckoo chick. By comparison, the average nesting success of the drongos was 58% during that same year (14 of 24 nests producing fledged young), which was higher than the rate for several other local open-nest passerines (about 40%). In an earlier study, Tarboton (1975) reported a breeding success rate of 38% (49 eggs producing 19 fledged drongos).

Host–parasite relations. The impact of this cuckoo on fork-tailed drongo breeding success would seem to be quite low, given the strong nest-defense behavior shown by the drongos and their apparent ability to recognize even those cuckoo eggs that are almost identical in appearance to their own. Tarboton (1986) judged that the drongo is nevertheless chosen as a host because it is common and because the drongos are very effective parents.

ORIENTAL CUCKOO

(Cuculus saturatus)

Other Vernacular Names: Blyth's cuckoo, Himalayan cuckoo, Indonesian cuckoo, saturated cuckoo, Sunda cuckoo.

Distribution of Species (see map 27): Palearctic from Russia east to Siberia, Mongolia, Korea, and Japan. Also in peninsular Malaysia, Sumatra, Borneo, Java, and the Lesser Sundas. Winters south to India, southeast Asia, Australia, and throughout the East Indies.

Subspecies

C. s. saturatus: Southern Himalayas to southern China. Winters to India, southeastern Asia and East Indies.

C. s. horsfieldi: Central and eastern Asia, Japan. Winters southeastern Asia, East Indies, and New Guinea region.

C. s. lepidus: Malaysia, Sumatra, east to Timor Island; nonmigratory.

C. s. insulindae (previously considered a race of *poliocephalus):* Borneo; nonmigratory.

Measurements (mm)

10–13″ (26–34 cm)

C. s. horsfieldi, wing, males 198–211 (avg. 210), females 191–209 (avg. 198); tail, males 159–175 (avg. 167), females 150–164 (avg. 156) (Cramp, 1985). Wing:tail ratio 1:0.8.

C. s. lepidus, wing, males 145–160, females 138–145 (Medway & Wells, 1976).

C. s. saturatus, wing, males 184–195 (avg. 190), females 174–181 (avg. 177) (Cramp, 1985).

Egg, avg. of *horsfieldi* 20.3 × 14.5 (range 19–21.5 × 13.7–15.2); *saturatus* 21.2 × 14 (range 20–25.4 × 12–16.2); *lepidus* 20.4 × 14 (range 18.8–21.5 × 13.5–14.7) (Schönwetter, 1967–84). Shape index 1.4–1.51 (= oval to marginally long oval). Rey's index: *horsfieldi* 1.96; *saturatus* 2.04.

Masses (g)

One male 105, one female 72 (Ali & Ripley, 1983). Spring and summer specimens from Siberia, males 91–128, females 75–89 (Cramp, 1985). Estimated egg weight of *saturatus* 2.2, of *horsfieldi* 2.3 (Schönwetter, 1967–84); also of *saturatus* 2.89 (Becking, 1981). Actual egg masses avg. ~1.9

MAP 27. Range of oriental cuckoo.

(Cramp, 1985). Egg:adult female mass ratio (*horsfieldi*) ~2.8.

Identification

In the field: This species is similar in plumage to the common cuckoo, and singing and concealing postures of these two species are nearly identical (fig. 33). The oriental cuckoo is most easily recognized by the male's breeding song, which over mainland Asia is a four-noted "hoop-oop-oop-oop" vocalization somewhat similar to the "hoo-poo" or "hoo-poo-poo" of a hoopoe. At close range, a soft preliminary and higher-pitched grace-note may be heard. In Borneo, Sumatra, and Java the song consists of a preliminary grace-note followed by two or three additional "hoop" notes. While wintering in Australia the species is rather quiet, but sometimes utters a subdued trill of three notes that are repeated in a rising crescendo. Harsh, repeated "gaaak" notes may be uttered while foraging. Females sometimes utter a rapid and loud quavering or bubbling call.

Both sexes are similar in appearance, and both have larger barrings on the flanks and underparts than occur in *canorus.* Rarely, females exhibit a rufous or hepatic plumage morph, and these birds resemble the common cuckoo's corresponding morph, but have barred rumps. The eye is whitish in adults, with a yellow eye-ring, and the feet are bright yellow. Immature individuals are similar to the hepatic plumage variant of adult females, but have less bright soft-part colors. In the carpal area of both immature birds and adults, the wrist feathers are pure white, rather than barred with gray,

FIGURE 33. Adult cuckoo social behavior, including concealing (A) and singing (B) postures of a common cuckoo, plus singing (C) and defensive gaping (D) behavior of the oriental cuckoo. After sketches in Cramp (1985).

thus helping to distinguish them from common cuckoos.

In the hand: This species and the common cuckoo are extremely similar in size and appearance. Generally, this species has a grayer chest but a darker back, so that the head seems paler. There is also somewhat broader barring on the underparts with barring extending over the thighs and vest, the under-tail coverts are blotched, not finely barred, with black, there is a more buffy or yellow tone to the underparts, and a clear white patch at the carpal joint. Males have a yellow iris, but females may have variable yellowish to brown irises. Both sexes have yellow to cream-colored eye-rings as adults. Females exist in a (typical) malelike gray morph and also rarely occur in a barred rufous or hepatic plumage morph. The gray morph females have coarser neck and breast markings (tinted with rufous) than the common cuckoo, and the hepatic morph females have considerable broader and blacker barring throughout. Likewise, juveniles exist in gray and hepatic morphs, which also on average have bolder barring that do the counterpart plumages of the common cuckoo. As with the common cuckoo, a white nape patch is usually (but not always) present in young birds. Nestlings have a vermilion or orange gape, a dark brown iris, and the edges of the bill are yellowish white (Cramp, 1985). Older birds have a brown bill, with the base of the lower mandible somewhat green-tinged. The iris color of immature birds is paler than in adults, and the eye-ring is green-cream rather than yellow or creamy.

Habitats

Hilly, wooded country in subtropical to temperate zones is favored, at elevations of 1500–3300 m in India and Nepal. These very high elevations probably mostly involve use of coniferous forests, but oak forest and oak–rhododendron forests are also used. The birds may range higher altitudinally than the Indian cuckoo and often occur on northern slopes and in conifers. However, more open forests and subtropical woodlands near streams seem to be preferred. In Japan the species occurs mostly in broadleaved foothill forests from about 600 m, but ranges to 2000 m in subalpine mixed forests. In New Guinea it is a wintering migrant and mainly occurs from sea level to 1500 m (but a presumably migrating specimen was also once found dead at 4400 m). On wintering areas of Australia it occupies open woodlands, gallery forests, scattered trees in open land, and pockets of rainforest. It also uses mangroves and coral cays during migration.

Host Species

Baker (1942) listed 18 host species, based on his collection of parasitized clutches, but 9 of these were of single species records. Only the Blyth's leaf-warbler ("Khasia crowned willow warbler") had more than six host records, and this single species accounted for about 40% of the total host records reported by Baker. In the northwestern Himalayas it appears that *Phylloscopus* warblers are nearly the only species parasitized, but there is a single record of parasitism of the rufous-bellied niltava. Farther east, in Malaysia and Indonesia, flycatcher warblers of the genus *Seicercus* are often parasitized (Roberts, 1991). A single record of parasitism from Malaysia involving the chestnut-crowned warbler, previously attributed to the drongo cuckoo, is also actually a record for this species (Becking, 1981). Eurasian host records include the arctic warbler, chiffchaff, willow warbler, Pallas' warbler, and olive-backed pipit (Cramp, 1985), and in eastern Asia the chiffchaff, arctic, greenish, and lemon-rumped warblers, plus the eastern crowned leaf-warbler are reputed hosts (Balatski, 1994). In Japan the eastern crowned leaf-warbler and the stub-tailed bush warbler are the major hosts on Honshu, but on Hokkaido (where lesser cuckoos are absent) it regularly parasitizes the Japanese bush warbler. At least 10 minor Japanese hosts are also known (Brazil, 1991).

Egg Characteristics

Eggs range from oval (or elliptical) to nearly long-oval (or long-elliptical) in shape, with a white or whitish buff ground color and speckles, spots, or fine lines of reddish brown, forming a ring around the more rounded end (Becking, 1981). These eggs are fairly close mimics of several *Phylloscopus* leaf warblers, such as the western crowned warbler, a common host in Pakistan (Roberts, 1991).

Breeding Season

In the Indian subcontinent breeding probably occurs from March to August, judging from the period of maximum vocalizations (Ali & Ripley, 1983). In Pakistan calling occurs from early May to early July (Roberts, 1991). Breeding in Nepal extends from March to August (Inskipp & Inskipp, 1991). In the Malay Peninsula advertisement calling occurs from January to July, and eggs have been found between February and May (Medway & Wells, 1976). In Japan breeding occurs from April to late June (Brazil, 1991).

Breeding Biology

Nest selection, egg laying. No detailed information exists. The small warbler hosts of this species are reflected in the relatively small size of the eggs laid by the cuckoo. Indeed, the host's nests are so small, and their access is often so restricted, that much speculation has gone on as to how the female cuckoo introduces her egg into such a tiny target. Rather than holding it in its bill, and thus dropping it into the nest, as has often been speculated, it is more likely that the egg is introduced directly from the female's cloaca, as in other cuckoos (Roberts, 1991).

Incubation and hatching. No information on the incubation period exists. It is known that the newly hatched nestling ejects the host's young or eggs from the nest (Roberts, 1991).

Nestling period. No information.

Population Dynamics

No information.

LESSER CUCKOO

(Cuculus poliocephalus)

Other Vernacular Names: Little cuckoo, small cuckoo.

Distribution of Species (see map 28): Asia from Pakistan northeast to China, Siberia, Korea, and Japan. Winters south to India, and also in southeastern Africa. (Earlier range attributions for the Sundas and Borneo are erroneous: such records apply to races of *saturatus* that were once considered part of this species.)

Measurements (mm)

10–11" (26–28 cm)

Wing, males (from India) 149–161 (avg. 154.5, $n = 6$), females 144–154 (avg. 149.4, $n = 10$). Tails, males (from India) 124–142 (avg. 134, $n = 6$), females 123–132 (avg. 128.4, $n = 10$). Measurements from African specimens are similar (Becking, 1988). Wing:tail ratio ~1:0.85.

Egg, avg. 21.4 × 15.5 (range 19–23.3 × 14–17) (Schönwetter, 1967–84). Shape index 1.38 (= oval). Rey's index 2.16 (Becking, 1981).

Masses (g)

Males 48–59 (avg. 54.2, $n = 3$), female 40 (Fry et al., 1988). Two males 48, 54; October weights of unsexed birds, 32–44 (Ali & Ripley, 1983). Unsexed birds 32–44 (avg. 40.1, $n = 10$) (Becking, 1988). Estimated egg weight 2.75 (Schönwetter, 1967–84). Actual mass of fresh eggs, avg. of 4, 2.89 (Becking, 1981). Egg:adult mass ratio ~6%.

Identification

In the field: This species closely resembles the common cuckoo, but is noticeably smaller, and its tail is often (but not always) the same color as the upper-tail coverts (rather than being darker than the coverts). The iris is brown, not pale yellow as in the common cuckoo, the tail has a noticeably darker tip, and the marginal coverts at the bend of the wings are gray (rather than clear white as in the Oriental cuckoo or lightly marked with brownish as in the common cuckoo). The adult female is malelike, but with some tawny tinges on the breast and sides of the neck. Females (or perhaps

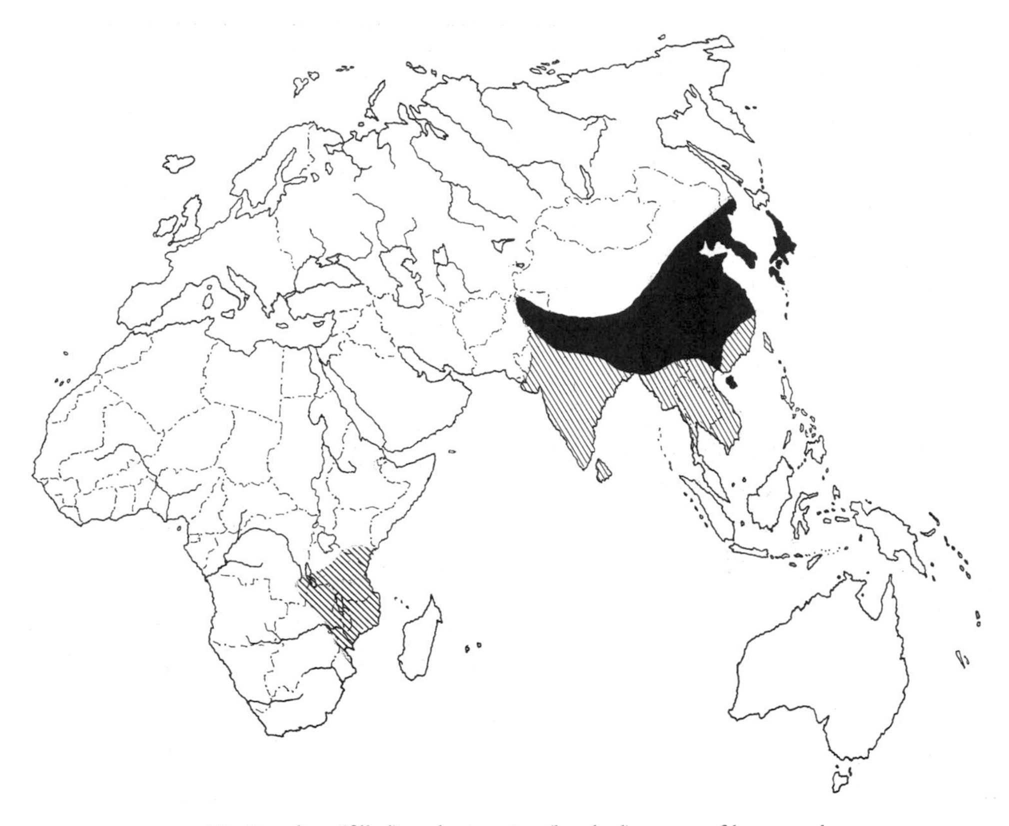

MAP 28. Breeding (filled) and wintering (hatched) ranges of lesser cuckoo.

subadults generally) also reportedly have a hepatic plumage morph that is more reddish brown dorsally, with blackish barring. Immature individuals likewise have plumages that are mostly edged or barred with tawny on the upperparts. The song of the breeding male is a distinctive series of five or (usually) six "kyioh" notes that seem to ascend and then descend the scale, with each note rising and falling in pitch, sometimes interpreted as "That's your choky pepper, choky pepper." The first two notes are brief and rising in scale, the third is often loudest, and the remaining notes accelerate while diminishing in volume. This phrase, which lasts about 1 second, may be uttered several times, with the average pitch gradually descending somewhat with each successive utterance. Its timbre is distinctly harsh, rather than soft and dovelike. Another interpretation is "who-who-whar," "-who-who-wha," with a stress on the prolonged third syllable, and sounding something like a cackling hen. Becking (1988) provided a sonogram of this vocalization. Females have a laughing or bubbling call similar to those of many other cuckoos and similarly associated with courtship.

In the hand: Except in southeastern Africa, where confusion with the Madagascar cuckoo might occur, the small size of this species (maximum wing length 171 mm and tail length maximum 150 mm), distinguishes it from the other grayish-above, barred-below cuckoos. In the absence of vocalizations (which are not likely to occur in wintering areas), in-hand distinction from the Madagascar cuckoo in Africa is possible using the latter's slightly larger size—the minimum wing length of *rochii* is 162 mm, which is the maximum length among all *poliocephalus* specimens reported by Fry et al. (1988) for Africa.

Habitats

This cuckoo is associated with wooded country, open scrub, and second-growth woodlands at elevations from about 1200 m to as high as 3360 m in India and from 1500 to 3660 m along the southern Himalayan slopes of Nepal. It occurs up to about 1800 m (or timberline) in Pakistan, and about 2000 m in the Malayan peninsula. In Japan it is mainly associated with broadleaved forests, usually to about 1200 m but sometimes to 1700 m, and rarely approaches the timberline zone.

Host Species

Baker (1942) provided a list of 71 parasitism records, involving 21 species (table 12). Of these, the brownish-flanked bush warbler ("strong-footed bush warbler") accounted for nearly one third of the total, and other bush warblers of this same genus (*Cettia*) made up most of the rest. All of the affected bush warbler clutches had been parasitized by closely mimicking "red" (brownish red) cuckoo eggs. However, some of the other putative hosts listed by Baker, such as willow warblers and leaf warblers, had been parasitized with "white" eggs of questionable identity (see below). In Sikkim the pale-footed bush warbler is a known host, and in the former USSR the principal host is the Manchurian bush warbler (Becking, 1981). In Japan the closely related Japanese bush warbler is likewise a major host, but the winter wren is also sometimes used. At least five other species have also reported as Japanese hosts (Brazil, 1991).

Egg Characteristics

Eggs of this species are broad ovals, with rounded ends, and with colors ranging from terra cotta red to deep chocolate brown. These eggs are close mimics of *Cettia* bush warbler eggs. Bush warblers lay eggs that average about 17–18 × 13–14 mm and are bright to deep chestnut in color, the more rounded end somewhat mottled with darker tones. Becking (1981) questioned whether the speckled white eggs identified by Baker (1942) as belonging to this species and affirmed by Cramp (1985) were actually produced by the lesser cuckoo. Becking believed that the identity of such eggs needed verification, although the eggshells conform in their ultrastructural characteristics with those of the noncontested brown type. In Japan this cuckoo evidently produces only reddish brown eggs that closely mimic those of a *Cettia* species (Nakamura, 1990). Roberts (1991) suggested that in Pakistan this cuckoo sometimes breeds at altitudes above the treeline, where the most common potential host is a *Phylloscopus* leaf warbler rather than any of the *Cettia* species.

Breeding Season

In the Indian subcontinent this species breeds from May to July (Ali & Ripley, 1983). In Pakistan calling occurs from late May to early August, possibly peaking in June (Roberts, 1991). In Nepal the breeding period is similar, extending from early May to late July (Inskipp & Inskipp, 1991). In Japan it breeds during June and July, which is later than the other Japanese cuckoos (Brazil, 1991).

Breeding Biology

Little reliable information is available, partly owing to the uncertainties associated with egg identification. The nests of bush warblers are deep cups or domed-over structures with lateral entrances near the top, placed near the ground. At least the latter type would be difficult to parasitize easily. There is no information on the incubation periods of the hosts, nor of the nestling stages of the cuckoo.

Population Dynamics

No information

MADAGASCAR CUCKOO

(Cuculus rochii)

Other Vernacular Names: Madagascan lesser cuckoo.

Distribution of Species (see map 29): Breeds throughout Madagascar; winters in eastern Africa.

MAP 29. Breeding (filled) and nonbreeding (hatched) ranges of Madagascan cuckoo.

Measurements (mm)

10–11″ (20–28 cm)

Wing, males 169–179 (avg. 167, n = 28), females 159–163 (avg. 160.5, n = 4). Tail, males 135–155 (avg. 144.3, n = 28), females 134–141 (avg. 138.1, n = 4) (Becking, 1988).

Egg, 18.5 × 14 (range 17–19.9 × 13.2 × 14.5) (Schönwetter, 1967–84). Shape index 1.32 (= broad oval).

Masses (g)

Males 64, 65 (Fry et al., 1985). Estimated egg weight 1.9 (Schönwetter, 1967–84). Egg:adult mass ratio ~3%.

Identification

In the field: Nearly identical in appearance to the Asian lesser cuckoo, but easily distinguished by its song. The male's breeding song is distinctive, and usually consists of four (sometimes three) "ko" or "kow" notes, with the first three on the same pitch, and the last usually noticeably lower. The usual sequence might be written as "ka-ka-ka′-ko." It has also been described as "ko-ko, ko′-ko" or "ko-ko, ko′-kof," and in an abbreviated version, "ko-ko, kof." The entire phrase lasts about 1 second, with each note starting high in pitch and slurring downward. It is mellower in timbre than that of the Asian lesser cuckoo.

In the hand: Distinction from the lesser Asian cuckoo in Africa has been described under that species and can be achieved by the differences in their wing and tail measurements. The wing coverts of the wrist joint in the Madagascan species are mostly white, with darker spotting or banding, but in the lesser Asian cuckoo the inner vane is blackish, and the outer one is white or whitish. Even in the hand females are scarcely distinguishable from males (perhaps being more buffy or rusty on the chest, as in the Asian species). Immature birds are generally browner than those of the lesser Asian cuckoo, with a brown head and strong barring on the underparts, dorsal areas, and upper wing coverts. No rufous morph is present in adult females (Becking, 1988).

Habitats

Forest edges and denser habitats in savanna are used by wintering birds in Africa. In Madagascar the birds are found from sea level to about 1800 m, in nearly any habitat that is even slightly forested or sometimes even only brush covered.

Host Species

The usual host of this species is the Madagascar cisticola. Less frequently, the northern jery and Madagascar paradise flycatcher are exploited (Landgren, 1990). The Madagascar swamp warbler is a fairly frequent host, and the souimanga sunbird is also sometimes used (Schönwetter, 1967–84).

Egg Characteristics

The eggs of this species are much like those of the lesser Asian cuckoo, but are slightly smaller. The usual ground color is white, but sometimes is light yellow or rosy-tinted. Dark sepia brown or reddish brown speckles and spots are present around the more rounded end.

Breeding Season

In Madagascar singing may be heard from August to April, with a peak from September to December (Landgren, 1990). Wintering birds are present in Africa mainly from April to September.

Breeding Biology

No specific information is available, beyond that presented above.

Population Dynamics

No information.

PALLID CUCKOO

(Cuculus pallidus)

Other Vernacular Names: Brainfever bird, grasshopper hawk, harbinger of spring, rain bird, semitone bird, storm bird, weather bird.

Distribution of Species (see map 30): Australia, including Tasmania. Migrates rarely in winter to New Guinea and the Moluccas.

Subspecies

C. p. pallidus: Eastern and southern Australia.

C. p. occidentalis: Western and northern Australia. Probably not a valid race (Hall, 1974).

Measurements (mm)

12–13″ (30–33 cm)

C. p. pallidus, wing, males 188–198, females 183–194 (Hall, 1974). Tail, adult 175 (Frith, 1974). Wing:tail ratio ~1:0.9.

C. p. occidentalis, wing, males 180–196 (avg. 187.8), females 179–194 (avg. 185.8) (Hall, 1974). Wing:tail ratio 1:1.0.

Egg, avg of *pallidus* 24.2 × 17.5 (range 22.5–26.4 × 13.5–14.7) (Schönwetter, 1967–84). Shape index 1.38 (= oval). Rey's index 1.76.

Masses (g)

Range of 25 males 59–119 (mean 83.9), of 14 females 58–100 (mean 85.3) (Brooker & Brooker, 1989b). Estimated egg weight 3.9 (Schönwetter, 1967–84), 3.8 (Brooker & Brooker, 1989b). Egg:adult female mass ratio 4.4%.

MAP 30. Breeding range (filled) and wintering or secondary breeding range (hatched) of pallid cuckoo.

Identification

In the field: This species is generally pale gray dorsally, with no barring on the grayish white underparts. The tail feathers are edged with white barring, and there is a white nape patch and white carpal wrist patch. The eye-ring is bright yellow to whitish yellow, and the lower mandible is grayish to (perhaps only in breeding birds) orange. There is a noticeable dark streak extending from the lores through the eye to the white nape patch. Juveniles are mostly strongly spotted above with dark gray and white and spotted with gray and white below. Somewhat older immature birds are more rusty-tinged on the upperparts, with a buffy nape patch, and the lower mandible has a dark yellow base. The song of the breeding male consists of a series of about eight loud, rather melancholy, whistles that rise in pitch by quarter-tones following an initial slight drop after the first note, and males also utter wild "crookyer" notes when chasing females. Females have an ascending call that is hoarser than that of males and consists of repeated "wheeya" notes. Females also utter single brassy whistling notes, and juveniles produce harsh begging calls.

In the hand: This is the only Australian cuckoo with entirely pale gray underparts and only slightly darker gray underparts. The iris is dark brown and surrounded by a bright yellow eye-ring. The black-eared cuckoo is similar but is much smaller and has a considerably darker area extending from the lores to the ears. Females are similar to males but are somewhat spotted and marked with chestnut and buff on the upperparts. Juveniles have broad white margins on the upperpart feathers and some darker markings on the face and neck. There is also some brown barring on the breast. Older immature birds and subadults are more tawny-colored, with pale rufous streaking and spotting on the crown, sides of neck, and back, as well as on the upper wing coverts. As the birds become older, they gradually lose the brown barring on the breast and the white mottling on the upperparts. The lower mandible is dark yellow in immature birds, and the mouth color is probably bright yellow to orange-red, as it is in adults.

Habitats

This species is associated with diverse habitats in Australia, ranging from arid semideserts through scrub woodlands, mangroves, gardens, paddocks, roadsides, and secondary growth to the edges of tropical forests, but forest habitats are avoided. It primarily favors fairly open country with some shrubs, including coastal dune scrub and tree-lined riparian vegetation in dry creek beds. It also extends into deserts where there are scattered shrubs.

Host Species

A list of 12 major biological hosts of the pallid cuckoo (host species individually composing at least 2% of the total records) is provided in table 13, based on the summary by Brooker & Brooker (1989b). These authors reported that 111 host species have been reported among 1052 records of parasitism. Of the total host list, 32 species are known to represent true biological (fostering) hosts, and 21 of these are honeyeaters. All of the major hosts are members of the honeyeater family Meliphagidae, and the egg pattern of this cuckoo closely matches those of some honeyeater species.

Egg Characteristics

The eggs of this species are oval, with a pinkish, pale flesh-colored ground color, and either have no darker markings or only a few dots (Brooker & Brooker, 1989b). The eggshells have mean weights of 0.24 g and mean thicknesses of 0.09 mm (Schönwetter, 1967–84).

Breeding Season

In southern Australia this species lays from August to January, with a peak in October or November. Moving north, the records become more scattered, and those from the northern half of the continent include every month of the year except April. Out of a total of 706 Australian breeding records, 74% are for the period September through November (Brooker & Brooker, 1989b).

Breeding Biology

Nest selection, egg laying. Essentially only cup-shaped nests are used by this species for laying; only 2% of the records of parasitism involve dome-shaped or cavity nests. The reduced host clutch size of parasitized nests indicate that a host egg is typically removed at the time of parasitism; at least five host species are known to bury the cuckoo's egg when it is deposited before they have laid their own. Almost invariably only a single cuckoo egg is deposited per host nest; in only 11 out of 843 parasitized nests were 2 (10 cases) or 3 (1 case) pallid cuckoo eggs reported to be present. Additionally, only rarely (two known cases) are the eggs of any other cuckoo species present in the nests parasitized by the pallid cuckoo (Brooker & Brooker, 1989b).

Incubation and hatching. The incubation period lasts about 12–14 days, and eviction of the host's eggs or young may occur as early as 48 hours after hatching. However, it may also occur as late as 5 days following hatching, depending on the depth of the nest and the size of the host's young (Brooker & Brooker, 1989b).

Nestling period. The nestling period is unreported, but fledglings may continue to be fed by their foster parents for as long as 6 weeks.

Population Dynamics

Parasitism rate. No information on the principal hosts is available. Marchant (1974) found only 4 of 565 nests of the willie wagtail to be parasitized.

Hatching and fledging success. No information.

Host–parasite relations. There appears to be a fairly high level and effectiveness of egg mimicry of honeyeater eggs, especially those of such host species as the yellow-tufted and singing honeyeaters. Relatively high levels of egg similarity exist with the majority of the known biological hosts (Brooker & Brooker, 1989b). There have been some instances of apparent egg rejection by a few species, including the willie wagtail, for which the level of egg mimicry is poor (Marchant, 1974).

DUSKY LONG-TAILED CUCKOO

(Cercococcyx mechowi)

Other Vernacular Names: Dusky cuckoo, Mechow's long-tailed cuckoo.
Distribution of Species (see map 31): Sub-Saharan Africa from Sierra Leone east to Zaire.
Measurements (mm)
13–14″ (33–36 cm)
Wing, males 128–143 (avg. 137), females 128–145 (avg. 135). Tail, males 148–194 (avg. 175), females 117–195 (avg. 164) (Fry et al., 1988). Wing:tail ratio 1:1.21–1.28.
Egg, no information.
Masses (g)
Males 52–60 (avg. 57, $n = 7$), females 50–61 (avg. 56, $n = 8$) (Fry et al., 1988).

MAP 31. Range of dusky long-tailed cuckoo.

Identification

In the field: Like the other long-tailed cuckoos (see fig. 27), this species has a tail length greater than half the total body length, and the plumage is generally dark brown above and heavily barred brown and white below. The tail is similarly barred with brown and lacks white notching or spotting except on the edges and tips of the tail. The under-wing coverts are also white. The sexes are monomorphic in all three species, and immature individuals of the three are also very similar. All three species additionally have three-noted songs, uttered emphatically. In this species the song sounds like "hit-hit-hit," with the notes all on about the same pitch. These phrases, lasting about 1 second, are repeated at regular (about 1-second) intervals, so that about 25 songs may be uttered per minute. There are other vocalizations, including a series of about 30 repeated whistled notes lasting about 10 seconds, and a rapid, clamorous jumble of bisyllabic notes.

In the hand: Like the other long-tailed cuckoos, this species has a tail length that averages more than the wing length and is uncrested. It may be distinguished from the barred long-tailed cuckoos by the fact that the back feathers of *montanus* are barred with rufous-brown and olive brown, and the outer webs of the greater wing coverts are similarly barred with rufous brown, whereas in adults of *mechowi* there is no such barring. Adults may be distinguished from the olive long-tailed cuckoo by the fact that in *olivinus* the back and rump are olive-brown, not dark grayish, and in*olivinus* the outer webs of the flight feathers are scarcely if at all spotted with rufous. The sexes are virtually identical as adults, although females may be slightly less barred below. Immature birds are strongly barred with rufous on the upperparts, thus closely resembling immature birds and adults of the barred long-tailed cuckoo.

Habitats

This species occurs in the lower story of lowland forests and tall second growth and in the undergrowth of dense vegetation such as occurs along water courses. It extends from near sea level up to at least 1830 m in Zaire.

Host Species

No proven host species are known, but the brown illadopsis is a probably host (Fry et al., 1988).

Egg Characteristics

No information.

Breeding Season

Little information is available, but breeding probably occurs during the rainy season. In tropical West Africa there are evidences of breeding in Cameroon from December to February. In eastern Zaire (Itombwe) there is similar breeding evidence for January–April plus September; also in northern Zaire (Ituri R.) for April–July. Birds in breeding condition have been reported from Angola during October and November (Fry et al., 1988).

Breeding Biology.

No information.

Population Dynamics

No information.

OLIVE LONG-TAILED CUCKOO

(Cercococcyx olivinus)

Other Vernacular Names: Olive cuckoo.
Distribution of Species (see map 32): Sub-Saharan Africa from Ivory Coast east to Zambia.
Measurements (mm)
13–14" (33–36 cm)
Wing, males 138–156 (avg. 145), females 126–148 (avg. 139). tail, males 139–182 (avg. 161), females 136–175 (avg. 153) (Fry et al., 1988). Wing:tail ratio 1:1.1.

MAP 32. Range of olive long-tailed cuckoo.

Egg, one egg 23 × 16.4 (Schönwetter, 1967–84). Shape index 1.4 (= oval).
Masses (g)
Males 64, 66 (Fry et al., 1988). Estimated egg weight 3.38 (Schönwetter, 1967–84). Egg:adult mass ratio 5.2%.

Identification

In the field: Compared with the *C. mechowi* this species is slightly more olive-toned on the upperparts and less heavily barred below; the tail is somewhat shorter and more heavily marked with white below (see fig. 27). The vocalizations include the male's primary song, a three-noted "ee-eye-owe" or "whi-whow-whow" that progressively drops in pitch with each repetition. It is repeated at short intervals, producing an average rate of 10 songs in 25 seconds, or about 25 per minute. The first note is sometimes so weak that only the last two may be heard, and the phrases are repeated numerous times. Another call consists of a long series of uniform "how" notes that increase gradually in volume, the sequence lasting about 10–15 seconds.

In the hand: This is the least barred of the three long-tailed cuckoos, with only faint rufous markings on the greater wing coverts and outer webs of the flight feathers in adults. The back and rump are dark olive-brown, and this same olive tone extends forward to the crown. The sexes are identical as adults, but immature birds can be distinguished from adults by their heavier rufous barring on their upper wing coverts and tail and are somewhat streaked rather than barred below.

Habitats

This species occurs mainly in fairly dense, unbroken forests but also in forest fragments and gallery forests, from sea level to 1500 m.

Host Species

No proven host species are known, but the rufous ant thrush is a possible host (Fry et al., 1988).

Egg Characteristics

The egg is reportedly pure white (Fry et al., 1988). No other information is available.

Breeding Season

Breeding probably occurs during the rainy season. In Angola breeding-condition birds have been reported for September and November, and a female with an oviducal egg was found in September (Fry et al., 1988).

Breeding Biology

No information.

Population Dynamics

No information.

BARRED LONG-TAILED CUCKOO

(Cercococcyx montanus)

Other Vernacular Names: Barred cuckoo.

Distribution of Species (see map 33): Sub-Saharan Africa from Zaire east to Kenya, Malawi, and Mozambique.

Subspecies

C. m. montanus: Uganda, Zaire, and Rwanda.

C. m. patulus: Kenya to Zambia and Mozambique.

Measurements (mm)

13–14" (33–36 cm)

MAP 33. Range of barred long-tailed cuckoo.

Wing, males 141–152 (avg. 147), females 146–148 (avg. 147). Tail, males 144–177 (avg. 159), females 163–174 (avg. 169) (Fry et al., 1988). Males of nominate *montanus* may have shorter wings (143–145) but longer tails (182–201) than those of *patulus* (Chapin, 1939). Wing:tail ratio 1:1.1.

Egg, one egg 21 × 15 (Schönwetter, 1967–84). Shape index 1.4 (= oval).

Masses (g)

Both sexes 60–68.5 (avg. 63.4, $n = 4$) (Fry et al., 1988). Four males 58–62.7 (avg. 60.8), one female 58.7 (Rowan, 1983). Estimated egg weight 2.5 (Schönwetter, 1967–84). Egg:adult mass ratio 3.9%.

Identification

In the field: Over most of its range (except in the Zaire–Uganda region), this is the only species of long-tailed cuckoo, so the very long tail (about half the overall length) and rather uniformly barred, brown plumage should be diagnostic (see fig. 27). White barring occurs on the outer rectrices, and barred white to buffy feathers are also present on the underparts and flanks. The male's songs include a three-noted (sometimes four-noted, rarely of five) "wit-wit-you" that is much like those of the other long-tailed cuckoos (and also of the red-chested cuckoo). Chapin (1939) stated that in this species the call is usually four notes, rather than three as in the two other long-tailed species, and resembles the phrase "see which fits best." This song is repeated almost without pause. It may be interspersed with or preceded by an extended (40-second) disyllabic series of "dee-u" or "you-too" notes. These are uttered at about 1-second intervals and are similar to the "how" notes of the olive long-tailed cuckoo.

In the hand: This is the most heavily barred species of long-tailed cuckoos; its remiges and greater wing coverts are always heavily barred with rufous. The sexes are alike as adults. Juveniles of all long-tailed cuckoos have buffy-tipped body feathers, and those of this species have strong throat streaking and extensive crescent-shaped and barred dark underpart markings. By comparison, juveniles of the dusky long-tailed cuckoo are more regularly barred below, and those of the olive long-tailed cuckoo are streaked below (Chapin, 1939).

Habitats

This species occurs mainly in elevations from less than 500 m to 1200 m, but sometimes to as high as 2800 m in montane forests. It also extends locally to lowland forests, forest-advances, miombe (*Brachystegia*)woodlands, and coastal thickets. It is generally found above the altitudinal levels typical of the dusky long-tailed cuckoo.

Host Species

No host species are yet proven, but the akalats, particularly the Sharpe's akalat, are probable hosts, as are broadbills such as the African broadbill (Fry et al., 1988).

Egg Characteristics

An egg laid by a captive bird was white, with a very faint band of red around the more rounded end. A second oviducal egg was entirely white (Fry et al., 1988).

Breeding Season

Relatively little information is available. In Malawi, Zambia, and Zimbabwe breeding probably occurs between December and March, with somewhat earlier (October to January or February) breeding for Tanzania and Mozambique. Evidence exists for February and March breeding in Kenya (Fry et al., 1988).

Breeding Biology

Little information is available. A female was observed mantling the nest of a Sharpe's akalat, and it evidently either removed or destroyed the akalat's clutch of two eggs. The egg laid by a captive individual closely resembled that of an akalat. A similar egg (which later hatched into a cuckoo of uncertain identity) was found in the nest of an African broadbill (Dean et al., 1974; Rowan, 1983).

Population Dynamics

No information.

BANDED BAY CUCKOO

(Cacomantis sonneratii)

Other Vernacular Names: Banded cuckoo, Ceylon bay-banded cuckoo (*waiti*), Indian bay-banded cuckoo (*sonneratii*).

Distribution of Species (see map 34): Asia from Pakistan east to China and south to Sri Lanka, Sumatra, Borneo, Java, Bali, and the Philippines.

Subspecies

- *C. s. sonneratii:* Pakistan, India, Burma, Thailand, and southern Indochina.
- *C. s. waiti:* Sri Lanka.
- *C. s. malayanus:* Northern and central Malayan Peninsula.
- *C. s. schlegeli* (= *fasciolatus*): Southern Malayan Peninsula, Sumatra, Borneo, Philippines.
- *C. s. musicus:* Java.

Measurements (mm)

8.5–9″ (22–23 cm)

C. s. sonneratii. Wing, both sexes 116–128; tail, both sexes 112–118 (Ali & Ripley, 1983). Wing, both sexes 116–133; tail, both sexes 123–130 (Delacour & Jabouille, 1931). Wing:tail ratio ~1:0.95–1.0.

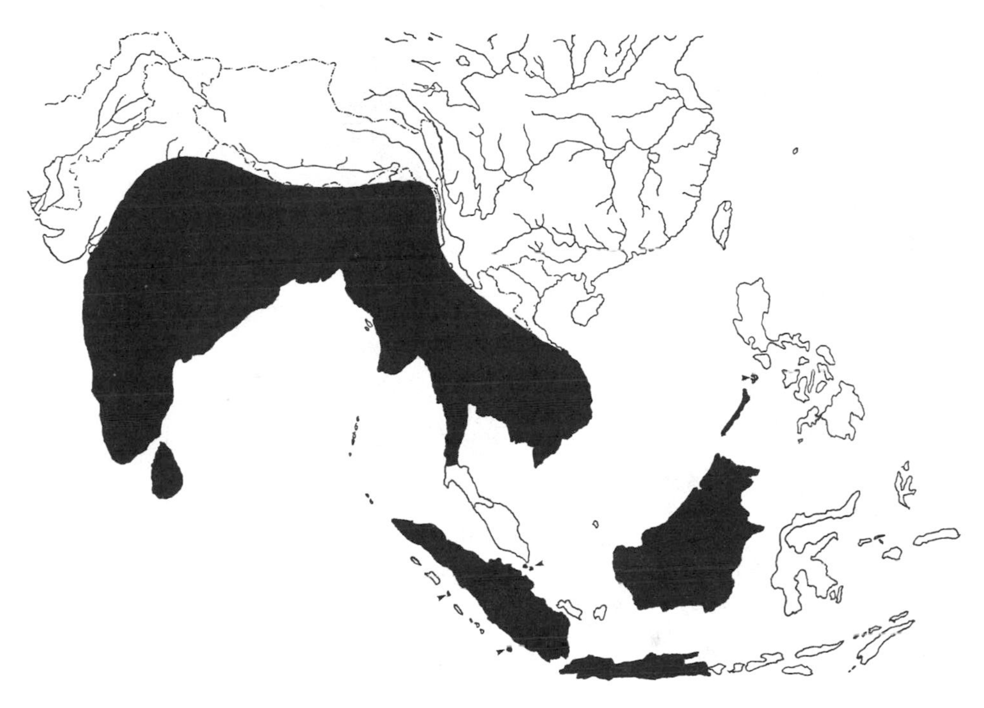

MAP 34. Range of banded bay cuckoo.

C. s. waiti. Wing, both sexes 121–126; tail, both sexes 110–111 (Ali & Ripley, 1983). Wing:tail ratio 1:0.8.

Egg, avg. of *sonneratii* 19.6 × 15.5 (range 17.7–20.8 × 14.4–16.5 (Schönwetter, 1967–84). Avg. of *musicus* 17.6 × 13.5, one egg of *schiegeii* 18.1 × 14.4, one egg of *waiti* 17.4 × 13.5 (Becking, 1981). Shape index 1.26–1.3 (= broad oval). Rey's index, *sonneratii* 2.39–2.42, *waiti* 2.71. (Becking, 1981).

Masses (g)

One unsexed bird, 32 (Dunning, 1993). Three unsexed adults 30–35 (avg. 33.7) (Becking, 1981). Estimated egg weights 1.8 (*musicus*) to 2.12 (*schlegeli*) (Becking, 1981); also (*sonneratii*) 2.45 (Schönwetter, 1967–84). Egg:adult mass ratio ~ 6%.

Identification

In the field: This small and inconspicuously brown-barred cuckoo has finely barred whitish underparts and a rather pale eyebrow stripe (fig. 34). Immature individuals have a less apparent eyebrow stripe and more buffy breast tones. The tail is fairly short and is barred with brown and rust, but with whitish edging and tips on the outer feathers. The male's typical song is a four-noted phrase sounding like "smoke, your-pepper" (also interpreted as "tee-tyup–tee-tyup," "wi-ti-tee-ti" and "yauk-hpa-kew-kaw"). It is characterized by having a plaintive tonal quality and a brisk or shrill enunciation, the first note being longer and the last three more run together. It closely resembles the "crossword puzzle" song of the Indian cuckoo in syllables and cadence, but is shriller and higher in pitch. Another typical vocalization is a series of clear whistles that rise progressively in pitch. This so-called cadence call often consists of two notes on the same pitch, followed by three more on a higher pitch, and finally three more on a still higher pitch. In Sri Lanka the birds reportedly utter a series of about five "whew" notes in sequence, each stanza

FIGURE 34. Profile sketches of six *Cacomantis* cuckoos: adult (A) and juvenile (B) of banded bay cuckoo; typical adult (C) and hepatic morph female (D) of plaintive cuckoo; typical adult (E) and hepatic morph female (F) of gray-bellied cuckoo; adult (G) and juvenile (H) of rusty-breasted cuckoo (also upper song pattern; lower diagram shows two songs of the nearly identical brush cuckoo); adult male (I) and juvenile (J) of chestnut-breasted cuckoo; adult male (K) and juvenile (L) of fan-tailed cuckoo. Morph-types of eggs, undertails, and typical song-phrase patterns are also shown.

a higher pitch than the last. The song stops abruptly when the bird seemingly runs out of breath or reaches the top of its vocal range.

In the hand. This is a strongly rufous-toned and heavily barred cuckoo and is similar in size and appearance to hepatic-morph females of the plaintive cuckoo. However, the bill of the banded bay is stouter and not so compressed, and the rectrices are narrower toward the tip and not so broadly tipped with white. The sexes are alike as adults. Immature birds are even more heavily barred with rufous than adults. Immature birds of the similar banded bay cuckoo are smaller, have upperparts a paler shade of rufous, are less clearly barred below, and the entire undersurface of the tail is distinctly and regularly barred (Deignan, 1945). Nestlings are distinctively striped rufous-red on their upperparts and have fine black barring on the throat and underparts. The feet of nestlings are a distinctive olive-green, a color that persists in juveniles and subadults (Becking, 1981).

Habitats

This cuckoo occurs in sparsely wooded to dense deciduous and evergreen woodlands from the Nepal terai foothills at about 150 m up to moderate elevations (rarely to about 2400 m) in the Himalayas and to about 1200 m in the Malayan highlands. In Sri Lanka the birds favor parklike habitats, including open jungles and sparsely cultivated areas around irrigation tanks and clearings. Singing males especially favor tall, dead trees or treetops that are somewhat exposed (Phillips, 1948).

Host Species

Baker (1942) provided a list of hosts of this species, based on his collection of parasitized clutches. However, Becking (1981) reported that some of these eggs were certainly misidentified, specifically those associated with babbler hosts. He believed that this cuckoo's primary host is the common iora, with a secondary dependence on minivets, including the orange minivet and probably also the little minivet. In Sri Lanka the common iora and orange minivet have reportedly been parasitized, and probably also the little minivet (Phillips, 1948; Becking, 1981).

Egg Characteristics

Becking (1981) described the eggs of this species as being broadly oval, with a white to pinkish ground color and with reddish-brown or purplish brown speckles and blotches and grayer underlying markings. One egg of the Sumatran race had olive-green blotches and gray spotting.

Breeding Season

In the Indian subcontinent breeding probably occurs from February to August; its primary host, the common iora, breeds mainly between April and July, and the minivets have similar breeding periods (Ali & Ripley, 1983). Calling in Sri Lanka suggests that most breeding occurs from February to April, although still-dependent birds have been seen as late as October (Phillips, 1948). In Nepal breeding extends from early February to late September (Inskipp & Inskipp, 1991). On the Malay Peninsula calling occurs between January and May (Medway & Wells, 1976).

Breeding Biology

Little reliable information exists, due to confusion of egg identification. In India the common iora builds a cuplike nest in shrubs or low trees and lays eggs that are pinkish white, with purplish brown blotches, averaging about 17.5–18 × 13–13.5 mm. The incubation period is about 14 days (Ali & Ripley, 1983). The iora's eggs average 1.71–1.89 g, or not significantly different from those of the cuckoo (Becking, 1981).

Population Dynamics

No information.

GRAY-BELLIED CUCKOO

(Cacomantis passerinus)

Other Vernacular Names: Gray-bellied plaintive cuckoo, gray-headed cuckoo, Indian plaintive cuckoo.

Distribution of Species (see map 35): Pakistan and India, wintering south to Sri Lanka.

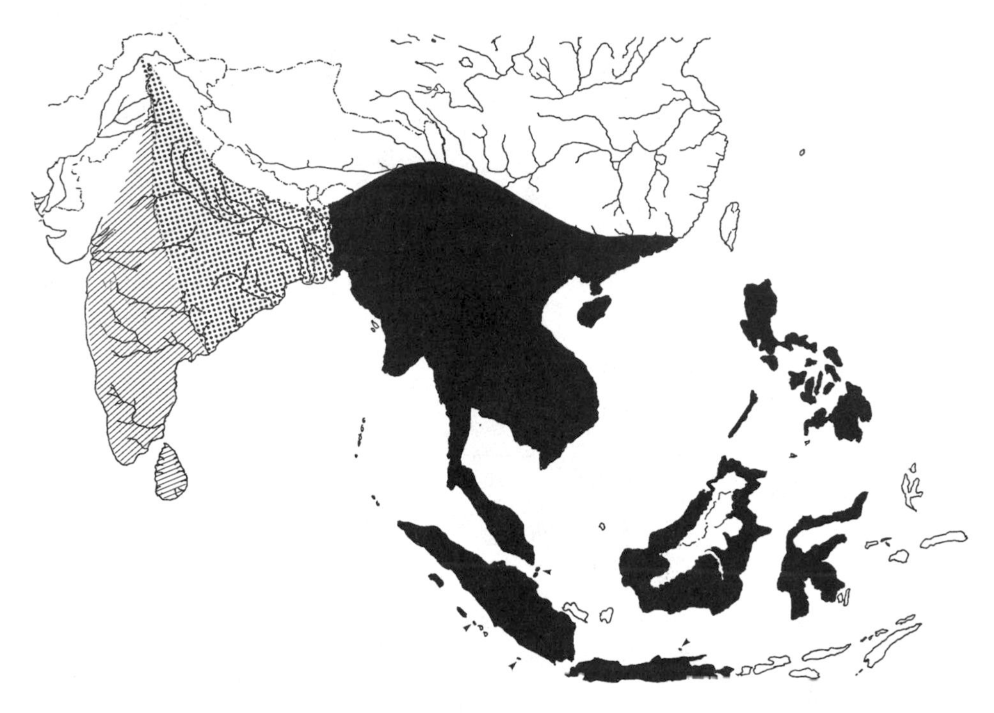

MAP 35. Range of plaintive cuckoo (filled), and breeding (shaded) plus nonbreeding (hatched) ranges of gray-bellied cuckoo.

Measurements (mm)

8.5″ (22 cm)

Wing, both sexes 113–120. Tail, both sexes 105–115 (Ali & Ripley, 1983). Wing:tail ratio ~ 1:0.95.

Egg, avg. 18.5 × 13.5 (range 16.1–21.2 × 12.1–14.2) (Schönwetter, 1967–84). Shape index 1.37 (= oval). Rey's index 2.14 (Becking, 1981).

Masses (g)

Avg. body weight (6 unsexed birds) 25.7 (Becking, 1981). Estimated egg weight 1.75 (Schönwetter, 1967–84), 1.48 (Becking, 1981). Egg:adult mass ratio 5.75–5.8%.

Identification

In the field: This rather small cuckoo is mostly grayish both above and below in adult males, with a blackish tail that is widely tipped and laterally barred with white (fig. 34). There is also a white patch at the base of the primaries on the underside of the wing, which is visible in flight. The adult female usually resembles the male, but may also sometimes exhibit a hepatic plumage morph (perhaps only a temporary subadult phase) that is strongly barred with rufous brown and closely resembles the barred rufous and dark brown plumage of immature birds. This rufous-dominated plumage also resembles the adult plumage of the banded bay cuckoo, but lacks the latter's whitish eyestripe. The male's song consists of a lilting series of three (or occasionally more) clear whistling "keveear-keveear-keevear" phrases that are uttered in a minor key and rise progressively in pitch. Other variations are "ka-weer, ka-wee-eer," "pee-pipee-pee, pipee-pee" or "pee-pipee-peepi, pipee-peepi." Another common sequence is a loud "wheeeh-whoo," followed by two more rapid "pe-ti-wear" or "peeter-peeter" phrases. This latter phrase somewhat resembles the "crossword puzzle" call of the Indian cuckoo but is higher pitched. Single "peeter" notes are also frequently uttered. As with other cuckoos, much calling occurs during relatively dark conditions.

In the hand: Like the closely related plaintive cuckoo, adult males have gray upperparts and a similar gray head and breast color, as well as a blackish and white-tipped tail that is barred with white on the outer feathers. This species is also grayish (not rufous) on the lower breast, flanks, and abdomen, distinguishing it from the more eastern "rufous-bellied" or Burmese plaintive cuckoo. Females of the typical plumage morph are nearly the same color as males, but with the slate gray interrupted by brown or gray. There reportedly also is a hepatic morph that is strongly barred with rufous and may be impossible to distinguish from the hepatic morph of the plaintive cuckoo. (It is quite possible that this "morph" is actually only a subadult plumage stage, as has been suggested for the plaintive cuckoo.) Immature birds are said to occur in a gray morph similar to the adult male, an intermediate morph, and a hepatic morph that closely corresponds to adult hepatic-morph females. This last plumage type may at times be nearly impossible to separate from the corresponding plumage of the plaintive cuckoo. However, the juveniles of the gray-bellied cuckoo are reportedly more brownish above, with chestnut (not rufous and black, as in the plaintive) markings on the crown, scapulars and upper back, and the underside is barred with black and white. The central rectrices have chestnut (not rufous) edge markings (Biswas, 1951). The validity of a species-level distinction between these two similar and doubtfully sympatric forms has recently been brought into question (White and Bruce, 1986).

Habitats

This species occupies forested habitats of the Indian subcontinent from the foothills (about 600 m in Pakistan) to as high as 1400 m (rarely to 2100 m) along the Himalayan slopes of Nepal and to about 1800 m in Pakistan. Savanna grasslands, village gardens, plantations, secondary forests, and open, scrub-covered hillsides are used. Sparse woodlands and open forests, rather than dense primary forests, are preferred habitats.

Host Species

A list of 10 host species reported by Baker (1942), and based on parasitized clutches in his collection, is presented in table 10. Becking (1981) noted that all known hosts are small warblers that are associated with rather open habitats, and most of them build dome-shaped nests with small side entrances or deep, purselike nests with very narrow, slitlike entrances, both of which seem difficult to parasitize. The zitting cisticola and common tailorbird are the host species most frequently represented in Baker's list, followed closely by the ashy prinia. Becking questioned the inclusion of the purple sunbird as a known host species.

Egg Characteristics

According to Becking (1981), the eggs of this species are polymorphic, with three primary phenotypic morphs associated with female gentes adapted to three different host groups. One of these is the chestnut-brown to mahogany egg morph that mimics eggs of the ashy prinia ("wren warbler"). Another egg morph has a blue ground color, with reddish, blackish-brown, or purple spots, lines, or blotches, and is evidently adapted to mimic the plain prinia ("common wren warbler"). The third is a light pinkish to bluish egg morph, with reddish brown blotches and spots around the more rounded end. This last morph is widely distributed geographically and ecologically and is generally adapted in pattern to resemble eggs of various tailorbirds and the zitting cisticola ("streaked fantail warbler"). All egg types have an oval shape, with a glossy to moderately glossy finish.

Breeding Season

Ali & Ripley (1983) report that the breeding season of the Indian subcontinent extends from June to September, synchronized with local host activity patterns. In Nepal the season is likewise from late May-early June to September (Inskipp & Inskipp, 1991). In Pakistan laying occurs between

March and September, also in close synchrony with host species (Roberts, 1991). Eggs in Baker's collection exhibited a distinct peak during July (58% of 53 records) (Becking, 1981).

Breeding Biology

Nest selection, egg laying. The act of egg laying has apparently never been observed in this species, which tends to select hosts that build nests with small and rather inaccessible lateral entrances. The eggs associated with tailorbird and *Cisticola* warbler nests are typically pinkish white or bluish white, with reddish brown markings similar to those of the hosts. The tailorbird hosts have eggs averaging about 16.5 × 11.5 mm, and an incubation period of 12 days; the zitting cisticola host lays white to pale bluish eggs of about 15 × 11–12 mm, with an incubation period of about 10 days. In southern India the common gens is one that lays mahogany-brown eggs in the nests of the ashy prinia. This host lays nearly identical eggs averaging 16 × 12 mm and has an incubation period of 12 days. The common prinia's eggs are pale blue, with numerous dark lines, spots, and blotches, averaging about 15.5 × 11.5 mm, with an incubation period of 11–12 days (Ali & Ripley, 1983). In most cases the cuckoo's eggs are only recognizable by being slightly larger than those of their host, and perhaps by then more rounded appearance. In spite of these similarities, Becking (1981) reported a high (20%) desertion rate by tailorbirds whose nests were parasitized.

Incubation and hatching. Little specific information exists. Whether the young cuckoo evicts its host's eggs and young is still unreported.

Nestling period. No information.

Population Dynamics

No information.

PLAINTIVE CUCKOO

(Cacomantis merulinus)

Other Vernacular Names: Burmese plaintive cuckoo, rufous-bellied cuckoo.

Distribution of Species (see map 35): Asia from eastern India and southern China south to Sumatra, Java, Bali, Borneo, Sulawesi, and the Philippines.

Subspecies

C. m. merulinus: Philippines, probably also Sulawesi.

C. m. celebensis: Sulawesi [questionably, according to White & Bruce (1986)].

C. m. lanceolatus: Java.

C. m. subpallidus: Nias Island (west of Sumatra).

C. m. threnodes: India, Malaysia, Sumatra, and Borneo.

C. m. querulus: Nepal and Assam to China and Hainan.

Measurements (mm)

8″ (21 cm)

C. m. merulinus, wing, both sexes 101–112 (White & Bruce, 1986).

C. m. querulus, wing, both sexes 104–122; tail, both sexes 99–119 (Delacour & Jabouille, 1931). Wing:tail ratio ~1:1.0.

C. m. threnodes, wing, both sexes 109–119; tail, both sexes 112–125 (Ali & Ripley, 1983). Wing:tail ratio ~1:1.05.

Egg, avg. of *querulus* 18.5 × 13.5 (range 17.5–19.8 × 12.2–13.8). Avg. of *lanceolatus* 18.2 × 13.2 (range 17.8–18.6 × 12.5–14.2). One egg of *celebensis* 19.5 × 14.8 (Schönwetter, 1967–84). Shape index 1.31–1.38 (= broad oval to oval). Rey's index: *lanceolatus* 2.18, *querulus* 2.27, *celebensis* 2.67.

Masses (g)

One unsexed bird, 26 (Dunning, 1993). One male of *threnodes,* 25 (Thompson, 1966). Estimated egg weights (*celebensis*) 1.47, (*lanceolatus*) 1.67, (*querulus*) 1.73 (Schönwetter, 1967–84). Egg:adult mass ratio ~6.5%.

Identification

In the field: Adult males closely resemble those of the gray-bellied cuckoo, but the flanks and underparts are buffy to rufous rather than gray (fig. 34). Females may resemble males or may sometimes exhibit a hepatic plumage morph that is mostly barred rufous and dark brown. Likewise, immature birds of the gray-bellied and plaintive cuckoo are similar, although those of *merulinus* are less whitish and more rusty-colored below. The male's "ascending" song is a cheerful three- or four-part sequence sounding like "tay-ta-tee" or "tay-ta-ta-tay" that is repeated and may increase in pitch and speed with each repetition. The first note is quite prolonged, lasting several (perhaps 4–5) seconds, the second lasts about 1 second, and the third or final note about 2 seconds. It may be repeated at about 5-second intervals or more irregularly. Another vocalization, the "cadence call," often follows. It is a series of two or three (but sometimes up to six) plaintive whistled notes uttered slowly and at the same pitch that are followed by a descending series of rapid and shorter notes, as in "pwee, pwee, pwee, pee-pee-pee-pee." A more extended variation consists of four slow notes (lasting about 2 seconds each, with longer intervals), a series of three or four more rapid double notes (each twice as fast and with shorter intervals), and a final prolonged and plaintive note (lasting up to about 3 seconds), the entire series forming a cadence that progresses down the scale. This sequence is usually rendered as "tee-tee-tee-tee-tita-tita-tita-tee" or as "pik, pik, pik, pik, pika-pika-pika-pika, peeeee." There is also a harsh screeching "tchree-tchree" call, as well as a call sequence of from two to four trilled and fading notes, sometimes rendered as "prrreee-prree-pre-pre."

In the hand: Adults normally can be distinguished from those of other (and locally sympatric) species by their more rufous underparts and flanks. Sexes are usually alike as adults, but a hepatic plumage morph occurs in females that is probably difficult to distinguish from the hepatic plumage morph of the gray-bellied cuckoo. However, this hepatic morph of females may simply represent a second immature plumage stage (Parkes, 1960; White & Bruce, 1986). Immature individuals have a gray (not brown, as in adults) iris, and a bright orange (not pink) gape. They also have extensive light rufous barring on the upperparts, with the lighter underparts barred and streaked with blackish brown markings. Juveniles are visually distinguished from those of the gray-bellied cuckoo on the basis of their rufous and black head, dorsal and central tail-feather barring (rather than being barred with brownish and chestnut). With increasing age the head barring is lost, the chin and throat become ashy, and the central tail feathers of older birds are barred only along their edges (Biswas, 1951).

Habitats

This species occurs in montane forest edges, lightly wooded or secondary forests, scrub jungle, gardens, and tea plantations. It extends from about 300 m to as high as about 2000 m in Nepal, Bhutan, and Sikkim. In Borneo and the Sundas the species reaches about 1200–1300 m elevation, but in the Philippines has been reported only at lowland elevations of no more than about 500 m.

Host Species

Hosts cited by Ali and Ripley (1983) for this species include three prinias (striated, hill, and gray-breasted), as well as the zitting cisticola and common tailorbird. Farther east in the species' range, it often parasitizes the yellow-bellied prinia (Smythies, 1960; Becking, 1981), and on Sulawesi there is a report of parasitism of the crimson sunbird (Schönwetter, 1967–84).

Egg Characteristics

The eggs of this species are more or less identical to those of passerines, at least in their measurements. The race *lanceolatus* reportedly has a white to pale greenish and brown-flecked egg morph that is adapted to the olive-backed tailorbird, as well as a mahogany-colored egg type that is adapted to the yellow-bellied prinia. The nominate race also has two egg morphs. One of these is an unspotted greenish egg that is adapted to parasitizing various species of cisticolas, tailorbirds, and prinias (such as the rufescent and graceful prinias). The other egg morph, with brown flecks on a white ground color, is adapted to parasitizing the striated prinia and those tailorbirds that lay whitish eggs (Schönwetter, 1967–84).

Breeding Season

Within the Indian subcontinent an April–August breeding season is typical (Ali & Ripley, 1983). In Burma much singing occurs during April (Smythies, 1953). In northern Thailand singing males may be heard from late February to mid-June (Deignan, 1945). Little information exists on breeding periods for Malaysia, the Sundas, and the Philippines.

Breeding Biology

No detailed information exists. This species is probably similar in its breeding biology to *passerinus,* which is likewise poorly studied.

Population Dynamics

No information.

RUSTY-BREASTED CUCKOO

(Cacomantis sepulcralis)

Other Vernacular Names: Gray-headed cuckoo, Indonesian cuckoo

Distribution of Species (see map 36): Malay Peninsula, Sumatra, Java, Lesser Sundas, Borneo, Sulawesi, and the Philippines. Often considered conspecific with the brush cuckoo.

Subspecies

C. s. sepulcraiis: Malay Peninsula, Greater and Lesser Sundas (including Belitung, Enggano, Simeulue, Bali), Borneo, and Philippines.

C. s. virescens: Sulawesi and nearby Molucca islands.

C. s. aeruginosus: Sula, Buru, Ambon, and Seram.

MAP 36. Ranges of rusty-breasted cuckoo (north and west of dashed line) and of brush cuckoo (south and east of dashed line).

Measurements (mm)

9″ (23 cm)

C. s. sepulcralis, wing, both sexes 112–120 (White & Bruce, 1986). Tail, both sexes 114–123 (U.S. National Museum specimens). Wing:tail ratio ~1:1.0.

C. s. virescens, wing, both sexes 113–120 (White & Bruce, 1986). Tail, both sexes 114, 124 (U.S. National Museum specimens). Wing:tail ratio ~1:1.0.

Egg, avg. 19.5 × 14.8 (range 17.2–21.9 × 13.2–16) (Schönwetter, 1967–84). Shape index 1.32 (= broad oval). Rey's index 2.22.

Masses (g)

Avg. of 14, both sexes, 33.4 (range 24.4–39.5) (Rand & Rabor, 1960). Estimated egg weight 2.25 (Schönwetter, 1967–84). Egg:adult mass ratio 6.7%.

Identification

In the field: Adults of this species closely resemble those of the plaintive cuckoo, but the rufous underparts extend forward to the chin, and the white outer tail barring is strongly suffused with a rusty color. There is also a more conspicuous yellow eye-ring. Females generally resemble males, but a hepatic plumage morph of adult or subadult females is presumably present. Immature individuals are barred dark brown and rusty brown above, and heavily barred with dark brown and white below. The male's typical song is a mellow single-note whistle that is repeated 10–20 times, becoming progressively lower in pitch, slower, and somewhat disyllabic toward the end. There is also a series of rising notes that is more rapid and jumbled than the corresponding "tay-ta-tee" vocalization of the plaintive cuckoo.

In the hand: Similar to the plaintive cuckoo, if somewhat larger (although with overlapping measurements), and generally darker (see fieldmarks above). Information on criteria associated with sex and age differences is still lacking.

Habitats

This species is associated with lowland forests in Malaysia, where it occurs in forest-edge, second-growth, and scrub habitats. In the Sundas it is found from sea level to 1200 m in woodland, forests, and cultivated areas, and in Sulawesi it ranged from lowlands up to about 1600 m, in similar diverse habitats. In the Philippines it occurs from coastal mangroves to forests of at least 2000 m elevation.

Host Species

Baker (1942) reported two parasitized clutches of the long-tailed ("Javan") shrike in his collection. Schönwetter (1967–84) stated that *Enicurus, Rhipidura, Culicicapa, Saxicola, Megalurus, Lanius,* and other genera are parasitized. He specificially mentioned the long-tailed shrike, the striated grassbird, the sooty-headed bulbul, and the white-crowned forktail as hosts.

Egg Characteristics

Eggs of this species range from yellowish white to pale orange, with small golden-red and grayish freckles and spots. Other eggs may have reddish or purplish spotting, perhaps according to the particular host.

Breeding Season

Little information is available, but females with enlarged gonads and fledglings have been reported for March and April in the Philippines.

Breeding Biology

No information.

Population Dynamics

No information.

BRUSH CUCKOO

(Cacomantis variolosus)

Other Vernacular Names: Fan-tailed cuckoo, gray-breasted brush-cuckoo, square-tailed cuckoo.

Distribution of Species (see map 36): From Timor, and Moluccas, the southern Philippines, and New Guinea south to southeastern Australia and southwestern Pacific islands (Bismarck Archipelago, Solomon Islands).

Subspecies

C. v. variolosus: Northern and eastern Australia, wintering to New Guinea, Moluccas.

C. v. tymbonomus: Timor Island (Lesser Sundas), wintering to New Guinea.

C. v. dumetorum: Northwestern Australia.

C. v. addendus: Bougainville (Solomon Islands).

C. v. macrocercus: New Britain, New Ireland, and Lihir Island (Bismarck Archipelago).

C. v. websteri: New Hanover Island (Bismarck Archipelago).

C. v. tabarensis: Tabar Island (Bismarck Archipelago).

C. v. blandus: Admiralty Island (Bismarck Archipelago).

C. v. infaustus: Western Papuan islands, northern New Guinea.

C. v. chivae: Biak Island (New Guinea).

C. v. oreophilus: Southern New Guinca.

C. v. fortior: Goodenough & Fergusson Island; D'Entrecasteaux Archipelago.

C. v. stresemanni: Ceram and Ambon (Amboina) Island.

C. v. virescens: Sulawesi, Tukangesi Island.

C. v. fistulator: Sulawesi.

C. v. everetti: Jolo, Tawitawi, Basilan, and Sanga Island (Philippines).

Measurements (mm)

7.5–9" (19–23 cm)

C. v. addendus, wing, males 120–126, females 119–124. Tail, males 124–140, female 123. Wing:tail ratio 1:1.13 (Amadon, 1942).

C. v. blandus, wing:tail ratio 1:1.02 (Amadon, 1942).

C. v. chivae, wing, both sexes 113–119 (Rand & Gilliard, 1969).

C. v. infaustus, wing, males 119–126 (avg. 123.4), females 116–124 (Rand, 1942a).

C. v. macrocercus, wing, males 124–131, female 128. Wing:tail ratio 1:1.04 (Amadon, 1942).

C. v. oreophilus, wing, males 114–122, females 116–121 (Mayr & Rand, 1937).

C. v. tymbonus. Wing, males 125, 131, females 123–125 (Mayr, 1944).

C. v. variolosus. Wing, males 117–125 (Diamond, 1972). Tail, adults 110–117 (Frith, 1977). Wing:tail ratio ~1:0.95.

C. v. websteri. Wing:tail ratio 1:1.06 (Amadon, 1942).

Egg, avg. of *oreophilus* 21 × 13.8; *macrocercus* 19.2 × 13.7; *variolosus* 18.2 × 14.5. Overall range 17.5–22.5 × 13–15.2 (Schönwetter, 1967–84). shape index 1.25–1.52 (= broad oval to long oval). Rey's index, *macrocercus* 1.88, *fistulator* 21.7, *variolosus* 2.2, *oreophilus* 2.23.

Masses (g)

Seven males of *variolosus* 32–36, 2 females 35, 40 (Diamond, 1972). Three adults males of *variolosus* 33.7–37.5; three adult females 30–36.4 (Hall, 1974). Eight males of *addendus* 34–42 (avg. 37.9) (Mayr, 1944). Range of 22 Australian males 31–49 (avg. 35.3), of six females 21.8–38 (avg. 32.2) (Brooker & Brooker, 1989b). Estimated egg weight, *macrocercus* 1.9, *variolosus* 2.0, *oreophilus* 2.1 (Schönwetter, 1967–84). Egg:adult mass ratio (*variolosus*) ~5.7%.

Identification

In the field: This small cuckoo's adult plumage is mostly uniformly brownish gray above, the head more uniformly gray, and pale rufous below (fig. 34). The tail is short and square-tipped, with a paler tip and buffy edge notching. Immature individuals are heavily barred with rust and brown above, strongly barred on the wing and tail feathers, and less strongly barred with white and brown below. The eye-ring is gray, and the feet are grayish pink. The usual male song is a repeated phrase of rising triple (or sometimes quadruple) notes, "Ph-ph-phew" (also interpreted as "where's the TEA", "sea to SEA," or "where's the tea Pete?"), that start slowly, have a lower and shorter middle syllable, and a loud and even higher-pitched final syllable. The phrase lasts slightly less than 1 second. These phrases gradually ascend in pitch become louder, faster, and more persistent with repetition, and they may be repeated

5–10 times in succession, in a somewhat "insane" manner.

Another song, often alternated with the first, consists of six to eight or more single, rather mournful, "peer" notes that usually gradually descend in scale and become louder, but sometimes remain at a constant pitch and volume. The individual notes are slightly upslurred and are spaced at nearly 1-second intervals, so that an eight-note phrase may be uttered in about 5 seconds. These phrases are often repeated at about 30-second intervals. Birds from Bougainville Island are said to have a different song phrase, of three to seven shrill notes uttered in either a rising or falling scale, and increasing in volume.

Immature birds are quite different from adults in appearance. They are mottled with brown and yellowish buff below and are dark brown with buffy barring above. They are more yellowish overall than young fan-tailed cuckoos and not so uniformly rufous below as young chestnut-breasted cuckoos.

In the hand: Similar to but slightly larger than the chestnut-breasted brush cuckoo (wing length usually >115 mm vs. usually <115 mm in *castanelventris*), and the upperparts are more olive, less slaty, and without rufous tones in the breast. In this species the sexes are alike as adults, but immature individuals are highly variable in appearance, although they usually have heavily barred brown underparts and upperparts barred with yellowish rufous. Immature birds of this species and the chestnut-breasted and fan-tailed cuckoos are all quite similar and are variably rufous-tinged, but in this species the upperparts are more strongly barred with pale rufous and the eye-ring is an inconspicuous gray rather than contrasting yellowish. The mouth color of adult males is orange to vermilion; females and immature birds have yellow to light orange mouth colors.

Habitats

The race *dumetorum* is mostly associated in Australia with open forests, shrubbery, and savanna woodlands, especially those near water. The race *variolosus* is more closely associated with open forests with dense canopies and/or dense understories. In New Guinea the species (several races) extends from sea level to about 1300 m and more rarely to 1800 m. It occurs in open habitats such as gardens, towns, hoop pine (*Casuarina*) plant-ings, and scattered trees, but also occupies heavier cover such as mangroves, second-growth woodlands, and forest-edge habitats. In Bougainville the race *addendus* occurs at elevations up to at least 1200 m.

Host Species

Brooker & Brooker (1989b) found 376 records of parasitism in Australia, involving 58 host species. Ten species were provisionally identified as biological (fostering) hosts. They include the five major host species listed in table 21, plus the rose robin, the satin flycatcher, the purple-crowned fairywren, the restless flycatcher, and the rufous fantail. Baker (1942) reported four parasitized clutches of the red-backed fairywren and the gray fantail. Known New Guinea hosts include the white-shouldered fairywren, lemon-breasted flycatcher, yellow-tinted and brown-backed honey-eaters, and probably the willie wagtail (Coates, 1985). Schönwetter (1967–84) also listed the emperor fairywren as a host in New Guinea, and he mentioned the black and olive-backed sunbirds as hosts of the insular race *macrocercus* on the Bismarck Archipelago.

Egg Characteristics

Brooker & Brooker (1989b) described the eggs of the Australian race as ranging from white to cream, sometimes marked on the more rounded end with spots and blotches of purplish brown or very lightly spotted with black. Those of other races appear to be quite similar. The eggshells of various races range from 0.12 to 0.14 g in weight and from 0.07 to 0.08 mm in thickness.

Breeding Season

In Australia there are breeding (egg) records for all months except June, but the records are seasonally concentrated between November and January, when 64% of 129 records involving biological hosts have been recorded. This temporal breeding concentration is most evident in southern Australia; in northern Australia there is a shift toward breeding between January and April (Brooker & Brooker, 1989b). In New Guinea the species breeds during the dry season, at the same time that the host fairywrens are breeding (Diamond, 1972). In the Philippines breeding females or unfledged young have been reported during March and April (Dickinson et al., 1991).

Breeding Biology

Nest selection, egg laying. There are apparently no observations of egg laying nor evidence of egg removal at the time of laying. Host clutch sizes in parasitized nests tend to average about one egg less than those of unparasitized nests, suggesting that egg removal does occur. Typically a single egg is laid per nest (1 in each of 291 nests, 7 nests with 2 eggs, and 1 with 3) (Brooker & Brooker, 1989b).

Incubation and hatching. The incubation period of one egg was less than 13 days. Within about 30 hours after hatching, the cuckoo nestling evicts any other eggs or young from the nest (Brooker & Brooker, 1989b).

Nestling period. In one nest a young cuckoo fledged in 17 days, and in another in about 19 days. Feeding of fledged young has been observed for at least 1 month after fledging (Brooker & Brooker, 1989b).

Population Dynamics

Parasitism rate. In 39 brown-backed honeyeater nests, the rate of parasitism was 26% (Miller, 1932).

Hatching and fledging success. No information.

Host–parasite relations. Little information is available. Breeding periods of the primary hosts (species of *Ramsayornis* and *Petroica*) rather closely coincide with those of the cuckoo, although some early-nesting, bar-breasted fantails and scarlet robins may escape parasitism. The presence of host-specific gentes and associated host-matching eggs has not been proven for this species, although some cases of strong similarity between cuckoo and host eggs have been found (Brooker & Brooker, 1989b).

CHESTNUT-BREASTED CUCKOO

(Cacomantis castaneiventris)

Other Vernacular Names: None in general English use.

Distribution of Species (see map 37): New Guinea.

MAP 37. Ranges of chestnut-breasted (filled) and Moluccan (shaded) cuckoos.

Also probably northern Australia's Cape York Peninsula, where breeding seems likely but is still unproven.

Subspecies

C. c. castaneiventris: Cape York Peninsula.

C. c. weiskei: Central and eastern New Guinea.

C. v. arfakianus: Northwestern New Guinea, western Papuan Island.

Measurements (mm)

7.5–9″ (19–24 cm)

C. c. arfaklianus, wing, adults 110–117 (Rand & Gilliard, 1968). Tail, both sexes 91–108 (U.S. National Museum specimens). Wing:tail ratio ~1:0.87.

C. c. castaneiventris, tail 131 (Frith, 1977).

C. c. weiskei, wing, males 110–119, females 107–114 (Diamond, 1972).

Egg, avg. 19 × 14.5 (range 18–20.8 × 14.4–14.7) (Schönwetter, 1967–84). Shape index 1.31 (= broad oval). Rey's index 2.50.

Masses (g)

Avg. of nine (both sexes) 34.9, range 25–38 (Dunning, 1993). Four males 32–38, three females 35–38 (Diamond, 1972). Estimated egg weight 2.14 (Schönwetter, 1967–84). Egg:adult mass ratio 6.1%.

Identification

In the field: This species has the richest chestnut underparts of any of the Australian or New Guinean cuckoos, and a conspicuous yellow eye-ring is present in adults. These birds are generally darker than the fan-tailed cuckoos and their outer tail feathers are not so strongly notched (fig. 34). Immature birds are unbarred and rich rusty brown and pale buffy to cinnamon below, generally resembling a pale version of the adults. The usual male call is a trilled, descending whistle much like that of the fan-tailed cuckoo, but shorter and repeated at approximate 1-second intervals. A second call is a slow, mournful-sounding three-note phrase taking about 2.5–3 seconds to complete. Its transliterated "seeei-to-sail" rendition is much like the "sea-to-sea" phrase of the brush cuckoo but is more prolonged. The first note is highest in pitch and longer and is followed by two shorter notes, the middle one lowest in pitch and briefest, and the third note intermediate in pitch and duration. These three-note phrases are repeated often, but unlike those of the brush cuckoo the phrases do not increase noticeably in pitch or speed.

In the hand: The rich chestnut underpart coloration distinguishes this species from others; it resembles a brighter version of the fan-tailed cuckoo, which has much duller chestnut underparts and darker above. Both species have conspicuous yellow eye-rings, but the outer tail feathers of the fan-tailed cuckoo are more strongly notched with white. They also differ in tail length (fan-tailed at least 140 mm) and wing length (fan-tailed at least 120 mm). The iris of adults is brown to yellowish brown, the eye-ring is slate gray, and the palate is bright orange, the mouth otherwise being grayish black. The legs and bill are also black to dark slate gray. Juveniles are uniformly cinnamon-rusty above, with little or no dark barring evident and unbarred pale cinnamon below. This generally cinnamon plumage tone of young birds, especially on their underparts, readily distinguishes them from the much browner and distinctly barred young of the fan-tailed cuckoo.

Habitat

This species is associated primarily with tropical rainforest interiors, but also occupies forest edges, forest clearings, secondary growth, and sometimes monsoon forest. In New Guinea it extends from the foothills up to about 1800 m, or occasionally to 2500 m, generally below the levels used by the fan-tailed cuckoo.

Host Species

Schönwetter (1967–84) listed three eggs of this little-known species from a nest of the large-billed scrubwren. No other hosts have been mentioned.

Egg Characteristics

The only available description (Schönwetter, 1967–84) of the eggs stated they are dull brownish white, with fine freckling of purplish brown spots organized as a wreath around the more rounded end. The eggshells had a mean weight of 0.11 g and averaged 0.06 mm thick.

Breeding Season
No information.

Breeding Biology
No information.

Population Dynamics
No information.

MOLUCCAN CUCKOO
(Cacomantis heinrichi)

Other Vernacular Names: Heinrich's brush cuckoo.
Distribution of Species (see map 37): Northern Moluccas (Halmahera, Bacan Island).
Measurements (mm)
7.5–9.5″ (19–24 cm)
Wing, 112–122; tail, 116–123 (Stresemann, 1931). Wing, males 118–123 (avg. 118.5, *n* = 4), females 111, 112. Tail, males 116–130 (avg. 123, *n* = 4); females 114, 116 (American Museum of Natural History specimens). Wing:tail ratio ~1:1.04.
Masses (g)
No information

Identification

In the field: This rare species is sympatric with the brush cuckoo and possibly also the rusty-breasted cuckoo, and it is impossible to distinguish the Moluccan cuckoo from these two in the field without better information about their comparative vocalizations and plumages than is currently available. Some minor plumage color differences between it and the brush cuckoo that would facilitate visual field identification are mentioned below. It distinctly resembles the rusty-breasted cuckoo (which occurs on the nearby Moluccas) in having a fairly long tail, rusty-colored underparts, yellow feet, and a barred immature plumage.

In the hand: According to Stresemann (1931), this species can be separated from the sympatric brush cuckoo by the Moluccan's shorter wing (usually <120 mm), its longer tail (>115 mm), its yellow (not brownish or greenish yellow) feet, its dark olive-brown (not grayish blue) upperparts, its dark reddish (not cinnamon) under-tail coverts, and a lighter rufous wash on the breast and abdomen. Distinguishing it from the rusty-breasted cuckoo perhaps is even more difficult, but the available descriptions do not allow for the establishment of diagnostic criteria.

Habitat

This species occurs in montane forests at altitudes of 1000–1500 m.

Host Species
No information.

Egg Characteristics
No information.

Breeding Season
No information, but breeding-condition birds have been collected in early October.

Breeding Biology
No information.

Population Dynamics
No information.

FAN-TAILED CUCKOO
(Cacomantis flabelliformis) [= C. pyrrophanus (Viellot)]

Other Vernacular Names: Ash-tailed cuckoo.
Distribution of Species (see map 38): New Guinea and Australia plus islands of southwestern Pacific (Solomon, Vanuatu)
Subspecies
C. f. flabelliformis: New Caledonia, Loyalty Island.
C. f. prionurus: Australia, including Tasmania.
C. f. excitus: New Guinea.
C. f. meeki: Solomon Island.
C. f. schistaceigularis: New Hebrides, Banks Island.
C. f. simus: Fiji Island.
Measurements (mm)
9.5–11″ (24–28 cm)

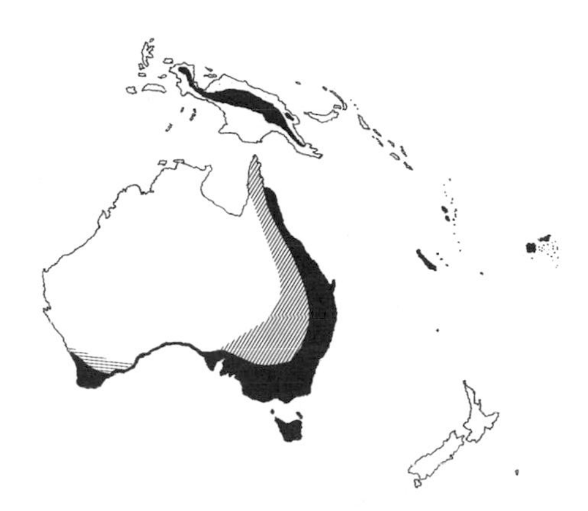

MAP 38. Primary (filled) and secondary or non-breeding ranges (hatched) of fan-tailed cuckoo.

C. f. excitis, wing, males 139–151 (avg. 143), females 142–143 (Rand, 1942a).
C. f. flabelliformis, wing, males 139–145, females 140–143. Tail, males 140–148, females 147–154 (Amadon, 1942). Wing:tail ratio 1:1.04.
C. f. prionurus, wing, 14 males 114–147 (Hall, 1974). Wing, female 138; tail, female 140 (Diamond, 1972). Tail, adult 145 (Frith, 1977). Wing:tail ratio ~1:1.0.
C. f. schistaceigularis, wing, males 129–140, females 130–139. Tail, males 135–144, females 133–148. Wing:tail ratio 1:1.03 (Amadon, 1942).
C. f. simus, wing, males 129–132. Tail, males 134–143. Wing:tail ratio 1:1.05 (Amadon, 1942).
Egg, avg. of *prionurus* 21.2 × 15.2; *flabelliformis* 19.9 × 13.2; *simus* 22.3 × 16. Overall range 19.5–23.3 × 12.8–16.4 (Schönwetter, 1967–84). Shape index 1.39–1.5 (= oval). Rey's index, *simus* 2.23, *prionurus* 2.30, *flabelliformis* 2.62.

Masses (g)

Range of 21 *prionurus* males 36–57 (avg. 44.1), of 8 females 40–50 (avg. 44) (Brooker & Brooker, 1989b). One female of *prionurus* 50 (Diamond, 1972). Males of *flabelliformis* 43.5–55 (avg. 48.2, *n* = 7) (Amadon, 1942). Estimated egg weight, *flabelliformis* 1.85, *prionurus* 2.57, *simus* 3.0 (Schönwetter, 1967–84). Egg:adult mass ratio (*prionurus*) 5.8% to (*flabelliformis*) 6.2%.

Identification

In the field: This mid-sized cuckoo is notable for its relatively long and heavily barred tail (fig. 34). Adults are mostly medium gray on the head and upperparts and pale cinnamon to buffy on the breast and underparts. A bright yellow eye-ring is also present. Immature individuals (presumably subadults) are quite different, with mottled grayish or brown markings on white underparts, a dark brown back with some rufous barring, and less conspicuous tail barring. Juveniles have the usual dark brown and white barring below and are mostly dark brown above, with much barring and notching of buff or cinnamon in the tail feathers. The male's usual song is a fast, mournful "peeer" trill that descends in pitch and is repeated several times. It is louder and the phrase lasts longer (2–3 seconds) than the similar call of the chestnut-breasted cuckoo. There is also a mournful whistle, "wh-phwee," with the preliminary syllable hard to hear unless one is close. Females reportedly utter shrill "preee-ee" notes.

In the hand: The relatively long tail (measuring nearly the same as the wing length), which is usually somewhat darker in color than the gray back and is strongly notched or banded with about seven white bars, distinguishes this bird from all the other Australian cuckoos except for the chestnut-breasted. The latter species is much brighter chestnut below and is somewhat smaller (wing length <120 mm vs. usually >120 mm in the fan-tailed). The sexes are similar as adults. Some island races have blackish to dark greenish olive upperparts. There is also a melanistic plumage morph in at least some races, in which the blackish underpart barring typical of juveniles is retained to varying degrees in adults. Juveniles are strongly barred below and lack the rufous underparts of adults but have tawny to rufous edging on the upperpart feathers and are generally dark brown to dark gray above. Their tail is also brown to blackish, the feathers usually barred and notched with white and pale rufous, but such barring is much reduced in melanistic individuals. Both sexes are mottled or barred with brown below. Older immature birds increasingly resemble adults, the young males gradually becoming fawn-pink on the breast. Young females are grayer than immature males, and have some dark barring present on the underparts. The gape color of adults is yellow-orange to red-orange, whereas that of juveniles and nestlings is yellow. Chicks are hatched with a flesh-colored skin that turns dark brownish black in a few days, and they have dark brown iris coloration and light horn-colored feet.

Habitats

This species is associated with open woodlands, low, dry rainforests, tropical woodlands, and woodland edges or second-growth forest. It also sometimes occurs in areas with sparse vegetation, but it generally favors denser habitats than the pallid cuckoo. In Australia it is mostly associated with open forests and woodlands, especially wooded ridges and mountain slopes. In New Guinea it mainly occurs from 1500 to 3000 m, sometimes extending from 1200 to 3900 m in forests and second growth. It mostly occurs above the altitudinal levels used by the two other resident *Cacomantis* species (brush and chestnut-breasted cuckoos).

Host Species

Brooker & Brooker (1989b) provided a list of 81 host species representing 662 records of parasitism in Australia. Of these, 17 species were provisionally identified as biological (fostering) hosts. The white-browed scrub wren and the brown thornbill (which is sometimes separated into two species) are perhaps the most important hosts and are parasitized throughout their ranges of sympatry with the cuckoo. Table 21 lists four additional species that represent records of parasitism exceeding 2% of the total. New Guinea hosts include the large scrub wren and the white-shouldered wren. Schönwetter (1967–84) lists the Fiji bush warbler as a host on the Fiji Islands and mentions several additional reported hosts of the nominate race. In New Hebrides the scarlet robin is a known fostering host (Amadon, 1942).

Egg Characteristics

Brooker & Brooker (1989b) described the eggs as rounded or elongated ovals ("oval" using this book's definition), with a dull white ground color and with spots and blotches of purplish brown that are sometimes organized into a distinct zone around the more rounded end. In spite of the large number of known Australian hosts, there is no indication there of egg polymorphism.

Breeding Season

In Australia there are egg records extending from July to January, but the majority (69%) of 469 records are for the period September–November. There is little indication of regional differences in egg dates (Brooker & Brooker, 1989b).

Breeding Biology

Nest selection, egg laying. All of the identified biological hosts of this species build enclosed rather than open-cup nests, although eggs are sometimes deposited in open-cup nests. The exact method of egg introduction into such nests is still unknown, but it is apparent that the cuckoo mush reach in and remove a host egg at the time of laying, as parasitized clutches are smaller on average (by about one egg) than unparasitized ones (Brooker & Brooker, 1989b).

Incubation and hatching. The incubation period of one egg was less than 13 days and 5 hours. Within 2 days of hatching, the young cuckoo evicts any host eggs or young (Brooker & Brooker, 1989b).

Nestling period. The nestling period lasts 16–17 days and is followed by a fledgling-dependency period of about 3 or 4 weeks (Brooker & Brooker, 1989b).

Population Dynamics

Parasitism rate. Little information is available. One early observer (McGlip, 1929) noted that scrub wren nests found in July and early August contained cuckoo eggs of this species or of *Chrysococcyx,* but that later nests were not parasitized. Yet, in New South Wales the two major hosts (white-browed scrub wren and brown thornbill) may breed early enough to have their first broods escape parasitism. In two different years the rate of parasitism of the yellow-throated scrub wren varied from 2% to 17% (Marshall, 1931; Brooker & Brooker, 1989b). Among 39 nests or fledged broods of brown-backed honeyeaters, at least 12 pairs had been parasitized (Miller, 1932).

Hatching and fledging success. No information.

Host–parasite relations. Nest desertion by the principal hosts, the white-browed scrub wren and brown thornbill, is said to be common following cuckoo parasitism.

LONG-BILLED CUCKOO

(Rhamphomantis megarhynchus)

Other Vernacular Names: Little long-billed cuckoo, little koel.

Distribution of Species (see map 39): New Guinea and nearby islands (Aru, Waigeu, Misol).

Subspecies

R. m. megarhynchus: New Guinea, Aru Island.

R. m. sanfordi: Waigeu Island (New Guinea).

Measurements (mm)

7″ (18 cm)

Wing (unsexed) 93–97, females to 101. Tail (unsexed), 69 (Rand & Gilliard, 1969). Wing:tail ratio ~1:0.7.

Egg, no information.

Masses (g)

No information.

Identification

In the field: This New Guinea endemic has a distinctive long and slightly decurved bill, and males have a bright red iris and eye-ring, contrasting with a blackish head (see fig. 37). Otherwise the birds are mostly brown, with no strong barring evident except perhaps on the outer tail feathers. Females are generally brighter and more cinnamon-colored, with finely barred underparts. Immature birds are brownish throughout, lacking the black head and bright red eyes of adults. The long, decurved bill is diagnostic for all age and sex categories. The male's usual song is a trill of descending and uniformly spaced notes that lasts about 4 seconds and may be repeated at approximate 5-second intervals.

In the hand: The relatively long (exposed culmen 21 mm), slender, and slightly decurved bill distinguishes this from all other New Guinea cuckoos.

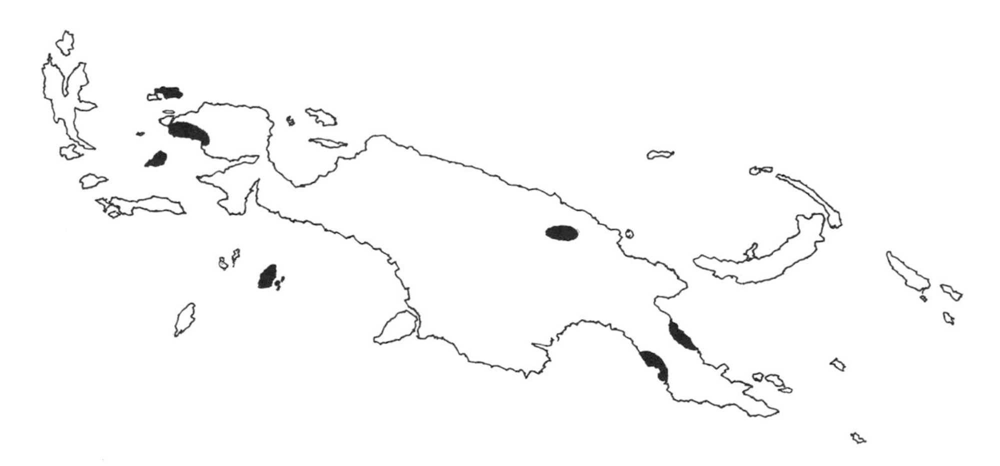

MAP 39. Range of long-billed cuckoo.

Habitats

This species occurs in rainforests, monsoon forests, forest edges, and secondary growth of the New Guinea lowlands.

Host Species

No information.

Egg Characteristics

No information.

Breeding Season

No information.

Breeding Biology

No information.

Population Dynamics

No information.

LITTLE BRONZE CUCKOO

(Chrysococcyx minutillus)

Other Vernacular Names: Australian bronze cuckoo, Gould's bronze cuckoo (*russatus*), Malaysian bronze cuckoo (*peninsularis*), Malaysian emerald (*peninsularis*), rufous bronze cuckoo (*russatus*), rufous-breasted bronze cuckoo (*russatus*), rufous-throated bronze cuckoo (*russatus*)

Distribution of Species (see map 40): Malayan Peninsula, Sumatra, Borneo, Java, Lesser Sundas, Sulawesi and nearby Moluccas, southern Philippines, New Guinea, northern and northeastern Australia.

Subspecies (arranged roughly from northwest to southeast)

C. m. peninsularis: Malayan Peninsula (previously known as *"malayanus"*), including Malaysia and extreme southeastern Thailand (but probably not elsewhere in Indochina).

C. m. albifrons: Western Java, northern Sumatra.

C. m. salvadorii: Babar Island (Lesser Sundas).

C. m. subspecies?: Timor Island. Probably a still-unnamed form in *russatus* complex (Parker, 1981).

C. m. cleis: Eastern and northern Borneo, where apparently sympatric with *aheneus.*

C. m. aheneus: Eastern and northern Borneo. Sulu archipelago, southern Philippines (part of *russatus* complex).

C. m. jungei: Southwestern and central Sulawesi, Madu and Flores Island (part of *russatus* complex).

C. m. misoriensis: New Guinea lowlands and neighboring islands (part of *russatus* complex).

C. m. poecilurus: Western Papuan islands, western and southwestern New Guinea. Considered a possible full species by Parker (1981).

C. m. minutillus: Northern Australia (Kimberly to Cape York Peninsula), also (probably only as migrants) on the Lesser Sundas, Moluccas, and New Guinea.

C. m. russatus: Cape York Peninsula (south to Bowen) and nearby islands, intergrading with *minutillus* on Cape York.

C. m. barnardi: Eastern Queensland (south of Bowen) and northeastern New South Wales, Australia.

Measurements (mm)

6″ (15–16 cm)

C. m. aheneus, wing, males 91–98, females 88–96. Tail males 60–64, females 60–64. Bill width at nostrils, males 5.1–6.2 (avg. 5.45), females 5–6.3 (avg. 5.7) (Parker, 1981).

C. m. barnardi, wing, males 94–105.5; females 93–101. Tail, males 62.5–70; females 59–68 (Ford, 1981). Wing:tail ratio ~1:0.7.

C. m. cleis, wing, males 89–93, tail 60–63. Bill width at nostrils 4–4.7 (avg. 4.45) (Parker, 1981).

C. m. minutillus, wing, males 89–98; females 88–97. Tail, males 58–66.5; females 57.5–64 (Ford, 1982). Wing:tail ratio ~1:0.7.

C. m. peninsularis, wing, males 91–97, females 83, 94. Tail, males 60–66, females 62, 63 (Parker, 1981).

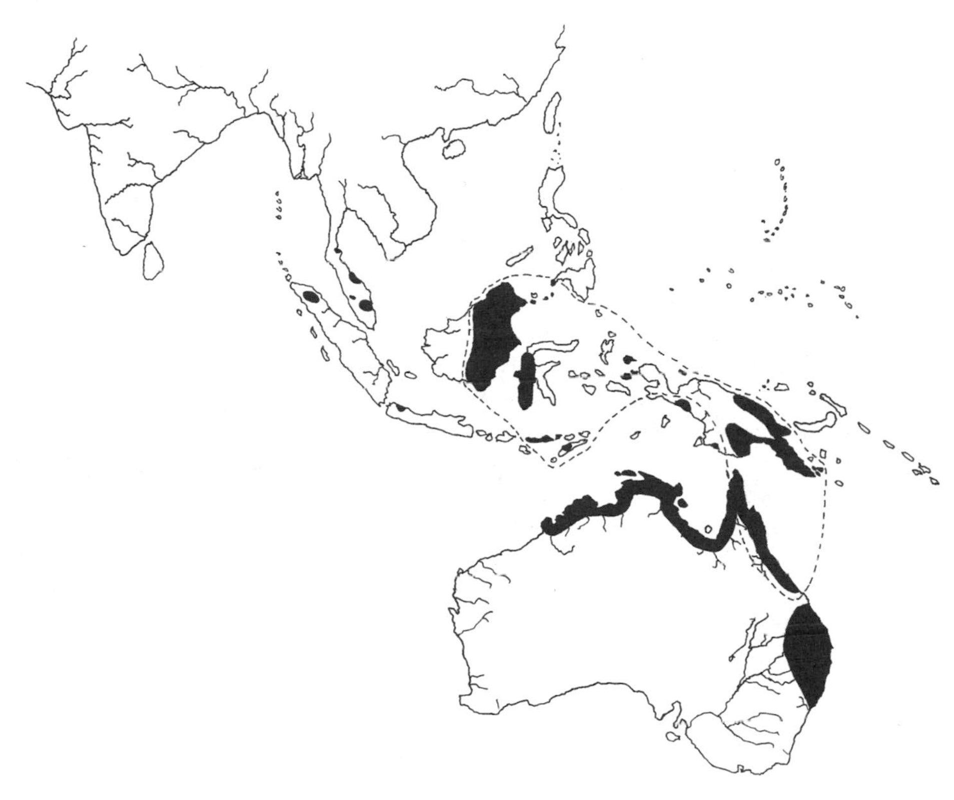

MAP 40. Range of little bronze cuckoo. Populations often attributed to form *russatus* are enclosed within the dashed line.

C. m. poecilurus, wing, males 87–96; females 90–97. Tail, males 61–67; females 59–65.5 (Ford, 1982). Wing, males 101, 101, females 96.5–100. Tail, males 65, 70; females 64–70 (Parker, 1981). Wing:tail ratio ~1:0.6–0.7.

C. m. russatus, wing, males 89–98; females 88.5–98. Tail, males 58–68; females 58.5–67 (Ford, 1982). Wing:tail ratio ~1:0.6–0.7.

All subspecies, wing, males 88–106.5, females 83–100. Tail, males 55–70, females 53–70 (Parker, 1981).

Egg, avg. of *peninsularus* 18 × 12.8; *poecilurus* 19.6 × 13.7; *russatus* 20.3 × 13.6; *minutillus* 18.9 × 13.7. Overall range 18.5–21 × 13.3–14.7 (Schönwetter, 1967–84). Shape index 1.38–1.49 (= oval). Rey's index (*poecilurus*) 2.24.

Masses (g)

C. m. aheneus, four males 17.5–21.1 (avg. 18.9), 2 females 17.5 (Thompson, 1966).

C. m. minutillus, five adult males 16.9–20.2 (avg. 18.4); two females 18.6, 18.7 (Hall, 1974).

Range of 14 Australian males (subspecies unstated) 14.5–20.2 (avg. 16.8), of 6 females 15.4–18.7 (avg. 17.4) (Brooker & Brooker, 1989b).

Estimated egg weight of *peninsularis* 1.5, *minutillus* 1.85, *poecilurus* 2.0, *russatus* 2.02, (Schönwetter, 1967–84). Egg:adult mass ratio (*minutillus*) 9.9%.

Identification

In the field: This is a small, bronze-green cuckoo with bright red eyes in adults, broad flank barring that does not extend across the white abdomen, and finer dark head barring that extends to the cheeks and sometimes to above the eyes (fig. 35). The eye-ring of males is consistently red; in females it varies from grayish green (especially the *minutillus* group) to yellow (especially in the *russatus* group). In the rufous-breasted (*russatus*) group the outer tail feathers are strongly tinged with rufous, and the barring on the head is less extensive, barely reaching the cheeks. Juveniles are paler than adults, lack red eyes, and have only slight barring on the flanks. In New Guinea the male's song is a descending series of five notes, with a slight delay after the third note. In Australian birds the usual song is a distinctive downward trill of four notes, "tew-tew-tew-teew". Or a preliminary long and downward inflected preliminary note may be followed by about five shorter and also downwardly inflected notes, the entire series lasting about 2 seconds. This song is apparently supplemented by a high-pitched "grasshopper-like" trill of uniform pitch, also lasting about 2 seconds.

In the hand: The close relationship between the *russatus* group of populations and the others has been discussed by many authors, and at least in Australia it appears to be impractical to try distinguish these two types consistently. Ford (1982) has summarized the differences in the most divergent individuals of nominate *minutillus* and *russatus.* Generally speaking, *russatus* is more rufescent overall than *minutillus,* especially on the breast, upperparts, and tail. Thus, *russatus* has a well-developed rufous tone on the breast and upper tail (versus little or no rufous in *minutillus*), considerable (vs. little or no) rufous on the two outermost rectrices, edges of the wing coverts, and the underwing coverts, broad (vs. narrow) ventral barring, strong (vs. weak) bronzy dorsal gloss, weak (vs. strong) sexual dimorphism, and weak (vs. strong) ear covert patterning. White flecking on the forehead and lores is more common in *minutillus,* and *russatus* usually has fewer black bars on the inner vane of the outermost (fifth) pair of rectrices. In *russatus* and *poecilurus* the eye-ring of females is likely to be yellow, tan, or orange, whereas in *minutillus* and *barnardi* it is gray, cream, or greenish. As for distinguishing *russatus* from the New Guinea form *poecilurus,* the latter has less rufous on the throat and is more pinkish-bronze above. However, *poecilurus* is notably variable in its tail coloration, the color and width of the ventral barring, the amount of rufous on the breast, and the amount of white on the face.

Females of all these forms have brown, rather than red, eyes, and sexual differences in plumage are more pronounced in the *minutillus* group than in *russatus.* Thus, females of the *minutillus* group are dull green on the crown and back, whereas the males have a glossy emerald green crown and a green to bronze-green back color. In contrast, in Borneo the two forms *cleis* (in *minutillus* group) and *aheneus* (in *russatus* group) may be sympatric without apparent interbreeding, and thus seem to act as good species. In Borneo, the width of the bill at the nostrils may be diagnostic (<5 mm for *cleis,* ≥ 5 mm for *aheneus*), and this interesting morphological difference may represent an example of character displacement (Parker, 1981). Immature birds of all forms are less bronzy above and are only faintly barred on the flanks below. They have dark brown eyes and inconspicuous eye-rings. The mouth color of immature birds is whitish to pale yellow. Among adults it is black, grading to pink in the throat.

Habitats

In New Guinea this species occurs from sea level to a maximum of 1400 m (mostly occurring under 500 m), and the three subspecies occurring there occupy such diverse habitats there as rainforests, monsoon forests, gallery forests, forest edges, mangrove swamps, secondary growth, and gardens. In Australia, nominate *minutillus* occurs in a wide variety of habitats, occupying rainforest edges, tall woodlands with heavy undergrowth, dense mangrove thickets, swamp woodlands, and

FIGURE 35 Profile sketches of eight Asian *Chrysococcyx* cuckoos: adults of two races (A = *plagosus,* B = *lucidus*) of shining bronze cuckoo; adults of little (C) and Gould's (D) bronze cuckoos; adult male (E) and female (F) of white-eared bronze cuckoo; adult (G) and juvenile (H) of Horsfield's bronze cuckoo, adult male (I) and female (J) of Asian emerald cuckoo; adults of black-eared (K) and (L) rufous-throated bronze cuckoos. Their eggs and undertail patterns are also shown.

tropical savanna gallery forests. However, also in Australia, *russatus* may be more prone to use open forests, tropical scrub, orchards, and gardens. In the Moluccas, *jungei* is found in dense bamboo thickets. In Borneo, swamp forests, mangroves, and open forests are used by *aheneus,* and secondary thickets, gardens and plantations in lowlands to about 800 m are preferred by *cleis.* In Java *albifrons* occupies swampy riverine forests, and the same race occurs at elevations of 500–1000 m in Sumatra.

Host Species

Major Australian host species, as summarized by Brooker and Brooker (1989b), are shown in table 13. They reported that 23 host species have been documented among a sample of 193 records of parasitism. Only four of these species were regarded as biological hosts, all of them species of gerygones. The largest number of host records were for the large-billed gerygone; this species and the fairy gerygone are major hosts in northern Australia. Two other gerygones, the mangrove and white-throated, are also probably hosts. In the Lesser Sundas, various species of gerygones, including the golden-bellied gerygone, are probably the chief hosts in the Moluccas (White & Bruce, 1986). On Java the golden-bellied gerygone is also parasitized, and the large-billed gerygone is parasitized by several races of bronze cuckoos that occur within its breeding range of northern Australia, New Guinea, and the Aru and Papuan islands (Schönwetter, 1967–84).

Egg Characteristics

The eggs of this species are elongated ovals, with a glossy surface that is bronze, olive-green, greenish brown or dark brown, and often with darker freckling at the more rounded end. The eggshells average 0.12 g and 0.08 mm in thickness (Schönwetter, 1967–84).

Breeding Season

In the Malay Peninsula eggs have been reported for March and August (Medway & Wells, 1976). Breeding for most insectivorous birds in Borneo, and thus presumably cuckoos as well, occurs between January and June (Parker, 1981). A total of 126 Australian breeding (egg) records include all the months except June and April, with the majority (53%) of records occurring between November and January. In north Queensland the breeding season is widely spread over 9 months (July–March), but in southern Queensland and northern New South Wales the breeding spread is only 5 months, from September to January (Brooker & Brooker, 1989b).

Breeding Biology

Nest selection, egg laying. Little information exists. Brooker & Brooker (1989b) reported that female cuckoos usually remove a host egg at the time of laying, since parasitized host clutches average about one egg less than unparasitized ones. In 147 parasitized nests, there was 1 cuckoo egg in 128, 2 in 16, and 3 in 3. Seaton (1962) reported seeing a cuckoo "carrying an egg in its bill" as it approached a yellow-breasted sunbird's nest. It then "clung to the side of the nest and, placing its head in the aperture, deposited the egg in the nest chamber" (p. 176). This remarkable account, if accurate, would help to explain how the eggs may be introduced into otherwise well-protected host nests.

Incubation and hatching. No information.

Nestling period. No information.

Population Dynamics

No information.

GREEN-CHEEKED BRONZE CUCKOO

[Chrysococcyx (minutillus) rufomerus]

Other Vernacular Names: Lesser Sundan bronze cuckoo. sometimes considered conspecific with the little bronze cuckoo and/or with the pied bronze cuckoo.

Distribution of Species (see map 41): Eastern Lesser Sundas (Kisar, Romang, Damar, Leti, Moa, and Sermata islands). Erroneously reported from Wetar.

Measurements (mm)

6″ (15–16 cm)

MAP 41. Ranges of white-eared (filled), pied (enclosed area), and green-cheeked (shaded) bronze cuckoos.

Wing, males 93–98.5 (avg. 95.87, n = 21), females 92, 95. Tail, males 62.5–72.5 (avg. 66.31, n = 19), females 67, 72. Bill width at nostrils, males 5–6.1 (avg. 5.68), females 5.8, 6 (Parker, 1981). Wing of females 92–97.5, tail 62–68.5 (Ford, 1982). Wing:tail ratio 1:0.69.

Egg, no information.

Masses (g)

No information, but must be similar to *minutillus,* considering their similar linear measurements.

Identification

In the field: This is often regarded as a subspecies of *minutillus,* and is very similar. However, at least breeding-condition birds of the genus *Chrysococcyx* occurring on the eastern Lesser Sundas can probably be safely attributed to this species. The identity of the breeders on Timor is uncertain. Both sexes resemble the little bronze cuckoo, with bright red eyes, barred flanks and underparts, and glossy greenish upperparts. Males are dark green above and are strongly barred with green below. Females are similar but are somewhat more brownish above and lack a bright red eye-ring, which instead is yellow. Typical vocalizations still remain to be described, as well as adequate descriptions of juveniles and females.

In the hand: Males are dark bronze-green above, with a slightly greener crown. No distinct white wing patch is present on the upper wing, but the secondary coverts are narrowly edged with white, and the white frosting on the head is confined to the lores and above the eyes. The face has a broad, dark green cheek smudge. The underparts are barred, and the tail resembles that of the pied bronze cuckoo in being blackish green, with all but the central pair of rectrices having white terminal spots or banding. The presence of narrow white edging on the upper-wing coverts, and dusky (rather than rufous-buff) undersides of most rectrices (all but the outermost pair) are said to be distinctive criteria for distinguishing this form from *minutillus.* However, a degree of intermediacy between the two types is especially evident among specimens from Roma, Letti, and Moa islands, as well as in a single specimen from Sermatta (Ford, 1982). However, Parker (1981) attributed the similarities existing between these two forms (namely, some rufous on the outer rectrices in a few specimens of *rufomerus*) to be the result of individual variation or perhaps the retention of a preadult characteristic, rather than indicative of possible hybridization or intergradation between them.

In its plumage traits *rufomerus* also closely resembles *crassirostris,* but in that species the tail is mostly blackish blue, whereas in *rufomerus* the tail is mostly blackish green, and the white edging of the wing-covert feathers also is greater in *crassirostris.* These two forms seem to differ in bill width as well (see measurements). In both species the eye-ring is vermilion red in males, and at least in this species the eye-ring is pale yellow in females (not known for *crassirostris*).

Habitats

No specific information on habitats is available, but these are probably much like those used by *crassirostris.* Habitats probably consist of coastal mangrove tangles where the host gerygone is found.

Host Species

The rufous-sided gerygone, a bird of coastal mangrove habitats, is the major or perhaps sole host of this little-known species (Parker, 1981). In addition to those islands known to be occupied by *rufomerus* (Kisar, Romang, Leti, Moa, Damar, and Sermata) or by *crassirostris* (Tanimbar and Kal), this host species also occurs on Kalaotoa, Madu, and Babar islands.

Egg Characteristics

No information.

Breeding Season

No information.

Breeding Biology

No information.

Population Dynamics

No information.

PIED BRONZE CUCKOO

[Chrysococcyx (minutillus) crassirostris]

Other Vernacular Names: Island bronze cuckoo. Sometimes considered conspecific with the little bronze cuckoo (e.g., Mayr, 1939; Peters, 1940). Also possibly conspecific with the previous species (Deignan & Amos, 1950; White & Bruce, 1985), but regarded by Parker (1981) as a distinct species.

Distribution of Species (see map 41): Tanimbar, Kur, and Kal islands (in eastern Banda Sea of eastern Indonesia).

Measurements (mm)

6″ (15–16 cm)

Wing, males 89–96.6 (avg. 92.5, $n = 3$), females 90–96.2 (avg. 92.3, $n = 6$). Tail, males 64–66.5 (avg. 64.83, $n = 3$), females 60.2–62.9 (avg. 61.25, $n = 6$). Bill width at nostrils, males 6, 6.6 (avg. 6.3), females 5.8–6.8 (avg. 6.25) (Parker, 1981). Tail lengths of 6 specimens at the American Museum of National History are longer (75–85 mm), but wing lengths are nearly the same (E. Levine, personal communication). Wing:tail ratio 1:0.7.

Egg, no information.

Masses (g)

No information, but unlikely to differ from *minutillus,* given their nearly identical wing lengths.

Identification

In the field: This species has a distinctive adult plumage which in males is dark blue and in females is emerald green on the upperparts; both sexes have unbarred underparts (fig. 36). The auriculars and area below the eyes are dark blue in males and are streaked with brown, gray, and white in females. The eye-ring is bright red in males and perhaps also in females (Parker, 1981). The blackish tail is tipped with white, and the outer vanes of the outer rectrices are also strongly edged with white, but little or no rufous is present. The upper-wing coverts have broad white edges forming a large white patch, and the upper-tail coverts are also edged with white. Juveniles have rufous edgings on the tail and upper-wing coverts, and the underparts are slightly barred on the flanks and breast. The lower surface of the tail is also strongly tinted with rufous. No information is yet available on the species' vocalizations, which are probably much like those of *minutillus.*

In the hand: The upperparts of males are glossy blackish blue, with some green shine; a variable white patch is present on the upper-wing coverts, and the flanks are weakly barred. The tail feathers are mostly blackish, with white tips or white barring; the central pair is uniformly blackish green. Females are more oil-green above and have a brownish-tinged crown. Males have a red eye-rim; that of females is unknown. Ford (1982) reported that a single specimen of *salvadorii* from Babar Island that he examined was geographically and morphologically intermediate between *crassirostris* and *rufomerus.* It is thus possible that these two populations will eventually be considered conspecific. However, Parker (1981) did not consider a single specimen adequate to provide sufficient evidence for making a decision and believed it might be an aberrant specimen or represent an unknown plumage stage.

Habitats

No specific information is available on habitats, but these are probably much like those described for *rufomerus* and presumably consist of mangrove tangles used for breeding by the host gerygone species.

Host Species

Only known to parasitize the rufous-sided gerygone. This host species has not yet been reported from Sorong, Halmahera, Ternate, Ambon, or Gorong islands. The pied bronze cuckoo has occurred on at least some of these islands, but these records are said to require confirmation (Parker, 1981).

Egg Characteristics

No information.

Breeding Season

No information.

Breeding Biology

No information.

Population Dynamics

No information.

SHINING BRONZE-CUCKOO

(Chrysococcyx lucidus)

Other Vernacular Names: Broad-billed bronze-cuckoo, golden bronze-cuckoo (Australia); greenback (Australia), shining cuckoo, whistler (New Zealand).

Distribution of Species (see map 42): Australia, New Zealand, islands of southwestern Pacific.

Subspecies

C. l. lucidus: New Zealand, Lord Howe and Norfolk islands, wintering mostly on the Solomon Island.

C. l. plagosus: Australia to Lesser Sundas and New Guinea.

C. l. layardi: New Caledonia; Loyalty and Santa Cruz Islands (Solomon Island).

C. l. aeneus: Banks Island (Vanuatu Island).

C. l. harterti: Rennell and Bellona islands (Solomon Island).

Measurements (mm)

7″ (17–18 cm)

C. l. harterti, wing, both sexes 90–95; tail, both sexes 60–64 (Mayr, 1932). Wing:tail ratio ~1:0.7.

C. l. layardi, wing, both sexes 96–101; tail, both sexes 66–73 (Mayr, 1932). Wing:tail ratio ~1:0.7.

MAP 42. Primary breeding (filled), secondary or migrant (hatched) and wintering (enclosed area) ranges of shining bronze cuckoo.

C. l. lucidus, wing, males 98–107; tail, males 67–70 (Oliver, 1955). Wing, both sexes 102–107. Tail, both sexes 66–70 (Mayr, 1932). Wing:tail ration ~1:0.7.
C. l. plagosus, wing, adults of both sexes 102.5–108; tail, both sexes 65–70 (Mayr, 1932). Wing:tail ratio ~1:0.6.
Egg, avg. of *plagosus* 18.3 × 12.9 (range 17.2–19.4 × 12–14.2); *lucidus* 18.9 × 13.1 (range 18–20.3 × 12.5–15.2) (Schönwetter, 1967–84). Shape index 1.42–1.44 (= oval). Rey's index, *lucidus* 2.47, *plagosus* 2.48.

Masses (g)

C. l. harterti, three males 19–20.5 (avg. 19.8), female 19 (Mayr, 1932).
C. l. lucidus, avg. of 19 (sexes unstated) 24.8 (range 21.9–27.5) (Dunning, 1993).
Range of 31 Australian males (subspecies unstated) 18–35 (avg. 24), 16 females 16–31 (avg. 22.9) (Brooker & Brooker, 1989b). Three males of *plagosus* 18–22.8 (avg. 19.9) (Hall, 1974).
Estimated egg weight of *plagosus* 1.66, *lucidus* 1.73 (Schönwetter, 1967–84). Egg:adult mass ratio (*lucidus*) 7.0%.

Identification

In the field: This small cuckoo resembles the other bronze cuckoos in that it is bronze-green above and barred dark and white below, the dark barring extending across the belly (fig. 35). The cheeks are only finely barred, and the eyes of males and females are dark brown (rather than bright red as in males of the little bronze cuckoo), with little or no white extending above the eye. The tail of adults is never tinted with rust color. The male's usual song in Australia is an extended series of upslurred whistles, "Su'wee, su'wee, su'wee. . .", sounding much like a person whistling to attract a dog, which may be followed by some downwardly inflected "peee-eerr" trills. In New Zealand the song has been quite differently described as resembling, "kui, kui, whiti-whiti ora, tio-o," which begins with upward-slurred "kui" or "whiti" notes that are followed by a few downward-slurred notes. The call note is a clear "tsui." Juveniles are less iridescent that adults, but do have distinct flank and underpart barring. They resemble the young of little bronze cuckoos but have more definite flank barring and no rust color present on the tail.

In the hand: Both sexes of this species are iridescent green above, with a more bronze cast in Australian birds and more green in the New Zealand population. Adults of both races have glossy bronze barring on the flanks that extends around the abdomen and reaches forward to the chin. The face is mostly finely barred, but in New Zealand the sides of the neck may be nearly immaculate, and white facial freckling extends farther forward on the forehead than in the Australian race. Females are somewhat more purplish bronze on the crown and nape than are males, and their abdominal barring is more bronzy. Immature birds (juveniles and subadults) are less bronze above than adults and have non-glossy, brownish barring on the flanks and breast. The mouth color of immature birds and females is yellow; in adult males it varies from fleshy or orange to grayish black, with a pink throat. Nestlings are pinkish orange on the shoulders and more grayish on the head and back at hatching, but the grayish color spreads and darkens with age. The mandibular flanges may be white (as in New Zealand) or bright yellow (southwestern Australia) (Gill, 1982; Brooker & Brooker, 1989a).

Habitats

In New Guinea this species occurs from sea level to nearly 200 m and occupies second-growth woodland, forest edges, scrub, savanna, mangrove tangles, gardens, casuarina (hoop pine) groves, and occasionally pine plantations. In Australia it is widespread in a variety of forest and woodland habitats, generally using rather denser habitats than Horsfield's cuckoo, but including rainforests, temperate forests, mixed woodlands, riverine forests, scrublands, golf courses, orchards, and gardens. In New Zealand it is mainly a forest dweller, including planted pine forests, but it also occurs in cultivated and residential areas wherever trees and shrubs are present.

Host Species

Brooker & Brooker (1989b) provided a list of 10 probable biological hosts of this cuckoo in Australia based on 909 records of parasitism involving a total of 82 possible host species. Eight of these species are listed in table 13; other species with smaller numbers of records are the splendid fairywren and the Tasmanian thornbill. Brooker & Brooker questioned whether fairywrens are important hosts and suggested instead that thornbills are the major hosts in all regions. In a separate study, Brooker & Brooker (1989a) reported that in Western Australia this species and the Horsfield's bronze cuckoo both parasitize one major host species, the western thornbill, but at differing rates. The shining bronze cuckoo largely concentrates on the yellow-rumped thornbill, but does not show egg mimicry (the domed nests are so dark that dark eggs may be desirable, and visual mimicry is unnecessary). However, Horsfield's cuckoo concentrates on the splendid fairywren, producing highly mimetic eggs. In New Zealand the gray gerygone is the shining bronze cuckoo's primary host (Gill, 1983). Additional native New Zealand genera that have been parasitized, not necessarily successfully, include *Petroica, Miro, Rhipidura, Mohoua, Anthornis,* and *Zosterops* (Schönwetter, 1967–84).

Egg Characteristics

The eggs of this species are oval in shape, with a dull surface. Australian eggs have a ground color of greenish white, bluish white, olive-green, olive-brown, or bronze, and typically unspotted. In Western Australia the eggs are olive-brown and do not mimic those of the primary thornbill hosts (Brooker & Brooker, 1989a). In New Zealand the eggs vary from greenish or bluish white to olive-brown or dark greenish brown and also do not mimic the eggs of the gray gerygone, the species' primary host. The dark egg color may be removed with a wet finger. The eggshell averages 0.095–0.1 g in mass and 0.06–0.065 mm in thickness (Schönwetter, 1967–84). The Rey's index is not suggestive of significant eggshell thickening in this cuckoo.

Breeding Season

In Australia there are egg records extending from June to January, but the great majority (77% of 637 records) cover the 3-month period September–November, plus a substantial number (12%) for December as well (Brooker & Brooker, 1989b). In a western Australian study, this species was found to lay for a 13-week period, starting in late August, and was well synchronized with breeding by the primary and secondary hosts (Brooker & Brooker, 1989a). In New Zealand the laying period occurs during November and December, during the austral spring (spring arrival typically occurs from August to October), and falling within the August–January overall breeding span of its host the gray gerygone (Oliver, 1955). Singing terminates there by early February. Other breeding cycles have not been so well documented.

Breeding Biology

Nest selection, egg laying. Brooker et al. (1988) videotaped one case of parasitism by this species. The laying occurred at 6:50 A.M. about an hour after sunrise, and required only 18 seconds. The cuckoo entered the nest, leaving her wings and tail partly exposed, and soon emerged backward, carrying a host egg, and without damaging the next. Only in 7 of 94 cases did females fail to remove a host egg, and on 13 occasions 2 eggs may have been removed. One egg is normally laid per host nest, but in a few cases (31 of 833) two cuckoo eggs have been found in a single nest (Brooker & Brooker, 1989b).

Incubation and hatching. The incubation period has been estimated for a variety of hosts; the shortest estimate is 13.5 days, and the longest is 15.5 days. This compares with, for example, the gray gerygone's mean incubation period of 19.5 days (Gill, 1983), the splendid fairywren's period of 13–14 days, and 18–20 days for thornbills (Brooker & Brooker, 1989a). The newly hatched cuckoos evict their host's eggs or nestlings when the cuckoo chicks are 42–56 hours old (Brooker & Brooker, 1989b) or at about 3–7 days (Gill, 1983).

Nestling period. Nestling periods have been estimated for several host species, with observed durations ranging from about 17.5 to 23 days. In four nests of yellow-rumped thornbills, the mean duration was 20.5 days. The subsequent period of postfledging dependence ranged from about 6 to 22 days in seven observed instances, but there are reports of feeding extending for as long as 28 days (Brooker & Brooker, 1989a, b).

Population Dynamics

Parasitism rate. Ford (1963) estimated a 16% parasitism rate for 113 yellow-rumped thornbill nests near Perth, and Gill (1983) reported a 55% parasitism rate for 40 gray gerygone nests in New Zealand. A study by Brooker & Brooker (1989b) in Western Australia revealed parasitism rates of 26% of 135 yellow-rumped thornbill nests, and 8% of 226 western thornbill nests, with substantial between-year differences evident.

Hatching and fledging success. The hatching success of this cuckoo on yellow-billed thornbill nests was 21 of 35 eggs (60%), of which 17 (81%) survived to near-fledging (14–15 days), resulting in an overall breeding success rate of 49%. The hatching success on western thornbill nests was 13 of 18 eggs (72%), of which 11 (85%) survived to this age, resulting in an overall breeding success rate of 61%. Of 20 parasitized yellow-billed thornbill nests in Ford's (1963) study, 12 cuckoos hatched, and all of these fledged (overall breeding success 60%). Of 23 cuckoo eggs studied by Gill (1982), 16 (70%) hatched, and 12 of the 16 survived to fledging (fledging success 75%, overall breeding success 52%).

Host–parasite relations. Considering the fairly high rates of parasitism on yellow-rumped thornbill nests and the similarly high breeding success of cuckoos with this host, the cuckoo must represent a significant factor in influencing the thornbill's productivity rates. Gill (1983) reported that there was a 32.9% breeding success among 70 gerygone eggs in unparasitized nests, versus a 1.9% success rate among 53 gray gerygone eggs in parasitized nests. Thus, there was a reduction of 31% in gerygone production in parasitized nests. Since there was a 55% incidence of parasitism, the overall effect of parasitism during the latter part of the nesting season was to produce a reduction in gerygone productivity of 17.1%.

HORSFIELD'S BRONZE CUCKOO

(Chrysococcyx basalis)

Other Vernacular Names: Bronze cuckoo, narrow-billed bronze cuckoo, rufous-tailed cuckoo.

Distribution of Species (see map 43): Australia (including Tasmania), wintering north to Indonesia and New Guinea.

Measurements (mm)

6.5″ (17 cm)

Wing (unsexed) 101, tail (unsexed) 69 (Rand & Gilliard, 1969). Wing:tail ratio 1:0.7.

Egg, avg. 18.1 × 12.7 (range 16.9–19 × 11.9–13.3) (Schönwetter, 1967–84). Shape index 1.42 (= oval). Rey's index 2.55.

Masses (g)

Range of 19 males 17.5–27.5 (avg. 21.2), of 10 females 18.7–27.5 (avg. 22.9) (Brooker & Brooker, 1989b). Adult males 18.8–20.2, adult females 22.2–27.5, immature birds of both sexes 21–22.2 (Hall, 1974). Estimated egg weight 1.55 (Schönwetter, 1967–84). Egg:adult mass ratio 7.0%.

Identification

In the field: This small bronze cuckoo resembles the shining cuckoo in that its eyes are dark brown rather than bright red, but its flank barring is weaker and incomplete below (fig. 35). It also has a distinct white stripe passing above and behind the eye, isolating an ear patch of bronze-green (not black, as in osculans) feathers. The outer tail feathers are barred blackish and white, with some rusty tinting in the outer feathers. The male's song is a series of whistled "prelll" or "tseeeeuw" notes with downward inflections, which average faster and higher in pitch than those of the black-eared cuckoo. Immature birds have no barring on the generally light gray flanks and underparts.

MAP 43. Primary breeding (filled), secondary or migrant (hatched) and wintering (enclosed area) anges of Horsfield's bronze cuckoo.

In the hand: The absence of barring on the central abdomen feathers help to identify this species. Its outer tail feathers are slightly tinted with brownish (especially the basal halves of the outer three rectrices), and its inner flight feathers and longer scapulars are also tipped with brown.

Juveniles generally resemble adults, but their underparts are unbarred or have only faint flank barring, and the head is less contrastingly patterned. Juveniles are paler and virtually unbarred below, thus closely resembling those of the larger black-eared cuckoo more than the other similar-sized bronze cuckoos. However, they are somewhat glossy greenish above, and like adults have rufous in the outer tail feathers and at the tips of their inner secondaries. Juveniles have bright yellow (initially) to creamy or yellow and mottled gray mouth colors; this becomes black in adults. Nestlings are hatched with flesh-pink shoulders and a more grayish head and back coloration, which darkens with age. The mouth lining of nestlings is yellow, and the mandibular flanges are white (Brooker & Brooker, 1986).

Habitats

In Australia this ecologically widespread species is usually associated with open savanna woodlands, bushy plains with scattered trees, shrub steppe, coastal scrub, and low spinifex grassland habitats of the arid interior. It also at times occurs in tropical rainforests, coastal saltmarshes, and mangroves, and in various residential or suburban habitats. In New Guinea, where perhaps it is only a migrant, this species occurs in savanna and scrub.

Host Species

Brooker and Brooker (1989b) reported 1555 records of parasitism involving 95 potential host species in Australia. They provisionally identified 28 of them as biological host species, of which 15 are listed in table 13. The majority (16) of the biological hosts are members of the thornbill family Acanthizidae, and five are malurid fairywrens. The locally breeding species of fairywrens are evidently the major hosts throughout Australia. The red-backed wren is most important in the north, the black-and-white fairywren in the interior, the splendid fairywren in the southwest, and the superb blue fairywren in the southeastern areas.

Egg Characteristics

The eggs of this species are oval and are white or pale pinkish white, with minute freckles, spots, and blotches of light reddish brown (Brooker & Brooker, 1989b). In western Australia this pattern is highly mimetic of the species' primary hosts. The eggshell averages 0.09 g in mass and 0.06 mm in thickness (Schönwetter, 1967–84).

Breeding Season

In Australia the breeding records extend throughout the entire year, but 75% of 1009 records are for September–November. This seasonal concentration is especially evident in southern Australia; farther north there are fewer records and they become less seasonally concentrated. In the dry interior there is an apparent peak in August, and a possible secondary spring peak, whereas in northern areas there are breeding records for all months but May (Brooker & Brooker, 1989b). In a Western Australian study it was found that laying began in late August, and breeding continued for up to 15 weeks (Brooker & Brooker, 1989a).

Breeding Biology

Nest selection, egg laying. This species lays in enclosed nests (76% of 1555 parasitism records) as well as open-cup nests (24%). On three occasions egg laying has been observed in the nests of splendid fairywrens (Brooker et al., 1988). In these cases the female cuckoo entered the nest through its small lateral opening, laid her egg within about 6 seconds, and retreated backward out of the entrance while carrying a host egg in her bill. In each case egg deposition occurred shortly after sunrise, and also shortly after the host female had laid. Judging from the reduction in the numbers of host eggs in parasitized nests, the cuckoo must almost invariably remove a host egg during the laying process. Rarely is more than a single egg of this species found in a parasitized nest (985 nests had 1 egg, 25 had 2, and 2 had 3 eggs). Three other species of *Chrysococcyx* have been observed to simultaneously parasitize host nests. Burying of the cuckoo egg has been observed in at least 10 species, and a change in nest location (including nest dismantling and reconstruction) has been observed in one (Brooker & Brooker, 1989b).

Incubation and hatching. The incubation period of 11 eggs in the nests of the splendid fairywren ranged from 11.8 to 13.5 days. Within about 30 hours after hatching, eviction of any host eggs or young occurs. Eggs of the splendid fairywren have a usual incubation period of 13–14 days, and those of the two thornbill hosts have incubation periods of 18–20 days (Brooker & Brooker, 1989a, b).

Nestling period. The nestling period of 10 cuckoos raised by splendid fairywrens ranged from 15–18 days, with a mean of 16.7 days, whereas in one western thornbill nest the nestling period was 20 days. Western thornbills usually have a 17- to 18-day fledging period, and those of the yellow-rumped thornbill 18–20 days, whereas splendid fairywrens have fledging periods of 10–12 days (Brooker & Brooker, 1989a).

Population Dynamics

Parasitism rate. Brooker & Brooker (1989a) reported for 1973–83 a parasitism rate of 17% for 332 nests of the splendid fairywren, and a 24% rate for 402 nests of this same species during a later period (1984–87). There was a 12% parasitism rate for 226 nests of the western thornbill.

Hatching and fledging success. Of 95 cuckoo

eggs laid in the nests of the splendid fairywren, 71 hatched (75% hatching success) and 41 fledged (59% fledging success), resulting in an overall breeding success of 43%. Among 26 eggs laid in the nests of the western thornbill, 22 hatched (85% hatching success), and 17 fledged (77% fledging success), resulting in an overall breeding success of 65% (Brooker & Brooker, 1989a).

Host–parasite relations. The impact of this cuckoo on its two primary hosts is considerable, judging from the data of Brooker and Brooker (1989b). Assuming a hatching success of 75% with the splendid fairywren and a parasitism rate of about 20%, the reduction in brood production for this species may be estimated as 15%. Correspondingly, with a hatching success of 85% with the western thornbill and a parasitism rate of 12%, the estimated reduction in host brood production would be 10.2%. In both the shining and Horsfield's bronze cuckoos, the breeding success appears to be somewhat higher with the secondary host (western thornbill) than with the primary fairywren and yellow-rumped thornbill hosts.

RUFOUS-THROATED BRONZE CUCKOO

(Chrysococcyx ruficollis)

Other Vernacular Names: Mountain bronze cuckoo, reddish-throated bronze cuckoo.

Distribution of Species (see map 44): Mountains of New Guinea.

Measurements (mm)

6.5″ (16 cm)

Wing, males 89–97 (avg. 93, $n = 4$), females 93, 96; tail, males 58–68 (avg. 63.4, $n = 4$), females 65.6, 68 (Parker, 1981). Wing (unsexed) 95–100, tail (unsexed) 67 (Rand & Gilliard, 1969). Wing:tail ratio ~1:0.7.

Egg, no information.

Masses (g)

No information

Identification

In the field: This species differs from the other New Guinea bronze cuckoos in its rufous-tinted throat, cheeks, and forehead. The eyes and eye-ring are not so conspicuously red as in the similar little and white-eared species. The usual song is a series of eight to nine downwardly inflected "tseew" whistles, uttered at the rate of about 2 per second. A single, downwardly slurred "tseew" may also be uttered irregularly.

In the hand: This species has much chestnut on the outer tail feathers and on the inner vanes of the flight feathers. The shining and little bronze cuckoos of the same region lack such well-developed brown tones on the face and throat and have less brown on the outer tail feathers. However, Mayr (1932) noted the similarity of this species' tail pattern to that of *C. lucidus* and suggested that they might be conspecific. The underparts of both species are strongly banded with broad glossy-green bars. The sexes are similar as adults, but in females the upperparts have a bright green to bluish green gloss, rather than a purplish bronze sheen. The immature plumages are still undescribed but are probably much like those of the shining bronze cuckoo.

Habitats

This species occurs from 1130 to 3230 m, but is usually found from 1600 to 2600 m, in forests, forest edges, secondary growth, and subalpine thickets. Relative to the other New Guinea breeding species of *Chrysococcyx*, this one occurs at the highest elevations, with the white-eared at intermediate elevations and the little bronze cuckoo at the lowest levels.

Host Species

No information.

Egg Characteristics

No information.

Breeding Season

No information.

Breeding Biology

No information.

Population Dynamics

No information.

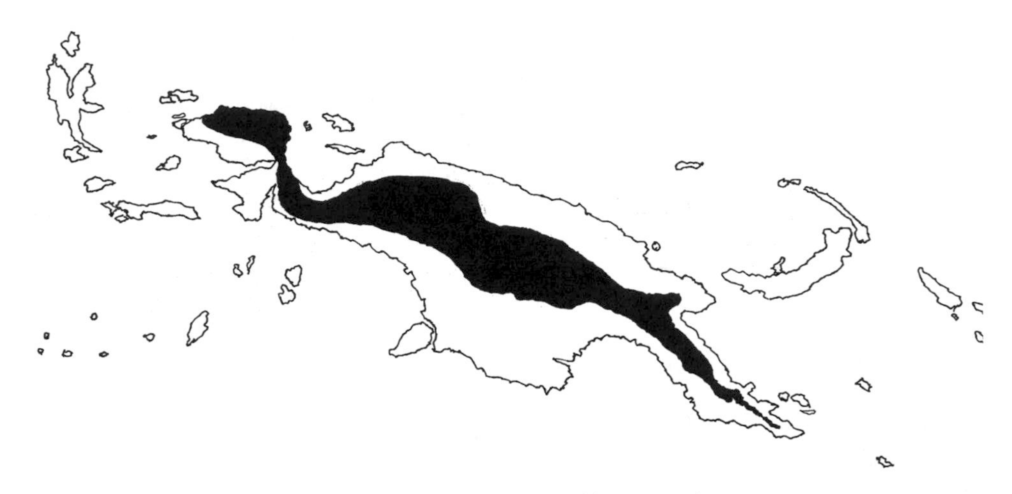

MAP 44. Range of rufous-throated bronze cuckoo.

WHITE-EARED BRONZE CUCKOO

(Chrysococcyx meyeri)

Other Vernacular Names: Meyer's bronze cuckoo, white-eared cuckoo.

Distribution of Species (see map 41): New Guinea and western Papuan Islands.

Measurements (mm)

6″ (15 cm)

Wing, male 90, female 92 (Diamond, 1972). Wing, males 88–92. Tail, male 65 (Rand & Gilliard, 1969). Wing, males 88–92, females 90–92 (Mayr & Rand, 1937). Wing:tail ratio ~1:0.7.

Egg, no information.

Masses (g)

Male 19.3, female 20.7 (Diamond, 1972). Avg. of 4 (sexes unstated) 19 (Dunning, 1993).

Identification

In the field: The conspicuous white patch, located behind the ear patch (auriculars), and separating the dark bronze-green of the shoulders from those of the nape and auriculars, is the best fieldmark (fig. 35). The birds are dark bronze-green above and barred below, but females have a rufous forehead (rather than green as in males), and adult males have a red eye-ring that is lacking in females. The male's song consists of a series of five to eight "peer" notes uttered at the rate of about one per second, the notes progressively dropping slightly in pitch before the last note, which is substantially lower in pitch than the others. Another distinctive call is a group of four pairs of downwardly inflected notes, the pairs alternatively rising and falling in pitch. Immature individuals have predominantly to entirely brown upperparts.

In the hand: No other species of *Chrysococcyx* has such a strongly emerald-green color above and barred white and glossy green below, with the white of the underparts extending up the sides of the neck as a semicollar. The outer rectrices are white, without chestnut tints, but there is some chestnut present on the basal two-thirds of the flight feathers. Females closely resemble males but average slightly larger (female wing-length to 95 mm; up to 92 mm in males), and the forehead of females is chestnut brown. In subadults the upperparts are entirely brown. The juvenal plumage is still inadequately known, but it may be brownish green above and whitish below, with some flank barring.

Habitats

This species occurs from near sea level (except in low flatlands) to about 2000 m. It is most common at intermediate elevations of about 800 m, occupying rainforests, monsoon forests, forest edges, tall secondary forests, clearings, and gardens.

Host Species
No information.

Egg Characteristics
No information.

Breeding Season
No information.

Breeding Biology
No information.

Population Dynamics
No information.

ASIAN EMERALD CUCKOO
(Chrysococcyx maculatus)

Other Vernacular Names: Emerald cuckoo.
Distribution of Species (see map 45): India, Tibet, and China south to Burma (Myanmar) and Thailand; wintering south to the Greater Sundas.

Measurements (mm)
6.5–7″ (17–18 cm)
Wing, both sexes 105–114; tail, both sexes 63–70 (Ali & Ripley, 1983). Wing, both sexes 103–114; tail, both sexes 63–70 (Delacour & Jabouille, 1931). Wing:tail ratio ~1:0.8.
Egg, avg. 17.6 × 12.3 (range 16.4–18.4 × 11.7–13.3) (Becking, 1981). Shape index 1.42 (= oval). Rey's index 2.44 (Becking, 1981).
Masses (g)
No adult weights of this species are available. Estimated fresh egg weight 1.45–1.5 (Becking, 1981).

Identification
In the field: Males of this species (fig. 35) are iridescent emerald green above and on the breast, with a red or orange eye-ring and a black-tipped, yellow bill. The flanks are barred with dark and

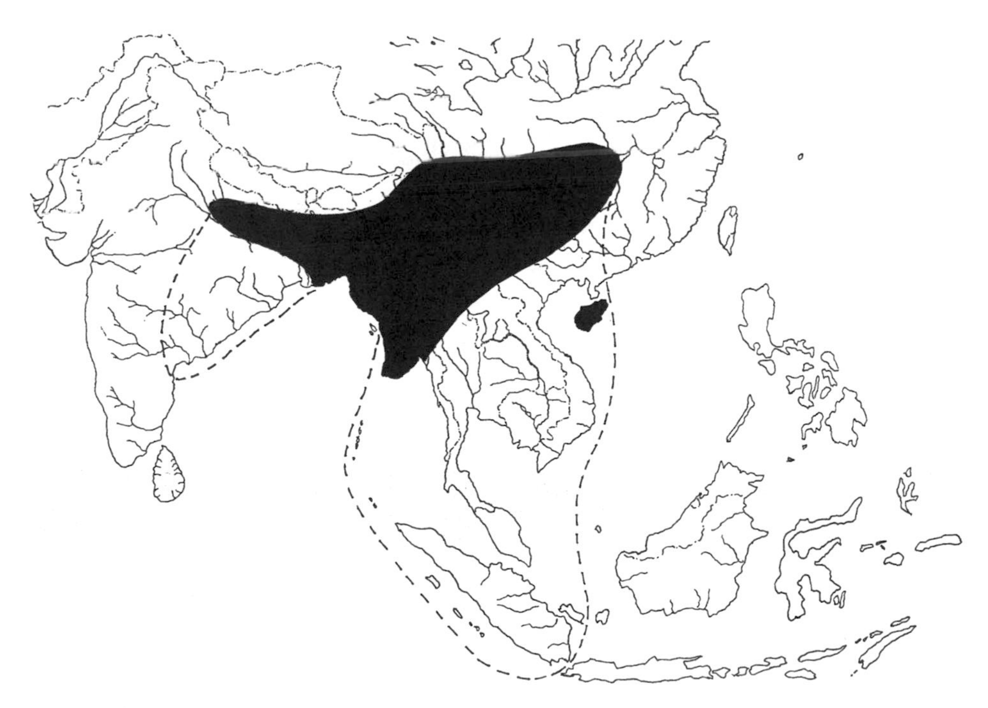

MAP 45. Breeding (filled) and nonbreeding (enclosed area) ranges of Asian emerald cuckoo.

white, and the tail is uniformly green. Females have a rather rusty brown head, an olive-brown back, and heavily barred underparts, the barring extending to the cheeks and slightly above the eyes. The bill is yellow, with a black tip. Immature birds of this species and the violet cuckoo have rufous heads with rather streaked crowns and may be difficult to distinguish. However, both adults and young of the emerald cuckoo have a white band at the base of the primaries on the underside of the wing, which is visible in flight. The male's song is not yet well described, but a quick, high-pitched rattle of five to six notes that descend slightly in pitch has been attributed to this species. Three ascending notes have also been described as the species' typical vocalization, as have various whistled twitters.

In the hand: The iridescent green plumage of males, with their white and glossy green barred underparts, and brown eyes with a coral-red eye-ring, are unmistakable. Females are notable for their rufous nape and crown, which becomes coppery green on the other upperparts, and the rufous outer tail feathers, which are also strongly marked with green and white. The female has a yellowish, rather than a bright orange, black-tipped bill and perhaps a paler eye-ring. The heavily barred juveniles may not be distinguishable from those of the violet cuckoo.

Habitats

This species occurs mainly in evergreen broadleaf montane forests and less often in second-growth and scrub habitats, at elevations up to 2400 m.

Host Species

Baker (1942) provided a list of 4 host species, based on 11 parasitized clutches in his collection. Three of these are listed in table 10; only a single record was listed for Gould's sunbird. Becking (1981) agreed that sunbirds of the genus *Aethopyga* (scarlet and Gould's) and the little spider hunter are the primary host species, although the eggs in Baker's collection probably belong to two different cuckoo species, the other one is probably the violet cuckoo. Those tentatively assigned by Becking to the emerald cuckoo are close mimics of the little spider hunter, their usual host.

Egg Characteristics

Eggs attributed to this species by Becking (1981) are long ovals, lack gloss, and are light buff to orange-tinted in ground color. They are covered with spots and specks of light brownish olive that form a distinct ring around the more rounded end. They closely match the eggs of the little spider hunter but are less rusty or fawn-colored and more olive-brown in tone. The shell weight averaged 0.089 g, as compared with 0.0751 g for the thinner-shelled but similar eggs of the violet cuckoo. Correspondingly, the two differ in their Rey's indexes (1.24–1.49, mean 1.34 for the violet, and 2.32–2.73, mean 2.44 for the emerald).

Breeding Season

In the Indian subcontinent, including Nepal, the breeding season is probably from the middle of April to the end of July. The little spider hunter mainly breeds from May to August in Assam and from December to August in southwestern India (Ali & Ripley, 1983).

Breeding Biology

No reliable information exists. Becking (1981) established that the eggs attributed by Baker (1943) to this species are those of two different cuckoo species. One type, smaller and more rounded, was judged by Becking to belong to the violet cuckoo. The other type, presumably of this species, is a close mimic of the little spider hunter's white to pinkish eggs, with reddish brown stipples, although the spots are slightly less russet than the spider hunter's typical eggs. The nests of spider hunters are cuplike and attached to the undersides of banana leaves or similar broadleaved plants. Their eggs average about 18 $\times$ 13 mm, and they are also sometimes parasitized by the Hodgson's hawk cuckoo. The other principal host, the scarlet sunbird, constructs an oval or pear-shaped nest, with a small opening near the top. Its eggs are whitish, with purplish to reddish freckles (Ali & Ripley, 1983). Host incubation periods are unknown, but it is likely that the host's eggs or young are evicted by the nestling cuckoos, since parasitized sunbird nests have been found to contain only nestling cuckoos.

Population Dynamics

No information.

VIOLET CUCKOO

(Chrysococcyx xanthorhynchus)

Other Vernacular Names: Asian violet cuckoo, Tenasserimese violet cuckoo (*limborgi*).

Distribution of Species (see may 46): India and Burma (Myanmar) to Indochina and south to Sumatra, Borneo, Java, and Philippines.

Subspecies

C. x. xanthorhynchus: India to southeast Asia, Borneo, and Java.

C. x. limborgi: Burma, northern Thailand.

C. x. bangueyensis: Banguey Island.

C. x. amethystinus: Philippines.

Measurements (mm)

6″ (16 cm)

C. x. limborgi, wing, males 104, 107 (Deignan, 1945).

C. x. xanthorhynchus, wing, both sexes 95–105; tail, both sexes 64–72 (Ali & Ripley, 1983). Wing, both sexes 98–104 (Medway & Wells, 1976). Wing:tail ratio ~1:0.7.

Egg, avg. of seven *x. xanthorhynchus* 17.2 × 12.5 (range 16.2–17.9 × 11.8–13.2 (Schönwetter, 1967–84). Avg. of eight *x. xanthorhynchus* 16.4 × 12.3 (Becking, 1981). Shape index 1.34–1.47 (= oval). Rey's index 1.34 (Becking, 1981).

Masses (g)

Two unsexed birds 19.8 and 22.1 (Becking, 1981). Estimated egg weight 1.4

MAP 46. Range of violet cuckoo.

(Schönwetter, 1967–84), 1.47 (Becking, 1981). Egg:adult mass ratio ~ 6.7%.

Identification

In the field: Males of this species are iridescent violet above and on the breast, rather than green as in the Asian emerald cuckoo, and are strongly barred with dark and white on the flanks. There is a bright red eye-ring. The bill is yellow, with a red base. Females resemble females of the Asian emerald, but are less rufous on the head and have a nearly all-black bill that becomes red at the base. Immature individuals resemble females, but are barred with dull rufous and brown above and lack red on the bill. The male's usual song consists of a musical and descending but accelerating trill; another vocalization that may be uttered during undulating flight is a repeated"kie-vik" note.

In the hand: Adult males are easily distinguished by their iridescent violet-purple upperparts and barred white underparts, white undertail coverts and white under-wing coverts. However, there is less white on the underside of the flight feathers than in the Asian emerald cuckoo. Females may be distinguished from those of the Asian emerald by the fieldmarks mentioned above and by the lack of a rufous-toned crown or rufous outer tail feathers. Immature birds of the two species are perhaps indistinguishable; both have a barred rufous and greenish brown pattern above and are barred with brown and white below.

Habitats

This cuckoo is associated with forest edges of lowlands, up to about 1500 m, and also occurs in secondary forests, orchards, and similar second-growth habitats.

Host Species

Baker (1942) listed five parasitized clutches in his collection, most of which involved the little spider hunter. The other four species were represented by single clutches. Becking (1981) has pointed out that the eggs of the Asian emerald cuckoo are nearly identical to those of the violet cuckoo, and thus easily confused. He agreed that the violet cuckoo primarily parasitizes the crimson ("yellow-backed") sunbird and the little spider hunter, but he described the other hosts mentioned by Baker as needing confirmation.

Egg Characteristics

Eggs provisionally identified by Becking (1981) as of this species are broad ovals (shape index 1.34), with faint to moderate gloss. Unlike eggs of *maculatus,* they are whitish buff to pink, profusely speckled and blotched with red, vinaceous, or violet, with secondary markings of olive-brown. A loose ring of dark markings sometimes is formed around the blunter end. Schönwetter's (1967–84) examples averaged slightly larger and were less broad (shape index 1.47). He described the eggs as highly variable in appearance, with a white ground color and a delicate yellow to rosy bloom, marked with various tones of brown and red. The shell weight of the eggs studied by Becking averaged 0.075 g, producing a fairly low Rey's index and presumably considerable resistance to breakage.

Breeding Season

No specific breeding information is available on the breeding chronology on this widespread but elusive species.

Breeding Biology

Little reliable information exists. Becking (1983) concluded that the emerald and violet cuckoos primarily parasitize the same host species (scarlet sunbird and little spider hunter). The similarities in egg types and the kinds of nests constructed by these two hosts were mentioned in the previous species account. Assuming Becking's identifications are correct, this species has slightly smaller and thinner eggs than those of the emerald cuckoo, and they are profusely speckled and blotched with violet to reddish markings (rather than spotted and speckled with brownish olive). The eggs of the presumed scarlet sunbird host averaged about 15 × 11 mm, or somewhat smaller than those of the little spider hunter.

Population Dynamics

No information.

BLACK-EARED CUCKOO

(Chrysococcyx osculans)

Other Vernacular Names: None in general English use.

Distribution of Species (see map 47): Australia; wintering north to New Guinea and the Moluccas.

Measurements (mm)

7.5–8″ (19–20 cm)

Wing (unsexed), 112. Tail (unsexed) 81 (Rand & Gilliard, 1968). Tail, 85–90 (Frith, 1977). Wing:tail ratio 1:0.77.

Egg, avg. 21.1 × 15.5 (range 19.5–22.4 × 14.5–17.5) (Schönwetter, 1967–84). Shape index 1.36 (= oval).

Masses (g)

Three males, range 26.7–34 (avg. 29.6); four females, range 27–34.5 (avg. 30.6) (Brooker & Brooker, 1989b). Estimated egg weight 2.75 (Schönwetter, 1967–84), also 2.6 (Brooker & Brooker, 1989b). Egg:adult mass ration ~8.7–9.2%.

Identification

In the field: This rather small and medium-gray cuckoo is readily recognized by the dark stripe that extends from the base of the bill through the eyes and continues through the ears. There is no flank barring, either in juveniles or adults, and very little barring on the grayish tail. The rump is noticeably paler than the back or tail. Juveniles closely resemble adults, but have a less conspicuous black facial stripe. The usual song is a series of downwardly inflected "peeeeeer" notes that gradually fade, and which may be repeated up to about eight times. A more animated "pee-o-wit-pee-o-weer" call has been heard when several males are interacting.

In the hand: Adults of this species are easily distinguished from the other Australian *Chrysococcyx* by the generally grayish brown overall dorsal color and the unmarked flanks and buffy brown to whitish underparts. The ear patch and lores are darker than the rest of the head or body plumage, and the tail feathers are tipped with white and have white barring on the outer pairs. In juveniles the underparts are pale gray, the ear patch and lores are brown and only slightly darker than the rest of the head, and the upperparts may be slightly darker than in adults, but with paler feather edging. A newly hatched young (probably but not definitely of this species) had a white gape, with yellow in the throat, and the skin was coal-black. Older nestlings may have more brownish gray skin. The legs, bill, and eyes range from dark brown to black in all age and sex categories from the juvenile stage onward.

MAP 47. Primary breeding (filled), secondary or migrant (hatched), and wintering ranges (enclosed area) of black-eared cuckoo.

Habitats

This species is associated with open (e.g., eucalyptus, sheeoak, mulga, mallee) woodlands, arid scrublands, lignum or samphire (glasswort) salt flats, drier coastal shrubby habitats, and open gallery forests of the Australian interior, but not extending into closed forests.

Host Species

Brooker & Brooker (1989b) listed 23 host species associated with 163 records of parasitism in Australia. Two of those hosts are listed in table 13; these are the only two species regarded by Brooker and Brooker as biological hosts of black-eared cuckoos. Of these, the redthroat was regarded as the most important host, although the egg color of the cuckoo matches the speckled warbler's egg somewhat more closely, and there are more records of parasitism for the speckled warbler.

Egg Characteristics

According to Brooker & Brooker (1989b), the eggs of this species are elongated oval ("oval" by this book's definition) and are a uniform, unspotted reddish chocolate brown, or almost the same color as those of the speckled warbler and redthroat hosts.

Breeding Season

Brooker & Brooker (1989b) reported on the seasonal distribution of 108 breeding (egg) records, which include all months except February and May. A majority (63%) of the records are for September–November; these records are mostly from southeastern Australia. Records from western Australia are more spread out temporally and extent mainly from June to December, with a possible peak in August.

Breeding Biology

Nest selection, egg laying. The best account of egg laying in this species is that of Chisholm (1973), who reported that four people observed a female cuckoo ward off the host speckled warblers as the cuckoo entered their globular nest. One of the warblers also entered, and both remained for about 7–8 seconds. Then both emerged, and the cuckoo flew away, the warblers in pursuit. The nest contained two warbler eggs (one broken) and one cuckoo egg. Based on the clutch sizes of parasitized nests, the cuckoo probably removes one of the host eggs at the time of laying (Brooker & Brooker, 1989b). The redthroat's eggs average about 19 × 14 mm, and those of the warbler average 17–19 × 15 mm. Both host species build enclosed and well-hidden nests on the ground or close to the ground.

Incubation and hatching. The incubation period is unreported; that of the hosts is about 12 days. The nestling cuckoo, while still naked and sightless, ejects the host's young or eggs from the next (Chisholm, 1973).

Nestling period. No information.

Population Dynamics

Parasitism rate. No information.

Hatching and fledging success. No information.

Host–parasite relations. Little information available. The high level of egg mimicry between this cuckoo and the speckled warbler makes distinguishing them extremely difficult; egg measurements and the more superficial pigment on the cuckoo's egg (which can be removed by rubbing with a wet finger) must be used for identification. There is no indication of egg rejection behavior for any of the reported host species. Ejection of the host young from the nest occurs, so host reproductive losses might be considerable.

YELLOW-THROATED CUCKOO

(Chrysococcyx flavigularis)

Other Vernacular Names: None in general English use.

Distribution of Species (see map 48): Western and central Sub-Saharan Africa from Sierra Leone to Zaire.

Measurements (mm)

8–8.5″ (20–22 cm)

Wing, males 94–99 (avg. 97), females 94–98.

MAP 48. Range of yellow-throated cuckoo.

FIGURE 36. Profile sketches of four African *Chrysococcyx* cuckoos: adult male (A) and female (B) of yellow-throated cuckoo; male (C) and female (D) of Klaas' cuckoo; male (E) and female (F) of African emerald cuckoo; male (G) and female (H) of dideric cuckoo.

Tail, males 65–80 (Fry et al., 1988).
Wing:tail ratio ~1:0.74.
Egg, no information

Masses (g)
Males 27.5–31, female 30 (Fry et al., 1988).

Identification

In the field: Within its rather limited West African range, this species is rare. Adult males are best recognized by the bright yellow chin and foreneck patch, contrasting with otherwise rather dark greenish brown upperparts and a finely barred greenish brown and buffy underparts (fig. 36). The outer rectrices are mostly white, with subterminal blackish barring, but there is little white elsewhere. Females lack the bright yellow throat of males, and instead are mostly bronze-brown above and faintly barred with dark and lighter brown below. Their outer tail feathers are mostly white, like those of the male. Immature individuals are quite female-like. The male's song is a clear whistle of 9–12 notes, all on the same pitch, the first note the longest, and the remaining notes accelerating and gradually diminishing in volume. The entire series lasts about 3 seconds and is repeated about eight times per minute. Another vocalization is said to be a series of short, sweet whistles that descend the scale slightly and can be heard over a few hundred meters. Other two-noted calls are also produced; one ("di-dar") resembles the first two notes of the dideric cuckoo's song.

In the hand: Males are easily recognized by their yellow underparts. Females resemble those of Klaas's cuckoo, but are distinctly barred with brown (not green) on the flanks, underparts, and throat. Immature birds resemble females, but the barring on the upperparts is more tawny, and less dusky. Like the Klaas's cuckoo the outer tail feathers of all sex and age groups lack brownish tints at their tips. The nestling stages remain undescribed.

Habitats

This cuckoo is associated with the canopies of lowland virgin forests but also occurs in forest edges, gallery forests, and thickly wooded savannas.

Host Species
No information.

Egg Characteristics
No information.

Breeding Season
No direct information. Birds with enlarged gonads have been recorded during July and December in Uganda (Fry et al., 1988).

Breeding Biology
No information.

Population Dynamics
No information.

KLAAS'S CUCKOO
(Chrysococcyx klaas)

Other Vernacular Names: Brown cuckoo, golden cuckoo, green cuckoo, white-throated emerald cuckoo.

Distribution of Species (see map 49): Sub-Saharan Africa from Senegal to Somalia and south to South Africa.

Measurements (mm)
7″ (18 cm)

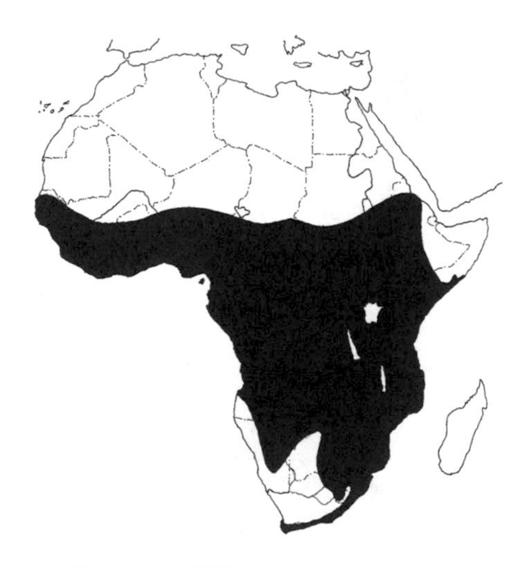

MAP 49. Range of Klaas' cuckoo.

Wing, males 98–108 (avg. 103), females 96–106 (avg. 102). Tail, males 69–80 (avg. 83), females 66–75 (avg. 73) (Fry et al., 1988). Wing:tail ration 1:0.7–0.8

Egg, avg. 18.9 × 12.8 (range 16.9–20 × 12.1–13.5) (Schönwetter, 1967–84). Shape index 1.48 (= oval). Rey's index 2.42.

Masses (g)

Males 21–31 (avg. 25.6, $n = 15$), females 28–34 (avg. 30.2, $n = 10$) (Fry et al., 1988). Estimated egg weight 1.65 (Schönwetter, 1967–84). Egg:adult female mass ratio 5.5%.

Identification

In the field: Adults of both sexes of this small cuckoo have almost entirely white outer tail feathers (distinguishing it from the African emerald and dideric cuckoos), and males are further distinguished by their pure white underparts with no flank barring as occurs in the dideric (fig. 36). Males are otherwise glossy green above, with a small white stripe behind the eye. Females are bronze-brown above and finely barred with brown and white below. They have a small white stripe behind the eye. Immature birds resemble females, but lack the white postocular stripe. The song of the male is a whistled bisyllabic phrase, "dee-da" (sometimes varied or extended to three or even four syllables, "may-ee-chee"), repeated four times in 3–4 seconds. A single-syllable "dew" note may be repeatedly uttered. Such songs may be repeated at intervals of about 15 seconds, for up to 30 minutes.

In the hand: Males of this species have immaculate white underparts (including the underwing coverts) and glossy green upperparts, including a patch of green at the sides of the chest, except for a white eye-stripe behind the eyes. The outer rectrices are white, with a bronze or green subterminal spot, and sometimes some dark barring, but no rufous tones. Females have upperparts variously barred with rufous, but they also have a white or pale postocular eye-stripe. Their outer tail feathers are mostly white, with no rufous tones. Immature individuals are scarcely distinguishable from adult females. They may be more brownish throughout than adult females and more barred dorsally, but their underparts are similarly barred with glossy green, and several of the upper-tail coverts have white outer webs. Their postocular patch is buffy, not white. Nestlings have bright orange gapes, and their naked skin is initially yellowish olive brown (not pink as in the dideric cuckoo), becoming slightly darker above. At 2 days the bill is blackish to horn-colored, rather than orange to red as in the dideric cuckoo. The skin color gradually darkens to deep blackish olive and is consistently darker than that of young emerald cuckoos (see following species account).

Habitats

This species favors riparian forests or forest edges, with moderately wooded miombe (*Brachystegia*) or acacia woodlands. It usually occurs in habitats less dense than those used by the African emerald cuckoo, but denser than those of the dideric. It often is found along forest edges, remaining high in the trees. It ranges up to about 1800 m.

Host Species

Thirty-nine known or highly probable hosts of this species were listed by Fry et al. (1988) and are summarized in table 20. Of these, ten or more records exist for the Cape crombec (13), greater double-collared sunbird (13), bronze sunbird (12), Cape batis (10), and yellow-bellied eromomela (10). Rowan (1983) listed 16 known biological hosts (those with nestling or fledglings) for southern Africa; in descending frequency they are the greater double-collared sunbird (13 records), the Cape crombec (8 records), Cape batis (7 records), and bar-throated apalis (7 records).

Egg Characteristics

The eggs of this species are oval and have a white, greenish white, or blue ground color, with freckles or speckles of brown, light rufous, or slate at the blunt end. Several gentes exist, and the eggs do not always match those of their hosts (Fry et al., 1988). The relatively high Rey's index suggests little shell thickening, and thus little resistance to breakage.

Breeding Season

Egg records for southern Africa (South Africa, Zimbabwe, Malawi, Namibia, and Angola) mostly fall within the period October–April (South African extremes September–July). Western African records (Senegambia, Liberia, Mali, and Nigeria) are mostly from March to November. Breeding in Zaire may be year-around, but mostly occurs during the rainy season from August to January. Ethiopia and Sudan records are for August–October. Those for East Africa seem quite variable. Eastern and coastal areas of Kenya have December–June records, while those from more western and northern areas are scattered from December through April, plus September (Fry et al., 1988).

Breeding Biology

Nest selection, egg laying. Little direct information available. Typically one egg is laid per nest, and evidently a host egg is removed at the time of laying, since a cuckoo has been seen carrying an egg and the clutch sizes of host eggs tend to be lower than normal for the species. Jensen and Clining (1974) reported "perfectly color-matched eggs" with both the dusky sunbird and the pririt batis. However, eggs of the host pririt batis average about 16 × 12 mm, and those of the dusky sunbird about 16 × 11 mm, so egg size alone can readily facilitate identification. It has been estimated that three or four eggs are laid on alternate days and that several clutches (totaling 20–24 eggs per season) may be laid (Fry et al., 1988).

Incubation and hatching. Various lines of evidence indicate that the cuckoo's incubation period must be at least 11–12 days, and no more than 14 days (Rowan, 1983). In one case, the cuckoo hatched 4 days before the host Cape batis' eggs, which have an incubation period of 17–18 days, and at 4 days of age, the cuckoo began to evict the host's young (MacLeod & Hallack, 1956). In another case, a chick hatched a day before the host dusky sunbird's egg, which has an incubation period of 13 days (Jensen & Clining, 1974).

Nestling period. There is one estimate of a 20- to 21-day nestling period for a partially hand-raised chick (Jensen & Jensen, 1969), which is close to an earlier estimate of at least 19 days. This hand-raised bird was independent 10 days later, but a bird that was tended by foster parents was independent about 25 days after leaving the nest.

Population Dynamics

Parasitism rate. On the basis of nest record card data from southern Africa, Payne & Payne (1967) estimated parasitism rates of 2.6% for 227 nests of the Cape crombec, 2.4% for 165 nests of the amethyst sunbird, and 1.4% for 209 nests of the bar-throated apalis. Jensen and Clining (1974) estimated overall parasitism rates over 4 years in a Namibian study as 8% for 84 nests of the dusky sunbird, 7% for 76 nests of the pririt batis, and lower percentages for two minor host species.

Hatching and fledging success. No information.

Host–parasite relations. No information available. In spite of the high degree of egg similarities between this cuckoo and its major hosts and associated apparent development of several host-specific gentes, the incidence of parasitism seems surprisingly low. If these parasitism rates are typical, the effects on host population must be relatively minor.

AFRICAN EMERALD CUCKOO

(Chrysococcyx cupreus)

Other Vernacular Names: Emerald cuckoo, golden cuckoo, green cuckoo.

Distribution of Species (see map 50): Sub-Saharan Africa from Senegal and Ethiopia south to South Africa.

Subspecies

C. c. cupreus: Mainland Africa and Bioko Island (Gulf of Guinea).

C. c. insularum: Islands in Gulf of Guinea.

Measurements (mm)

9″ (22 cm)

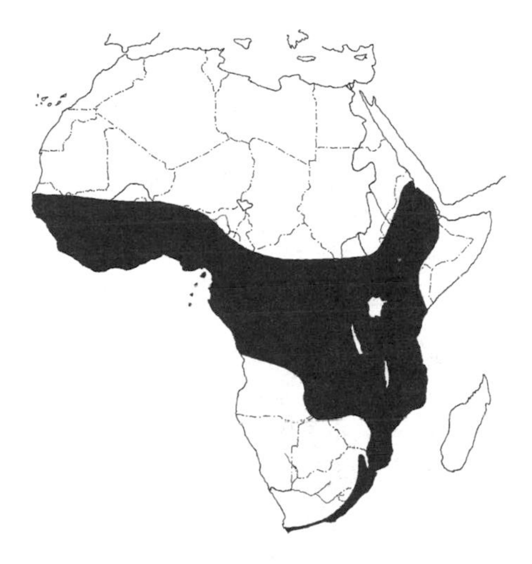

MAP 50. Range of African emerald cuckoo.

C. c. cupreus, wing, 110–120 (avg. 114), females 105–116 (avg. 112). Tail, males from South Africa 83–95 (avg. 90), males from Ethiopia 109–133 (avg. 121) (Fry et al., 1988). Wing:tail ratio 1:0.8.

C. c. insularum, tail, males 84–102 (Fry et al., 1988).

Egg, two *cupreus* eggs 19.1–21.8 × 15–18 (avg. 20.45 × 16.5) (Fry et al., 1988). Shape index 1.24 (= broad oval). Rey's index 3.6.

Masses (g)

Males 33–46 (avg. 38.3, n = 12), females 30–41 (avg. 36.7, n = 12) (Fry et al., 1988). Avg. of 32, both sexes, 37.7, range 30–46 (Dunning, 1993). Estimated egg weight 2.8. Egg:adult mass ratio 7.4%.

Identification

In the field: The adult male has a unique pattern of being emerald green throughout, except for bright yellow flanks and underparts (fig. 36). The tail is fairly long and graduated, with the outer feathers barred and tipped with white. The under-wing coverts and bases of the flight feathers are also white in both sexes; in the dideric and yellow-throated cuckoos this area is barred in both sexes, and in the Klaas's cuckoo only the male has an immaculate white under-wing lining. Females are distinctly barred with green and brown dorsally and strongly barred with white and green below, but they too have much white on the outer tail-feathers as well as their under-wing surfaces. Immature individuals resemble adult females, but are more heavily barred on the head and anterior upperparts. The male's usual song is a four-syllable "diyou, du, di," with the first doublet note strong, the middle note lower pitched and weaker, and the final note also stronger. This series lasts about 1.5–2 seconds and is repeated indefinitely every 2–3 seconds. A rapid series of "ju" notes is sometimes also uttered.

In the hand: Females and immature birds of this species can be distinguished from immature birds of Klaas's cuckoos by the fact that they never have more than a narrow margin of white on the green outer and upper-tail coverts, whereas immature birds of Klaas's cuckoos have several upper-tail coverts with white outer webs. Both can be distinguished from females of the yellow-throated cuckoo by their glossy green, rather than brown, barring of the underparts. Immature birds are very femalelike, with extensive dorsal and ventral barring, and young males tend to be more glossy overall than young females. Nestlings have a whitish upper mandible, an orange gape, and initially have skin that is pinkish yellow below to mauve above, but by a few days it becomes yellowish brown to violet-black. The similar Klaas's cuckoo chick is smaller, its skin is darker, and its bill is smaller and darker, rather than yellow to orange. Older and feathered nestlings are very hard to distinguish, but those of the emerald cuckoo are slightly larger and have their outer upper-tail coverts entirely green or bronze, with no more than a narrow white fringe, whereas those of the Klaas's have the outer webs white. The young emeralds are also heavily barred below with bronze-green (rather than more lightly barred with bronze-brown) and lack a whitish ear patch typical of young Klaas's cuckoos.

Habitats

This cuckoo occurs in evergreen forests, fairly densely wooded savannas, gallery forests, and similar woodlands. It extends to about 800 m of elevation, and the birds are more common in forests than in savannas, where they perch high in the trees.

Host Species

Fry et al. (1988) listed 18 probably hosts for this species, but noted that confusion with the Klaas's cuckoo frequently occurs. These host species are listed in table 20; only the common bulbul, with 12 records, and the Sao Tome weaver, with 20 records, had more than 10 records of parasitism. Rowan (1983) listed a few possible, but no definite, hosts for southern Africa.

Egg Characteristics

The eggs of this species are broad oval. Various color morphs reportedly exist, but possible confusion with dideric eggs makes some of these egg types questionable. One type from South Africa had a ground color of white, pinkish, or pale blue, with a pattern of brown freckles and blotches around the blunter pole. Another (from Gabon) is rose-red to rose-salmon, with a circle of red speckles (Brosset, 1976), and a third (from Sao Tome) is bluish gray, with a pattern of brown spots and blotches around the blunt end (de Naurois, 1979; Fry et al., 1988). Oviducal eggs have been white or white with sparse purple specklings, and another authenticated egg was white with brown freckles and blotches (Rowan, 1983).

Breeding Season

Records of breeding in South Africa are from October through January. Malawi records are for September–November; both Angola and Mozambique have February records. In West Africa the few available records are scattered throughout the year, but in Zaire they are more concentrated, from May through September. Western parts of East Africa have April–July records, but more eastern areas are from January to May, plus September–October (Fry et al., 1988).

Breeding Biology

Nest selection, egg laying. Little information is available. It is known that the cuckoo removes at least one host egg at the time of laying; in all but one of 20 nests of the Sao Tome weaver, an egg was removed (de Naurois, 1979; Fry et al., 1988).

Incubation and hatching. The incubation period is no longer than 13 days. As in other members of this genus, the newly hatched cuckoo soon evicts any eggs (but perhaps not always the young) of the host species (Jensen & Jensen, 1969; Fry et al., 1988). De Naurois (1979) doubted that a cuckoo chick could evict any eggs or young from the purse-shaped nest of the Sao Tome weaver.

Nestling period. The nestling period is 18–20 days, and the fledgings remain with their foster parents for as long as 2 weeks (Fry et al., 1988).

Population Dynamics

No detailed information exists. De Naurois (1979) found 20 parasitized nests among more than 100 nests of the Sao Tome weaver.

DIDERIC CUCKOO

(Chrysococcyx caprius)

Other Vernacular Names: Barred emerald cuckoo, bronze cuckoo, diederick cuckoo, didric, golden cuckoo, green cuckoo.

Distribution of Species (see map 51): Sub-Saharan

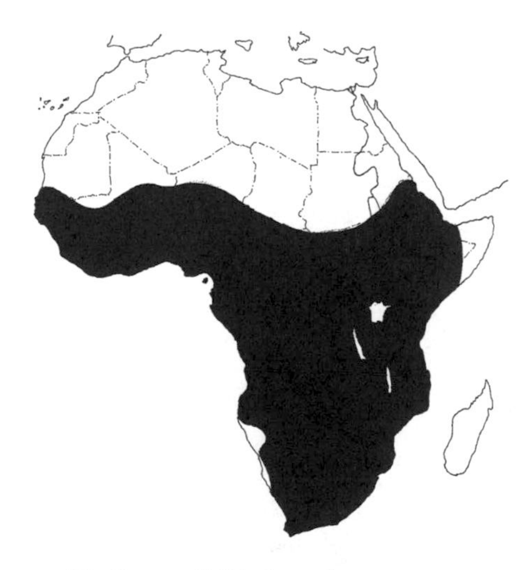

MAP 51. Range of dideric cuckoo.

Africa from Mauritania and Somalia south to South Africa.

Measurements (mm)

7.5" (19 cm)

Wing, males 111–124 (avg. 116), females 102–122 (avg. 116). Tail, males 73–86 (avg. 79), females 75–90 (avg. 80) (Fry et al., 1988). Wing:tail ratio 1:0.7.

Egg, avg. 21.5 × 14.8 (range 20–24.2 × 13.8–15.9) (Schönwetter, 1967–84). Shape index 1.45 (= oval). Rey's index 1.67.

Masses (g)

Males 24–36 (avg. 29, $n = 24$), females 29–44 (avg. 35, $n = 14$) (Fry et al., 1988). Estimated egg weight 2.55 (Schönwetter, 1967–84). Actual egg weight 3.1 (Chalton, 1991). Egg:adult female mass ratio 7.3%.

Identification

In the field: Males of this species are dark emerald green above, with a white stripe in front of and behind the red eye and orange-red eye-ring, and a dark malar stripe extending down toward the white throat (fig. 36). The upper flanks are barred with green, but the remaining underparts are white. The outer tail feathers are barred and tipped with white, and the inner webs of the flight feathers are also strongly barred with white. Females are generally similar in pattern, but more buffy below, less intensely iridescent above, and have brown or reddish brown rather than red iris color. Immature birds resemble females but have rufous crowns. The male's song is a series of high-pitched, multiple-syllable (five to seven notes) whistles lasting about 2 seconds. It is variously rendered as "day-dee-dee-deric," "dee-dee-dee-di-di-ic," or "deea-deea-deedaric," with the middle notes becoming louder and higher in pitch, and the final ones declining. The phrase, "Oh dear, Dad did-it" approaches the typical rendering. This series is repeated after brief intervals, resulting in an average of four songs in 9 seconds. Females have a plaintive "deea" note they may use to respond to males.

In the hand: Females and immature birds of this species can be distinguished from those of the three other African species of bronze cuckoos by the presence of streaking and spotting on the throat and foreneck and the absence of barring on the back, throat, and foreneck. Immature dideric cuckoos are very rufous, especially on the head. They resemble the female dorsally, but in addition to being more rufous on the crown, their underpart markings are more longitudinally streaked, spotted, or blotched, rather than barred. Juveniles resemble those of other *Chrysococcyx* young, but are distinctive in their white spotting on the wing coverts and by the irregular spotting or blotching (not barring) on their underparts. Their outer tail feathers are mostly dark, with white blotching or barring. Nestlings have a pinkish skin color initially, which turns blackish within 48 hours. The gape and bill color of newly hatched nestlings is a distinctive vermilion to orange-red, but the bill soon darkens and becomes blackish red while the gape remains bright reddish. This distinctive bill color persists for at least 18 days after fledging.

Habitats

This species is associated with a wide array of rather open, scrubby habitats, including thorn (acacia) savannas, forest edges, clearings, gardens, steppe, and semidesert habitats, but is absent from evergreen forest.

Host Species

Fry et al. (1988) listed more than 30 species as known hosts of this species. Table 20 includes seven of these species. At least in South Africa, the commonest hosts are the red bishop (245 records), the masked weaver (219 records), and the cape sparrow (118 records). Other possibly important hosts include the cape weaver (40 records) and the red-headed weaver (26 records). Regionally significant hosts may include the African golden weaver (main host on Zanzibar), the Heuglin's masked weaver (common host in Mali), the Vieillot's black weaver (common host in Zaire), and the black-necked weaver (common host in Nigeria). Rowan (1983) lists 24 biological hosts (those observed with nestlings or fledglings present) in southern Africa; in descending frequency they are the red bishop

(131 records), cape sparrow (73 records), masked weaver (72 records), and Cape weaver (14 records).

Egg Characteristics

The eggs of this species are oval, with highly variable colorations that may include up to 10 host-specific gentes (Colebrook-Robjent, 1984), but at least consist of three distinct gentes and their associated host groups (cape sparrow, red bishop, and *Ploceus* weavers such as the masked weaver). One common egg morph (adapted to the red bishop) is blue, with or without dark green spots. Eggs adapted to the masked weaver are also blue, as are those adapted to the cape weaver and Bocage's weaver. Another frequent gens morph, often found in cape sparrow nests, is bluish white, with fine and sparse brown spotting. Blue eggs with fine purplish speckling are associated with and evidently adapted to mimicking the red-headed weaver (Fry et al., 1988; Rowan, 1983).

Breeding Season

In South Africa the breeding season is from October to March, while in Zimbabwe and Malawi egg records span from October to April. The available records in Zambia are distributed from August to April. West African areas (Senegambia, Ghana, Mali, Nigeria) are well scattered between January and November, and those from East Africa are similarly well spread, from December to August (Fry et al., 1988).

Breeding Biology

Nest selection, egg laying. It is believed that egg laying is done without any assistance on the part of a mate. After silently watching the host colony for 30 minutes or longer, the female may fly directly to the nest in the face of any nest defense put up by the host, even among colonial nesters such as the red bishop. There is no clear evidence that the female cuckoo usually removes a host egg at the time of laying. However, some egg stealing (and subsequent egg eating) certainly does occur, perhaps before the cuckoo's actual egg laying. Egg stealing from nests other than hosts may also occur (Rowan, 1983). It is likely that laying occurs on alternate days and that females may be able to retain an egg in the oviduct for periods of up to a day. It has further been estimated that from 16 to 21 eggs might be laid by a single female over a breeding season of 10 weeks (Payne, 1973b).

Incubation and hatching. Incubation periods of 10–12 days (2 days ahead of its host) and of 11–13 days have been determined, and other less precise estimates of 11 or 12 days have been made. Chalton (1991) estimated a 9–10.5 incubation period. Only in a few cases (1 of 74 nests studied by Reed, 1968) have two cuckoo eggs been found in the same nest. By the second or third day after hatching the young cuckoo begins to evict other eggs or nestlings from the nest, and this eviction behavior may persist to the fifth or sixth day. On those rare occasions (5 cases among 74 studied by Reed, 1968) when the host young hatched before the cuckoo, this eviction may not be successful, and in such cases the cuckoo may fail to survive or possibly both the cuckoo and the host young may survive (Rowan, 1983).

Nestling period. A nestling period of 20–22 days was determined by Reed (1968) for two nestlings hosted by the red bishop and the cape sparrow. Similar estimates of 20 and 22 days have been made, as well as some longer periods of up to 26 days for the cape sparrow (Rowan, 1983) and a shorter period of 18.5–19 days for a spectacled weaver host (Chalton, 1991). The period of post-fledging dependency was estimated by Reed as 25 days for young tended by masked weaver hosts, 18 and 32 days for cuckoo chicks tended by cape sparrows, plus periods of 17 and at least 25 days for those hosted by red bishops. It is likely that the cuckoo chicks not only learn to recognize the calls of their foster species but also are able to mimic the juvenile hunger calls of the host species effectively, which differ considerably among the masked weaver, red bishop, and cape sparrow hosts (Reed, 1968).

Population Dynamics

Parasitism rate. Using nest record data from southern Africa, Payne & Payne (1967) estimated overall parasitism rates of 8.5% for 648 cape sparrow nests, 6.6% for 1173 southern masked weaver nests, 3.8% for 472 cape weaver nests, 3.4% for 295 village weaver nests, and 2.4% for 3735 red bishop and 174 lesser masked weaver nests. Several other species had somewhat lower rates of parasitism. Hunter (1961) observed a reduction of parasitism incidence from 18% to 4% during two successive years in a southern masked weaver colony. Reed (1968) reported a 25% parasitism rate among 52 nests of red bishops, but noted large variations in rates associated with different times and locations. Jensen and Vernon (1970) observed major seasonal and year-to-year variations in parasitism incidence of red bishops, but collectively found that about 9% (75 of 847 nests) were parasitized, which compares well with about 10% (32 of 324) of nests for various *Ploceus* hosts in Natal and the Transvaal (Rowan, 1983). Jackson (1992) reported a rate of parasitism of less than 1% for 645 nests of the northern masked weaver. Craig (1982) reported only 3 losses to *Chrysococcyx* cuckoos in a sample of 438 red bishop eggs.

Hatching and fledging success. No direct information.

Host-parasite relations. It would seem that locally or seasonally this cuckoo can exert strong effects on the fecundity rates of important hosts such as the red bishop. Both Reed (1968) and Jensen and Vernon (1970) found seasonal or local rates of red bishop parasitism to exceed 50% at times, which could have major implications for this species potential fecundity. Even at a fairly conservative estimate of a 10% parasitism rate, the effects might be biologically significant. Reed (1968) observed that, of 74 nests with cuckoo chicks, 48 of these had cuckoos as sole occupants, and in another 10 nests originally having eggs or chicks of the host, the cuckoo was later found alone. Thus a nearly complete loss of host productivity would be typical among parasitized nests, although rare instances of one or more host chicks surviving in the presence of a cuckoo chick have been reported. However, brood parasitism is likely to put a normally multibrooded hen or pair out of production for the rest of the breeding season due to the long period of fledgling dependency (Reed, 1968).

WHITE-CROWNED KOEL

(Caliechthrus leucolophus)

Other Vernacular Names: None in general English use.

Distribution of Species (see map 52): New Guinea.

Measurements (mm)

12–14″ (30–35 cm)

Wing, males 169, 175 (Diamond, 1972).

Wing (unsexed) 166–176, tail (unsexed) 159

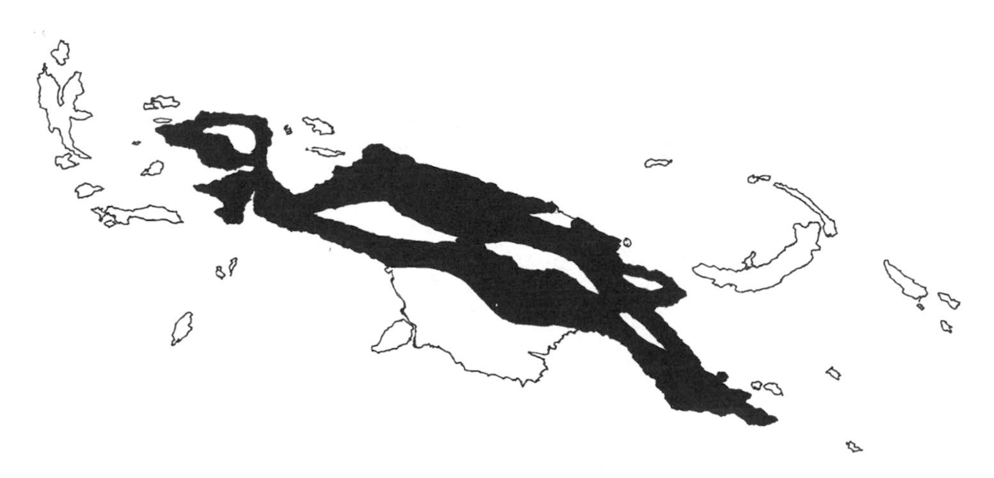

MAP 52. Range of white-crowned koel.

(Rand & Gilliard, 1969). Wing, males 166–175 (avg. 171) (Mayr & Rand, 1937). Wing:tail ratio ~1:1.

Egg, no information.

Masses (g)

Males 113, 125 (Diamond, 1972). Three males 110–125 (avg. 116) (Dunning, 1993).

Identification

In the field: This medium-sized cuckoo is almost entirely black, except for a white central crown stripe and white-tipped tail features (fig. 37). Immature individuals are more strongly edged and barred with white. The male's song consists of a series of three or four descending, whistled "too" notes, with the last the most prolonged. These phrases are repeated many times, each series uttered at a slightly higher pitch and more emphatically, until the sequence finally ends with some excited "week!" notes. The birds also utter loud, rolling "ka-ha-ha-ha" calls of three or four notes that resemble human laughter.

In the hand: Adults of both sexes are rather easily recognized in the hand by the white crown stripe on an otherwise black head. Immature birds may be somewhat barred below with white, but possibly younger birds have brownish tinges to the upperparts and patchy white crown stripes.

Habitats

Associated primarily with the middle levels and canopies of mature forests, but it extends to forest edges, secondary growth, and isolated tall trees, mainly from the lowlands to 1740 m.

Host Species

No information.

Egg Characteristics

No information.

Breeding Season

No information.

Breeding Biology

No information.

Population Dynamics

No information.

DRONGO CUCKOO

(Surniculus lugubris)

Other Vernacular Names: Indian drongo cuckoo (*dicruroides*), Ceylon drongo cuckoo (*stewarti*).

Distribution of Species (see map 53): Asia from India and southern China south and east through the Greater Sundas, Philippines, and northern Moluccas.

Subspecies

S. l. lugubris: Java and Bali.

S. l. velutinus: Baslin, Jolo, Mindanao, Samar, Tawitawi, Leyte, and Bohol (Philippines).

S. l. chalybaeus: Luzon, Mindoro, Negros, and Gigante (Philippines).

S. l. minimus: Palawan, Balabad, and Calauit (Philippines).

S. l. musschenbroeki: Sulawesi.

S. l. barussarum: Malaysia, Sumatra, and Borneo.

S. l. stewarti: Sri Lanka and southern India.

S. l. dicruroides: North-central India to Indochina and southern China.

Measurements (mm)

9″ (23 cm)

S. l. dicruroides, wing, both sexes 135–148; tail, both sexes 128–152 (Ali & Ripley, 1983). Wing, both sexes 129–147; tail, both sexes 106–133 (Delacour & Jabouille, 1931). Wing:tail ratio ~1:1–1.13.

S. l. stewarti, wing, males 127–128, female 130; tail, males 136–146 (Ali & Ripley, 1983). Wing:tail ratio ~1:0.9.

Egg, avg. of *dicruroides* 22.8 × 16.4, *stewarti* 19 × 14.5, *lugubris* 20.3 × 15. Overall range 17.5–23.5 × 13.8–17.5 (Schönwetter, 1967–84). Shape index 1.31–1.35 (= broad oval). Rey's index 2.24 (Becking, 1981).

Masses (g)

Avg. of 10 (both sexes) 35.7, range 32.6–39 (Dunning, 1993). Adults of both sexes of *dicruroides* 30–43.6 (avg. 36.2, $n = 28$) (Becking, 1981). Estimated egg weight of *stewarti* 2.0, *lugubris* 2.4, *dicruroides* 3.3

FIGURE 37. Profile sketches of four endemic East Indian cuckoos: juvenile (A), female (B), and adult male (C) of long-billed cuckoo; female (D) and adult male (E) of dwarf koel; immature (F) and adult (G) of white-crowned koel; adult of black-billed koel (H).

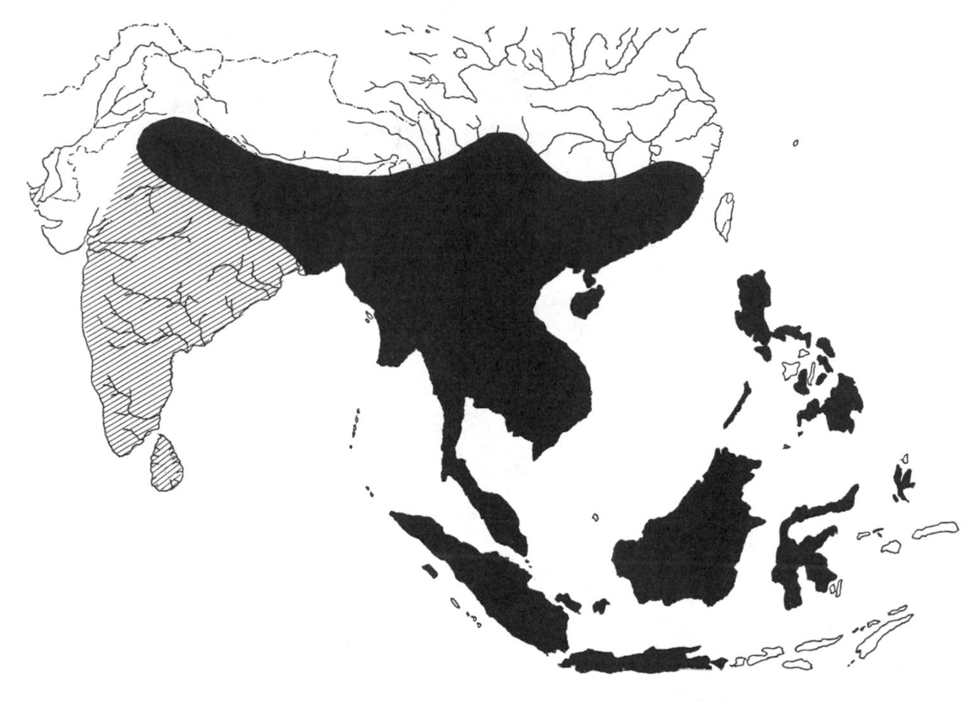

MAP 53. Breeding (filled) and wintering (hatched) ranges of drongo-cuckoo.

(Schönwetter, 1967–84). Egg:adult mass ratio (*dicruroides*) 9.1%.

Identification

In the field: This rather small cuckoo is readily distinguished by the drongolike forked tail of adults (see fig. 6). Adults of both sexes are almost entirely black, except for some white barring on the under-tail coverts and outer rectrices. Immature birds are more extensively barred with white in these areas, and the head, breast, and upperparts are additionally heavily spotted with white. Adult females differ from males in having yellow, not brown, eyes. The species' most typical song is a series of five to eight evenly spaced and whistled notes that ascend the scale gradually. This series may be preceded by a higher introductory note and typically is repeated monotonously over and over, with intervals of a few seconds. A second utterance is a rapidly trilled and ascending series of notes that ends with about three descending notes. Yet another vocalization is a shrill version of the "brain-fever" call of hawk cuckoos, as well as a loud, clear "whee-wheep," the second note of higher pitch.

In the hand: The somewhat forked tail is unique to this species of cuckoo, and the outwardly similar true drongos lack the zygodactyl feet and rounded nostrils of cuckoos. Instead, their nostrils are largely covered by forehead feathers, and rictal bristles are present. Female drongo cuckoos are not readily distinguished from males, and immature birds have a nonglossy uniformly blackish plum-age, with many small white spots at the tips of the feathers. This distinctive plumage is acquired by nestling birds and does not appear to mimic any particular host species. The mouth interior of young birds is bright vermilion.

Habitats

This is a widespread woodland species, ranging from about 200–1800 m in India and Nepal and occurring in open secondary forest, plantations, orchards, and occasionally in dense evergreen jungle.

Host Species

Becking (1981) has established that many of the eggs that Baker (1942) attributed to the banded bay cuckoo are actually of this species. These eggs were all associated with babbler hosts, especially the Nepal fulvetta (or "Quaker babbler"), and the dark-fronted (or "black-headed") babbler. In Myanmar the birds reportedly parasitize various shrikes, bulbuls, the white-crowned forktail, and the striated grassbird, in addition to the questionable exploitation of drongos. In Java the species is known to parasitize the Horsfield's babbler, less frequently the gray-cheeked titbabbler, and still less often (one record) the brown-cheeked fulvetta (Becking, 1981). In Malaya a fledging was observed being fed by adults of the striped tit-babbler.

Egg Characteristics

According to Becking (1981), the eggs of this species are broad oval, with a white to pinkish ground color and usually with heavy streaks and blotches of red and purple. Sometimes the eggs are more faintly marked. In Java the nominate race is polymorphic as to egg color, with one egg morph that mimics its major host the Horsfield's babbler, having a rosy ground color, with bluish gray flecks and light chestnut brown flecks, clouds, and scroll-like markings. Other Javan hosts include the gray-cheeked tit-babbler, for which the cuckoo's eggs are adaptively white with brown flecks, and the crescent-chested babbler, for which the parasite's eggs are white (Schönwetter, 1967–84).

Breeding Season

In the Indian subcontinent the breeding season probably extends from March to October (but perhaps from January to March in Kerala), when the birds are most vocal and gonadal enlargement is most evident (Ali & Ripley, 1983).

Breeding Biology

Nest selection, egg laying. Remarkably little is known of this fairly common species of cuckoo, and most of the information that exists is unreliable, as it is based on Baker's erroneous identifications of this cuckoo's eggs. In spite of assertions to the contrary, there is no evidence that drongo cuckoo eggs are ever laid in the nests of drongos. Becking (1981) has effectively shown that the "*Pycnonotus* type" eggs that Baker (1942) attributed the banded bay cuckoo are in fact mostly eggs of the drongo cuckoo, although some are authentic pycnonotid (bulbul family) eggs. No observations of egg laying are available.

Incubation and hatching. No information exists on the incubation period or on hatching behavior. It is apparent that shortly after hatching the young cuckoo must evict host eggs or young from the nest, since even at early nestling stages only single cuckoo nestlings have been found in nests (Becking, 1981).

Nestling period. No information.

Population Dynamics

No information.

DWARF KOEL

(Microdynamis parva)

Other Vernacular Names: Black-capped cuckoo, black-capped koel, little koel.

Distribution of Species (see map 54): New Guinea.

Subspecies

M. p. parva: Southwestern and eastern New Guinea.

M. p. grisescens: Northern New Guinea.

Measurements (mm)

8″ (20–21 cm)

Wing, male 102, females 100, 104 (Diamond, 1972). Wing (unsexed) 104–115; tail (unsexed) 93 (Rand & Gilliard, 1969). Wing, males 107, 110 (Mayr & Rand, 1937). Wing:tail ratio ~1:0.8.

Egg, no information

Masses (g)

Male 40; females 40, 49 (Diamond, 1972).

Identification

In the field: This small cuckoo is mostly brown, but males have a blackish head and malar stripe and bright red eyes (fig. 37). Females are brown almost throughout, but have reddish-brown eyes. The bill

MAP 54. Range of dwarf koel.

is short and unusually stout for a cuckoo. Two song types are known. One consists of a long series of slightly upslurred notes uttered at the rate of about one per second and lasting 30 seconds or more. The other is a more rapid series of down-slurred notes that gradually rise in pitch until they reach a plateau and then continue at a constant pitch.

In the hand: The relatively short bill (culmen 17 mm) that is also robust and shrikelike, is distinctive, as is the bright red (males) to reddish hazel (females) iris color of adults. Because of these traits the birds are not very cuckoolike in appearance, and they also lack the tail-barring or spotting that usually is present in cuckoos. The males are easily distinguished from adult females by their mostly black head and malar stripe and their red rather than reddish brown to hazel iris color. Adult females and immature birds of both sexes are mostly brown to grayish brown, but adult females are more distinctly barred on the breast and flanks.

Habitats

This species occurs from the New Guinea lowlands near sea level to about 1450 m, mainly in the canopies of tall rainforests and monsoon forests, but it also uses forest edge habitats and tall trees in gardens.

Host Species

No information.

Egg Characteristics

No information.

Breeding Season

No information.

Breeding Biology

No information.

Population Dynamics

No information.

ASIAN KOEL

(Eudynamys scolopacea)

Other Vernacular Names: Black cuckoo, common koel, Indian koel, koel.

Distributions of Species (see map 55): Asia from Iran and Pakistan east to southeast Asia to Sri Lanka, the Greater Sundas, Moluccas, Philippines, and New Guinea.

Subspecies

E. s. scolopacea: India, Sri Lanka, and Nicobar Island.

E. s. chinensis: Indochina, western and southern China.

E. s. harterti: Hainan Island.

E. s. dolosa: Andaman Island.

E. s. simalurensis: Simalur and Babi Island (western Sumatra).

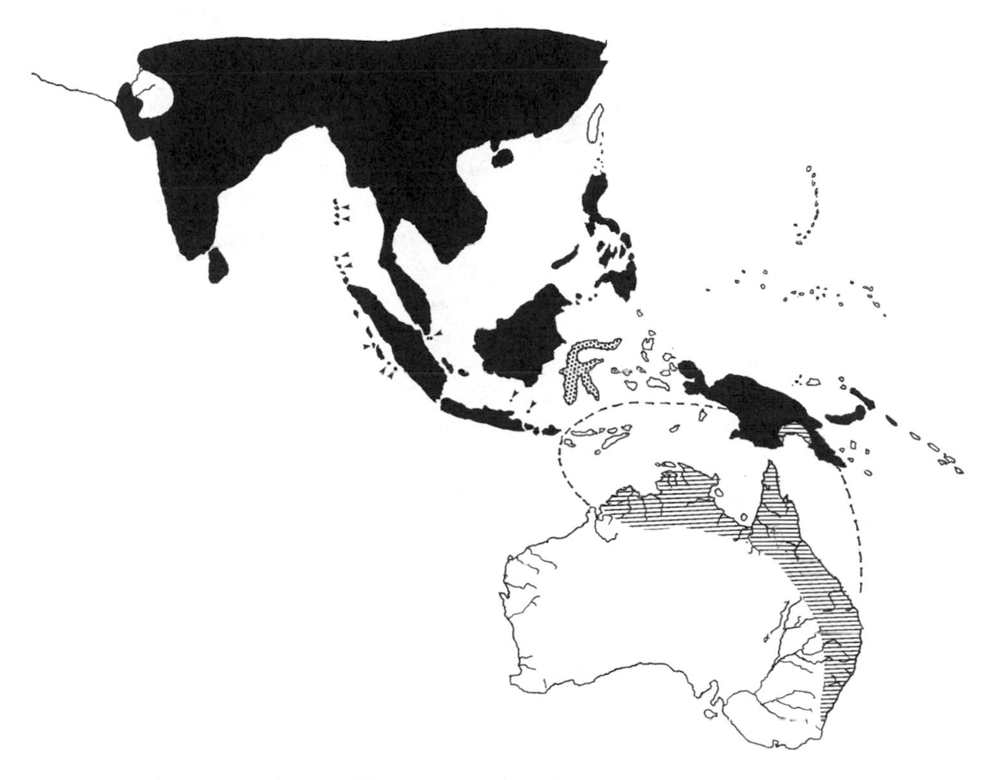

MAP 55. Breeding ranges of Asian (filled), black-billed (shaded) Australian (hatched) koels, plus wintering range of Australian koel (enclosed area).

E. s. malayana: Western Indochina, Malaysia, Sumatra to Flores Island.

E. s. everetti: Sumba to Timor and Roma Island, Kei Island (Lesser Sundas).

E. s. mindanensis: Philippines (reported from more than 40 islands); also Palawan, Sangir, and Talaut Island (includes *paraguena*).

E. s. frater: Calayan and Fuga islands (Philippines).

E. s. corvina: Northern Moluccas.

E. s. salvadorii: Bismarck Archipelago.

E. s. alberti: Solomon Island.

E. s. rufiventer: New Guinea and western Papuan islands.

E. s. minima: Southern New Guinea.

Measurements (mm)

16″ (42 cm)

E. s. dolosa, wing, males 203–235, females 201–216; tail, males 189–221, females 184–197 (Ali & Ripley, 1983). Wing:tail ratio ~1:0.9.

E. s. everetti, wing, males 199–203, females 194–210. Tail, males 187, 206, females 177–200 (avg. 183.8) (Mayr, 1944). Wing:tail ratio 1:0.87.

E. s. rufiventer, wing, male 185 (Diamond, 1972). Wing (unsexed) 180–196 (Rand & Gilliard, 1969). Wing, males 176–196; tail, male 188 (Mayr & Rand, 1937). Wing:tail ratio ~1:1.

E. s. scolopacea, wing, males 182–205, females 179–203; tail, males 186–205, females 171–189 (Ali & Ripley, 1983). Wing:tail ratio ~1:1.

Egg, avg. of *scolopacea* 30.9 × 23.2 (range 28–34.4 × 21.6–24.6); *chinensis* 32.5 × 24.2; *salvadorii* 39 × 26 (range 38.5–39 × 25.5–26.5); *malayana* 33.8 × 25.5 (overall range 30–37 × 25–26) (Schönwetter, 1967–84). Shape index 1.33–1.5 (= oval). Rey's index (*scolopacea*) 1.09.

Masses (g)

E. s. mindanensis, males 133–231, females 191.5, 244.9 (Rand & Rabor, 1960).

E. s. scolopacea, 10 males 136–190 (avg. 167) (Ali & Ripley, 1983).

Avg. of 11 (various subspecies, both sexes) 238 (range 192–327) (Dunning, 1993).

Estimated egg weight of *scolopacea* 9.0; *chinensis* 10.2; *malayana* 11.8; *salvadori* 14.0 (Schönwetter, 1967–84). Egg:adult mass ratio (*scolopacea*) 5.4%.

Identification

In the field: This rather large cuckoo is entirely glossy black in males, the plumage contrasting with a pale ivory to greenish or horn-colored bill and bright red eyes (fig. 38). Females and young are strongly spotted and barred with rufous and buff; the head has a distinct pale malar stripe and the tail is strongly barred with rufous and dark brown. Females usually have a rufous-brown head (rather than being mostly black, as in *cyanocephala*) plus bright red eyes like those of males. Immature birds are similar but have distinctly barred backs, brown eyes, and grayish buff (not greenish) bills. However, in some races such as *everetti,* the throat color of females may be entirely black or streaked with black and rufous, or there may be two broad black malar stripes separated by a rufous stripe in the middle of the throat. These traits suggest that intermediacy exists in female plumages between these two questionably distinct species.

Many different vocalizations are produced. The usual song of the males, and perhaps also of females, consists of a series of upslurred "couel" (also described as "cooee" or "you're-ill") notes. These are often uttered singly or in groups and are the basis for the koel's English vernacular name. These notes are usually uttered in an extended series at a rate of about two notes per 3 seconds, with the phrases gradually becoming louder and higher before suddenly terminating. A second song type of males is a series of paired or bisyllabic warbling "wuroo" notes that resemble water-bubbling sounds. They are uttered at a rate of about four notes per second, which gradually rise and then plateau in pitch. Other single-noted or quickly repeated calls also are produced by males, including a rapid series of up to eight falsetto and brief "dulli" notes, and a rising series of up to 10 high-pitched and nasal whistles.

In the hand: Adult males of this and the other two koels are unique in being entirely black, with red eyes. Males of this species have ivory to greenish, rather than black, bills that should distinguish them from those of the black-billed koel. Distinction from the Australian species may not always be possible. Females of the two species are more readily distinguished (see account of the Australian koel), but some are rather intermediate between the brown-headed condition typical of the Asian koel and the much more blackish head and upperpart markings found in Australian birds. Juveniles resemble the brownish females, but are more distinctly barred, rather than spotted, on the upperparts, and have dark brown rather than red eyes. Juvenile females are barred on the tail and underparts and are slightly browner above, whereas young males have some chestnut-buff on their wing coverts, producing a spotted shoulder pattern. Juvenile females are more sooty and grayish black dorsally than are adult females, which apparently represents a host-mimesis adaptation favoring a crow-like dorsal aspect, whereas ventrally they more closely resemble adult females. Both adults and juveniles have crimson-red gapes that are used in threatening situations and for food-begging, respectively. Newly hatched young are initially pinkish red, but soon become black-skinned. Their first emerging feathers are black, tipped with white (usually) or reddish fawn. Fledglings are mostly black on the head and mantle and have a blue-black bill, thus continuing to somewhat resemble their crow hosts during the postfledging period, at least until their postjuvenal molts occur.

FIGURE 38. Profile sketches of three Australasian cuckoos: adult (A) and juvenile (B) of channel-billed cuckoo; juvenile (C) and adult (D) of long-tailed koel; male (E), female (F) and juvenile (G) of Australian koel. A male (H) and female (I) of the Asian koel are also illustrated. Morph-types of their eggs are shown at enlarged scale.

Habitats

This is a widespread, lowland-adapted species, ranging from sea level to about 1500 m. It is found in secondary tropical and subtropical forests, plantations, gardens, forest edges, and sometimes primary forests, but occasionally it extends into savannas. It is often associated with fruiting trees.

Host Species

Baker (1942) provided a list of three host species for the nominate race, four host species for the Malayan race, and one host (black-collared starling) for the Chinese form of this species. Collectively, the house crow is probably the most frequently exploited host, with the large-billed crow/jungle crow species complex representing important secondary hosts. The blue magpie is certainly also a significant host species within its range. Mynas evidently serve as hosts in some areas, such as on Palawan Island (Dickinson et al., 1991).

Egg Characteristics

According to Schönwetter (1967–84), two discrete egg morphs are produced by the nominate race of this species. One of these, the *Corvus* or crow-raven type, has a dull grayish green ground color, with medium-sized, dark sepia or olive brown and dull gray markings, the egg mostly being of dark overall color tone. The second morph, which Schönwetter called the *Urocissa* (blue magpie) type, has a yellowish brown ground color that is marked with darker reddish brown flecks but is of lighter overall appearance.

Breeding Season

In the Indian subcontinent the breeding season extends from March to August but is concentrated from May to July, depending on the hosts' breeding cycles. Over much of India the nominate race of the host house crow breeds mainly from April through June, but in Kerala it mainly breeds from March to May. The Sri Lankan race breeds mainly during June. The other major host, the large-billed crow/jungle crow species complex, also breeds over an extended period, but mainly from January to March (Himalayas), March to April (northern India), or May to July (Sri Lanka) (Ali & Ripley, 1983). In Pakistan laying is mostly during June and July (Roberts, 1991).

Breeding Biology

Nest selection, egg laying. Observations made by Dewar (1907) and by Lamba (1963, 1975) indicate that the male may participate in egg-laying behavior by flying up to the host's nests and advertising his presence by crowing. The host birds typically attack the male koel, leaving the nest unguarded long enough for the female to visit the nest and deposit an egg. Frequently more than one koel egg is present in the nest, and it is likely that individual females may lay more than one egg in the same nest. There is no clear evidence that a host egg is removed or destroyed by the visiting koel, although this seems quite possible. It has also been suggested that destruction of the host egg may not necessarily occur at the time the koel egg is laid, but may occur later (Dewar, 1907). Among 24 nests observed by Lamba, in 19 cases the koel egg was laid after the host's first egg, in 3 cases after there were 2 host eggs, and in 2 cases laying occurred after there were 3 host eggs.

Incubation and hatching. The incubation period of three eggs was 13 days (Lamba, 1963). It has also been estimated at 13–14 days, as compared with 16–17 for house crow hosts and 18–20 for jungle crows.

Nestling period. According to Lamba (1963), the koels typically hatch a few days before their crow hosts, but fledge at about the same time, namely, at 3–4 weeks of age. They continue to be fed by their hosts for more than 2 weeks after nest departure (Roberts, 1991).

Population Dynamics

Parasitism rate. Little information is available. In seven of eight house crow nests observed in Pakistan and in which young survived to fledging, there were koels present. In one nest there were four koels and no crows, and in four nests one crow and one koel each. In one nest there were two crows and one koel, and in another there were two koels and one crow. In the eighth (presumably unparasitized) nest, four crows were brought to fledging (Roberts,

1991). Evidently 3 of 20 house crow nests (15%) were parasitized in Lamba's (1963) study.

Hatching and fledging success. Lamba (1963) noted that 44 house crows (and three koels) fledged from 81 crow eggs that were laid in 20 nests, representing a high host breeding success rate of 54%. Because the number of koel eggs initially present remains unknown, the koel's breeding success cannot be determined from this information.

Host–parasite relations. The ability of the host to rear young in the presence of koel nestlings reduces the impact on its reproductive potential, although Lamba (1963) stated that it is rare for more than one crow to survive when a single koel nestling is present in the nest and doubted that any could survive in the presence of two koels.

BLACK-BILLED KOEL

(Eudynamys melanorhyncha)

Other Vernacular Names: Moluccan koel; sometimes considered conspecific with the Asian koel.

Distribution of Species (see map 55): Sulawesi and nearby Moluccas.

Subspecies

E. m. melanorhyncha: Sulawesi, Tongian, and Peling Island.

E. m. facialis: Sula Island (Moluccas).

Measurements (mm)

est 15–16" (139–142 cm)

Wing, both sexes 183–214 (White & Bruce, 1986); tail, both sexes 173–215. Wing, males 202–214 (avg. 209.3, $n = 3$), females 195–200 (avg. 198.3, $n = 3$). Tail, males 193–215 (avg. 204.3, $n = 3$), females, avg. 177.7 ($n = 3$) (American Museum of Natural History specimens).

Egg: One *melanorhyncha* egg 38.6 × 24.2 (Schönwetter, 1967–84). Shape index 1.59 (= long oval). Rey's index 1.00.

Masses (g)

No information on body weights. Estimated egg weight 12.3.

Identification

In the field: This is the only koel occurring on Sulawesi and the nearby Molucca Islands, so the fieldmarks provided for the closely related Australian and Asian koels should apply to this similar species. The bill is black in adults of both sexes, which should distinguish it from the other koels (fig. 37).

In the hand: In addition to the black bill, this species differs from the similar Asian koel in having a more rounded wing, with the ninth primary shorter than the fourth (vs. longer than the fifth), and the eighth shorter than the sixth (vs. being the longest). Females exist as three different but possibly intergrading plumage morphs. These include a malelike glossy black type (but with blue-green rather than blue-violet iridescence), another type with rusty and blackish barred underparts and a gray or blackish throat and chest, and a third type with a streaked rusty-colored and black crown, a barred black and rusty-colored back, and streaked fawn and blackish underparts (Bruce & Wright, 1986).

Habitats

This species uses habitats similar to those of the Asian and Australian koels; it reportedly occurs from the Sulawesi lowlands to 1500 m elevation, in open woodlands, humid forests, riparian woodlands, towns, and farmlands.

Host Species

No specific information is available on the hosts of this species, but mynas are the presumed hosts. Schönwetter (1967–84) suggested that the myna genus *Gracula* is a host. The only known egg is much like those of the hill myna, and mynas are also reportedly hosts of the closely related if not conspecific Asian koel. It has also been suggested that mynas of the endemic Moluccan genus *Streptocitta* might be hosts.

Egg Characteristics

Schönwetter (1967–84) listed a single egg specimen for this species, which had an eggshell weight of 0.93 g and a shell thickness of 0.16 mm. It is surprisingly long relative to its width for a cuckoo egg (shape index 1.59) and is bright bluish green, with sparse and coarse markings of brown and violet green.

Breeding Season

Little information is available, but laying reportedly occurs in early March (Bruce & Wright, 1986).

Breeding Biology

No information.

Population Dynamics

No information.

AUSTRALIAN KOEL

(Eudynamys cyanocephala)

Other Vernacular Names: Australasian koel, black cuckoo, cooee, Flinders cuckoo, rainbird. Sometimes considered conspecific with the Asian koel.

Distribution of Species (see map 55): Australia and perhaps southern New Guinea, where wintering occurs.

Subspecies

E. c. cyanocephala: Northeastern Australia.

E. c. subcyanocephala: Northwestern Australia, wintering in New Guinea: Possibly resident in Trans-Fly region of southern New Guinea.

Measurements (mm)

15.5–18″ (39–46 cm)

E. c. cyanocephala, wing, male 218; female 210. Tail, male 203, female 199 (Mayr & Rand, 1937). wing:tail ratio ~ 0.9.

E. c. subcyanocephala, wing, males 200–219 (avg. 209, $n = 6$); females from New Guinea 199–215 (avg. 205, $n = 10$), females from Australia 194–227 (Rand, 1941).

Egg, avg. 33.3 × 23.6 (range 30–35.3 × 22.1–26) (Schönwetter, 1967–84). Shape index 1.41 (= oval). Rey's index (*cyanocephala*) 1.05.

Masses (g)

range of 15 males (subspecies unspecified) 120–254 (avg. 215), of 15 females 167–290 (avg. 234) (Brooker & Brooker, 1989b). Estimated egg weight 10.2. (Schönwetter, 1967–84). Egg:adult mass ratio 4.5%.

Identification

In the field: Adult males of this large cuckoo are all black, with red eyes and grayish bills and legs (fig. 38). Females are heavily barred and spotted with white, but are generally more blackish than those of the more widespread Asian koel, especially on the upper head and malar areas. According to Rand (1941), females from Australia's Northern Territory and southern New Guinea are entirely black on the crown and nape, whereas those from New South Wales and southeastern Queensland have crown feathers and a malar stripe that are conspicuously streaked with rufous. Juvenile birds (until about 3 months of age) are mostly barred rufous and dark brown, with a clear rufous crown and darker fuscous or blackish stripes through the eye and in the malar area. Many different vocalizations are produced, one of which is a male song of repeated "koo-el" or "koo-ee" notes that soon rise to a frantic climax and abruptly terminate. There is also a series of falsetto "quodel-quodel-quodel . . ." calls, and repeated, rising "weir-weir-weir" notes of a slightly "insane" quality. Several other diverse calls have been described; some are similar to those described for the Asian koel.

In the hand: In Australia this species is unlikely to be confused with any other cuckoo; males are the only all-black cuckoos with red eyes. Females can be distinguished from those of long-tailed koels by their shorter < 220 mm) tails, a dark malar stripe, and a rather uniformly blackish upper head color. Juveniles resemble adult females but have a facial stripe from the lores to the ear region, with a crown that is cinnamon-rufous. The iris color of juveniles is dark brown, rather than the adult red condition. Their mouth color is bright reddish orange, and they have bluish gray legs and deep buff-colored bills. Newly hatched and still-naked young have similar bright orange-pink mouth coloration.

Habitats

This is a lowland species occupying rainforests, monsoon forests, dense gallery forests, and other woodland habitats, preferring denser woodlands such as rainforests to more open habitats. Gallery forests near water, and especially forests with fruiting trees, are preferred habitats.

Host Species

Brooker & Brooker (1989b) provided a list of 21 species representing 196 records of parasitism in Australia. Six of these species were classified as biological hosts, four of which are listed in table 21. The two excluded species, for which fewer than 10 records each exist, are the helmeted friarbird and the silver-crowned friarbird. In northern Australia the major host is probably the little friarbird, and in southern parts of the range the primary hosts are the noisy friarbird, the figbird, and the magpie lark.

Egg Characteristics

Brooker & Brooker (1989b) described eggs of this species as tapered oval, with a pinkish buff ground color, sparingly to sometimes moderately spotted and blotched with chestnut and purplish brown, especially around the more rounded end. Schönwetter (1967–84) described two different egg morphs, one being smaller (32.3 × 25.0) and broader (shape index 1.29, or broad oval) from Cape York, and another that is larger (36.2 × 24.6) and less broad (shape index 1.47, or oval). Brooker & Brooker's measurements are closer to the former category.

Breeding Season

In Australia the egg records extend over a 7-month period (September to May), but 78% of the total 124 available egg dates fall between November and January. In southern Australia egg-laying is somewhat earlier (October to January) than in the north (November to February) (Brooker & Brooker, 1989b).

Breeding Biology

Nest selection, egg laying. In contrast to the Asian koel, this species parasitizes birds that are generally smaller than itself, and thus the females probably have little difficulty fending off nest defenders. They deposit their eggs in host nests that are cup-shaped and fairly accessible. In one early observation (North, 1895) a female was seen sitting on an olive-backed oriole nest (that had previously contained three oriole eggs) for 30 minutes before leaving. The nest then contained three host eggs and one koel egg, so a host egg had not been removed. However, more recent observations (Gosper, 1964; Crouther, 1985) and the depleted clutch sizes of parasitized host nests suggest that a host egg is probably often removed or destroyed by the visiting cuckoo. Typically only a single parasitic egg is laid per nest; of 125 parasitized nests, 120 had single koel eggs, and the remainder had 2 parasitic eggs present per nest (Brooker & Brooker, 1989b).

Incubation and hatching. The incubation period is probably 13–14 days, as in the Asian koel (Gosper, 1964; Crouther, 1985). A common host, the magpie lark, has an incubation period of about 16 days. In contrast to the Asian koel, in which eviction behavior of the much larger host eggs and/or young is apparently absent, in this species it has been well documented for such medium-sized host species as the figbird and the little friarbird. This eviction behavior occurs when the young are about 24–48 hours old (Gosper, 1964; Crouther, 1985).

Nestling period. The nestling period has been reported to range from 18 to 28 days (Gosper, 1964; Crouther & Crouther, 1984). Young koels are cared for and fed by their hosts for at least 2–3 weeks after fledging (Gosper, 1964).

Population Dynamics

Little information exists. Gosper (1964) judged that a pair of koels might have a breeding territory encompassing that of five pairs of magpie larks. He judged that 53 days may be required from egg laying to independence for the koel, and that up to three koels per season might be raised by a pair of magpie larks.

LONG-TAILED KOEL

(Eudynamys taitensis)

Other Vernacular Names: Long-tailed cuckoo, Pacific long-tailed cuckoo.

Distribution of Species (see map 56): Breeds in New Zealand; winters widely across southwestern Pacific, mainly in Polynesia and Micronesia.

Measurements (mm)

15–16″ (38–40 cm)

Wing, males 188–195; tail, males 230–250 (Oliver, 1955). Wing:tail ratio ~1:1.3.

Egg, avg. 23 × 17.4 (range 22.5–23.5 × 17–18) (Schönwetter, 1967–84). Shape index 1.32 (= broad oval).

Masses (g)

Avg. of four (unsexed) 126, range 111–140 (Dunning, 1993). Estimated egg weight 13.7 (Schönwetter, 1967–84). Egg:adult mass ratio 10.9%.

Identification

In the field: In New Zealand this is the only long-tailed cuckoo (the tail accounts for more than half the total length) breeding on the islands, and thus it is easily recognized (fig. 38). In Australia it occurs with and closely resembles females of the Australian koel, but it is browner and more heavily barred with pale rufous to buffy throughout. The underparts are white and streaked or spotted with brown rather than barred with brown. Its commonest call is a loud, shrill whistle or screeching "zzwheesht" that may be uttered by night as well as during the day. Another vocalization and a possible male song is a rapid, ringing, and prolonged series of "zip" notes, or a loud "rrrp-pe-pe-

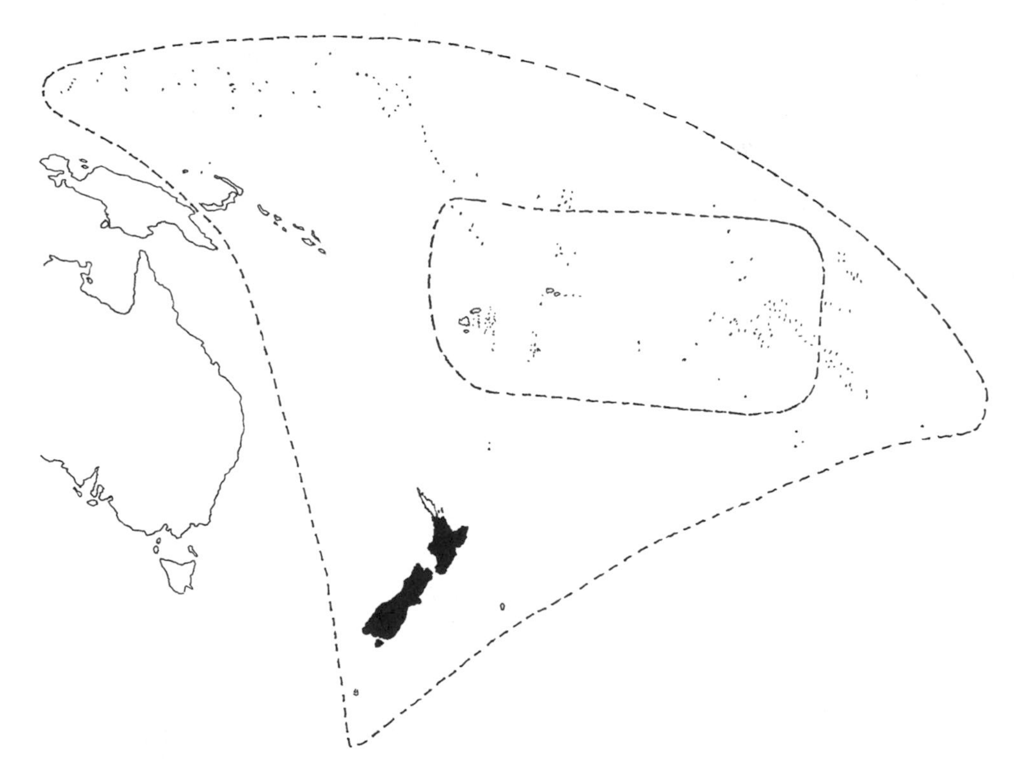

MAP 56. Breeding range (filled), plus primary wintering range (smaller enclosed area) and peripheral wintering range (large enclosed area), of long-tailed koel.

pe-pe . . . ," with the preliminary ringing "rrrp" syl-lable sometimes uttered independently of the sharp "pe" notes.

In the hand: The very long tail (>220 mm and longer than the wing), and heavily barred rufous and dark brown upperpart coloration provide for easy species identification. Unlike the other koels, the iris color of adults is light brown to yellow, and the bill is also pale brown. Females closely resemble males but are somewhat more rufous and slightly smaller. Juveniles have conspicuous white spots on the upperparts and, unlike the rather lightly streaked underparts of adults, have considerable dark striping or spotting below on otherwise rufous rather than white underparts and face.

Habitats

This species is associated with forest-canopy habitats in its New Zealand breeding grounds, but on nonbreeding areas it is often found in lower and more open vegetation, including the scrubby vegetation typical of many sandy islands.

Host Species

Oliver (1955) reported that the major hosts of this species are the whitehead on the North Island and the pipipi ("brown creeper") on the South Island. Additional known hosts are the yellowhead, the South Island tomtit, the South Island robin, the silvereye, and the introduced greenfinch and song thrush.

Egg Characteristics

The eggs of this species were described by Oliver (1955) as ovoid (= broad oval) and creamy white, with spots and blotches of purplish brown and gray, especially near the more rounded end. This corresponds well to the description accepted by Schönwetter (1967–84), who noted that the shell is somewhat glossy and also is relatively thick and hard (Rey's index not available).

Breeding Season

This species reportedly lays its eggs during November and December, during the austral spring. Its major host, the whitehead, mainly breeds from October to December; the pipipi breeds during November and December, as does the yellowhead, and the silvereye similarly breeds from September to January (Oliver, 1955).

Breeding Biology

Nest selection, egg laying. Little detailed information is available. Although it has been suggested that the egg might be inserted into the nest by the koel carrying it in its bill, there is no evidence for this, and it may be presumed that it is laid directly in the nest.

Incubation and hatching. The incubation period has not yet been established with certainty, but is probably about 16 days. Incubation periods of the major hosts include 18 days for the whitehead, 17–21 days for the pipipi, and about 21 days for the yellowhead. It is now known that the newly hatched long-tailed koel evicts its host's young or eggs (McLean & Waas, 1987), and thus far it is the only member of this genus known to exhibit such behavior.

Nestling period. McLean (1982) estimated a 21-day nestling period for this koel, as compared with a 17.4-day average for the whitehead.

Population Dynamics

Parasitism rate. On Little Barrier Island, the only certain host species is the whitehead. There the overall rate of parasitism was judged by McLean (1988) to be 16.5%, with a higher (37.5%) rate at altitudes above 250 m, and a lower (5.4%) rate at altitudes below 250 m, based on observations of fledglings and nestlings. The parasitism rate did not vary between years, nor were between-year breeding success rates of the host whiteheads directly related to brood parasitism.

Hatching and fledging success. No information.

Host–parasite relations. Although McLean (1988) reported that the reproductive success rate of the whitehead host varied significantly between years during a 2-year study, this variation could not be attributed to brood parasitism effects.

CHANNEL-BILLED CUCKOO

(Scythrops novaehollandiae)

Other Vernacular Names: Fig hawk, flood bird, giant cuckoo, hornbill, rainbird, stormbird, toucan.

Distribution of Species (see map 57): Sulawesi and Moluccas to Lesser Sundas (probably only as a wintering migrant), New Guinea (probably a local breeder but mainly a wintering migrant), and Australia (migratory breeder).

Measurements (mm)

24–26″ (60–67 cm)

Wing, males 330–350, female 316. Tail, male 264 (Rand & Gilliard, 1969). Males, wing 331–342 (avg. 338), females 322–341 (avg. 334) (Rand, 1942a). Wing:tail ratio ~1:0.8.

Egg, avg. 40.7 × 28.9 (range 38–46.2 × 26.6–32) (Schönwetter, 1967–84). Shape index 1.41 (= oval). Rey's index 1.15.

Masses (g)

Range of five males 535–655 (avg. 604), of five females 560–777 (avg. 623) (Brooker & Brooker, 1989b). One immature male 566; three females 560–632 (avg. 592.3) (Hall, 1974). Estimated egg weight 18.2 (Schönwetter, 1967–84), also 19.4 (Brooker & Brooker, 1989b). Egg:adult mass ratio 3.1%.

MAP 57. Breeding (filled) and wintering ranges (enclosed area) of channel-billed cuckoo.

Identification

In the field: This large cuckoo (the largest of parasitic cuckoos), with its yellow, toucanlike bill, is easily recognizable. Adults have bright red eyes and eye-rings, with similar scarlet red facial skin extending down the lores to the base of the bill (fig. 38). Females resemble males but are noticeably smaller. Immature birds have a dull grayish brown facial skin, their head and neck are pale buff, and their upperpart feathers are also tipped with buffy brown rather than black. The species' diverse vocalizations include various loud, booming "korrk, orrk, orrk" or, "graaah-graah" notes, plus repeated screeching and screaming calls. One of these apparent song phases begins with a long, loud squack, followed by a series of similar notes that descend in pitch but increase in speed. There are also other single-noted and ascending nasal "wark" or "oik" screams that are usually uttered in flight or while perching, and often may be heard during the nighttime.

In the hand: The very large, slightly grooved bill (culmen 85 mm in males, 75 mm in females) serves to identify this species immediately. Immature birds lack the bright soft-part colors typical of adults (grayish brown facial skin, olive-brown iris, pale yellow eye-ring, and pale grayish horn to reddish brown bill color) and have buff-tipped feathers, as noted above. Nestlings are initially naked (hatchlings of the usual crow and pied currawong hosts are downy), with bronze-colored skin and a pinkish red mouth coloration, which is similar to the pinkish white mouth color of adults. Fledgings have deep buff to golden or rufous feather markings on the lighter parts of the body, and the bill is mostly dark blackish brown, but paler toward the tip. There is no bare skin around the eye, but a dark line of bare skin extends from the bill to the eye (Goddard & Marchant, 1981).

Habitats

This species is usually found in forests at the canopy level, often in figs or other fruit-bearing trees, and favors eucalyptus forests and rainforests. It also occupies forest edges, savannas, woodlands, and partially cleared forests, from near sea level to about 1200 m elevation.

Host Species

Brooker & Brooker (1989b) reported that 9 host species were associated with 138 records of parasitism in Australia. Of these, various corvid and cracticid taxa were identified as biological hosts, including several species of crows (71 records), the pied currawong (46 records), and the Australasian magpie (9 records). Goddard & Marchant (1981) also listed 9 host species, with a total of 78 records of parasitism. Known biological hosts among the crows include the Torresian and little crows and the Australian raven; the forest raven may also be a biological host.

Egg Characteristics

Brooker & Brooker (1989b) described this species' eggs as "swollen oval" (= oval, as defined here) and varying in ground color from dull white to pale reddish brown. The surface markings are spots and blotches of light to medium brown, which vary in quantity from few to moderate. The very low Rey's index suggests that a thick, hard shell is typical. The only host species that the channel-billed cuckoo's eggs mimic closely is that of the pied currawong, whose eggs can scarcely be distinguished from those of the cuckoo (Goddard & Marchant, 1981).

Breeding Season

Australian breeding (egg or nestling) records for this species are few and seem to fall within October–January (Brooker & Brooker, 1989b). No information on reproductive seasonality is available for New Guinea. There breeding is still unproven but is likely to occur, at least in southern New Guinea and the Bismarck Archipelago, where the Torresian crow is common.

Breeding Biology

Nest selection, egg laying. Eggs are deposited directly into the rather large, open-topped nests typical of most host species. The female probably simply drops them in while standing on the nest rim, inasmuch as the host's eggs are often damaged. Destruction of or preying on the eggs of the host has also been reported for several species. Although up to as many as eight channel-bill eggs have been seen in a single nest, most commonly there is only one per nest. However, of 61 nests that contained both channel-bill and host eggs, 29 (47.5%) had more than one channel-bill egg present (18 nests had 2 eggs, 5 had 3, and the remainder had 4–8). Goddard & Marchant (1981) estimated a mean of 2.5 channel-bill eggs present in parasitized nests of various *Corvus* species and 1.7 in those of the pied currawong. Various estimates of 12–25 eggs laid per season have been made; based on indirect evidence the estimated egg-laying interval is about 48 hours (Goddard & Marchant, 1981).

Incubation and hatching. The incubation period is still unknown. It is also not yet known whether the host's young and eggs are evicted from the nest or if the host nestlings simply starve, but the former is possible, inasmuch as the host species' young are usually gone from the nest within a week of hatching. As many as five channel-bill nestlings have been reported occupying a single little crow's nest. Of 14 observations of channel-bill nestlings in various host's nests, 12 nests contained only channel-bill nestlings and the other nests contained channel-bill nestlings in addition to one (one case) or two (one case) host chicks (Brooker & Brooker, 1989b).

Nestling period. The nestling period is believed to require 17–24 days (Goddard & Marchant, 1981). There is an additional period of post-fledging dependence that is still of uncertain length but which may be about 1 month.

Population Dynamics

Parasitism rate. No information.

Hatching and fledging success. No information.

Host–parasite relations. At least some times the host species manages to rear one or more of its own young in the presence of the parasite. Salter (1978) mentioned a case in which a pair of crows managed to raise two young of their own species in addition to a single channel bill.

9

AMERICAN GROUND CUCKOOS

Family Neomorphidae

The ground cuckoos of the New World are a relatively small assemblage of mostly tropical species, of which 11 species in five genera have been accepted by Sibley and Monroe (1990). Among these, only the greater roadrunner has been studied in any great detail, and this is also the only species of the group with a range extending north of Mexico. Five of the other species are limited to South America, three species occur both in Middle and South America, and the remaining two are limited to Middle America.

Friedmann (1933) suggested that "the ancestral cuculine stock that reached the Americas brought with it a tendency toward parasitism" (p. 533), or, more probably, in the New World cuckoos parasitism developed independently of that occurring in the Old World. Whereas in the Old World the parasitic cuckoos are believed to be the most highly specialized, in the New World the species most like the Old World social parasites are members of such non-parasitic genera as *Coccyzus.* Following this argument, Friedmann suggested that the brood parasitic ground cuckoos of the New World are not members of one of the "higher" groups of cuckoos but rather are relatively primitive types. Friedmann observed, for example, that the striped cuckoo lays a primitive type of nonpigmented egg, but it might also be noted that the two species of *Dromococcyx* lay spotted or otherwise patterned eggs.

This argument as to the relative primitive or advanced status of the American ground-cuckoos seems to be rather nonproductive, especially given the rudimentary state of knowledge concerning not only the ground-cuckoos, but other New World cuckoos that are nonparasitic. These include the members of the genera *Crotophaga* and *Guira,* some representatives of which are communal nesters (Skutch, 1954, 1976; Cavalcanti *et al.,* 1991). Such communal nesting behavior represents at least one potential route toward social parasitism, as has been noted earlier. Additionally, the "nonparasitic" yellow-billed cuckoo has at times been observed to parasitize the nests of other species (Bent, 1940; Wiens, 1965).

STRIPED CUCKOO

(Tapera naevia)

Other Vernacular Names: Brown cuckoo, crespin, four-winged cuckoo, tres pesos.

Distribution of Species (see map 58): Mexico south to southern Brazil and northern Argentina, also Trinidad.

Subspecies

T. n. naevia: Northern South America and Trinidad.

T. n. excellens: Panama to southeastern Mexico.

T. n. chochi: Southern Brazil and northern Argentina.

Measurements (mm)

10–11″ (26–29 cm)

T. n. excellens, wing, males 108–117.5 (avg. 112.4), females 104–112 (avg. 108.2) Tail, males 148–165 (avg. 157.7), females 140–162 (avg. 146.2) (Wetmore, 1968). Wing:tail ratio 1:1.35–1.4.

T. n. naevia, wing, males 103–112 (avg. 108), females 99–106 (ffrench, 1991).

Egg, avg. of *naevia* 21.3 × 16.4 (range 19.8–23.5 × 15.4–17.2). *chochi* 21.4 × 15.8 (range 19.8–23 × 15.3–16.1) (Schönwetter, 1967–84). Avg. of *naevia,* 22 × 15 (Friedmann, 1933). One egg of *excellens,* 23.43 × 16.46 (Kiff & Williams, 1978). Shape index 1.29–1.35 (=broad oval). Rey's index, *naevia,* 1.75; *chochi* 1.54.

Masses (g)

Avg. of 10 of both sexes, 52.1 (Dunning, 1993). One female, 47 (Sick, 1993). Males 40–50, females 41–53 (Haverschmidt, 1968). One male 41, one female 41 (ffrench, 1991). Estimated egg weight 2.87 (Schöwetter, 1967–84); actual weights of *naevia* 3.1–3.7 (Haverschmidt, 1968), 3.4 (Sick, 1993). Egg:adult mass ratio 5.5%.

MAP 58. Range of striped cuckoo.

Identification

In the field: This is a medium-sized, forest-dwelling cuckoo with a long tail, a back that is streaked with black, and conspicuous blackish alulas (fig. 39). The tail is grayish brown, with buff or white feather tips. Immature individuals have a blackish crown and blackish barring on the neck.

FIGURE 39. Profile sketches of the parasitic ground-cuckoos: adults of striped (A), pavonine (B) and pheasant cuckoos (C). Also shown is the striped cuckoo's alula-spreading posture (D, after a sketch in Wetmore, 1968), a nestling threat-gaping (E, after a photo by Haverschmidt, 1961), and bill structures of nestling (F) and adult striped cuckoos (G, after sketches by Sick, 1993).

The usual vocalization is a whistled, usually bisyllabic (but sometimes monosyllabic or trisyllabic) "sa-see" (the basis for "cres-pin," "wife-sick," and other onamatopoeic names), with the second syllable accented and a half-tone higher in pitch. The song is metallic in timbre and is monotonously repeated for long periods. Slud (1964) described its song in Costa Rica as a highly ventriloquial, metallic whistle that has the second note a step higher than the first. At times a third, still higher note is added, and occasionally a somewhat lower fourth note as well. Another variation is a series of four whistles that ascend the scale, followed by a fifth descending note so that a complete song phrase becomes "pee-pee-pee-pee′-dee." In Suriname the usual song is a three-note "pee-pee-de," but it varies from two to five syllables. In Trinidad the three-note version of call has the first two notes on the same pitch, and the last higher, and evidently is the basis for its "Trinity" vernacular name. Another song there consists of four or five uniformly spaced and pitched notes, followed by one or two evenly spaced but fainter notes. The songs are usually uttered at about 5- to 10-second intervals, for minutes on end. The bird raises and lowers its crest each time it vocalizes and also often lowers its alulas and wings white vocalizing (fig. 39D).

In the hand: This species is rather easily recognized by its shaggy crest and unusually large and conspicuous black alular feathers (the basis for its vernacular name "four-winged cuckoo"). The sexes are similar if not identical as adults, but juveniles and immature birds have black vermiculations on the foreneck and breast, as well as yellow spots on the feathers of the upperparts (Sick, 1993). Young nestlings are naked, with a yellow-orange gape and pinkish skin. Within a week the skin has turned a violet color, and by 10 days feathers are sprouting (Haverschmidt, 1961).

Habitats

This species is found in scrubby and thickety fields, forest edges, and shrubby woodlands with scattered trees. Boglike habitats are sometimes also used. It occurs from wet lowlands near sea level up to about 1400 m.

Host Species

A list of at least 20 reported host taxa of this species is provided in tables 30 and 31, based on various literature sources including Friedmann (1933), Sick (1953, 1993), and Salvador (1982). Too few records are currently available to judge which of these species might be the most significant and universal hosts, but they probably consist mostly of spinetails. In Costa Rica the usual hosts include *Synallaxis* spinetails, *Throthorus* wrens, and *Arremonops* sparrows (Kiff & Williams, 1978; Stiles & Skutch, 1989). In Trinidad *Certhiaxis* and *Synallaxis* spinetails are known hosts (ffrench, 1991). The commonest host in Argentina in Spix's spinetail (Friedmann, 1933).

Egg Characteristics

The eggs reportedly range in color from pale blue or greenish blue to white. Haverschmidt (1968) stated that of 20 Suriname eggs, 9 were white, 8 were bluish green, and 3 were bluish white. Eggs in Trinidad also range from white to bluish or greenish, and blue eggs have likewise been observed in Panama. Many of this species' known hosts lay white eggs, such as the furnariid spinetails and the plain wren, but the rufous-and-white wren lays plain blue eggs.

Breeding Season

There are few records of breeding, but in Panama singing begins in January, reaches a peak by the end of that month, and lasts at least until June (Wetmore, 1968). In Suriname eggs can be found nearly throughout the year. Haverschmidt (1968) lists 20 seasonal records, with 13 of these from December to April and 7 from June to October. In Trinidad singing occurs mainly from December to April, but breeding activity has been reported for nearly all months between March and October. Thus, as in Suriname, some reproduction perhaps occurs throughout the entire year in Trinidad (ffrench, 1991). Argentina breeding

TABLE 30 Reported Host Species of the Neotropical Ground Cuckoos[a]

Well-documented or major hosts	Probable or minor hosts
STRIPED CUCKOO (52.1 g; mean host mass 47%)	
A. Spherical stick nests	A. Pendant nests
Stripe-breasted spinetail (M, 28%)	Tody-flycatchers (~60–125%)
Spix's spinetail (M, 24%)	B. Cavity nests
Plain-crowned spinetail (M, 35%)	Cinclodes (~60–100%)
Pale-breasted spintail (M, 28%)	Earthcreepers (~60–100%)
Yellow-chinned spinetail (FH, 28%)	C. Roofed adobe nests
Sooty-fronted spinetail (29%)	Horneros (~60–110%)
Chotoy spinetail (35%)	
Azara's spinetail (32%)	
Rufous-breasted spinetail (33%)	
Common thornbird (47%)	
Red-eyed thornbird (47%)	
Greater thornbird (74%)	
B. Spherical/retort-shaped grass nests	
Tody-tyrants (M, ca 50%)	
White-headed marsh tyrant (27%)	
Rufous-and-white wren (M, 31%) and other *Thyrothorus* wrens	
C. Cavity nests	
Buff-crowned foliage gleaner (51%)	
D. Open-cup nests	
Black-striped sparrow (M, 67%) and other *Arremonops* sparrows	
PHEASANT CUCKOO (84.5 g; mean host mass 26%)	
A. Spherical grass nests	
Tody-tyrants (~30%)	
Pied water-tyrant (14%)	
B. Basketlike pendant nests	
Barred antshrike (33%)	
PAVONINE CUCKOO (43.2 g; mean host mass 26%)	
A. Enclosed pendant nests	
Ochre-faced tody flycatcher (13%)	
Drab-breasted bamboo tyrant (16%)	
Eared pygmy tyrant (12%)	
Tody tyrants (~60%)	
B. Basketlike pendant nests	
Plain ant vireo (30%)	

[a]From various sources including Friedmann (1933) and Sick (1993). Major hosts are indicated by M, and a known fostering host by FH. Mean adult host masses (as percentages of mean adult cuckoo mass) are also shown.

records range from late October to late January, during the austral spring (Friedmann, 1933).

Breeding Biology

Nest selection, egg laying. Little information is available. Most of the nests known to be used by this species are difficult of access, and questions have arisen as to how the cuckoo is able to enter such nests. Sick (1953, 1993) observed nests of *Certhiaxis* that had been broken into near the incubation chamber and suggested that this is the cuckoo's means of access. However, direct observations of egg laying are still lacking. The data summarized by Friedmann (1933) on this species, and other more recent observations, suggest that single-egg parasitism is the usual rule, but Haverschmidt (1968) noted that two-egg cases of parasitism may be found "fairly often."

Incubation and Hatching. The incubation period lasts 15 days, as compared with the usual 18-day period of a common Suriname host, the plain-crowned spinetail. The nestlings of this host disappeared soon after they had hatched, although the long nest entrance made it unlikely that they had been evicted by the cuckoo. Possibly they simply starve to death (Haverschmidt, 1961), but it is more likely that they are killed by the young cuckoo, which has a pincherlike bill with a sharp point and an associated ability to kill nest-mates (Sick, 1993; Morton& Farabaugh, 1979). In any case, their bodies are probably removed by the parents, since they are generally not found below the nest.

Nestling period. The nestling period is 18 days, although at this age the young bird is still unable to fly well (Haverschmidt, 1961).

Population Dynamics

Parasitism rate. Little information exists. Haverschmidt (1968) reported that 14 of 21 nests (67%) of the stripe-breasts spinetail nests he found in Suriname were parasitized.

Hatching and fledging success. No significant information.

Host–parasite relations. Little information exists. Sick (1953, 1993) reported that the host owners of a *Certhiaxis* nest that had been broken into by a cuckoo immediately begin to repair the damage. It seems likely that, if parasitism rates are high, considerable damage to a host species could be done by this cuckoo, both in terms of nest damage by visiting cuckoos and reduced fecundity as a result of parasitism.

PHEASANT CUCKOO
(Dromococcyx phasianellus)

Other Vernacular Names: None in general English use. *Distribution of Species (see map 59):* Mexico south to Paraguay, Bolivia, and Argentina.

Subspecies

D. p. phasianellus: Tropical and subtropical South America, except for Colombia.

D. p. rufigularis: Mexico to Colombia.

Measurements (mm)

13-15" (33–39 m)

Wing, males 163–176 (avg. 167.8), females 160–176 (avg. 168). Tail, males 162–203 (avg. 185.4), females 177–208 (avg. 192.6) (Wetmore, 1968). Wing, both sexes 159–189; tail, both sexes 193–234. The mean wing length was 170.4 for males and 173.5 for females; the mean tail length was 219.4 for males and 224.8 for females (Ridgway, 1916). Wing:tail ratio 1.29–1.3.

Egg, avg. 24.6 × 15.8 (range 23.3–25.6 × 14.5–16.9). (Schönwetter, 1967–84). Shape index 1.56 (= long oval). Rey's index 1.77.

Masses (g)

Ave. of 4 unsexed birds, 84.5 (Dunning, 1993). One unsexed individual 90 (Smithe, 1966). Estimated egg weight 3.3. Egg:adult mass ratio 3.9%.

Identification

In the field: This is a medium-sized, forest-dwelling cuckoo with a bushy, cinnamon-colored crest, a pale eye-stripe on an otherwise brown head, and a very long tail (fig. 39). It is mostly dark brown above, with cinnamon to rufous barring, and white to buff below. The upper breast and throat are streaked with dark brown. The long tail feathers are mostly covered by white-tipped coverts, and the rectrices are also narrowly tipped with white. Its vocalizations are evidently similar to, but are lower in pitch than, those of the pavonine cuckoo. The typical call is a double-noted (sometimes three- or four-noted) whistle of successively higher-pitched notes (similar to that of a striped cuckoo) that is followed by a tinamoulike trilling note or a tremulo. In the usual call uttered in Central America, the second note is higher than the first, and the third note is quavering, "whoo-hee-whe-rrr" (Smith, 1966). In Brazil the birds are said to utter a series of whistling notes that either

MAP 59. Range of pheasant cuckoo.

ends in a "eweerrew" tremulo, or, if the bird is more excited, progressively ascends the scale "eww–eww–dew–rew" (Sick, 1993).

In the hand: The crested condition and long, rather filmy and spotted upper-tail coverts identify this as a *Dromococcyx* cuckoo. In this species the wing length is greater than 150 mm. The sexes are alike as adults, with yellowish eyes and yellow-green eye-rings that also extend backward as bare skin over the ears. Immature birds differ from adults in lacking white tips on the rectrices, but they have conspicuous buffy tips on the wing coverts. They also are sooty brown, rather than rufous, on the crown. Iris (perhaps more brownish) and soft-part color may also differ in immature birds, but this is not yet certain. Immature birds also reportedly differ from young of the pavonine cuckoo in having conspicuous buffy postocular stripes and in exhibiting buffy tips on their greater wing coverts. Young nestlings are still undescribed.

Habitats

This is a little-observed and woods-adapted lowland species that extends into the subtropical zones and perhaps locally into the lower montane forest, at elevations up to 800 m in Costa Rica.

Host Species

Three taxa of host species are listed in table 31, based on the still limited available literature, such as Sick (1993). In Costa Rica the usual hosts evidently include flycatchers of the genus *Myiozetetes* and also flatbills, presumably the eye-ringed flatbill. These forest-edge species range in mass from about 23 to 40 g (or about 30–50% of adult cuckoo mass) and build cuplike, roofed, or retort-shaped nests (Stilles and Skutch, 1989).

Egg Characteristics

Wetmore (1968) has summarized the available information on this species' eggs. The most reliable record is from an oviducal egg that measured 25.2 × 14.3 mm (shape index 1.76), was faintly buff in ground color and lacked gloss, and had scattered and irregular dots of rugous to dull grayish rufous. A similar egg, found in the nest of a pied water tyrant, was 23.3 × 16 mm (shape index 1.45), with a pale reddish ground color and small reddish brown markings. Another recently described oviducal egg was whitish, with reddish brown dots on the more rounded half. The mean measurements given by Schönwetter and cited earlier are from only three eggs, including the first two mentioned here.

PAVONINE CUCKOO

(Dromococcyx pavoninus)

Other Vernacular Names: None in general English use.

Distribution of Species (see map 60): Tropical South America from Colombia, Venezuela, and the Guianas south to Paraguay, southern Brazil, and extreme northeastern Argentina.

Subspecies

D. p. pavoninus: Tropical South America, except northern Venezuela.

D. p. perijanus: Northern Venezuela.

Measurements (mm)

11–11.5″ (28–29 cm)

Wing, both sexes 137–139.7; tail, both sexes 139.7–172.7 (Ridgway, 1916). Wing:tail ratio ~1:1.12.

Egg, avg. of 4, 21.5 × 14.8 (range 21.2–22 × 14.4–15.2) (Schönwetter, 1967–84). Shape index 1.45 (= oval). Rey's index 2.45.

Masses (g)

Two unsexed birds, 40.5 and 45.9 (Dunning, 1993). One adult 48 (Sick, 1993). Estimated egg weight 2.6 (Schönwetter, 1967–84). Egg:adult mass ratio 5.8%.

Identification

In the field: Over much of its range, this species occurs sympatrically with the pheasant cuckoo, and the two species probably cannot be easily distinguished in the field. This species is somewhat smaller than the pheasant cuckoo and lacks dark spotting or streaking on the lower neck and upper breast (fig. 39). It has a mostly rufous-brown head, with a cinnamon crest and a pale eye-stripe, similar to that of the pheasant cuckoo. Likewise, its extremely long upper-tail coverts are white-tipped. Its distinctive song is a repeated series of four to five whistled notes, sounding like "ew-i, ew, ew" or "ew-i, ewi-i," and of the same timbre as that of the striped cuckoo. The first syllable of each couplet or phrase is lower than the following syllable or syllables. Neunteufel (1951) has diagrammed the call in dot-dash form (·---·---··), and described it as sounding like "yasy-yatere."

In the hand: The smaller size (wing <150 mm) separates adults of this species from the pheasant cuckoo. The sexes cannot be distinguished externally, but immature birds lack the white terminal spots on the adults' upper-tail coverts and have sooty brown rather than cinnamon-colored crowns. Young birds additionally differ from those of the pheasant cuckoo in that they have inconspicuous grayish postocular stripes and have rather broad buffy brown streaks on the upper-wing coverts.

MAP 60. Range of pavonine cuckoo.

Habitats

This species is found in forest edges and dense secondary woodlands of lowland tropical forests.

Host Species

Five taxa of birds are listed as probable hosts in table 31, based on the relatively scant literature currently available, especially Sick (1993).

Egg Characteristics

Schönwetter (1967–84) mentions four eggs that he examined and ascribed to this species. Three of these had a pinkish white ground color, with thick spots, scrawls, and scribbles of bright purple. The fourth one had a white ground color, with loose and light streaking present. A fifth egg has been described as measuring 22.2 × 15 mm and having a creamy white ground color and covered with small yellowish flecks (these perhaps representing stains from nest materials).

Breeding Season
No specific information.

Breeding Biology
No information.

Population Dynamics
No information.

TABLE 31 Reported Host Species of the African Parasitic Finches[a]

Primary or unique hosts	Minor or questionable hosts
Parasitic Weaver	
Black-chested prinia	Pectoral-patch cisticola (2)
Tawny-flanked prinia	Winding cisticola (1)
Zitting cisticola	Croaking cisticola (1)
Desert cisticola (5)	Singing cisticola (1)
Tinkling cisticola (3)	Wing-snapping cisticola (1)
Rattler-grass cisticola (2)	
Village Indigobird	
Red-billed firefinch	Bronze munia (1)
Jambandu Indogobird	
Zebra waxbill	Black-bellied firefinch
Baka Indigobird	
Black-throated firefinch	
Variable Indigobird	
African firefinch (except *codringtoni*)	Common waxbill (2)
Peters' twinspot *(codringtoni)*	
Dusky Indigobird	
Jameson's firefinch	
African firefinch	
Pale-Winged Indigobird	
Bar-breasted firefinch *(wilsoni)*	Common waxbill (1)
Brown firefinch *(incognita)*	
African quailfinch *(nigeriae)*	
Brown twinspot *(camerunensis)*	
African firefinch *(camerunensis)*	
Steel-Blue Indigobird	
Black-cheeked waxbill *(delameriei)*	
Red-rumped waxbill	
Straw-Tailed Whydah	
Purple grenadier	
Queen Whydah	
Common grenadier	Scaly weaver (3)
	Black-chested prinia

(continued)

(1)

TABLE 31 *(continued)*

Primary or unique hosts	Minor or questionable hosts
Pin-Tailed Whydah	
Common waxbill (widespread host)	Bronze munia (2)
Zebra waxbill (possible host in Natal)	Swee waxbill (2)
	Crimson-rumped waxbill (2)
	Red-collared widowbird (1)
	Magpie munia (1)
	Fawn-breasted waxbill (1)
	Black-rumped waxbill (1)
	Orange-cheeked waxbill (1)
	Streaky seedeater (1)
	African golden-breasted bunting (1)
	Tawny-flanked prinia (1)
	Piping cisticola (1)
Northern Paradise Whydah	
Green-winged pytilia ("red-lored" races only)	
Togo Paradise Whydah	
Red-faced pytilia	
Long-Tailed Paradise Whydah	
Red-winged pytilia	
Eastern Paradise Whydah	
Green-winged pytilia (excepting "red-lored" races)	
Broad-Tailed Paradise Whydah	
Orange-winged pytilia	

[a]Host list based largely on Friedmann (1960), but with some updating, especially based on studies by R. Payne. Minimum number of probable but not certain valid host records are indicated for

10

AFRICAN PARASITIC FINCHES

Family Passeridae

The only obligate parasites among the finches and sparrows of the world are found in Africa, where the approximately 16 species of indigobirds and whydahs of the genus *Vidua* ("viduine finches") all occur as well as the single species of parasitic weaver or so-called cuckoo finch.

The geographic breeding ranges of the viduine finches encompass all of sub-Saharan Africa (fig. 40), but the most species-rich regions (supporting seven to nine species per 5° latilong quadrants) consist of subequatorial habitats extending from Nigeria east to the upper Nile Valley and Rift Valley. Another species-rich region occurs in southeastern Africa, where seven species often occur in the general area of Botswana, Zimbabwe, and northern South Africa.

Friedmann (1960) reviewed the probable phyletic relationships of the viduine finches and weaver finch at length. He concluded that the parasitic weaver is perhaps related to such rather typical plocied genera as the bishops and widowbirds (*Euplectes* spp.), although the structure of its sternum is somewhat aberrant with regard to this group, and one of its sternal features suggests possible affinities with the buffalo weavers (*Bubalornis* and *Dinemellia*). More recently, Sibley and Monroe (1990) placed the parasitic weaver in linear sequence between the bishop–widowbird group and the grosbeak weaver (*Amblyospiza*), all within the subfamily Ploceinae (family Passeridae).

Friedmann (1960) proposed that a phyletic relationship exists between the viduine finches and their hosts, the estrildine finches, thus supporting an idea that had been advanced earlier by Chapin (1917). Chapin has suggested that the viduine finches branched off from an ancestral estrildine line now most closely represented by the African quailfinch (*Clytospiza*). He

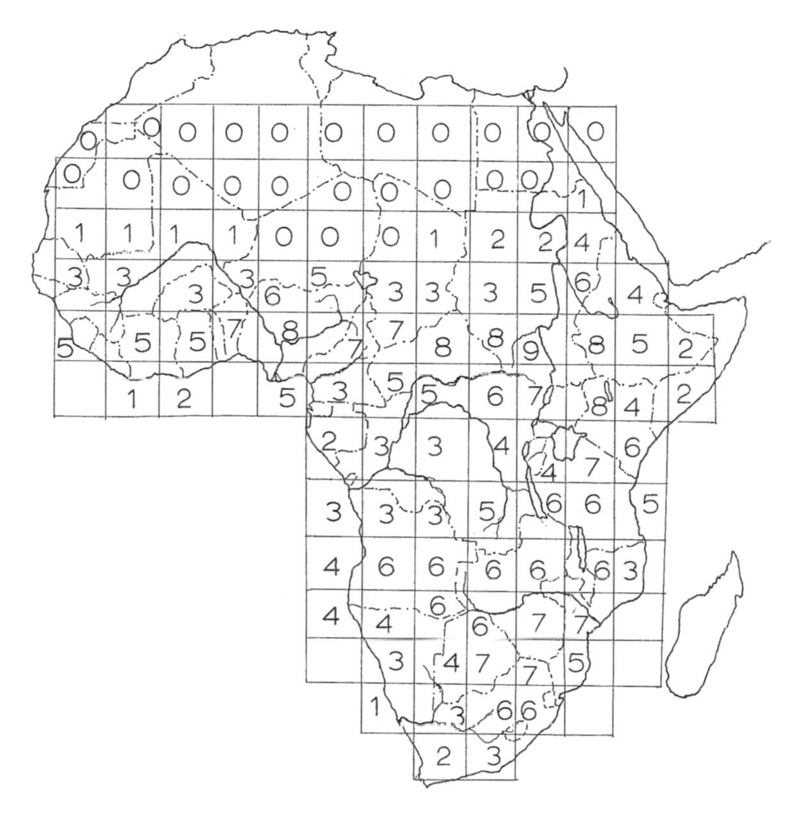

FIGURE 40. Species-density map of parasitic viduine finches in Africa, by 5° latilong quadrants.

also believed that the short-tailed forms (subgenus *Hypochera*) that are commonly called indigobirds or combassous are more primitive than the long-tailed species (subgenera *Vidua* and *Steganura*), of which the latter subgenus is the most highly differentiated in male plumage traits and exhibits the most complex male display behavior, including aerial display flights.

Friedmann (1960) believed that the close similarities in the mouth markings of viduine and estrildine finches, and also their similarities in juvenal plumages, can be interpreted as a result of close phyletic relationships rather than reflecting selection for host mimicry on the part of the viduine parasites. Although a general level of mouth and palatal similarity between the two groups may indeed be the result of shared ancestral traits, the extreme degrees of species-specific similarity exhibited between almost every known host–parasite pair can only be interpreted as being the result of direct selection pressures favoring host mimicry, as was first suggested by Nuenzig (1929) and later supported by Southern (1954). Friedmann (1960) interpreted the white egg color of both host and parasite as the result of phyletic relationship, rather than evolved similarities.

Friedmann (1960) was unable to determine if the viduines are monogamous or polygamous (but he noted that monogamy is the usual estrildine condition) and observed that in estrildine and ploceine finches, incubation may be performed by the female alone, by both sexes, or largely by the male, with no clear patterns evident that might lead to the evolution of brood parasitism as an adaptive mode of reproduction. He suggested that an endocrine imbalance or change might have been the basis for a shift from a nonparasitic to a parasitic mode of reproduction. He also suggested that the evidence for such a possible endocrine "lag" might include the fact that viduines do not appear to breed before their second year of life. Because of the lack of any structures or habits directly deleterious to the host young or eggs, Friedmann believed that the development of parasitism on the part of the viduines is a relatively recent phenomenon. However, one might also argue that the lack of direct mortality to species-specific hosts is actually a highly derived or specialized condition, a position that was advocated earlier by Southern (1954), and might easily be supported by the available facts.

PARASITIC WEAVER

(Anomalospiza imberbis)

Other Vernacular Names: Cuckoo finch, cuckoo weaver

Distribution of Species (see map 61): Sub-Saharan Africa from Sierra Leone to Kenya and south to South Africa.

Subspecies

A. i. imberbis: Southern Africa north to Kenya.

A. i. macmillani: Ethiopia.

A. i. butleri: Western Africa from the Congo to Sierra Leone.

Measurements (mm)

5″ (13 cm)

Wing, males 64–71 (avg. 68). Tail, males 40–46 (avg. 43.5) (Friedmann, 1960). Wing, both sexes 56–73. Tail, both sexes 41–44 (McLachlin & Liversidge, 1957).

Egg, 17–18 × 12.75–13 (Friedmann, 1960); 17–17.3 × 12.5–13 (McLachlin & Liversidge, 1957). Shape index ~1.35 (= broad oval or oval).

Masses (g)

Males 18–21 (avg. 19.8, $n = 8$). Females 19–21 (avg. 19.6, $n = 6$) (Williams & Keith, 1962). Avg. of 6 females 21.4; unsexed adults 23–26 (avg. 24, $n = 4$) (Maclean, 1984). Estimated egg weight 1.59 (Payne, 1977b). Egg:adult mass ratio ~13.2%.

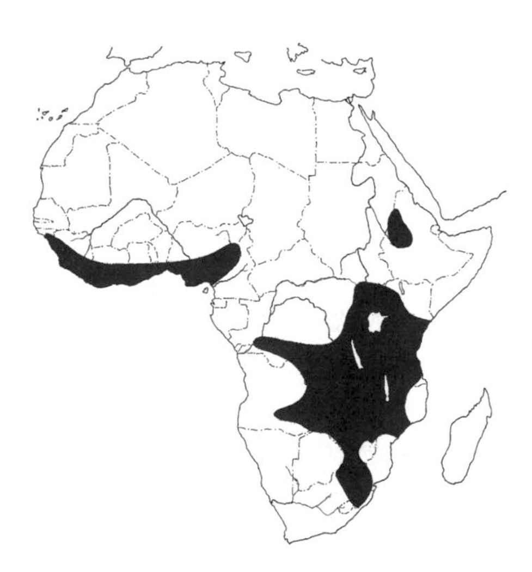

MAP 61. Range of parasitic weaver.

Identification

In the field: This is a rather chubby, short-tailed finch, with the males mostly having bright yellowish plumage and a heavy black bill (fig. 40). Females are browner above and more buffy below, as are young birds, but the rather large, stubby bill shape is evident in these birds also. Immature birds generally resemble adult females, but their flanks are streaked with black, and the lower mandible is yellowish. Nonbreeding males are duller than breeding birds, but a gradual brightening before breeding occurs as result of wearing away of the duller barbule surfaces, exposing the bright yellow feather interior. The vocalizations of males include "squeaky," "chattering," or "tittering" calls uttered during flight, and the courtship song is a similar squeaky "tsileu-tsileu-tsileu." One description states that the principal components of the song are thin, high, sibilant "tissiwick" and "tissiway" phrases, the former rising on the last syllable, and the latter falling in pitch. Another call is a deliberate "dzi-bee-chew" that is less sibilant and more three-syllabled than the song proper (Williams & Keith, 1962).

In the hand: Easily recognized by the short, stubby bill (culmen length 12.5–14 mm) and predominantly yellow plumage in adults of both sexes. Adult females may be distinguished by their whitish rather than yellow underparts. Immature individuals have buffy flanks that are streaked with black and have yellowish present on the lower mandible. In adults the bill is brownish black (females) to blackish (breeding males), but there may some buffy gray, flesh-white, or pinkish tones at the extreme base of the lower mandible. Palatal papillae or other special mouth markings are lacking in nestlings, but the mouth is flesh-colored, and the tongue is purplish pink. The interior of both mandibles is bright yellow, and there is pale yellow mandibular flange (or "wattle") at each commissural junction. The upper mandible of young birds is sepia-colored and the lower one ochre with a sepia tip (Benson & Pitman, 1964). Feathered nestlings may be readily recognized by the presence of upperpart feathers conspicuously margined with tawny buff, whereas their cheeks, throat, and breast are uniformly buffy.

Habitats

Open grasslands and lightly wooded grasslands, especially near water, are preferred habitats.

Host Species

A list of 11 host species is shown in table 31, based mostly on Friedmann (1960). This list includes three probable primary fostering hosts (black-chested and tawny-flanked prinias, zitting cisticola), four hosts with at least two parasitism records, and four that appear to have at least one reliable parasitism record. Vernon (1964) listed eight host species (eight cisticolas, two prinias) for Zimbabwe, with the largest number of records (six) for the zitting cisticola, followed by four records each for the desert cisticola and black-chested prinia, and three records for the tawny-flanked prinia. He also listed one species (wing-snapping cisticola) not previously recorded as a host. The average adult masses of the zitting cisticola and black-chested prinia are less than 10 g (Dunning, 1993), or roughly half that of the weaver.

Egg Characteristics

The eggs are white, pale blue, or pinkish, with brown, reddish brown, and violet markings. The eggs of the zitting cisticola host average 15 × 11 mm and are white to bluish, with fine red to brown spotting. The eggs of the pectoral patch cisticola are about 16 × 10.5 mm and of the croaking cisticola about 18.4 × 11.9 (Vernon, 1964). Those of the black-chested prinia average about the same, 16 × 11.5 mm, and have a more blue or blue-green ground color, with blotches and sometimes scrolls of various darker colors, especially at the more rounded end. Assuming a mean weight of 1.07 g for the zitting cisticola's egg (Maclean, 1984), the parasite's egg averages nearly 50% larger than its cisticola host's eggs, and slightly larger than those of the prinia.

Breeding Season

In southern Africa breeding occurs from September to March (spring to autumn), when host species that are dependent on fresh grass are breeding (Ginn et al., 1989). In the Congo Basin area the birds breed during the rainy season, when warblers are also nesting (Chapin, 1954). In East Africa breeding-condition birds have been collected during December and January, and on Pemba Island (off Tanzania) eggs are laid from September to January. Nestlings have been seen in Kenya during May, and fledged birds during June and July. In Ethiopia, nestlings have been seen later in the year, during August and November.

Breeding Biology

Nest selection, egg laying. Judging from limited information, nest structure plays no clear role in nest choice, but habitat may, as moist meadows seem to be a favored location for parasitizing nests. However, as Friedmann (1960) mentioned, there is no information on the number of eggs laid by a female or on the interval between successive eggs in a laying sequence. The female either removes or consumes a host egg when depositing her own (Vernon, 1964). There is no clear evidence as to possible egg mimicry. The presence of two parasitic weaver eggs or young in the same nest has been documented (Parkenham, 1939), but the incidence of such multiple parasitism is unknown. Payne (1977a) estimated the "clutch size" of this species as 2.9 eggs, with the eggs being ovulated one per day on successive days and with laying occurring the day after ovulation. Of 21 reports, 13 are of single eggs or chicks present in the nest, and 8 are of 2 eggs or chicks (Vernon, 1964).

Incubation and hatching. The incubation period is unknown, but it is not more than 14 days

FIGURE 41. Sketches of an adult male (A) and female (B) parasitic weaver, plus eggs of the weaver (left) and its prinia host (right). Also shown is a black-chested prinia feeding a juvenile weaver (C, after a photo by K. Newman, 1971). Shown below are gape patterns: parasitic weaver (D); an adult (left) and a nestling cisticola (E); a nestling prinia (F); nestling and adult prinias (G) (mostly after Swynnerton, 1916).

(Vernon, 1964) and is presumably similar to that of its *Cisticola* and *Prinia* hosts, namely, about 11–13 days. It is also not known whether any nestling/egg eviction occurs, although this seems unlikely, as two parasitic weaver chicks have been found occupying the same nest.

Nestling period. The fledging period is 18 days (Vernon, 1964), or distinctly longer than the usual 11–15 days typical of *Cisticola* and *Prinia* host species. One nest having a chick whose feathers were just appearing on March 19 left the nest on April 4, representing a minimal fledging period of about 16 days, and probably closer to 3 weeks. Friedmann (1960) summarized the information on the nestling stage and found at least one instance of a black-chested prinia nest having a young parasitic weaver and a surviving prinia nestling. Likewise, in one nest of the winding cisticola, two nearly fledged parasites and one host were observed, but in parasitized nests of the pectoral-patch cisticola, host young were never seen (Cheesman & Sclater, 1935). (These parasitic young were specifically attributed to *Vidua* by the authors, but were regarded as those of the parasitic weaver by Friedmann.) The nestling parasitic finch lacks any specific gape or tongue markings indicative of host mimicry (the host cisticolas and prinias have paired dark tongue spots, as shown in Fig. 41), and in contrast to the viduine and emberizine finches, the chick's begging posture is not the nearly inverted head position typical of these two groups (see Fig. 16), but rather a normal perching position similar to that of cuckoos or cowbirds (fig. 41). A postfledging dependency period of 10–40 days is typical (Vernon, 1964).

Population Dynamics

No significant information.

VILLAGE INDIGOBIRD

(Vidua chalybeata)

Other Vernacular Names: Green indigobird, Neumann's combassou (*neumanni*), purple indigobird, red-billed firefinch indigobird, Senegal combassou (*chalybeata*), South African indigobird, steel-blue widowfinch, variable widowbird (*amauropteryx*).

Distribution of Species (see map 62): Widespread in Sub-Saharan Africa, from Senegal east to Ethiopia and south to South Africa.

Subspecies

V. c. chalybeata (= *aenea*): Western Africa from Senegai and Sierra Leone east to Mali.

V. c. neumanni: Mali to Sudan.

V. c. ultramarina (= ionestii): North-central Africa from Chad to Sudan and Ethiopia.

V. c. centralis (= orientalis): Zaire to Kenya and south to Namibia, Zambia, and Tanzania.

V. c. okavangoensis: Angola, northwestern Botswana, and Namibia. First described by Payne (1973a).

V. c. amauropteryx: South-central and southern Africa from Angola, Zambia, and Tanzania south to South Africa.

Measurements (mm)

4.5″ (11–12 cm)

V. c. amauropteryx, wing, males 62–69, females 61–65. Tail, males 35–41, females 36–42 (Payne, 1973a).

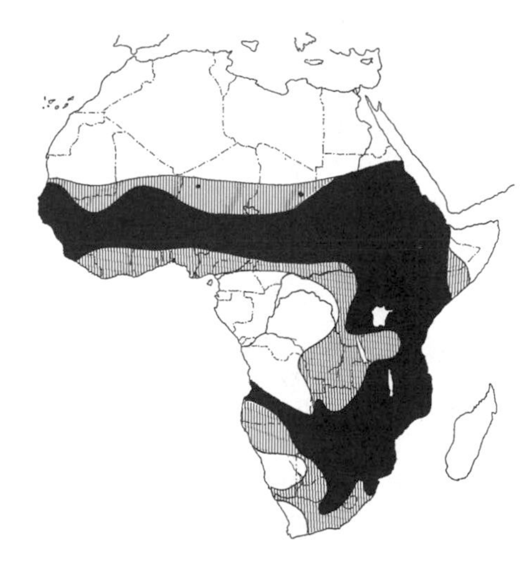

MAP 62. Ranges of village indigobird (filled) plus red-billed firefinch host (hatched plus filled).

V. c. centralis, wing, males 64–70, females 63–68. Tail, males 34–41, females 34–38 (Payne, 1973a).
V. c. chalybeata, wing, males 59–65, females 58–62. Tail, males 36–40, females 34–37.5 (Friedmann, 1960).
V. c. neumanni, wing, males 58–65, females 58–61. Tail, males 34–41, females 35–38 (Payne, 1973a).
V. c. ultramarina, wing, males 57–70, females 59–65 (Friedmann, 1960). Wing, males 60–65, females 59–62 (Payne 1973a).
Egg, avg. of *chalybeata* 15.1 × 11.8 (range 14.5–15.7 × 11.4–12.2); *amauropteryx* 15.3 × 12.3 (range 14.7–15.9 × 11.9–12.9). One egg of *ultramarina* 15 × 11 (Schönwetter, 1967–84). Shape ratio 1.24–1.36 (= broad oval).

Masses (g)

Avg. of 51 females, 13.2 (Payne, 1977a). Range of 64 males, 11.1–15.1, of 51 females, various races, 11.6–14.5 (Payne, 1973a). Avg. of 12 of both sexes, 12.5 (Dunning, 1993). Both sexes 11–15.2 (avg. 12.8, n = 19) (Maclean, 1984). Estimated egg weight of *ultramarina* 0.95, *chalybeata* 1.1, *amauropteryx* 1.2 (Schönwetter, 1967–84). Egg:adult mass ratio ~8.5%.

Identification

In the field: Like other viduine finches, females and non-breeding males generally cannot be safely identified in the field by plumage traits. Males in breeding condition range from iridescent blue to dark purplish black across their entire range, and have varied foot, bill, and wing colors (fig. 42). In West African populations the males have black to dark brownish black flight feathers (primaries and secondaries), orange feet, and white to grayish or light brownish bills. In East African birds the feet of breeding adults are orange-pink, the bills are pinkish white, and the flight feathers are browner (less blackish) than in West Africa. In southeastern Africa the feet are bright red, the bill is either distinctively orange-red (east and south of Victoria Falls) or is white to pinkish (to the west and north), and the flight feathers are medium to dark brown. Females have bill, foot, and flight feather colorations that are similar to those of males in their respective populations.

The males' songs include clear, whistled "wheeet-wheet-wheetoo" notes and the mimicked song of the red-billed firefinch. This song consists of two to six soft, upslurred fluty notes that often drop slightly in pitch toward the end of the series. Payne (1982, 1990) has provided sonograms of the village indigobird's mimetic and nonmimetic vocalizations; he stated that males have three or more mimic songs, whereas the firefinch has a single song type. Nicolai (1964) stated that vocal mimicry of the host species includes not only the male firefinch's song, but also its distance call, contact notes, nest calls, and the begging calls of young birds. According to Payne, each male sings throughout the day from a specific tree, and males also perform courtship hovering before females, but do not exhibit courtship head-swinging or aerial dive displays. Like other indigobirds, rather harsh chattering calls (at the rate of 8–16 notes per sound) are commonly uttered, and these show no apparent interspecific differences. In this species many such chatters may be followed by buzzy notes or complex mixtures of harsh and whistled notes that vary rapidly in pitch. Each male may have about 12 or more nonmimetic song variations, plus 6 or more mimetic ones. Payne (1982) reported that females are attracted to playbacks of mimetic song recordings of their own species, but not to those of others, and respond rather weakly to playbacks of the nonmimetic song types.

In the hand: Nestlings of this species (and their host) have a distinctive combination of an orange buccal cavity, a pastel yellow horny palate, a blue commissural junction, and white commissural tubercles with blue bases. Like other indigobirds, there are five black palatal spots, arranged with three large spots forming a semicircle in front on the horny palate and two smaller ones behind, and one on each side of the choana. These spots persist for about 30–40 days after hatching (or un-

FIGURE 42. Profile sketches of breeding male village (A), straw-tailed (B), steel-blue (C), and pin-tailed whydahs (D). Also shown are male pin-tailed whydahs performing upright display (E), wing-shaking (F), and hovering before a female (G). After sketches in Shaw (1984).

til parental independence is attained and the bill turns reddish). Then the palatal colors fade, and the commissural tubercles regress. The black palatal spots gradually become smaller, eventually either entirely disappearing or persisting in adults as gray points (Payne, 1982). However, in contrast, palatal markings of the host species often persist into adulthood. The nestling gape patterns of this species and its red-billed firefinch host are shown in fig. 9.

Adults of this species average slightly smaller in wing length than other indigobirds except for the pale-winged, and adult females tend to have brighter orange feet than do other West African indigobirds. Breeding males of the nominate West African race are mostly green-glossed, with black flight feathers and remiges, a white bill, and reddish orange feet. In *neumanni* (western and interior subequatorial Africa) the breeding males are more bluish, and their feet are orange, but they are otherwise similar. In *ultramarina* (north–central Africa) the plumage of males is mostly purplish, the bill is white, and the feet are reddish. In *amauropteryx* (southeastern Africa) the males are greenish blue to purplish blue, and bill and feet are similarly and distinctively salmon-pink to orange. In these birds the flight feathers are medium brown, not black. In *okavangoensis* (Okavango region) the males are greenish blue to bluish, the wing and tail feathers are medium to dark brown, and the feet are red but the bill is white. Distinguishing adult males from the dusky and variable indigobirds in areas of sympatry in South Africa is possible by observing their differences in foot color; red-footed in this species, white-footed in the dusky and variable. Farther north in Zimbabwe and Malawi, the village and dusky indigobirds are somewhat more distinct in their plumages, although some interbreeding may occur. Additionally, some interbreeding between the village and variable indigobirds probably occurs in the southern Congo region (Payne, 1973a).

Habitats

This species prefers brushy country but avoids deserts, humid woodlands, and forests. It occupies cultivated areas and gardens as well. Its fostering host, the red-billed firefinch, favors dry areas, including acacia-dominated savanna, especially inundation-zone woodlands. It also uses second-growth brushy areas, distributed areas adjoining cultivated areas, and the edges of relatively dry riparian woodland.

Host Species

Host species are listed in table 31. They include the red-billed firefinch (*L. senegala*), which is certainly the primary, if not the exclusive, host (Payne, 1982). There is also a possible record of parasitism for the bronze munia (Friedmann, 1960). The red-billed firefinch has a mean adult weight of 8.3 g, as compared with 13.2 g for the indigobird, representing an adult mass advantage of nearly 60% for the indigobird.

Egg Characteristics

The eggs are pure white and probably not distinguishable in size from those of other viduine finches, although their averages are the smallest of any of the viduine finches so far measured. They may be somewhat smaller than those of the host, which average 13.5 × 10.2 mm (shape ratio 1:1.32) in South Africa (Maclean, 1984). In Zambia the host eggs similarly average 13.1 × 10.8 mm (shape ratio 1:1.21), and those of the indigobird average 15.4 × 12.0 mm (shape ratio 1.28). Thus the host egg has nearly identical shape ratios but averages considerably smaller in its linear measurements. The estimated masses of fresh eggs of host and parasite are also significantly different (1.27 vs. 0.84 g), representing a substantial difference of about 50% greater egg mass in the indigobird (Payne, 1977a).

Breeding Season

In southern Africa this species primarily breeds during the austral spring (South Africa) and farther north its breeding coincides with the end of the rainy season and subsequent early dry season. In South Africa breeding of both host and parasite occurs from about November to April, probably peaking in January and February. In Zimbabwe parasitized firefinch broods have been seen from January to April (Ginn et al., 1989), although the host species has a virtually year-around breeding period there. In Kenya laying occurs during May and June, and in Nigeria during July (Payne, 1973a). Payne (1973a:180) illustrates a generalized breeding season for southern Africa lasting from December until about September. Approaching equatorial Africa, breeding occurs over a progressively longer period, primarily by lasting into later months, and within two degrees of the equator breeding may continue throughout the year. North of the equator most breeding of hosts and parasites occurs between July and December, representing a half-year displacement from the breeding cycles typical south of the equator. In Senegal, the host firefinch breeds from August to May, and especially during the period October–December, immediately after the rainy season, when seed supplies and relatively cool temperatures are at an optimum (Morel, 1973).

Breeding Biology

Nest selection, egg laying. So far as is known, only the red-billed firefinch represents a biological host of this species. Its nest is usually roofed-over, but at times may be somewhat cup-shaped. According to Morel (1973), the host firefinch is usually remarkably tolerant of the indigobird; the incubating host adult permits the indigobird to lay its eggs on the host's back while it is sitting in the nest. Host eggs are usually not removed at the time of parasitism, probably because the continuous incubation behavior of the adults means that the host eggs are out of sight (below the incubating adult) at the time of laying by the indigobird.

Egg-rejection behavior by the host has also not been observed. Morel reported a mean season-long host clutch size of 3.5, (241 unparasitized clutches averaged 3.5; 133 parasitized ones averaged 3.4, plus 2.2 parasitic eggs). Multiple parasitism is known to occur fairly frequently (in 45% of 133 parasitized nests found by Morel). As many as 6 indigobird eggs were deposited in a single host nest, but only rarely (8.2% of 133 parasitized nests) were more than four indigobird eggs present. Payne (1977a) estimated that female indigobirds probably lay a mean of 2.98 eggs per laying cycle, at a rate of 1 egg per day, and have about one laying cycle per 10-day period. He suggested that the clutch size in indigobirds has evolved to match the maximum number of nestlings that a host firefinch pair can successfully rear. He also estimated that a female village indigobird might lay up to 26 eggs per breeding season (Payne, 1977a).

Incubation and hatching. Morel (1973) believed that the presence of parasitic eggs in a nest results in a "super-stimulus" to brooding behavior by the hosts, with the result that nest losses during incubation are lower (45.7% vs. 56.3%) for parasitized nests than for parasite-free nests. The incubation period has been reported as 11 days (Olsen, 1958) and 10–11 days (Payne, 1977a); that of the host species is also 11–12 days (Goodwin, 1982). Morel concluded that hatching success of the parasite is related to both the relative synchrony of hatching by host and parasite and by the host's own relative hatching success. She stated that, the more parasitic eggs that are present in a nest, the greater the degree of hatching synchronization and hatching success. At the time of hatching she found the mean brood size of nonparasitized nests to be 3.3 and that of parasitized nests to be 3.1 host plus 1.7 parasite chicks. There is no hostility shown by the nestling indigobirds toward other host or indigobird nestlings, which is not surprising inasmuch as there is a fairly high probability that one or more of the young sharing their nest may be their own siblings. Additionally, an increased number of brood members may be desirable to the degree that they may stimulate more foraging activity by parents or help provide shared metabolic heat (Payne, 1977a).

Nestling period. The nestling period lasted 18 days in one instance, with the red-billed firefinch as a host (Olsen, 1958). An 18-day fledging period is also typical of the host species, followed by an approximate additional 8 days of postfledging dependency (Goodwin, 1982). The indigobird's corresponding postfledging dependency period is still uncertain but probably occurs at about 30 days (Payne, 1977a). Apparently the nest-mates of both host and parasite interact as a single family unit until they have all reached independence (Morel, 1973).

Population Dynamics

Parasitism rate. In a large sample of 374 host nests from Senegal, Morel (1973) reported a 36% parasitism rate. In a smaller sample of 31 host nests from Zambia, there was a 42% rate of parasitism (Payne, 1977a).

Hatching and fledging success. In Morel's (1973) study, the average breeding success (percentage of nests producing one or more fledged young) was about 28%. Thus, unparasitized nests had a 35% success rate, and parasitized nests had about an 18% success rate. Among successful nonparasitized nests, the average number of firefinches fledged was 2.8, whereas among successful parasitized nests it was 2.1 firefinches (plus 1.3 indigobirds). With regard to parasitic breeding success, Morel (1973) found that the percentage of indigobird eggs that hatched and subsequently produced fledged young ranged from 17–20% in nests having one to four parasitic eggs, but two nests with more than four indigobird eggs produced no offspring. However, posthatching survival by indigobirds was significantly related to the number of parasitic eggs present; fledging success averaged 13–14% in nests with one or two indigobird eggs, but only 6–8% in nests with three or four indigobird eggs. The overall rate of indigobird breeding success (percent survival from egg to fledging) varied from 16.6% to 20% in nests having one to four indigobird eggs present; this range of success being statistically nonsignificant. The mean number of host young fledged from nonparasitized nests was 2.6, as compared with 2.1 host young (plus 1.3 parasites) fledging from parasitized nests. Among the 133 nests that were parasitized, a total of 232 eggs were laid. Of these, 75 parasitic young were hatched in 42 nests. A total of 41 indigobirds were subsequently fledged from 31 of these nests, producing an overall egg-to-fledging breeding success rate of 17.6%. By comparison, the 241 unparasitized firefinch nests (containing 854 eggs) produced 243 fledged firefinch young, representing a breeding success rate of 28%. The 462 firefinch eggs in the parasitized nests produced 133 fledged firefinches, representing a nearly identical breeding success rate of 29%.

Host–parasite relations. Morel's (1973) data suggest a negligible impact of the indigobird on the host firefinch's reproductive success. She believed that the most deleterious effects of parasitism occurs when indigobird eggs are added to firefinch clutches of at least four, but that with smaller clutches the presence of the additional eggs and nestlings are not measurably harmful. Indeed, nests with parasitic eggs present had a better hatching success than nonparasitized clutches, although this apparent hatching advantage was counterbalanced by a poorer rate of host fledging success. Additionally, the total brood size at fledging was larger in parasitized nests than in nonparasitized ones (3.5 vs. 2.8 fledglings, including 1.3 indigobird young), evidently because of the "super-stimulus" brooding effect of the larger families on host parents. The reduction in average host brood size in parasitized nests (2.1 firefinch young fledged, representing a 25% reduction from the 2.8 fledglings typical of nonparasitized nests), was evidently compensated by the inexplicably higher hatching success rates of parasitized nests (56% hatch rate) versus nonparasitized nests (45% rate), resulting in a nearly identical breeding success rate for parasitized (29%) versus unparasitized (28%) nests. Some host pairs may nest as many as five times per year, but they usually nest four times, and may rear up to as many 14 young to fledging. Thus, the overall negative effects of indigobird parasitism on firefinch productivity must be rather limited overall and is at least partly balanced by the seemingly beneficial effects mentioned above.

Payne (1977a) noted that the probability of avoiding destruction by predation of at least one of three eggs laid by a female indigobird in a single nest is 37.8% (or equal to the probability that the host nest will survive predation). However, there is a much lower probability of losing all three nests to predation (24.1%) if a female indigobird's three-egg clutch is deposited in separate nests. Thus "scatter-laying," rather than laying all the eggs in a single host nest, provides the greatest statistical likelihood for breeding success by the indigobird.

JAMBANDU INDIGOBIRD

(Vidua raricola)

Other Vernacular Names: Goldbreast indigobird. Recently described (Payne, 1982): previously confused with the variable, baka, and pale-winged indigobirds (especially *nigeriae*).

Distribution of Species (see map 63): Sub-Saharan Africa from Sierra Leone to Sudan. According to Payne & Payne (1994), known from Sierra Leone, Ghana, Nigeria, Cameroon, Zaire, and Sudan.

Measurements (mm)

4.5″ (11–12 cm)

Wing, males 61–67 (avg. 64.05, $n = 19$) (Payne & Payne, 1994); tail, male 39 (Payne, 1982).

Egg, no information.

Masses (g)

One male, 11.8 (Payne, 1982).

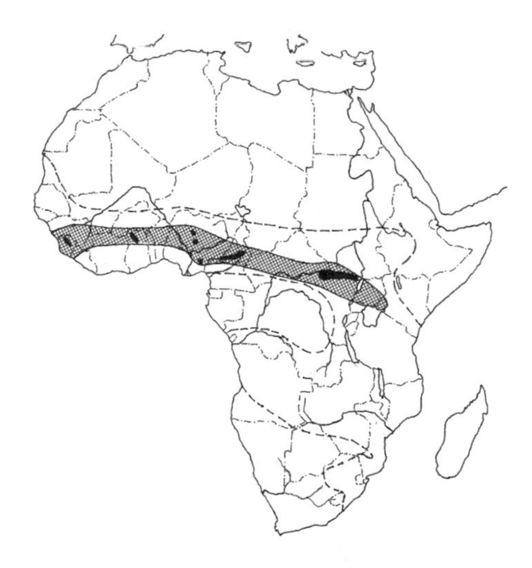

MAP 63. Ranges of jambandu indigobird (filled) plus reputed host black-bellied firefinch (cross-hatched). The range of zebra waxbill, another possible host, is also indicated (dashed line).

Identification

In the field: The breeding male plumage is glossy green (more commonly) or bluish, with pale brown flight feathers. Males (and adult females) also have a whitish bill and foot color ranging from grayish orange to reddish gray. These traits overlap with those of the baka, variable, and the "*nigeriae*" form of the pale-winged indigobird, and thus song (mimicry of host-specific firefinches) must be used to distinguish these sibling species when identifying adults in the field. Payne (1982:22–25) has provided sonograms of various vocalizations of this species (alarm, contact, etc.) and its then-presumed host the black-bellied firefinch. More recently, Payne and Payne (1994) have provided comparative sonograms of this species' vocalizations and those of the zebra waxbill.

In the hand: Nestlings (host and parasite) have a species-specific combination of a purplish white to reddish lilac buccal cavity, a purplish-white or purplish-gray horny palate, violet-red to bluish gray commissural junctions, and blue commissural tubercles. Five black palatal spots are present, and there two distinct gape papillae on either side, ranging from white to pale blue, with a dark blue to black intervening area. A similar palate pattern occurs in young of the zebra waxbill host, in which there are two white to blue papillae on the upper mandible and one blue papilla on the lower one (Payne & Payne, 1994). Older juveniles, females, and nonbreeding indigobird males may be impossible to identify to species by plumage traits alone, and their palatal traits begin to fade soon after fledging.

Habitats

Open, brushy country similar to that used by other indigobirds is preferred. Habitats used by its firefinch host include savannas, grasslands, and cultivated areas.

Host Species

The black-bellied firefinch was until recently considered the most likely fostering host of this species (Payne, 1982). Its mean adult weight is 11.8 g (Dunning, 1993). However, Payne & Payne (1994) have recently reported that males of this indigobird mimic the vocalizations of the considerably smaller zebra waxbill, which is thus the more probable host. A single unsexed specimen of the zebra waxbill weighed 7 g (Dunning, 1993), and Payne & Payne (1994) give the species' adult weight as 6–7 g.

Egg Characteristics

No information exists. The egg of the zebra waxbill averages about 14 × 10 mm.

Breeding Season

Breeding in Cameroon corresponds with the end of the rainy season, with males singing during October and November and females with oviducal eggs taken during October.

Breeding Biology

Little information exists. Payne (1982) provided some information on ecology, sympatry, and host species distribution, but no information on the egg-laying and brood-rearing phases is available. The zebra waxbill's incubation period is 11–12 days, and the young fledge in 18–21 days (Goodwin, 1982).

Population Dynamics

No information.

BAKA INDIGOBIRD

(Vidua larvaticola)

Other Vernacular Names: Bako Indigobird. Recently described (Payne, 1982); previously confused with the jabandu, variable, and pale-winged indigobirds (especially *camerunensis*).

Distribution of Species (see map 64): Sub-Saharan Africa from Guinea-Bissau to eastern Sudan and extreme western Ethiopia.

Measurements (mm)

4.5" (11–12 cm)

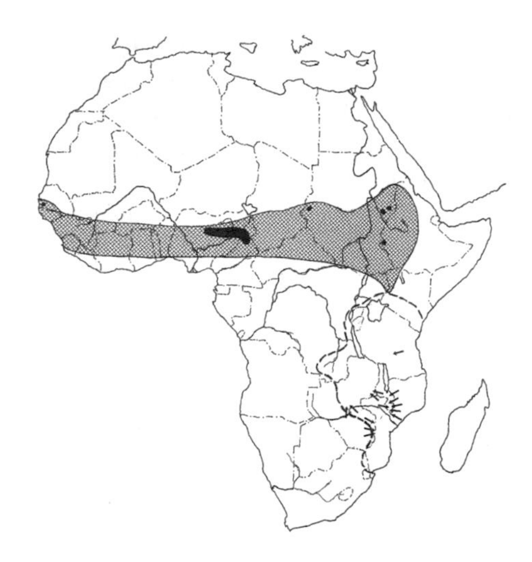

MAP 64. Ranges of baka indigobird (filled) plus black-throated firefinch (cross-hatched) and Peters's twinspot (dashed line) hosts. Locality records for *V. (funerea) codringtoni* are indicated by arrows (see also map 65).

Wing, males 64–69 (avg. 66.7, n = 16) (Payne & Payne, 1994); tail, male 40 (Payne, 1982).

Egg, no information.

Masses (g)

One male 13 (Payne, 1982).

Identification

In the field: Breeding males are typically glossed with blue or (less commonly) bluish green and have light brown flight feathers. Adults of both sexes also exhibit whitish bills and whitish mauve to grayish flesh or light purplish feet (Payne, 1982). Breeding males with blue iridescence cannot be distinguished from the "*camerunensis*" form of the pale-winged indigobird, which is of doubtful taxonomic validity. Furthermore, the variable iridescence of breeding males is of little or no value in species identification (Nicolai, 1972). Immature birds, adult females, and nonbreeding males cannot be identified to species in the field. This species parasitizes the black-throated (or "masked") firefinch, and mimicry of that species' song is perhaps the best fieldmark. Payne (1982:36–37) has provided comparative sonograms of several of this indigobird's vocalizations and those of its firefinch host, including begging calls, alarm calls, and two-parted "whee-hew" slurred whistles uttered at the rate of about four notes per second.

In the hand: Nestlings (host and parasite) have a distinctive combination of a orange buccal cavity, a pale yellow horny palate, blue-black commissural junctions, and a pair of blue commissural tubercles. There is also a ring of five black spots on the palate (Payne, 1982). These palatal traits may also be useful for identifying recently fledged juveniles (until about 1 month of age), but are nearly identical to those of the host species's nestlings. Male traits are mentioned above, but the plumages of immature birds, females, and even breeding males may not always be adequate for achieving species identification consistently.

Habitats

Brushy areas, gardens, and woodland edges are all used by this species. Its host firefinch species (at least of the eastern race *larvata*) favors grassy savanna woodlands, bamboo thickets, and grassy banks of woodland streams, at elevations of 1000–1500 m. The more widespread race *vinacea* also favors bamboo thickets (Goodwin, 1982).

Host Species

This indigobird is known only to parasitize and to be fostered by the black-throated firefinch (Payne, 1982). This host species has a mean adult weight of 9.6 g (Dunning, 1993). This adult mass is considerably less than that of its parasite, which has an approximate 35% weight advantage, based on limited available information (weight of one male).

Egg Characteristics

No information is available, although females with oviducal eggs have been collected. The eggs of its host species are white and average 16.5 × 11.4 mm, representing an estimated average mass of 1.12 g (Schöwetter, 1967–84).

Breeding Season

In Nigeria breeding by this indigobird closely coincides with that of its host species, which has been reported to nest there during July and August. Associated singing behavior by male indigobirds begins at the end of the rainy season and extends from about July through September. Laying females have been collected during August and September. Recently fledged young have been seen as late as December. West African breeding records for the host species are from August through September (Payne, 1973:180). Little other specific information on breeding periodicity is available.

Breeding Biology

No significant information exists on this rather recently described species; Payne (1982) has summarized what little is known of its ecology and behavior.

Population Dynamics

No information.

VARIABLE INDIGOBIRD

(Vidua funerea)

Other Vernacular Names: Black indigobird (or widowfinch), brown-winged dusky combassou (or indigobird), brown-backed firefinch indigobird, Codrington's indigobird (*codringtoni*), dusky indigobird, funereal indigobird, gala indigobird (*sorora*), green widowfinch (*codringtoni*), plateau indigobird (*maryae*), twinspot indigobird (*codringtoni*), white-footed indigobird (northern races).

Distribution of Species (see maps 64, 65): Widespread in Sub-Saharan Africa, probably from Sierra Leone east to Sudan, and south to South Africa. (Uncertainties of species limits make an accurate range description impossible to provide, given available information.)

Subspecies (including some possibly distinct biological species)

V. (f.) marvae: Initially described (Payne, 1982) from the northern plateau of Nigeria. Recently

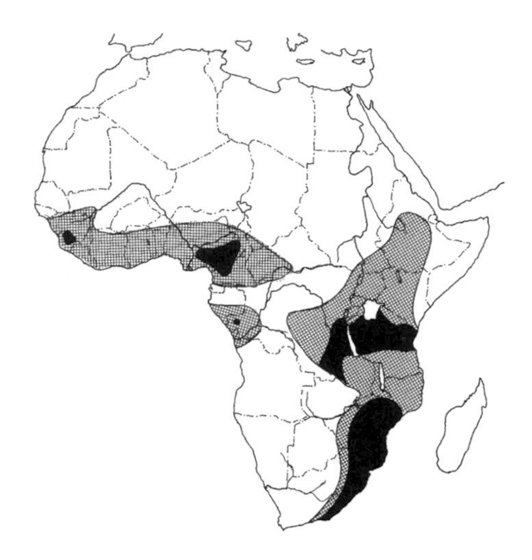

MAP 65. Ranges of variable indigobird (filled) plus African firefinch host (hatched). See also map 64 for locality records for *V. (f.) codringtoni.*

listed by Payne & Payne (1994) as a distinct species.

V. f. sorora: Originally described (Payne, 1982) from Cameroon. Probably extends from Sierra Leone east to Sudan, but there are few records.

V. (f.) nigerrima: Angola, Zambia, southern Zaire, and Malawi. Possibly represents a species distinct from *funerea* (Nicolai, 1967).

V. (f.) codringtoni: Tanzania, Zambia, Zimbabwe, and Malawi (see map 64). Originally described in 1907, but until recently included within *chalybeata* (e.g., Traylor, 1968). Considered a distinct species by Payne et al. (1992a,b) and so-listed by Sibley and Monroe (1990). The race *lusituensis* (Payne, 1973) has been recently reidentified as a synonym of *codringtoni* (Payne et al., 1992).

V. f. funerea: Transvaal, Swaziland, and south to Cape province.

Measurements (mm)

4.5″ (11–12 cm)

V. f. funerea, wing, males 65–71, females 63–68. Tail, males 38–44, females 36–42 (Payne, 1973). Wing, males 65–71.5 (avg. 68.6, $n = 6$), females 64–69 (avg. 66.2, $n = 6$) (Maclean, 1984).

V. f. nigerrima, wing males 61–70, females 65–66. Tail, males 34–42 (Payne, 1973). Wing, males 67–71 (avg. 68.5, $n = 15$) (Payne et al., 1992).

V. (f.) maryae, wing, males 66–69 (avg. 67.75, $n = 4$). (Payne & Payne, 1994).

V. (f.) codringtoni, wing, males 65–69, females 64–66. Tail, males 37–44, females 39–41 (Payne, 1973). Wing, males 66–70 (avg. 67.87, $n = 15$) (Payne et al., 1992).

Egg, one of *funerea,* 14.9 × 12.3 (Schönwetter, 1967–84). Shape index 1.21 (= broad oval).

Masses (g)

V. f. funerea, 7 males 14–16.5 (avg. 15.2), 17 females 12–16.1 (avg. 14.1) (Maclean, 1984).

V. f. nigerrima, 14 males 12–13.8 (avg. 12.81) (Payne, et al., 1992).

V. (f.) codringtoni, 15 males 12.2–14.1 (avg. 13.04) (Payne et al., 1992).

Estimated egg weight 1.24. Egg:adult mass ratio 8.9%.

Identification

In the field: This species is well named, for it is perhaps the most variable of the indigobirds, with no obviously consistent geographic pattern of plumage and soft-part color variation. The bill is whitish in adults throughout its range, but foot and flight feather colors and male breeding plumage iridescence vary geographically. The plumage color of breeding males in eastern and southern Africa may be glossed with bluish-purple (Angola, South Africa), bluish (Zambia, Zimbabwe), or greenish (southern Rift Valley highlands). In West Africa breeding males may also be greenish (Nigeria), blue (Cameroon), or more purplish (lower Congo basin). In one Nigerian race (*maryae*) the brown flight-feathers are edged with light brown, the bill is white, and the feet are white or purplish, whereas in a Cameroon race (*sorora*) the wings and bill are similarly colored but the feet are light orange. In the biologically distinctive taxon *codringtoni* the males vary from glossy green to blue, have blackish wings, a white bill, and bright orange-red feet, and females have a white bill, orange feet, and grayish breast plumage. In southern Africa the bill is likewise white in adults, but the feet may vary from bright red (South Africa) to orange (Zululand northward) or pinkish (Angola, northern Zambia, and Malawi). Additionally, in southern Africa the flight feathers of both sexes may range from dark brown to pale brown, but the distinctive orange to red foot color and white bills of the more southern populations may help to distinguish them from other indigobirds in that region. Thus, in southeastern Africa adult village indigobirds have red bills and feet, the dusky indigobirds have whitish bills and feet, and the variable indigobirds have white bills but orange to reddish feet. Adults of the taxonomically puzzling form *codringtoni* also have white bills and orange to reddish feet like those of variable indigobirds, and breeding males similarly vary in color from green to blue.

At least the nominate race of this species is a parasite of the African firefinch. Males of the host species have a large repertoire of vocalizations, including some complex and songlike trills that are mimicked by the male indigobirds (Nicolai, 1967a). Males sing from prominent perches, and the song is said to be linnetlike. Payne has published sonograms of mimetic vocalizations of two indigobird forms (*nigerrima* and *codringtoni*) that he originally assigned (1973, 1982) to this species, plus those of the African firefinch. However, Payne et al. (1992a,b) have since identified the Peters' twinspot as a taxon-specific host of the form *codringtoni* and have provided sonograms of its calls and songs, plus those of the indigobird.

In the hand: Nestlings of this species (except perhaps for those of *codringtoni*) have a pink or pinkish white buccal cavity, a pale yellow horny palate, a blue-black commissural junction, and blue commissural tubercles. The gape markings of the host species are extremely similar. In chicks of both host and parasite, five black palatal spots are present (three in front, plus two smaller ones behind on each side of the choanal opening), and the tongues of the host and parasite are also similarly marked with black (Payne, 1973, 1982). However, in the twinspot host species of *codringtoni,* there are three black spots on a yellow palate and two light yellowish (not blue) swellings on each side of the gape, so it is possible that the form of indigobird will be found to have a similar pattern (Payne et al., 1992). Immature indigobirds may be impossible to identify to species by plumage or soft-part traits alone, especially in West Africa, but their palatal colors may be adequate for identifying juveniles up to about 1 month old. As described above, traits of breeding birds are highly variable geographically, and such birds can probably be easily identified only in those areas of southern Africa where their bill color is distinctively orange to red.

Habitats

Brush and woody habitats are preferred by this species, including rocky, wooded hillsides, abandoned fields, and similar shrubby plant communities, up to about 1800 m elevation. Its host species prefers forest edges, thick woody cover along streams, mixed grass and thorn-tree scrub habitats, and similar combinations of grass and low woody cover. It favors relatively moister and ranker situations that does its relative the Jameson's firefinch and avoids both dense forests and open savannas (Goodwin, 1982).

Host Species

The African firefinch is certainly the primary fostering host of this species' nominate form. A few records of other possible hosts, perhaps reflecting erroneous indigobird identification, exist. They include the common waxbill and, much less reliably, the red-billed firefinch. This latter host record may apply to the village indigobird (Friedmann, 1960). Payne et al. (1992) have recently concluded that the Peters's twinspot is the host of the form *codringtoni,* which thus appears to act as a separate biological species and is locally sympatric with the *nigerrima* form (pale-footed and purplish blue) of the variable indigobird and with the red-billed, red-footed, and blue-plumaged *chalybeata.*

Egg Characteristics

Little specific information exists about this species' eggs, beyond the fact that they are white and evidently differ little if at all in size from those of other indigobirds (see village indigobird account). Besides Schönwetter's measurements of a single egg (14.9 × 12.3 mm), Friedmann (1960) mentioned another egg measuring 15.2 × 12.3 mm, but noted that better-authenticated information on this species' eggs is still needed. The eggs of the host African firefinch species similarly average about 15 × 11.4 mm, with a mean mass estimate of 1.02 g, so such measurement differences probably would not serve to consistently distinguish host eggs from parasite eggs. Payne (1977a) reported an estimated fresh weight of 1.33 g, as compared with an estimated 1.16 g for the host firefinch, representing a 15% greater egg mass in the indigobird. The mean adult weight of the firefinch species is 10.2 g, so the parasite also has an average adult mass about 30% greater than its host.

Breeding Season

Breeding at the southern end of this species' range occurs from about November to May, terminating as the dry season approaches, in common with the breeding cycles of the host species. Closer to the equator its season is more restricted and is concentrated from about February or March to June or July, also in general synchrony with its host (Payne, 1973:180). The host species has been found nesting from November to June in South Africa, from January to May in Zambia, from February to June in Malawi, and from March to May in Mozambique (Goodwin, 1982). North of the equator, the host firefinch breeds during June in eastern Angola, from August to December in the northern Congo (Goodwin, 1982), and during November in Cameroon (Payne, 1982:48).

Breeding Biology

Nest selection, egg laying. Payne (1977a) estimated that a mean of 2.96 eggs (range 1–4) are laid per laying cycle in this species and that an average of 1.8–2.5 eggs are laid over a 10-day period, based on examination of ovaries.

Incubation and hatching. No information.

Nestling period. No information.

Population Dynamics

No information.

DUSKY INDIGOBIRD
(Vidua purpurascens)

Other Vernacular Names: Pink-backed firefinch indigobird, purple combassou, purple indigobird, purple widowfinch. Sometimes considered as composing part of the variable indigobird species (e.g., Traylor, 1966, 1968).

Distribution of Species (see map 66): Sub-Saharan Africa from Kenya and Angola south to South Africa.

Measurements (mm)

4.5″ (11–12 cm)

Wing, males 62–70, females 63–67. Tail, males 35–44, females 38–43 (Payne, 1973). Wing, males 66–70 (avg. 67.87, $n = 23$) (Payne et al., 1992a). Wing, males 68–73 (avg. 69.8, $n = 39$), females 65–69 (avg. 66.7, $n = 1$) (Maclean, 1984).

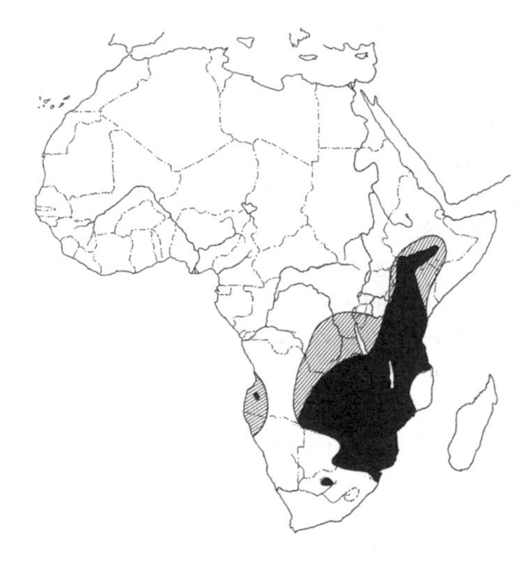

MAP 66. Ranges of dusky indigobird (filled) plus Jameson's firefinch host (hatched).

Egg, avg. 15.5 × 12.2 (range 15.4–15.6 × 12–12.4) (Schönwetter, 1967–84). Shape index 1.27 (= broad oval).

Masses (g)

Avg. of 21 adults, 13.4 (Payne, 1977). Range of 43 males 11.6–15, 17 females 12.7–15.5 (Payne, 1973). Males 11.8–13.7 (avg. 12.82, *n* = 21) (Payne et al., 1992a). Both sexes 11.4–14.6 (avg. 13.3, *n* = 13) (Maclean, 1984). Estimated egg weight 1.21 (Schönwetter, 1967–84). Egg:adult mass ratio 9.0%.

Identification

In the field: Across their entire southern African range, the breeding males of this species have a purplish to bluish purple (never green and rarely blue) overall plumage sheen, a white or pale pink (not red) bill, whitish to pale pink (not orange or red) feet, and have pale brown (not medium brown to blackish) flight feathers. Except south of the Limpopo River (where adult have distinctive orange feet), they cannot be visually distinguished from individuals of the sympatric variable indigobird, which also has pinkish white feet. Females, nonbreeding males, and young birds often cannot be identified to species in the field, but breeding females have bill and foot colors much like those of the males, which may help in identification.

The mimicked song of the Jameson's firefinch is a tinkling, canarylike trill. One of the firefinch's courtship songs consists of a plaintive, whistling "feeeeee" that is repeated three or four times. Payne (1973:70–73) has provided sonograms of some of this species' vocalizations and those of its firefinch host. He noted that in Zimbabwe this indigobird species (and its firefinch host) has a song distinct from that of the variable indigobird (and its corresponding host), but in Malawi the two indigobirds appear to intergrade, and their mimetic songs are similar. Payne (1980) also provided some sonograms of a probable hybrid of this species and the eastern paradise whydah. According to Payne, male dusky indigobirds perform courtship hovering, and individual males also sing from single display tree sites throughout the day, but they do not perform aerial dive displays, nor do they engage in courtship head-swinging.

In the hand: The nestling gape pattern of this species and its Jameson's firefinch host are very similar. Nestings have five large blackish palatal marks (three in front, two behind) on a grayish to pinkish horny palate. They also have a yellow tongue, pinkish commissural junctions, and a pair of silvery white tubercles above and below each of the commissures. The gape of the Jameson's firefinch is almost identical in both color and pattern (Nicolai, 1974), but minor differences in tongue patterning might exist (Payne, 1973:55). Juveniles are streaked above, buffy on the breast, and have whitish abdomens. The bills of juveniles are gray to grayish brown above and white below, and their feet are light gray to creamy gray. Adult females have similar light brown to whitish bills, and their foot color is flesh-white to pale purplish. Breeding males range from bluish to purple or even purplish black in general plumage, with brown flight-feathers and rectrices, a white bill, and whitish to purplish white feet. Adults of either sex of this species can often be identified in the hand by their pale, whitish feet and their fairly large size.

Habitats

This species' habitats are not well described, but it is apparently similar to the other indigobirds in preferring open, brushy country, especially thornveld. Its host, the Jameson's firefinch, likewise prefers thickets and grassy tangles in thornveld, especially where coarse grasses and bushes are to be found, such as at the edges of drier riparian thickets.

Host Species

Host species are listed in table 26. The Jameson's (or 'pink-backed') firefinch is certainly this species' primary fostering host. The reported host records for the African or "brown-backed" firefinch may have resulted from observer confusion with the variable indigobird.

Egg Characteristics

In common with other indigobirds, the eggs are white, like those of their host species. The mean adult host weight is 8.8 g (Dunning, 1993), as compared with a 13.4 g mean adult weight of the parasite, representing a mass advantage of 50% for the latter. Payne (1977a) reports an estimated mean fresh egg weight of 1.13 g for this indigobird, as compared with a host egg weight of 1.04 g, representing a roughly 12% greater egg weight and volume for the parasite.

Breeding Season

In southern Africa the host species' breeding season extends from about December to as late as September, and especially from December to April, corresponding generally to the wetter season. In South Africa host breeding is centered in May, and in Zambia and Malawi it extends generally from January to July (Payne, 1973:180). In Zambia this indigobird's gonads are correspondingly active during the period January–May (Benson et al., 1971). Slightly south of the equator, breeding by the indigobird is spread throughout the year, but somewhat north of the equator there is evidently mid-year (May, June) breeding (Payne, 1973:180).

Breeding Biology

Nest selection, egg laying. Payne (1977a) estimated that females lay an average of 2.96 eggs per laying cycle or clutch and that a maximum of four eggs may be laid in such a cycle, presumably at the rate of one per day.

Incubation and hatching. No information.

Nestling period. No information.

Population Dynamics

No information.

PALE-WINGED INDIGOBIRD

(Vidua wilsoni)

Other Vernacular Names: Bar-breasted firefinch indigobird (*wilsoni*), brown twinspot indigobird (*camerunensis*), Cameroon combassou (*camerunensis*), Nigerian combassou (*"nigeriae"*), quail-finch indigobird (*nigeriae*), violet widowbird (*incognita*), Wilson's dusky combassou, Wilson's indigobird, Wilson's widowfinch.

MAP 67. Ranges of pale-winged indigobird (filled) plus host bar-breasted firefinch (hatched). Ranges of other known or probable hosts are not shown.

Distribution of Species (see map 67): Sub-Saharan Africa from Senegal to Ethiopia and south to Zaire, Caprivi, and northeastern Namibia.

Subspecies

V. w. wilsoni: Senegal and Guinea-Bissau to Zaire, Sudan, and Ethiopia. Includes *lorenzi,* according to Payne (1982), and considered a separate species ("bar-breasted firefinch indigobird") by Payne & Payne (1994).

V. w. incognita: Probably northeastern Namibla, Caprivi, northwestern Zimbabwe, adjacent Angola, and also parts of Zambia and southern Zaire. Possibly considered a distinct species (Nicolai, 1972), but Payne (1982) has tentatively placed it within *wilsoni.*

V. w. nigeriae. Earlier regarded by Payne (1985) as a *nomen dubium,* but more recently Payne & Payne (1994) have designated birds from northern Cameroon that mimic the African quailfinch as representing this taxon, which they considered as specifically distinct and have called the "quail-finch indigobird." Reported only from Cameroon (Garoua) and Nigeria (Kiri).

V. w. camerunensis. Earlier considered by Payne (1985) as a *nomen dubium,* but more recently Payne & Payne (1994) have designated birds mimicking the brown twinspot as representing this taxon and considered it as specifically distinct. Reported only from Cameroon (Tibati and Meng).

Measurements (mm)

V. w. wilsoni, wing, males 63–67, females 60–61 (Friedmann, 1960). Wing, males 60–65 (Payne, 1982). Wing, males 62–69, females 60–61. Tail, males 32–38, females 33–34 (Bannerman, 1949).

V. w. nigeriae, wing, males 63–66 (avg. 64.25, *n* – 8) (Payne & Payne, 1994).

V. w. camerunensis, wing, males 62–67 (Payne & Payne, 1994).

Egg, one egg, 15.4 × 12.1 (Schönwetter, 1967–84).

Shape index 1.27 (= broad oval).

Masses (g)

Five males of *wilsoni* 13–14 (avg. 13.4). Thirteen males of *camerunensis* 13–14 (avg. 13.38).

One male of *nigeriae* 13 (Payne, 1973).

Estimated egg weight 1.18 (Schönwetter, 1967–84).

Egg:adult mass ratio 8.8%

Identification

In the field: In breeding males of the typical purple-glossed form, the brown flight feathers are edged with light brown, the bill is grayish white, and the foot color may be pale purplish, pinkish, or light gray. Two other male plumage variants are sometimes included in this species, namely the green-glossed "*nigeriae*" and the blue-glossed "*camerunensis*" types. The more greenish birds also have pale brown-edged flight feathers, but the bluish-tinted birds may have somewhat darker brown flight feathers (see "In the hand," below). Immature birds, adult females, and nonbreeding pale-winged indigobird males cannot be distinguished from those of other indigobirds, but adult females have bill and foot colors similar to those of the breeding males.

Songs of probable host species include mixtures of high metallic and low nasal notes (in the bar-breasted firefinch), as well as repeated chirping strophes or series of interspersed close and distant contact calls (in the closely related brown firefinch, a probable additional host) (Nicolai, 1972; Goodwin, 1982). Recently Payne and Payne (1994) have reported that some mimicry also occurs with the African quailfinch and with the brown twinspot and regard these populations as constituting distinct biological species.

In the hand: Nestling birds have a distinctive reddish lilac to pinkish buccal cavity and horny palate with a white to dark blue commissural junction and a white bill flange that lacks distinct tubercles. Five black palatal spots are also present (Payne, 1982). Payne (1973:55) has provided sketches of the gape patterns of adults of typical *wilsoni* plus those of *"camerunensis"* and *"nigeriae,"* which exhibit a few minor pattern variations in tongue and palatal spotting. Breeding males vary from green through blue to purple in their plumage iridescence. The flight feathers are pale brown in the purplish birds (typical *wilsoni*) and in the most greenish ones (*nigeriae*), but are slightly darker in the bluish ones (*camerunensis*). Bill color varies from white to pinkish white, and foot color ranges from whitish to pinkish purple in all of these variant types. At least in some areas of West Africa, females of this species can be separated from those of the village indigobirds by their purplish white (not orange) feet and their somewhat grayer upperparts (Payne, 1973).

Payne concluded in 1985 that the available descriptions and reference specimens of the *"camerunensis"* and *"nigeriae"* races of the pale-winged indigobird cannot be certainly identified to species, nor can they be distinguished morphologically from his two previously (1982) described similar species (jambandu and baka indigobirds). However, Payne and Payne (1994) have recently reported that males of *"nigeriae"* from northern Cameroon mimic the African quailfinch, and they have called this possibly specifically distinct taxon the "quailfinch indigobird." They additionally reported that males of at least one population of birds they classified as *"camerunensis"* mimic the songs of the brown twinspot. Other populations that they assigned to *camerunensis* are song-mimics of the black-bellied firefinch and of the African firefinch. However, Traylor (1966, 1968) considered *nigeriae* as a race of the variable indigobird and regarded *camerunensis* as a synonym of *nigeriae.* Additionally, Nicolai (1972) has urged the species recognition of *incognita,* based on its host-specific adaptations and host mimicry of male song types.

Habitats

Like the other indigobirds, this species is found in open, brushy areas, including those of second-growth woodlands, forest edges, abandoned fields, and similar transitional grassland–woodland environments. Its hosts, the bar-breasted, African, and brown firefinches, also occupy diverse habitats, ranging from disturbed sites near human habitations to thick brushy cover near streams. The brown twinspot favors forest edges, savannas with stands of tall grasses and bushes, and clearings or cultivated areas where grasses and other natural cover are also available. The African quailfinch is found among open grasslands and sometimes even marshy habitats, favoring short-grass environments and sites with alternating areas of tufted grasses and bare sand or soil (Goodwin, 1982).

Host Species

Host species are listed in table 26. Of these, the bar-breasted firefinch is probably the major fostering host, at least of the nominate form *wilsoni* (Payne, 1973). It appears that the brown firefinch is the specific host of the problematic indigofinch taxon *incognita,* both of which perhaps represent sibling species (Nicolai, 1972). However, the brown firefinch appears to be geographically allopatric with the bar-breasted firefinch, and their male advertising songs seem to be identical, suggesting that they (and thus their parasitic counterparts) should perhaps be regarded as conspecific (Payne, 1982). However, the brown firefinch is now increasingly regarded as a species distinct from the bar-breasted (Sibley & Monroe, 1990; Goodwin, 1982). Other probable estrildine hosts of taxa here regarded as part of the present species include the African quailfinch (host of *nigeriae*), the brown twinspot (host of *camerunensis*). and the African firefinch (putative host of *camerunensis).* The African firefinch is also the putative host of the indigobird taxon *maryae* (Payne & Payne, 1994), which is here considered part of the variable indigobird. The mean adult masses of these known or putative hosts are: bar-breasted firefinch 9.0 g, African firefinch 10.2 g, African quailfinch 12.9 g, and brown twinspot 15.1 g (Dunning, 1993).

Egg Characteristics

The eggs of this species are poorly documented (measurements are available only for a single egg); they are white and indistinguishable from those of all other *Vidua* species. Approximate egg measurements of the known or presumptive hosts are: African quailfinch 15 × 10 mm (estimated mass 8.3 g), bar-breasted firefinch 14 × 11 mm (estimated mass 0.92 g), brown firefinch 16 × 10.5 mm (estimated mass 0.95 g), and African firefinch 15 × 11.5 mm (estimated mass 1.1 g). Thus, the indigobird may have a slightly to considerably larger egg than any of these presumptive hosts, based on this single indigobird egg's measurements.

Breeding Season

The breeding dates for this species are mainly from June to December for the nominate form *wilsoni* (Payne, 1973:180). This general period represents the last part of the rainy season and the early dry season, which typically occurs in sub-equatorial areas north of the equator during the second half of the year. Thus, in Nigeria the rains peak during July or August to September, singing by the indigobirds occurs in October and November, and the firefinch host species nests from at least July to November. In coastal Ghana, where rains are heaviest during May and June, both indigobird and the host firefinch's breedings have been observed from April through September. Breeding is probably more delayed in northern Ghana, where the period of heavy rainfall occurs later in the year (Payne, 1982). The brown twinspot and African quailfinch also breed at the start of the dry season.

Breeding Biology

Nest selection, egg laying. Little information is available. Payne (1977a) estimated that the mean number of eggs laid per laying cycle is 3.43 eggs (range 2–4, n = 7), with an average of 2.5 eggs laid per 10 days.

Incubation and hatching. No information

Nestling period. No information.

Population Dynamics

No information.

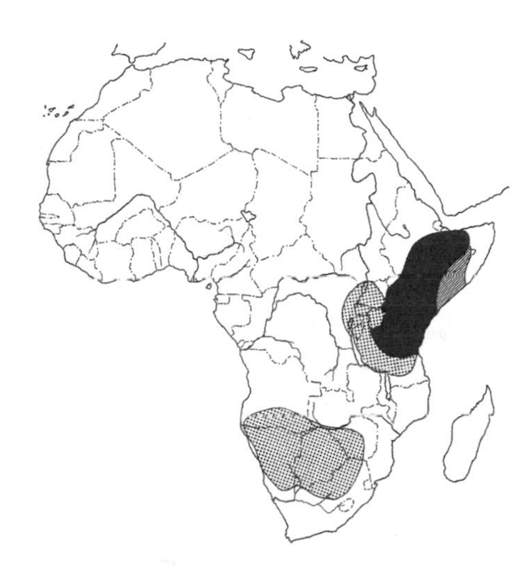

MAP 68. Range of steel-blue whydah (filled) plus ranges of hosts black-cheeked waxbill (shaded) and red-rumped waxbill (hatched).

STEEL-BLUE WHYDAH

(Vidua hypocherina)

Other Vernacular Names: Blue widowbird.

Distribution of Species (see map 68): Sub-Saharan Africa from Sudan and Somalia south to Tanzania.

Measurements (mm)

Female 4″(10 cm; males 12″(30 cm Wing, males 59–67.5 (avg. 64), females 63–67. Tail 40–47 (avg. 44), with the four ornamental rectrices of breeding males up to 205 (Friedmann, 1960). Wing, males 63–67, females 60–65 (Mackworth-Praed & Grant, 1960).

Egg, no quantitative information. Judging from a photo showing host and parasite eggs (Nicolai, 1989), the whydah's eggs must measure about 17 × 12.

Masses (g)

No body or egg weights available. Judging from their wing lengths, adults probably average 12–14. The eggs probably weigh 1.35, and have a shape index of 1.4 (oval), judging from a photo.

Identification

In the field: In breeding plumage, the bluish- to purplish-black males are easily identified by their four long, black central tail feathers, which are slender and widen somewhat toward their tips (fig. 42 and 43). The bill is black and the feet are grayish to grayish brown. The male's songs include a sustained soft warble. The song of its host, the black-cheeked waxbill, mostly consists of thin "teeh-heeh" or "fwooee" notes (Goodwin, 1982). Nicolai (1964) reported that no unquestionably mimicking host phrases are evident in the vocalizations of the parasite. Females and nonbreeding males are easily distinguishable in the field from indigobirds by their white under-wing coverts and inner webs of their flight feathers. They more closely resemble the corresponding plumages of the pin-tailed whydah, but are whitish, rather than buffy, on their underparts.

In the hand: The nestling gape pattern of this species and one of its two hosts, the black-

FIGURE 43. Sketches of adult female (A) and breeding male (B) steel-blue whydah, compared with an adult of its black-cheeked waxbill host (C). A juvenile head profile (D), gape pattern (F), and nestling (H) of the host waxbill are also shown, together with corresponding views (E, G, I) of the whydah. After photos in Nicolai (1989).

cheeked waxbill, are shown in fig. 43. Both species have three (or four fused into three) palatal spots arranged in a semicircle on the upper palate. The tongue is also marked basally with black, and there are poorly developed flanges at the commissural junctions. Both host species have essentially identical palatal markings (Nicolai, 1989). Juveniles of this species have a smaller bill than do the quite similar young of the pin-tailed whydah. Females and nonbreeding males also resemble those of the pin-tailed whydah, but have a brownish or grayish (not reddish) bill, and the white in the tail is limited to the edges and tips of the feathers.

Habitats

Dry brushveldt habitats, ranging from sea level up to about 1350 m elevation are favored. Its black-cheeked waxbill host's preferred habitat is dry acacia thornbush, and the red-rumped waxbill host similarly occupies arid thorn scrub, up to about 1400 m elevation (Goodwin, 1982).

Host Species

Host species are listed in table 31. Fostering hosts include the black-cheeked waxbill and the red-rumped waxbill (Nicolai, 1989). One or two other possible hosts have been mentioned by Friedmann (1960). The adult mass of the black-cheeked waxbill averages about 8 g (Dunning, 1993), but the corresponding parasite weight is still unknown.

Egg Characteristics

Nicolai (1989) illustrated the eggs of this species and its two host species but did not provide any measurements. The eggs of hosts and parasite are white, and those of the parasite in each case are noticeably larger and somewhat more rounded than the host species' eggs. Eggs of both waxbill hosts average about 14.5–15.2 × 11–11.5 mm (McLachlin & Liversidge, 1957; Mackworth-Praed & Grant, 1960) and have an estimated mass of about 1.0 g. Based on Nicolai's photographs, the whydah would appear to have eggs averaging about 17 × 12 mm, which represents a mean difference of about 30% greater egg volume and mass of the parasite's egg over that of the host.

Breeding Season

In Kenya and Tanzania breeding-condition birds have been reported between February and June. In Tanzania its host is known to breed from February to April, and both host species have breeding seasons associated with the rainy season (Nicolai, 1989).

Breeding Biology

Nest selection, egg laying. Payne (1977a) estimated that the usual number of eggs laid in a laying cycle is 3.33, based on counts of ovulated follicles in laying birds. Nicolai found both one-egg and two-egg "clutches" of whydah eggs present in four-egg clutches of the red-rumped waxbill. The presence of such complete four-egg clutches of the waxbill host in doubly parasitized nests suggests that no egg removal or egg destruction occurs during laying by the whydah.

Incubation and hatching. No information on the incubation period is available, but that of the host species is 12 days and presumably matches that of the parasite.

Nestling period. No definite information on the nestling period of the whydah is available, but the host waxbill has a nestling period of 22 days, and they begin self-feeding at 32 days (Goodwin, 1982). Nicolai (1989) reported that the young that he reared in captivity under Bengalese finches had similarly become independent by 4 weeks of age.

Population Dynamics

No information.

STRAW-TAILED WHYDAH

(Vidua fischeri)

Other Vernacular Names: Fischer's whydah.

Distribution of Species (see map 69): Sub-Saharan Africa from Sudan and Somalia south to Tanzania.

Measurements (mm)

Female 4″ (10 cm); males 11″ (28 cm)

Wing, males 64–71 (avg. 68). Tail, males 44–48, the four lengthened ornamental rectrices of breeding males as long as 190 (Friedmann, 1960). Wing, males 65–71, females 61–65 (Mackworth-Praed & Grant, 1960).

Egg, avg. 15.7 × 12.6 (range 15.7 × 12.5–12.7)

(Schönwetter, 1967–84). Shape index 1.25 (= broad oval.

Masses (g)

One unsexed bird, 13.6 (Dunning, 1993). Estimated egg weight 1.31 (Schönwetter, 1967–84).

Egg:adult mass ratio 9.6%.

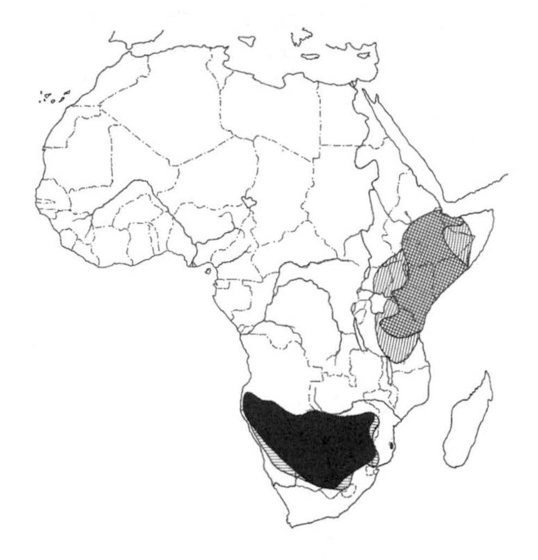

MAP 69. Ranges of queen whydah (filled) plus host common grenadier (horizontal hatching), and of straw-tailed whydah (hatched) plus host purple grenadier (vertical hatching).

Identification

In the field: Males in breeding plumage are easily identified by the presence of four pale yellow, straw-colored central tail feathers that extend well beyond the others (figs. 41 and 43). The crown and underparts are also yellow; the upperparts are blackish. Males have a four-note song that no doubt mimics that of the purple grenadier. The latter species is said to have a nine-noted song that is mainly used as a contact call between mated pairs and a male-limited song that begins with a soft crackling and buzzing and ends with a trill.

In the hand: The nestling gape patterns of this species and its purple grenadier host are illustrated in (figs. 9 and 44). There are three black palatal spots, the horny palate is mostly pale bluish to golden yellow, and the tubercles at the commissural junction are bright blue (Nicolai, 1974). Immature individuals are tawny-colored from the head to the flanks and breast and are streaked with brown and dusky on their upperparts, with narrow buffy feather edgings. They also have white underparts and a blackish bill. Adults of both sexes have reddish bills that probably are brighter in breeding males, which also have orange feet. Females closely resemble those of the pin-tailed whydah, but these two species have nonoverlapping distributions.

Habitats

Thorny scrub, rather heavy bushveld, and the vicinity of tree-lined marshes are this species' favored habitats, but it sometimes extends into steppelike and near-desert habitats. The species' altitudinal range is from lower than 300 m to about 1500 m. Its host, the purple grenadier, also favors thick thorn scrub but at times occurs in more open busy habitats.

Host Species

The purple grenadier is believed to be the primary or sole fostering host of this species (Nicolai, 1974). It has a mean adult weight of about 13 g, or only slightly less than that of its parasite.

Egg Characteristics

Like the other *Vidua* species, the eggs are white and oval. The eggs of its host species average 15.5 × 12.2 mm (estimated mass 1.27 g), representing a mean mass difference of about 3% in favor of the parasite.

Breeding Season

Breeding in Tanzania and Kenya occurs at least from March to May and probably occurs during June in southern Ethiopia. Its host breeds in southern Kenya during March and April and in northern Tanzania between December and February (Goodwin, 1982).

Breeding Biology

Nest selection, egg laying. No information.

Incubation and hatching. The incubation period is still unreported, although 12–13 days is likely, considering this is the period typical of grenadiers, or at least the violet-eared grenadier (Goodwin, 1982).

Nestling period. The nestling period of this species was reported to be 16 days (Nicolai, 1969). A 16-day fledging period is also typical of the

FIGURE 44. Sketches of breeding male (A), juvenile (C), and nestling gape patterns (E) of the straw-tailed whydah, compared with the corresponding features of its purple grenadier host (B, D, F). Also shown is a male straw-tailed whydah displaying to a female (G). After sketches and photos in Nicolai (1969).

violet-eared grenadier, although that of the purple grenadier is not yet reported (Goodwin, 1982).

Population Dynamics

Parasitism rate. Nicolai (1969) reported that 11 out of 15 host nests (73.3%) that he found in Tanzania were parasitized.

Hatching and fledging success. No information.

Host–parasite relations. No information.

QUEEN WHYDAH

(Vidua regia)

Other Vernacular Names: Shaft-tailed whydah.

Distribution of Species (see map 69): Sub-Saharan Africa from Namibia, Botswana, and Mozambique south to South Africa.

Measurements (mm)

Females 4.5″ (11 cm); males 12–13.5″ (31–34 cm)

Wing, males 70–75 (avg. 72.5, $n = 17$), females 68–70. Tail, males 37–42, the four lengthened rectrices of breeding males 210–243, females 37–43 (Maclean, 1984).

Egg, avg. 16 × 12.5 (range 15.3–17.2 × 11.4–13.3) (Schönwetter, 1967–84). Shape index 1.28 (= broad oval).

Masses (g)

Avg. of 10 females, 15.7 (Payne, 1977). Avg. of 4 unsexed birds, 13.8 (Dunning, 1993). Two males 15, 15.4, one female 14.6 (*Ostrich,* 1974). Males 11.9—15.7 (avg. 14.5, $n = 4$), females 14.5–15.2 (avg. 13.8, $n = 3$) (Maclean, 1984).

Estimated egg weight 1.31 (Schönwetter, 1967–84).

Egg:adult mass ratio 9.5%.

Identification

In the field: Breeding males can be instantly recognized by the presence of four very long, ornamental black tail feathers that widen and become narrow-spatulate toward their tips (fig. 45). The bill is red, there is a black crown and upperparts, and the nape, sides of head, and remaining underparts are all golden yellow except for a darker breast-band. Females resemble those of the pin-tailed whydah, but have a less well-developed dark eye-stripe and ear patch and have a dark reddish bill. The species parasitizes the common grenadier (or "violet-eared waxbill") and mimics its songs, which are said to be twittering or "zizzling," and end in fluty tones. Nicolai (1964) stated that this whydah mimics not only the song but also the excitement calls, excitement phrases, greeting notes, nest calls and "rage" calls of its host species. Females closely resemble other whydahs in plumage, but have reddish bills and feet, at least during breeding, thus visually distinguishing them from all species but the allopatric straw-tailed whydah.

In the hand: Nestlings closely resemble their host species, the common grenadier, but juvenile whydahs have no blue on the rump area. The nestling waxbill has a mostly orange palate, and the tongue is also orange or yellow, crossed with a black band that is lacking in the parasite. The palate has the usual five-spot pattern, but the two lower spots are either small or absent (Payne, 1970; Goodwin, 1982). Additionally, the whydah has a narrower bill (nostrils 3 mm apart, rather than 4 mm), its crown is dull brown (not reddish-brown), its upperpart feathers are edged with light brown (not uniformly brown), its tail is dark brown (not black), its underparts are mainly white (not mainly light brown), and its brown tarsus has seven scales (not bluish, with six scales) (Skead, 1975). When the young waxbill is only 24–35 days old, a rapid molt occurs, so that the colorful adult feathers appear on the head, at which time the juvenile parasites are doubtless quite distinct from their hosts.

Habitats

Preferred habitats are dry, grassy areas in savanna thornveld. The birds are especially attracted to open, grassy and barren areas around stockyards. The common grenadier host species similarly prefers dry

FIGURE 45. Sketches of adult female (A), breeding male (B), and fledgling (C) of the queen whydah, as compared with an adult female (D), breeding male (E), and fledgling (F) of the common grenadier host. Sexual differences in the juvenile female (G) and male (H) of the common grenadier are also shown. Partly after Nicolai (1967).

thorn scrub, thorn tangles near streams, and sometimes dry river beds well away from water.

Host Species

Host species are listed in table 31. The common grenadier is believed to be the primary fostering host. Friedmann (1960) listed six other possible hosts, of which at least the records for the scaly weaver and black-chested prinia seem to be acceptable, the others hosts appear questionable. The mean weight of the common grenadier is 10 g (Dunning, 1993), so the parasite has an approximate 60% weight advantage over its host.

Egg Characteristics

The eggs of this species are white and much like those of their host species, the latter having eggs averaging 15.7 × 12 mm. Such measurements represent an approximate 13% greater estimated mean volume for the parasite than that of the host.

Breeding Season

Breeding in southern Africa occurs mainly early in the year, from December to May in the Transvaal, during February in Botswana, and during April in Namibia. Its host also breeds there during January–May (McLachlin & Liversidge, 1957;

Maclean, 1984; Ginn et al., 1989). In northern Transvaal the waxbill host may breed as late as June, but in Zambia it breeds during January and February (Goodwin, 1982).

Breeding Biology

Nest selection, egg laying. Skead (1975) observed a female entering a waxbill nest that already contained four host eggs and two parasitic eggs, plus the incubating male. The nest was entered briefly by the waxbill female, but this bird soon left. The whydah also left after 45 seconds. A second female whydah appeared about an hour later, and apparently pecked at the female waxbill that was then incubating. This female whydah soon entered the nest, but departed after only about 30 seconds. Two eggs were deposited during this brief observation period. Although it has been suggested that for each parasitically laid egg, one of the host's is removed (Maclean, 1984), this is in contrast to the usual situation among parasitic whydahs, and seems doubtful in view of Skead's observations. He found one parasitic egg in four of five waxbill nests, and five in the remaining one. This last nest also contained four waxbill eggs (the modal clutch size), suggesting that host eggs are rarely if ever removed by the whydah.

Incubation and hatching. The incubation period is not known, but that of its host is 12–13 days (Goodwin, 1982).

Nestling period. The nestling period is not known, but that of its host is 16 days (Skead, 1975).

Population Dynamics

Parasitism rate. Skead (1975) reported that 5 of 15 waxbill nests found in central Transvaal had been parasitized.

Hatching and fledging success. No information.

Host–parasite relations. No information.

PIN-TAILED WHYDAH

(Vidua macroura)

Other Vernacular Names: Pintail widowbird.

Distribution of Species (see map 70): Sub-Saharan Africa from Senegal and Sudan south to South Africa.

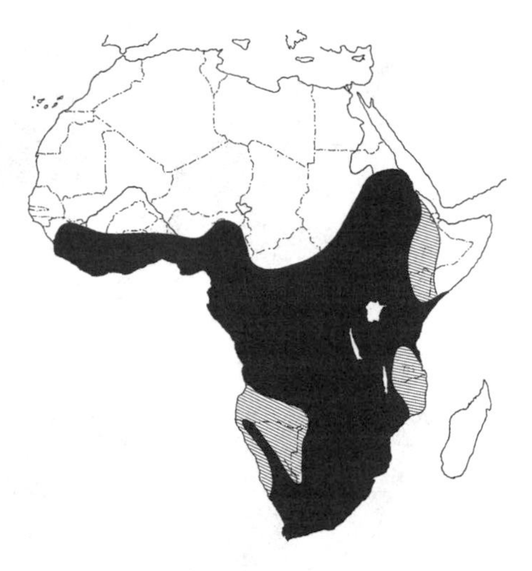

MAP 70. Ranges of pin-tailed whydah (filled) plus host common waxbill (hatched).

Measurements (mm)

Females 4.5–5″ (12–13 cm); males 10–13.5″ (25–34 cm)

Wing, males 64.5–75; tail, males 44–51.5, the four longest rectrices of breeding males 235–340 (Friedmann, 1960). Wing, males 69–79 (avg. 73.5, n = 29), females 64–71 (avg. 67, n = 14).Tail, males 47–52, the longest rectrices 163–264; females 43–50 (Maclean, 1984). Wing, males 69–76, females 63–67. Tail, males 42–47, the longest rectrices 180–260; females 38–50 (Bannerman, 1949).

Egg, avg. 15.8 × 11.9 (range 14.5–17.2 × 11–12.4) (Schönwetter, 1967–84). Shape index 1.33 (= broad oval).

Masses (g)

Avg. of 22 females, 14.4 (Payne, 1977). Males 14.1–18.7 (avg. 15.9, n = 16), females 13.8–15.9 (avg. 14.5, n 26). Estimated egg weight 1.31 (Schönwetter, 1967–84). Egg:adult mass ratio 6.9%.

Identification

In the field: Breeding males can be instantly recognized by their four lengthened and black central tail feathers, which are all uniformly narrow and tapering to a sharp tip, rather than spatulate as in the shaft-tailed whydah (fig. 42). The upperparts are mostly glossy black and the underparts white. The male's legs are grayish black, and its bill is pink to bright red, even in the femalelike nonbreeding plumage. Females are similar to but smaller than males in nonbreeding plumage and are less strongly streaked above. Females also have brownish to grayish, rather than blackish, legs. Females have either a blackish upper mandible (when breeding) or a more reddish brown one (when nonbreeding). Females also reportedly have a more contrasting (more "sparrowlike" or more like that of paradise whydahs) head pattern than do female queen whydahs.

Males display in a hovering, dancing flight before females, uttering "tseet-tseet-tseet" notes. This song is probably a repetition of the species' usual simple call-note, but is uttered in a rapid series of 5–15 notes, with only slight modulation. According to Nicolai (1964), it is not yet possible to associate any of this species' vocalizations with those of their host species as potential examples of mimicry.

In the hand: Newly hatched nestlings of this species resemble those of their primary host, the common waxbill, but they are more mauve-colored and are covered with down (rather than being pinkish and virtually naked). Nestling gape patterns are similar in both, including a spot on the lower mandible and two dark spots on the tongue. However, the palate of the waxbill has a circle of six spots and a central seventh spot, whereas that of the whydah has a circle of five spots. Juveniles are femalelike but have buffy feather-edgings and horn-colored (not pinkish horn to salmon) bills and also lack white in the tail (rather than having white on the inner webs). The chestnut and buff tones of adults are more grayish in young birds. Nonbreeding males are generally more rufous-colored than females and have stronger blackish striping on their upperparts.

Habitats

A variety of habitats are used, including savannas and grasslands, but this species prefer open, grass-dominated areas with scattered trees or bushes, especially those near wet areas. This species may also be found in forest clearings and along the edges of tropical rivers. It additionally occurs in cultivated areas and suburban gardens. It ranges altitudinally from sea level to about 2250 m, but is most common below 1800 m.

Host Species

Host species are listed in table 31. Of these, the common waxbill (*E. astrild*) is certainly the primary fostering host, but the zebra waxbill (*E. subflava*) may be a significant host in Natal (Friedmann, 1960). Payne (1977b) listed the orange-cheeked waxbill as an additional host species, and later suggested (1985) that the black-rumped waxbill, Anambra waxbill, and fawn-breasted waxbill may also serve as local foster species. Maclean (1984) listed secondary hosts as including the bronze munia, zebra waxbill, red-billed firefinch, swee waxbill, piping cisticola, and tawny-flanked prinia. Such species would all seem to be minor, or even accidental, hosts.

Egg Characteristics

The mean egg measurements given above for this species are considerably larger than the average measurements of its common waxbill host (13 × 10 mm), and the eggs also differ slightly in surface texture, so it is probable that they can be distinguished by these traits. The estimated fresh weights of the two species are 1.34 g vs. 0.87 g, respectively (representing a mean weight advantage of about 33% for the parasite egg relative to the host), so this method may also serve as a practical means of helping distinguish them. The mean adult weights of the common waxbill is only 7.5 g (Dunning, 1993), so the parasite has an approximate 90% weight advantage over the host.

Breeding Season

Breeding by this whydah occurs in southern Africa between October and March, through the austral summer (Ginn et al., 1989). In the Cape region its host waxbill similarly breeds from September to January, but farther north in South Africa it breeds from November to April. In Malawi, Mozambique, and Zimbabwe breeding by the host also occurs from November or later to April. In East Africa (Kenya, Uganda) the host's season is complex (March–May in Uganda, November–January and March–July in Kenya). In West Africa host breeding occurs during or after the fall rainy season (September–November in Sierra Leon and Cameroon). In the Congo Basin of Zaire the whydahs breed during the rainy season. In the northern Ituri forest and the grasslands around Lake Albert, the males are thus in breeding plumage from about May to November. However, south of the Ituri forest, the breeding plumage occurs between October and June. Therefore, with the crossing of the equator there may be a fairly distinct shift of breeding periods (Chapin, 1954). The host waxbill likewise breeds in the northeastern Congo (race *occidentalis*) from August to November and in the southeastern Congo (race *cavendishi*) from February to April (Mackworth-Praed & Grand, 1973). In Uganda there may also be two breeding seasons, with male whydahs molting during opposite seasons in two different regions of the country (Chapin, 1954).

Breeding Biology

Nest selection, egg laying. Payne (1977a) estimated the average number of eggs produced during a laying cycle as 3.11 (range 2–4).

Incubation and hatching. The incubation period is unreported. that of the host species is 11–12 days (Goodwin, 1982).

Nestling period. The nestling period is about 20 days (Maclean, 1984), as compared with 17–21 days in the host waxbill species (Goodwin, 1982).

Population Dynamics

No information.

NORTHERN PARADISE WHYDAH

(Vidua orientalis)

Other Vernacular Names: None in general use. Sometimes included (often with *interjecta* and *togoensis*) as part of the broad-tailed paradise whydah (Mackworth-Praed & Grant, 1973). Also sometimes included as part of the eastern paradise whydah (Payne, 1971, 1985, 1991).

Distribution of Species (see map 71): Africa from Senegal east to northern Ethiopia in the Sub-Saharan sahel zone.

Subspecies

V. o. orientalis: Chad to northwestern Ethiopia (Eritrea).

V. o. kadugliensis: Southern Sudan (Kordofan).

V. o. aucupum: Senegal to northern Nigeria and western Chad.

Measurements (mm)

Female-like birds 5.5" (14 cm); breeding males

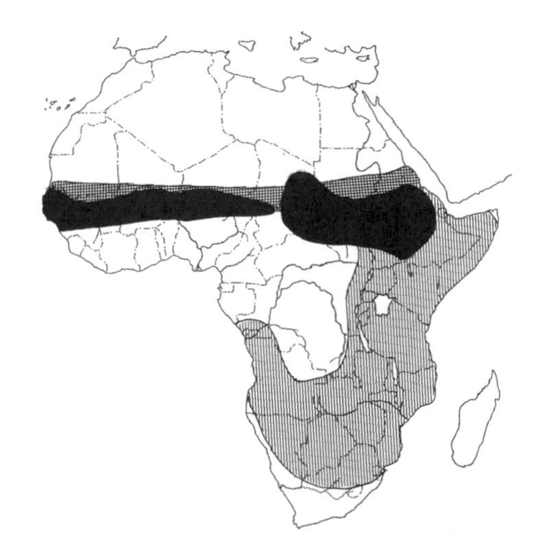

MAP 71. Ranges of northern paradise whydah (filled) plus host green-winged pytilia. The host's indicated range includes that of the parasitized race *citerior* (cross-hatched) plus the ranges of several races parasitized by the eastern paradise whydah but apparently not by the northern (hatched).

10–12″ (25–31 cm)

V. o. aucupum. Wing, both sexes 73–76 (Mackworth-Praed & Grant, 1973); males 72–80 (avg. 75.54, $n = 62$) females (including some *orientalis*) 71–74 (avg. 73, $n = 11$), (Payne, 1991). Longest pair of rectrices in breeding males <275 (Friedmann, 1960); 230–255 long × 25 wide; the next rectrix pair 55; female 52 (Bannerman, 1949).

V. o. orientalis. Wing, males 73–80 (avg. 77.43, $n = 68$), females (including some *aucupum*) 71–74 (avg. 73, n 11) (Payne, 1991); males 74–80, females 71–73 (Bannerman, 1949). Longest pair of rectrices in breeding males ~200 (Mackworth-Praed & Grant, 1973); 195–272 long × 32 wide (Bannerman, 1949). Width 24–30 (Friedmann, 1960). The next rectrix pair 53–55; females 51–53 (Bannerman, 1949).

Egg, range 18–19 × 13.5–14 (avg. 18.5 × 13.75, $n = 3$) (Bannerman, 1949). Shape index 1.34 (oval).

Masses (g)

No information on body mass, but linear measurements suggest an average adult mass of about 16–18. Estimated egg mass 1.86.

Identification

In the field: Except at the easternmost end of its range, this species is not in contact with any other paradise whydahs, so breeding males should be easily recognized by their dual-length ornamental tail feathers and by the combination of a black head, chestnut nape, a rufous breast, and buffy underparts (fig. 46). In areas where the eastern paradise whydah also occurs (along the eastern edge of this species' range, in northern Ethiopia), the eastern whydah may be recognized by its long, gradually tapering ornamental tail feathers, rather than possessing the uniformly broadened tail feathers of the northern paradise whydah. Possible contacts with the long-tailed paradise whydah may occur along the southeastern edge of the northern's range, in the Central African Republic and extreme southern Sudan. All species of paradise whydahs probably perform advertising displays involving tail-feather exhibition while they are perched, and some species also display while in flight above their territories. Their vocalizations are usually quite weak and inconspicuous. Nicolai (1964) has provided comparative sonograms of call-notes ("wit" calls and two-syllable whistles) of all three races of this species and the red-lored form of the green-winged pytilla, all of which are similar. This host species has a distinctive male song that includes a series of "veet"notes, a phrase of whistles that vary in pitch, and two final and short, gurgling phrases.

In the hand: Breeding males can be readily identified by the fieldmarks mentioned above. Females, nonbreeding males, and immature birds may not be readily distinguishable from those of other paradise whydah species, although young of this species are said to be a paler earth-brown than are those of the eastern paradise whydah; adult females and nonbreeding males are also paler tawny above, with narrower mantle streaking. Males are more brightly colored in overall plumage than females and more strongly striped. Juveniles have pale rufous feather edges and dark brown bills (adult females also have brownish bills, but the bills of adult males are blackish).

Habitats

Open acacia savanna is the preferred habitat, where scattered trees provide convenient perching and display sites. Dry thorn woodland is also the favored habitat of the host pytilla species, but it extends into semideserts and cultivated areas having interspersed bushes or thorn scrub patches (Goodwin, 1982).

Host Species

An arid-adapted, pale-colored and red-lored taxon (*citerior*) of the green-winged pytilia (or "common waxbill") group is the only known biological host of the *aucupum* race of *V. orientalis.* This red-lored form and perhaps some closely related populations (such as the poorly distinguished taxon sometimes recognized as *clanceyi*) are also

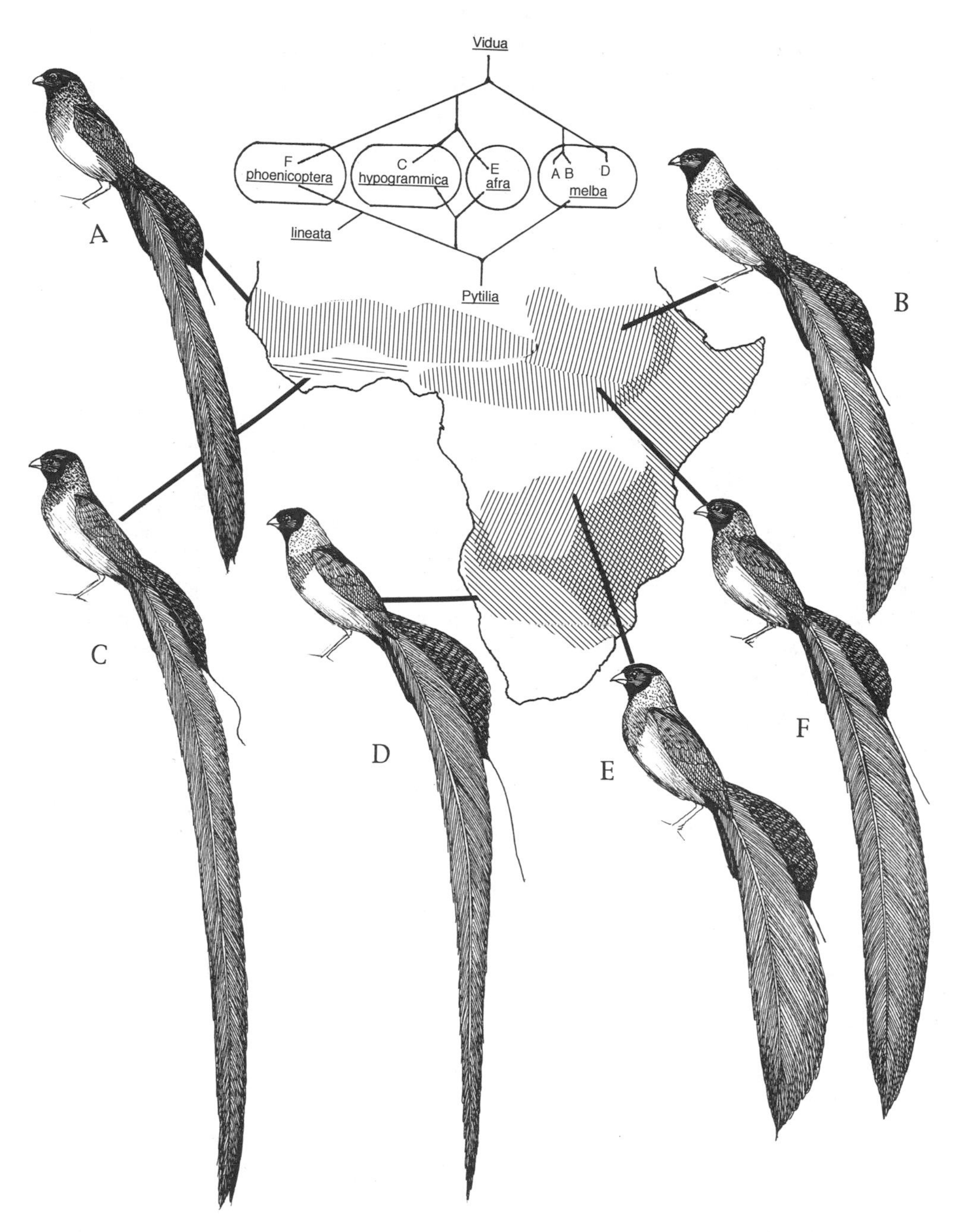

FIGURE 46. Geographic distributions and male breeding plumages of the paradise whydahs, including the northern paradise whydah (A = *aupicum* race, B = *orientalis* race), plus the Togo (C), eastern (D), broad-tailed (E), and long-tailed (F) paradise whydahs. Shown above are postulated co-phylogenies of these whydahs (adapted from Nicolai, 1977) and their pytilia hosts (after Goodwin, 1982).

the probable hosts of the other two races of *V. orientalis.* Observations by Nicolai (1964, 1977) suggest that *citerior,* which (if liberally interpreted taxonomically) ranges from Senegal and Guinea east to the Eritrean region of Ethiopia, might be best considered a distinct species separate from the various gray-lored populations of *melba* that extend all the way from southern Sudan and Ethiopia south to South Africa. Weights of *citerior* are not available, but those of the South African population of the green-winged pytilia average 15.1 (females) to 15.5 (males) (Skead, 1975), or only slightly less that the probable average adult weight of the whydah.

Egg Characteristics

Little information on the eggs of this species is available. The eggs of its pytilia host species average about 15 × 12 mm (= broad oval, and about 1.2 g mass). Those eggs attributed by Bannerman (1949) to this whydah are distinctly larger and have an approximate 50% greater estimated mass.

Breeding Season

The breeding season of the host pytilia species occurs during August and September in Nigeria, or during the latter part of the rainy season. In the Sudan, host breeding records extend from October to February, and a second round of breeding evidently occurs from May to July, which no doubt must roughly correspond with the breeding periodicity of the parasite.

Breeding Biology

Nest selection, egg laying. No information.

Incubation and hatching. The incubation period is still unknown. It is probably close to the 12- to 13-day incubation period reported for the host pytilia species (Goodwin, 1982).

Nestling period. No information. The nestling period of the host species (at least of *afra)* is 21 days, which is followed by a 14-day postfledging dependency period (Goodwin, 1982).

Population Dynamics

No information.

TOGO PARADISE WHYDAH

(Vidua togoensis)

Other Vernacular Names: None in general English use; sometimes included within a common expanded species as the broad-tailed paradise whydah, usually together with the long-tailed paradise whydah (Mackworth-Praed & Grant, 1973), or considered as part of the northern paradise whydah (Traylor, 1968).

Distribution of Species (see map 72): Sub-Saharan Africa from Sierra Leone or the Ivory Coast east to at least Nigeria, and probably to northern Cameroon and southern Chad (Payne, 1985).

Measurements (mm)

Female-like birds 5.5″ (14 cm); breeding males to 15″ (38 cm)

Wing, males 74–78 (avg. 76.6, $n = 15$) (Payne, 1991). Longest rectrices of breeding males ~325 (Mackworth-Praed & Grant, 1973); 290–360 (Bannerman, 1949). Width of longest rectrices (flattened) <30 mm

MAP 72. Ranges of Togo paradise whydah (filled) plus host red-faced pytilia (hatched). Arrows show specimen localities for breeding-plumaged *togoensis,* but still unproven breeding.

(Payne, 1985). Length of next pair 52 (Bannerman, 1949).

Egg, no definite information. One probable egg found in the nest of the only known host species measured 17.2 × 13.3 mm (shape index 1.29).

Masses (g)

No information on body mass, but wing-length data suggest an average adult mass of about 18–20. The estimated mass of one egg was 1.6.

Identification

In the field: Within this species' rather small range, it is evidently the only breeding paradise whydah, although not far to the north (as in Burkina Faso), the northern paradise whydah also occurs (fig. 45). There is still disagreement as to the range limits of these two species, but Payne (1985) observed the Togo paradise whydah as far west as Kabala, Sierra Leone. Breeding males of both species have a rather chestnut-toned nape color in addition to a black head, a rufous breast, and buffy underparts. Males also differ somewhat in tail lengths, with the Togo's being longer. There is also some local contact with the long-tailed paradise whydah, as in Liberia and probably northern Cameroon; there the differences in the ornamental tail feather width (narrower in the Togo) and underpart color (more two-toned in the long-tailed) help to distinguish breeding males (Payne, 1985). The song of the Togo paradise whydah's host, the red-faced pytilia, is a repeated "vee-vee-vee" that is rather intermediate in form to the songs of the oranged-winged and red-winged pytillias and which the paradise whydah effectively mimics. There is no display flight in this species, and its display posturing generally resembles that of the long-tailed paradise whydah (Nicolai, 1964, 1977). Nonbreeding males, females, and immature birds are probably not distinguishable in the field from those of other paradise whydah of western Africa, such as the northern paradise whydah, and breeding males cannot be distinguished with certainty but have somewhat longer elongated tails than any of the other potentially occurring species in this same general region.

In the hand: This species reportedly parasitizes the red-faced (also "golden-winged" or, in Nicolai's terminology, "yellow-winged") pytilia, and the nestlings of the two are probably similar to one another (and they also resemble the young of the closely related orange-winged pytilia, which is parasitized by the broad-tailed paradise whydah not far to the south and east). However, from an early age, host and parasite can be distinguished by the parasite's tendency to open its bill to a much greater degree when begging than does the host. Additionally, the parasite and host differ in their palate characteristics in that the blue-violet signal markings of the mouth are larger and more elongated in the host than in the parasite. The palates of the host and parasite are also both covered with small papillae that probably stimulate parental feeding (Nicolai, 1977). As noted above, immature birds and females of the Togo and northern paradise whydah are probably not distinguishable, even in the hand. Breeding males of this species are best separated from the northern and long-tailed paradise whydahs by the Togo's somewhat longer but narrower ornamental tail feathers (at least 290 mm long, and no more than 30 mm wide).

Habitats

This species is apparently similar to the other paradise whydahs in preferring brushy grasslands or savannas. It probably can also be found along woodland edges and in secondary growth or among areas of derelict cultivation, where the host species is likely to occur.

Host Species

The only known fostering host of this species is the red-faced pytilia (Nicolai, 1977). It occupies savanna woodlands and overgrown cultivated areas and nests in lower-elevation sites such bushes, shrubs, and trees. Its range is slightly overlapping with but mostly just south of the red-winged pytilia, which is parasitized by the long-tailed paradise whydah. The adult mean body mass of the host is 14.8 g (Dunning, 1993); the whydah's body mass is not yet known, but judging from its wing measurements, it is certainly considerably larger than its host.

Egg Characteristics

No definite information is available on the eggs of this species, which no doubt are identical in appearance to those of the other paradise whydahs. A whydah egg found in the nest of the red-faced pytilia in Nigeria measured 17.2 × 13.3 mm but was attributed to the long-tailed paradise whydah (Serle, 1957), a species not known to parasitize the red-faced pytilia. Males collected by Serle in the same area had ornamental tail feather measurements (length 276–303 mm, maximum rectrix width 28–32 mm) that were within the range of the Togo paradise whydah. Eggs laid by the host species are about the same size as those of the other pytilias (14.8–15.5 × 11.5–11.9 mm, estimated weight 1.4 g). This paradise whydah thus appears to have eggs averaging about 15% greater in mass and volume than those of its pytilia host.

Breeding Season

This species' breeding season is evidently concentrated during the dryer period immediately after the rainy season, when ripening seeds become easily available and breeding conditions are especially suitable (Nicolai, 1977). Its host, the red-faced pytilia, reportedly nests during the dry season, which roughly extends from October to January over much of its range. In Nigeria, where the rainy season occurs from May to October, nesting by the pytilia has been reported during November and January. There the male paradise whydahs molt from their femalelike plumages into breeding plumages between mid-September and mid-October and are ready to breed with the onset of the dry season (Serle, 1957, Nicolai, 1977).

Breeding Biology

Nest selection, egg laying. Nicolai observed parasitism of red-faced pytilia nests in eastern Nigeria, where three parasitized nests were found. These nests had one, two, and four parasitic eggs, in addition to the host's respective clutches of three, four, and four eggs. This eight-egg combined clutch suggests that no host eggs are removed or destroyed at the time laying by the parasite.

Incubation and matching. The incubation period averaged 11.75 days, as compared with 12–13 days for the host pytilia. In a nest containing four parasitic eggs and four host eggs, the host eggs all hatched over a 3-day period. However, the parasitic eggs hatched over a 7-day period, suggesting that they may have been laid by a single female whydah. In another nest containing two parasitic eggs and four host eggs, the host eggs hatched over a 2-day period and the parasitic eggs hatched over a 3-day period (Nicolai, 1977).

Nestling period. The nestling period for this whydah requires about 15–16 days, as compared with 17–19 days for the host pytilia (Nicolai, 1977).

Population Dynamics

Parasitism rate. Three of the four pytilia nests (75%) found by Nicolai (1977) contained parasitic eggs.

Hatching and fledging success. Little information exists. Nicolai (1977) removed the eggs he found from their host nests, and these were subsequently hatched and reared in captivity by foster Bengalese finch parents.

Host–parasite relations. No information.

LONG-TAILED PARADISE WHYDAH

(Vidua interjecta)

Other Vernacular Names: Congo paradise whydah, Nigerian paradise whydah; West African broad-tailed paradise whydah, Uelle paradise whydah. Sometimes also included in a common species with the broad-tailed paradise whydah (Mackworth-Praed & Grant, 1973) or included as part of the northern paradise whydah (Traylor, 1968).

Distribution of Species (see map 73): Sub-Saharan Africa from northeastern Zaire and southeastern Sudan west to at least as far as

MAP 73. Ranges of long-tailed paradise whydah (filled) plus host red-winged pytilia (hatched). Arrows indicate specimen locations of breeding-plumaged long-tailed whydahs, but still unproven breeding. The isolated Ethiopian pytilia population (hatched) is sometimes considered a separate species.

Nigeria. Breeding-condition birds have also been reported from much farther west, including Ghana, Ivory Coast, Mali, and Guinea. Since the host species extends west to Guinea, whydah breeding in these in these regions is also likely (Payne, 1985, 1991).

Measurements (mm)

Female-like birds, 5.5″ (14 cm); breeding males to 14″ (36 cm). Wing, males 78–79 (Chapin, 1954); males 76–82 (avg. 78.05, n = 39); females 74–76 (avg. 75, n =3) (Payne, 1991). Longest rectrices of breeding males 260–298 (Chapin, 1954); 284–298 (Friedmann, 1960): 290–304 (Payne, 1991). Width of longest rectrices, flattened 30–40 (Payne, 1985), unflattened 26 (Payne, 1991). The next-longest rectrix pair 63–64 (Bannerman, 1949).

Egg, one egg attributed to this species was 17.2 × 13.3 (Serle, 1957), but perhaps belonged to *togoensis.* Shape index 1.29 (= broad oval).

Masses (g)

One non-breeding male 20 (Payne, 1991). Estimated egg weight 1.6 (Schönwetter, 1967–84).

Identification

In the field: The range of this species lies immediately to the south of that of the northern paradise whydah (fig. 45), which is extremely similar to it in all plumages, and possible intermediates have been recorded from Ndele, Central African Republic. It may also be in limited contact with the eastern paradise whydah at the easternmost limits of its range, in the vicinity of the White Nile, southeastern Sudan, where a specimen of intermediate plumage has been collected, suggesting that local hybridization may be occurring. It is apparently also in sympatric contact (but not known to hybridize) with the Togo paradise whydah, since the long-tailed paradise whydah has been reliably reported as far west as northern Ghana, southern Mali, and Guinea (Payne, 1985, 1991). In such areas, males of the long-tailed paradise whydah should be identified by their wider and shorter ornamental tail feathers, darker napes, and a two-toned underpart appearance, as the maroon breast extends farther back toward the abdomen than in the Togo paradise whydah. Breeding males of the long-tailed paradise whydah are also difficult to distinguish visually from the northern paradise whydah, but the hindneck of this species is somewhat darker and browner than in the northern.

The song of this species' red-winged (or "aurora") pytilia host consists of a repeated series of rattling notes that are followed by a long, drawn-out and croaky whistle (Nicolai, 1964). This species' song is mimicked by the paradise whydah (Payne, 1991), but a close comparison of their respective vocalizations still has not been made.

In breeding plumage the male's bill is black and its feet are mostly dark gray, with some pinkish tints. Females and young of other paradise whydahs potentially occurring, in the same region as this species are probably impossible to distinguish in the field, but females have a dark line extending back behind the eye (as also occurs in broad-tailed and northern paradise whydahs), rather than curving down in a C-like pattern behind the ear, as in the eastern paradise whydah (Payne, 1971). Payne (1991) has recently described the female and immature male plumage of this species, noting that the most distinctive features of the adult female are its pastel-reddish legs and light orange bill. By comparison, females of the eastern paradise whydah and broad-tailed paradise whydah have gray feet and gray to blackish bills. The nonbreeding male plumage apparently does not differ in pattern from that of other paradise whydahs. However, nonbreeding males resemble females in their reddish bill and leg coloration, which may aid in field identification.

In the hand: This species parasitizes the red-winged pytilia, and the nestlings of the two are probably similar, but detailed descriptions of plumages and gape patterns are still lacking for both nestlings and juveniles. Juveniles probably have less strongly defined feather markings and pale feather edgings than do adults, as is the case in other related species of paradise whydahs. Adult females and males in nonbreeding plumage are not readily distinguishable from those of the other paradise whydahs by wing or other similar linear measurements; the wing measurements overlapping with those of all species except for the larger broad-tailed paradise whydah (Payne, 1991). Males in breeding plumage have shorter ornamental rectrices than those of the Togo paradise whydah (the longest pair usually <300 mm, rather than usually >300 mm, and 30–40 mm in maximum flattened rectrix width, rather than 20–30 mm). They also have much less gradually tapering, elongated rectrices than those of the eastern paradise whydah and have darker hindnecks than those of the northern paradise whydah.

Habitats

This species probably occupies habitats similar to those of the other paradise whydahs, namely, grasslands with scattered trees or tall bushes, as well as forest edges or woodland clearings, and similar transitional grassland–woodland habitats. Brushy pastures and the borders of cultivated ground are said to be favored habitats.

Host Species

The only known fostering host of this species is the red-winged pytilia (Nicolai, 1977). Its habitats include open woodlands, savannas, woodland edges, bamboo thickets, and cultivated areas around villages. A small and separate population of the red-winged pytilia in the Blue Nile region of Ethiopia (shown as cross-hatched area on map 73) is not yet known to be parasitized by any paradise whydah. This still-unstudied but potential host probably is best regarded as a distinct species, the "red-billed" or "striped" pytilia (Nicolai, 1968; Goodwin, 1982). The red-winged pytilia probably averages about 15.5 g in adult body mass (Dunning, 1993), giving the whydah an approximate 30% mass advantage.

Egg Characteristics

The eggs of this species are still not known with certainty. However, they probably are similar to those of the red-winged pytilia host, whose eggs have mean measurements of 15.5 × 12.5 mm (range 15–16.7 × 12–13.2 mm) and an estimated egg mass of 1.33 g. The latter figure compares with an estimated 1.6 g mass in an egg that has been attributed to this parasite (see account of Togo paradise whydah), suggesting an approximate 12% greater mean egg mass in the whydah.

Breeding Season

Breeding probably occurs during the drier season or latter part of the year in areas north of the equator and over much or all of the year in the areas more closely approaching the equator. Nests of the host pytilia have been found in Nigeria during November, December, and February.

Breeding Biology

No information.

Population Dynamics

No information.

EASTERN PARADISE WHYDAH
(Vidua paradisaea)

Other Vernacular Names: African paradise whydah.

Distribution of Species (see map 74): Sub-Saharan Africa from Angola, Zaire, and Somalla south to South Africa.

Measurements (mm)

Female-like birds 5.5–6″ (14–15 cm); males 13–15″ (33–38 cm)

Wing, males from Sudan to Somalia, 74–81 (avg. 77.91, $n = 54$), females 71–76 (avg. 74.36, $n = 11$); males from Kenya to South Africa 76–81 (avg. 77.66, $n = 98$), females 93–79 (avg. 76.11, $n = 11$) (Payne, 1991); males from Zaire 79–84 (avg., 81), females 78–82 (Chapin, 1954). Longest rectrices of breeding males 245–344 (Chapin, 1954); 270–342 long, 24–34 wide (Payne, 1980); 255–315 (Maclean, 1984).

Egg, avg. 17.8 × 13 (range 17–18.4 × 12–14.1) (Schönwetter, 1967–84); also 17.5–19.5 × 13–14 (avg. 18.2 × 14, $n = 3$) (Maclean, 1984). Shape index 1.37 (= oval).

Masses (g)

Males, 20.2–22 (avg. 21.2, $n = 5$), Avg. females 21.5, $n = 26$) (Payne, 1977a). Five males 20.2–22 (avg. 21.2), two females 19.9, 20.2 (*Ostrich* 45:192). Unsexed adults 18.9–21.4 (avg. 19.9, $n = 6$) (Maclean, 1984). Estimated egg weight 1.58 (Schönwetter, 1967–84). One fresh egg 1.6 (Skead, 1975). Egg:adult mass ratio 7.35%.

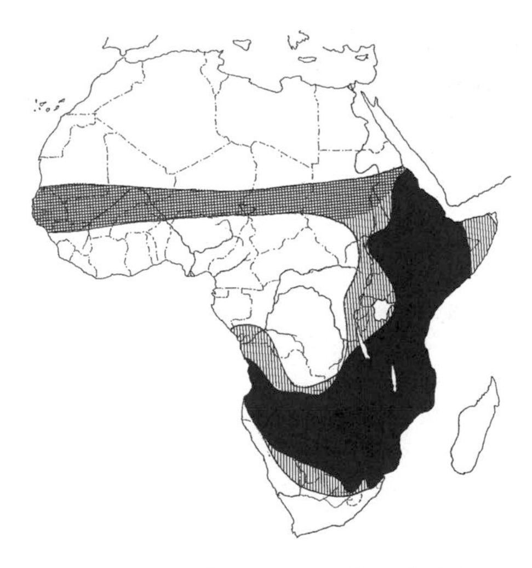

MAP 74. Ranges of eastern paradise whydah (filled) plus host green-winged pytilia (hatched, except for race *citerior*, which is cross-hatched).

Identification

In the field: This is a broadly ranging species that is widely sympatric with the broad-tailed paradise whydah over much of eastern and southern South Africa, from Angola east through Zambia and Zimbabwe to Mozambique and southern Tanzania (fig. 45). There are perhaps some more limited contacts with the northern and the long-tailed paradise whydahs near the northern limits of the eastern's range in western Ethiopia and southern Sudan, respectively. Nonbreeding males, females, and immature individuals of all of paradise whydahs are often indistinguishable in the field, but the longest ornamental tail feathers of breeding males of this species are noticeably longer and more tapering than are those of the broad-tailed paradise whydah. Both species perform prolonged territorial display flights, with their rounded median ornamental tail feathers raised vertically above the others. Vocalizations of these two species are apparently similar, but display postures while vocalizing distinctive (fig. 47). The song of this species is reported to consist of a series of about three monosyllabic and drawn-out sounds, partly shrill and partly euphonious, alternately ascending and descending in pitch. The song of the pytilia host species is prolonged, lasting up to 16 seconds, and begins with a note sounding like a drop of water landing on water, followed by gurgling and trilling phrases, and ending with three fluty notes. Two representative and nearly identical sonograms of the initial portion of the host's and parasite's songs are shown in fig. 48. Payne (1980) has also

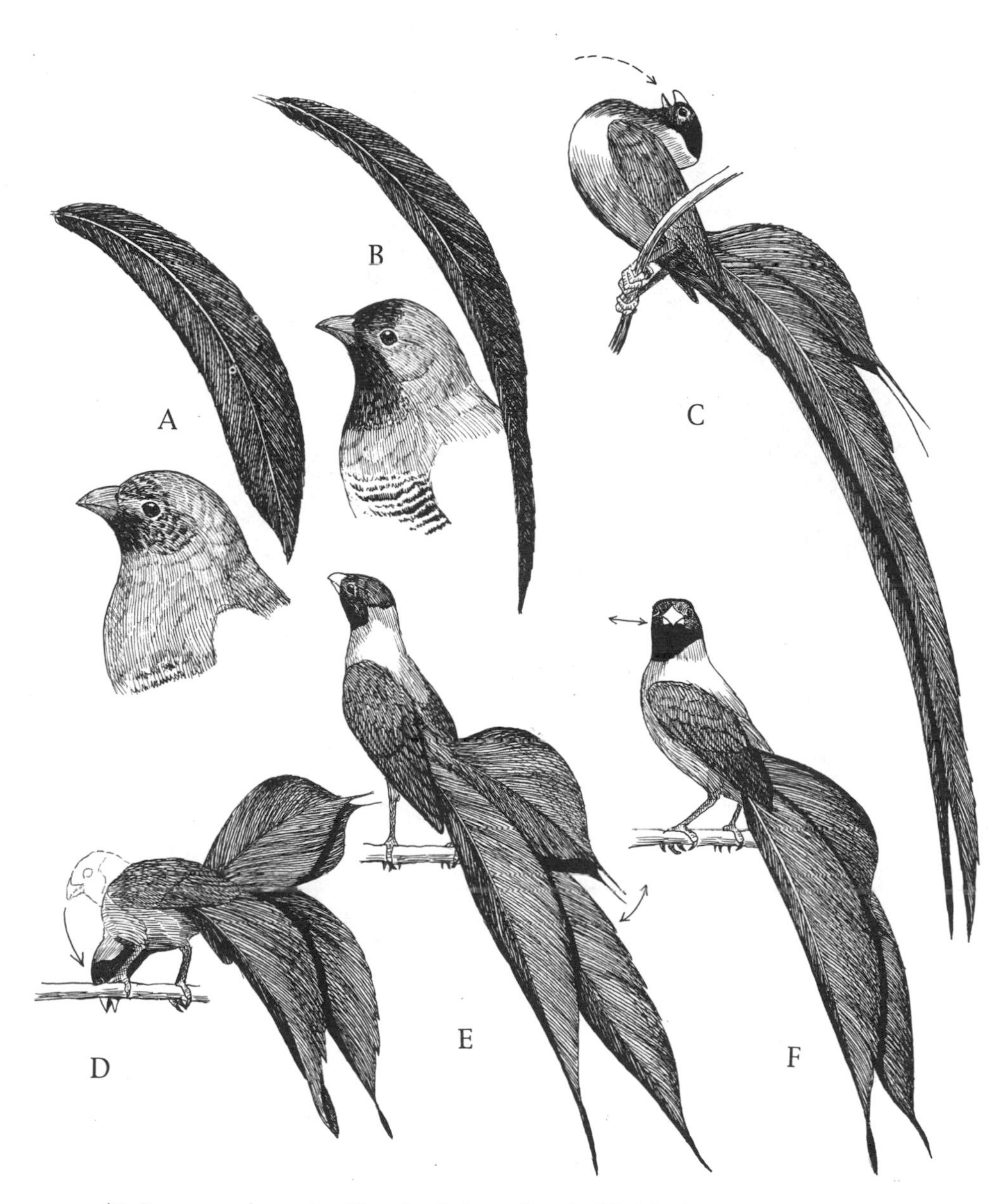

FIGURE 47. Longest male rectrix of broad-tailed paradise whydah (A), shown above its orange-winged pytilia host and beside that of the eastern paradise whydah and its green-winged pytilia host (B). Perched male display postures of the eastern (C) and broad-tailed paradise whydahs (D–F) are also shown. After sketches in Nicolai (1969).

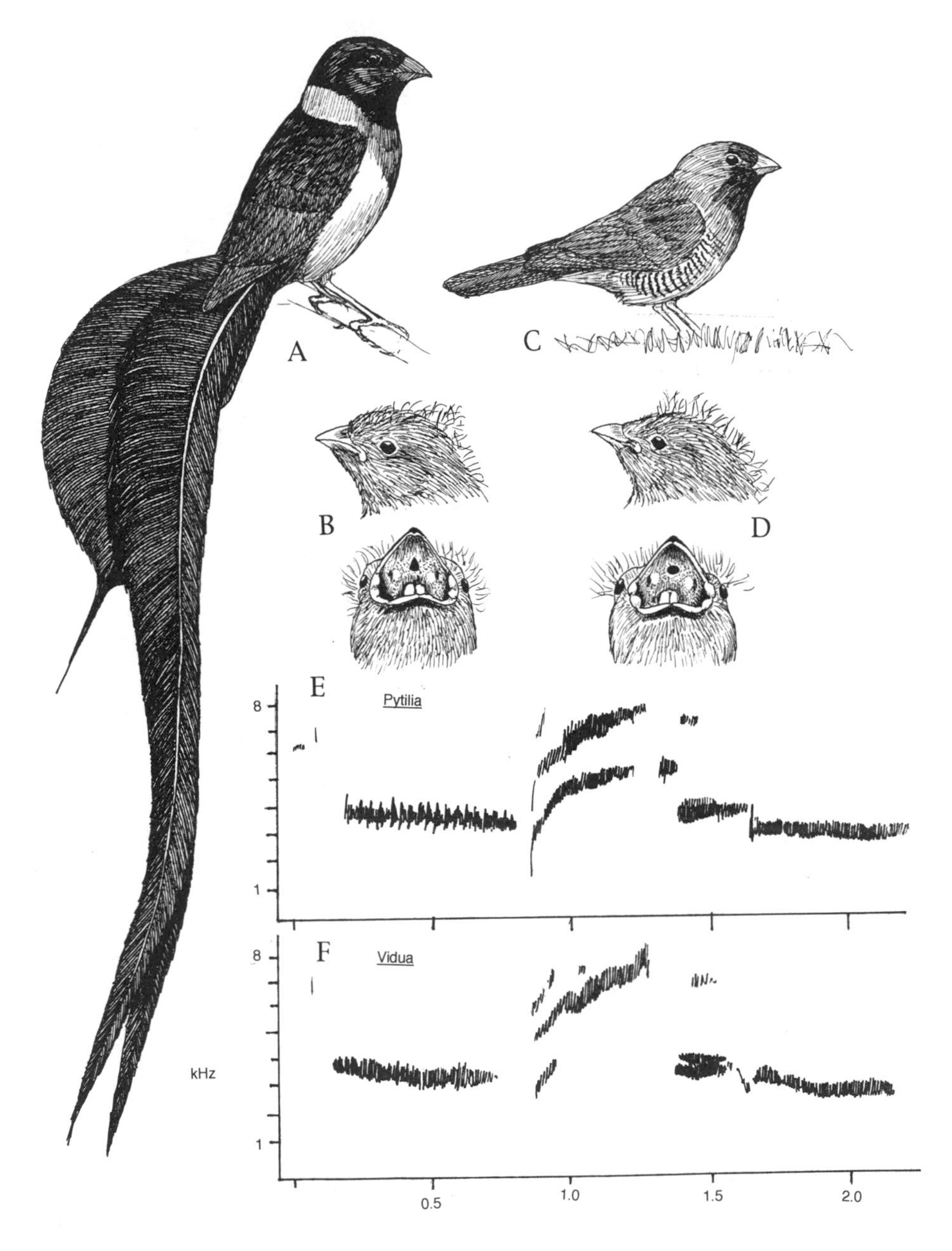

FIGURE 48. Sketches of breeding male (A) and nestling (B) of eastern paradise whydah, compared with an adult (C) and nestling (D) of its green-winged pytilia host. Also shown is a sonogram of a 2-second phrase of the pytilia's advertising song (E), with the whydah's version below (F). After Nicolai (1969).

provided representative sonograms of the pytilia's song as compared with the eastern paradise whydah. The paradise whydahs produced some long introductory whistles, followed by a downslurred note, a shorter rising whistle, and a long series of varied short notes, ending with two downslurred whistles.

In the hand: The gape pattern of nestlings of this species and its green-winged pytilia (or "melba finch") host are shown in figs. 8 and 47. There is a single black median palatal mark, surrounded by a grayish pink horny palate with paired bluish violet, wartlike enlargements, and two silvery white tubercles at the commissural junctions of the bill. The pale tongue is dark tipped, and the lower part of the mouth is blackish. Newly hatched nestlings of the two differ in that the whydah is more dark skinned, has a more grayish (not sandy-white) down, has a more conical and broader bill, and the upper pair of tubercles at the commissural junction are larger than the lower ones (not the same size). In older chicks about to leave the nest, the whydah averages larger than those of the host, is more grayish brown (not uniform olive-gray), lacks the reddish rump of the host chick, and has six (not eight) tarsal scales (Skead, 1975). Juvenile whydahs resemble the host species in their general plumage traits, and their upperpart plumage is correspondingly somewhat greenish toned. By 8 weeks they have acquired the stripped brownish upperparts, prominent head striping, and whitish underparts typical of adult female and nonbreeding male whydahs (Nicolai, 1964, 1974, 1980).

Habitats

Open, mixed woodlands and savannas, especially acacia savannas, from near sea level to about 2100 m elevation are favored. Most numerous below 1500 m, it is especially abundant in open savannas with scattered acacia trees or shrubs. Its host species is mainly found in dry thornbush that is more arid than habitats typically used by the orange-winged pytilia (and parasitized by the broad-tailed paradise whydah), and so the two paradise whydahs are separated ecologically.

Host Species

The only known fostering host of this species is the green-winged pytilia (Nicolai, 1974, 1977). It occupies dry acacia thornbush, open woodlands, semideserts with thorny scrub cover, and cultivated areas with scattered bushes and thorn scrub over a large portion of equatorial and subequatorial Africa. In the sub-Saharan sahel zone, the local red-lored form of the pytilia host is parasitized by the northern paradise whydah rather than by this species, and perhaps this arid-adapted type of pytilia should be recognized as a distinct species (Nicolai, 1964, 1974). The mean adult mass of the green-winged pytilia is 12.8 g (females) to 13.4 g (males) (Skead, 1975), so the whydah has an approximate 60% greater adult mass.

Egg Characteristics

Payne (1977a) noted that the mean estimated egg weight of this species is 1.63 g, as compared with 1.41 g in its host pytilia species, representing a mean difference an approximate 15% greater egg mass in the whydah. The mean linear measurements of the pytilia's eggs range from about 15 × 12 mm in East Africa to 16.4 × 12.5 mm (range 14.7–17.3 × 11.6–13.5 mm) in South Africa, representing a shape index of 1.25–1.31 (= broad oval). Skead (1975) reported mean measurements of 18 × 14 mm (shape ratio 1.28) for the whydah versus 16.7 × 12.5 mm (shape ratio 1.34) for the host in the Transvaal region. He stated that the parasite's eggs may be easily distinguished from host eggs by their larger size and different shape.

Breeding Season

In southern Africa males are in breeding plumage from late November to May or early June, and they mainly breed from January until June. Its pytilia host species similarly breeds mainly from February to June. Somewhat farther north, as in Zambia and Malawi, the birds are in breeding plumage from January or February to July, and in Kenya from about October to March. In Ethiopia the males are in breeding plumage from May to December or sometimes as late as February or March. In all cases the breeding season is probably closely synchronized with the rainy season, and the corresponding breeding season of the host pytilia species. In South Africa this breeding mainly occurs during February and March (extremes November to June), peaking after the heaviest rains and extending into the dry season. The overall host breeding period in Zambia, Malawi, and Mozambique extends from January to June, and in Tanzania, Kenya, and Uganda from at least March to May (but probably also during winter) (Fry et al., 1988). In Nigeria host breeding occurs during August and September, in Sudan it extends from October to February and from May to July, and in Ethiopia it occurs during May and June (Goodwin, 1982).

Breeding Biology

Nest selection, egg laying. Nicolai (1969) reported on 19 pytilia nests containing parasitic eggs, and on 15 additional pytilia nests with parasitic nestlings. Among the nests with parasitic eggs, the number of host eggs ranged from 1 to 5 (mean 3.63). Two unparasitized pytilia nests had four eggs. Skead's (1975) South African data suggest a similar mean host clutch size of 4.1 eggs. Skead further reported finding one whydah egg in each of 11 parasitized nests, two parasitic eggs in each of seven nests, and three pytilia nests with three whydah eggs (mean whydah "clutch," 1.6 eggs). In Nicolai's study area, the number of whydah eggs present per parasitized nest ranged from 1–5 (mean "clutch" 2.3 eggs). A maximum combined clutch of 10 eggs (5 of each species) was present in one nest, suggesting that host eggs are rarely if ever destroyed or removed by whydahs during laying. Payne (1977a) estimated on the basis of ovarian dissections that the usual number of eggs laid by this whydah species per laying cycle averages 3.42, (range 3–4, n = 21), and that about 22 eggs might be laid by a single female during a breeding season.

Incubation and hatching. The incubation period is still unknown, but that of its host pytilia lasts 12–13 days (Goodwin, 1982).

Nestling period. The nestling period lasts 16 days. The fledging period of the pytilia host is 21 days, and the postfledging dependency period lasts another 14 days (Goodwin, 1982). Among 15 parasitized nests containing young, the number of surviving pytilia nestlings varied from 1 to 4 (mean 2.4), and the number of whydahs from 1 to 3 (mean 1.7) (Nicolai, 1969). Thus, in both species the number of surviving host chicks was diminished from the mean clutch size (as observed in other parasitized but unhatched nests) by about 25–35%. The mean age of nestlings at the time of these counts was 9.7 days, or well on their way to fledging and probably beyond their period of highest mortality.

Population Dynamics

Parasitism rate. Nicolai (1969) reported that 13 of 15 pytilia nests found in Tanzania were parasitized during one breeding season and that 34 out of 36 nests were parasitized 3 years later, resulting in an overall parasitism rate of 92%. These are among the highest rates of brood parasitism reported for any avian species. Skead (1975) reported a 28% parasitism rate (21 parasitized nests, presumably from a total sample of about 75 pytilia nests) in central Transvaal.

Hatching and fledging success. No direct information is available, but as noted above, Nicolai (1969) found that the mean number of nestlings (all ages) in 15 parasitized nests averaged 1.7 whydahs and 2.4 pytilias, suggesting a fairly high rate of hatching and fledging success for both parasite and host. Among 4 parasitized nests having chicks at least 10 days old, the average number of surviving young was 1.5 parasitic and 2.6 host chicks. As many as six surviving chicks (both species combined) were present in this group of older young. All of the nests in this group contained at least two surviving host chicks, suggesting that starvation or other means of elimination of the host by the whydah is probably uncommon.

Host–parasite Relations. The hatching and fledging success data of Nicolai (1969) as summarized above suggest that little damage is done to host species productivity by the presence of the whydah, even when two or more whydah chicks are present.

BROAD-TAILED PARADISE WHYDAH
(Vidua obtusa)

Other Vernacular Names: Chapin's paradise whydah.

Distribution of Species (see map 75): Sub-Saharan Africa from Angola, Zaire, and Tanzania south to Zimbabwe and Mozambique, occasionally reaching northern Transvaal.

Measurements (mm)

Female-like birds 5.5–6" (14–15 cm); breeding males 12–14" (31–36 cm)

Wing, males 81–87 (avg. 83.3, $n = 137$), females 77–84 (avg. 79.73, $n = 11$) (Payne, 1991). The longest rectrices in breeding males 176–222 long × 35–37 wide (Chapin, 1954); 175–228 long × 33–41 wide (Payne, 1980).

Egg, avg. 17.9 × 13 (range 17.6–18.2 × 13–13.1) (Schönwetter, 1967–84). Shape index 1.38 (= oval).

Masses (g)

Avg. of six females 19.5 (Payne, 1977a). Estimated egg weight 1.59 (Schönwetter, 1967–84), 1.64 (Payne, 1977a). Egg:adult mass ratio 8.2%

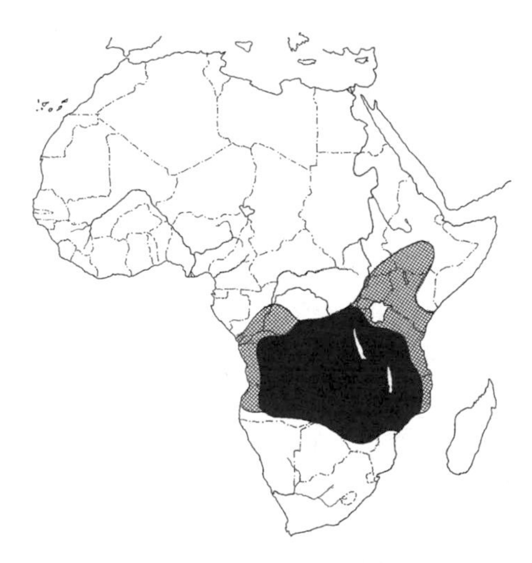

MAP 75. Ranges of broad-tailed paradise whydah (filled) plus host orange-winged pytilia (hatched).

Identification

In the field: Occurring broadly over the southern half of Africa, this species is in extensive sympatric contact with the eastern paradise whydah over about half of its total range, or from Angola east to the Rift Valley of Tanzania and southward across Zambia and Zimbabwe to Mozambique (fig. 46). The longest ornamental tail feathers of breeding males in this species are widest below their mid-point (fig. 47A). These feathers remain fairly wide almost to their tips, which are thus somewhat rounded rather than pointed in shape, and the length of the elongated tail is less than twice that of the body. In the eastern paradise whydah, these ornamental tail feathers are nearly three times the length of the body, and they taper gradually toward their tips. Unlike the eastern paradise whydah, males of this species do not perform display flights (Nicolai, 1977), and perched display postures of the two species may be distinctively different (fig. 47C–F). Nicolai (1964) has provided sonograms of comparative vocalizations of this species and those of its host the orange-winged pytilia. The host species' "whoooeee" call is one that is readily mimicked. Females of the broad-tailed species have relatively pale heads and horn-colored or slightly pinkish bills, whereas those of the eastern paradise whydah have somewhat darker head markings and dark gray to blackish bills, which might allow for field identification of females (and perhaps nonbreeding males) under idea conditions. Young birds may be impossible to identify in the field.

In the hand: Newly hatched young closely resemble those of the host pytilia from the first day, when both are partly covered by loose and sparse grayish down. The palatal markings are also similar between parasite and host. At least many of the orange-winged pytilia host chicks observed by Nicolai had no dark spot in the middle of the reddish upper palate such as occurs in its close relative the green-winged pytilia and which had earlier been reported as typical of this species as well. Rather, they exhibited only paired, wartlike, violet spots on either side of the palate, so perhaps there is some variation in this feature with some young whydahs possibly also having single median palatal spots. The plumages of nestling birds also closely resemble those of the host. Immature birds, females, and nonbreeding males are probably not readily distinguishable from those of the eastern paradise whydah except by such minor plumage differences as those noted above for adult females. Immature birds also closely resemble adult females, but they probably have some buffy feather edges and less distinct upperpart patterning. Females are nearly identical to males in nonbreeding plumage. They perhaps have more brownish (less blackish) head striping and paler rectrices than nonbreeding males, but these apparent differences seem rather subjective and somewhat speculative. Nonbreeding males of the broad-tailed species are more streaked and spotted with blackish on the breast than are those of the northern paradise whydah and also may have a narrower median crown stripe, but these reputed minor plumage differences are still questionable (Friedmann, 1960). Breeding males of the broad-tailed and eastern paradise whydah can be easily distinguished in the hand; the longest rectrices of the broad-tailed paradise whydah are 33–41 mm in maximum width and are so more than about 230 mm in length versus 24–34 mm wide and at least 245 mm long in the eastern paradise whydah. There are perhaps also some minor mensural differences for distinguishing adult females (and nonbreeding males) of these two species, with the broad-tailed whydah having slightly longer wing measurements (Payne, 1971, 1991).

Habitats

Open woodlands, such as the miombo (*Brachystegia*) woodlands of southern Africa, are favored habitats, but this species is also widespread in the acacia savannas of eastern Africa. Where it occurs with the eastern paradise whydah, the broad-tailed is more likely to be found in woodland habitats. Perhaps this is related to its host species' preference for tangled thornbreaks near water, but the host also uses open woodlands, gallery forest edges, and similar mixtures of grasses, bushes, and trees.

Host Species

The only known fostering host species is the orange-winged pytilia (Nicolai, 1977). This species has a body mass averaging 14–15 g, placing it at a substantial mass disadvantage relative to the parasite.

Egg Characteristics

he eggs of this species are white and their average measurements (17.9 × 13 mm) are slightly larger than those of their host pytilia (avg. 16.5 × 12.5 mm). Their estimated respective mean fresh egg weights are 1.64 g and 1.42 g (Payne, 1977a), representing an approximate 15% greater estimated mean egg mass in the whydah.

Breeding Season

In the southeastern Congo Basin, this species develops its breeding plumage in early February and retains it until late July (Chapin, 1954). The host pytilia similarly breeds during April and May in the southeastern Congo. The pytilia also breeds from January to May in Zambia and Zimbabwe, from March to June in Malawi, from April to June in Tanzania and Zanzibar, and probably during June in southern Ethiopia (Goodwin, 1982).

Breeding Biology

Nest selection, egg laying. Little information is available. Nicolai (1969) was unable to locate any host species' nests during his studies. Payne (1977a) determined from ovarian examinations of two birds that in each case three eggs had been laid per laying cycle.

Incubation and hatching. The incubation period is still undetermined, but that of its host species is 12–13 days (Goodwin, 1982).

Nestling period. The nestling period of this whydah is unknown, but that of its host species last 21 days, with an additional 14-day period of postfledging dependency (Goodwin, 1982).

Population Dynamics

No information.

11

PARASITIC COWBIRDS

Tribe Icterini

The cowbirds are a group of six species of passerines that are part of a larger group of "advanced" New World passerines of uncertain taxonomic status that includes approximately 95 species and 23 genera of cowbirds, blackbirds, oropendolas, caciques, and orioles. Orians (1985) regarded the group (which he collectively called "blackbirds") as a distinct family Icteridae. No single collective vernacular name fits this rather diverse group of birds, but the orioles, caciques, and oropendolas make up the largest single component group. One tropical oriole species is the troupial (so called because of its coloniality and sociality), and the term "troupials" is also sometimes used an inclusive vernacular name for the oriole-, cacique-, and oropendolalike birds. In recent years the entire group has increasingly been reduced by taxonomists from familial rank (Icteridae) to subfamilial (e.g., American Ornithologists' Union, 1983) or sometimes (as adopted here) even tribal rank (e.g., Sibley & Monroe, 1990), and this procedure allows for the use of convenient catch-all vernacular term "icterines" to refer to the cowbirds and their relatives.

Sibley and Monroe (1990) regarded the icterines as one of five tribes of birds in the subfamily Emberizinae, these in turn being placed with the larger family Fringillidae. Following tradition, they placed the cowbirds in generic sequence between the typical blackbirds (*Euphagus*) and the bobolink (*Dolichonyx*). Orians (1985) likewise placed the cowbirds in the same linear sequence but regarded them as being most similar to the hypothesized finchlike ancestral type. American Ornithologists' Union (1983) currently places the cowbirds within the tribe Agelaiini, together with the blackbirds, meadowlarks, and grackles, which is tribally separated from and sequentially placed between the monotypic bobolink tribe (Dolichonychini) and that

Female lazuli bunting and nestling brown-headed cowbird. After a photo by P. B. Witherspoon (in Bailey & Niedrach, 1965).

of the oropendolas, caciques, and orioles (Icterini). Structurally, the cowbirds and other icterines differ from the sparrowlike birds in that they lack rictal bristles at the bases of their bills. Friedmann (1929) stated that the cowbirds may be further characterized by their relatively short, stout bills, the bills having small nostrils with dorsal operculums and with feathering reaching their posterior margins. In this regard the giant cowbird is somewhat of an exception, as its bill is fairly long and closely approximates the configuration typical of some caciques.

Of the six species of cowbirds, five are obligate brood parasites, and the sixth, the bay-winged cowbird, raises its own young but nests almost exclusively in the nests of other species. Such nest-takeover behavior represents a trait that also sporadically occurs in various other icterine species. Friedmann (1929) regarded the bay-winged cowbird as behaviorally and structurally the most primitive cowbird type, characterized by a "female" type of coloration in both sexes, no courtship display, and a distinctive type of song that is acoustically simple and uttered by both sexes. He visualized the other species of cowbirds as forming a progressive series of parasitic stages. The screaming cowbird was considered a direct offshoot of ancestral bay-winged cowbird stock that was originally nonparasitic but became progressively parasitic as the males' weakened territorial instincts became disconnected with the egg-laying instincts of the females. The entirely parasitic shiny cowbird was regarded as having its "parasitic habit very poorly developed" because it "no longer knows how" to build a nest and because of its "wasteful" egg-deposition behavior. The bronzed cowbird was regarded as a direct offshoot from relatively primitive cowbird stock. The giant cowbird was regarded simply as a "large edition" of the bronzed cowbird and an extension of its phyletic line. The brown-headed cowbird was regarded as relatively efficient in its egg-deposition behavior, and thus was considered as being more advanced than the shiny cowbird. An alternative scenario has recently been proposed by Lanyon (1992), who suggested on the basis of DNA studies of the mitochrondrial gene for cytochrome-b that host specificity represents the primitive condition in cowbirds, rather than being the derived situation. Lanyon also concluded that the genus *Molothrus* as currently constituted is polyphletic, since the giant cowbird's placement on the phylogram fell between the screaming cowbird and the three remaining brood-parasitic species. The bay-winged cowbird was not included in this phylogram, and Lanyon stated that his DNA evidence does not provide any information as to its proper placement with respect to the other cowbirds.

Parasitic Cowbirds

SCREAMING COWBIRD

(Molothrus rufoaxillaris)

Other Vernacular Names: None in general English use.

Distribution of Species (see map 76): South America from Bolivia, Paraguay, and Brazil south to Uruguay and Argentina.

Measurements (mm)

7–8" (18–21 cm)

Wing, males 108–115 (avg., 112), females 104–105 (avg. 104.7). Tail, males 77–82 (avg., 79.2), females 79–82 (avg. 80.3) (Friedman, 1929). Wing:tail ratio 1:0. 70–0.77.

Egg, avg. 23 × 18 (21–23 × 17–19) (Friedmann, 1929). Shape index 1.28 (= broad oval). Rey's index 1.29, also 1.15–1.5, avg. 1.28 (Hoy & Ottow, 1964).

Masses (g)

Four males 56–65; females 38, 57 (Dunning, 1993). Four males 61–66 (avg. 63.25), five females 48–52 (avg. 50.2) (Fraga, 1979). Range (unsexed) 47–63 (Sick, 1993).

MAP 76. Range of screaming cowbird (filled), plus the additional range of bay-winged cowbird host (dashed line).

FIGURE 49. Profile sketches of both sexes of adult brown-headed (A), bronzed (B), shiny (C), and giant cowbirds (E), plus an adult and nestling of the screaming cowbird (D). Egg morphs, dorsal bill outlines and outer primary vane configurations (A & B only) are also shown.

Estimated egg weight 3.64 (Schönwetter, 1967–84). Egg:adult female mass ratio 7.6%.

Identification

In the field: Within its rather limited South American range, this species can sometimes be distinguished by its chestnut under-wing coverts and axillaries, although these are usually visible only when the birds are in flight. It is often found near the slightly smaller bay-winged cowbird, which has black-tipped wings that otherwise are chestnut-colored, and a somewhat shorter tail. Both sexes also resemble the shiny cowbird, but they are larger, have longer tails but shorter, more stubby, bills, and have less iridescent plumage (fig. 49). The juvenal plumage is much browner than that of adults, exhibits varying amounts of faint streaking on the underparts, and is more like that of the brown-headed cowbird. The male performs a song-spread courtship display by spreading his wings and tail horizontally and bowing forward while uttering a harsh "tsi-LIT-chech." His head is stretched forward, not tilted down to touch the throat, and his body is not stretched vertically upward, as seen in the shiny cowbird. Other harsh notes are also produced at other times.

In the hand: This species is easily distinguished by its chestnut brown under-wing coverts and axillarics, which are first acquired with the first winter plumage. Apart from their slightly smaller wing, tail, and bill measurements (culmen from base under 16 mm, rather than 16–18 mm), females cannot readily distinguished externally from males. In both species the mouth interior is reddish, the mandibular flanges are white, and the bill is pinkish yellow (screaming) or pinkish orange (bay-winged). However, the bill of the bay-winged has a darker pigmented area around the egg-tooth that is lacking in the screaming cowbird, and the skin of the nestling bay-winged cowbird is orange, whereas the skin of newly hatched screaming cowbirds is pink to pale pink (Fraga, 1979). Juveniles are virtually identical to those of their host the bay-winged cowbird, with both having much rufous-chestnut edgings on their greater wing coverts and flight feathers. However, during the postfledging period, young screaming cowbirds soon darken to black, whereas those of the bay-winged cowbirds remain dark brownish for the first 2–3 months (Fraga, 1979). As they molt their juvenal feathers and increasingly acquire their first winter plumage, black feathers begin to appear on the head and body.

Habitats

This species occupies open country having only scattered trees, especially cattle ranches, from about sea level to 1000 m, in tropical to temperate climates. It is also common around freshwater marshes.

Host Species

The only known biological host species of the screaming cowbird is the bay-winged cowbird. There are a few possible records of eggs being laid in the nests of other species (Friedmann, 1929; Fraga, 1984).

Egg Characteristics

The eggs are usually broad oval in shape but are highly variable in color, ranging from reddish, through bluish, greenish, grayish, and yellowish to white, without any tendency toward host mimicry (Sick, 1993). The surface markings range from grayish brown to purplish brown. How & Ottow (1964) reported that around Salta, nearly all eggs were white to bluish white and bluish green in ground color, but around Rio de Janeiro more than half were of other colors. The surface markings range from reddish brown (in most eggs) to brown or (rarely) greenish brown, and the underlying spots are grayish. The eggs of the bay-winged cowbird host average 24 × 18 mm (range 21–26 × 16.5–20 mm, mean shape index 1:33), with grayish brown to purplish brown markings that are usually more scattered and sharply defined. Additionally, the screaming cowbird's eggs are harder to pierce than are those of its host the bay-winged (Friedmann, 1929). The mean egg mass of the screaming cowbird is about 12% less than that of the bay-winged, but the screaming cowbird's mean shell mass (0.32 g) and shell thickness (0.127 mm) are either equal to the shell weight or somewhat greater than the shell thickness of the bay-winged cowbird eggs (Rahn et al., 1988). The Rey's index of the bay-winged eggs correspondingly averages 1.62, as compared to 1.28–1.29 for the parasite (Hoy & Ottow, 1964).

Breeding Season

In the Lerma Valley of central Argentina this species breeds during the wet season, which typically begins in January or February. Most birds breed there during March, and the bay-winged cowbird's nesting period is also concentrated between mid-February and mid-March. The screaming cowbird's nesting season is generally synchronized with that of the bay-winged, but may at times begin earlier, and then they are forced to lay in old, empty nests (Hoy & Ottow, 1964).

Breeding Biology

Nest selection, egg laying. This species is essentially a single-host parasite, laying its eggs exclusively in the nests used (but usually not built) by bay-winged cowbirds. The eggs are deposited at daily intervals, and bay-winged cowbird eggs are neither destroyed nor removed at the time of laying. However, the host often evicts alien eggs from the nest, at least those that are deposited before its own egg-laying has begun. The usual clutch size of the host cowbird is five eggs, which are usually also laid at daily intervals (Friedmann, 1929). A significant percentage (about 15%) of the screaming cowbird's eggs are deposited before the first host eggs are laid (Fraga, 1986); these eggs are regularly ejected by the host. Typically, from 6 to 20 screaming cowbird eggs may be present in a single bay-winged cowbird's nest, and up to as many as 12 females may parasitize a single nest, according to Hoy & Ottow (1964). (Friedmann had believed that only a single female screaming cowbird might parasitize a bay-winged's nest and that perhaps only five eggs, representing a single clutch, are laid per season by any such female, but both of these conclusions seem unlikely in view of the recent observations of Hoy and Ottow.) In at least three cases, Hoy and Ottow (1964) noted that two eggs were laid in a nest by the same cowbird female, and in one case a single female probably had deposited three. Sometimes more than one bay-winged cowbird will lay in the same nest as well; in one nest a female bay-winged had an egg added from each of two other bay-winged females, plus the usual addition of parasitic screaming cowbird eggs. These authors reported finding a nest with 5 bay-winged cowbird eggs, plus 14 of the screaming cowbird, many of the latter having been thrown out of the nest-cup by the owners.

Incubation and hatching. The incubation period is 12 (occasionally) or 13 (usually) days, or about the same as the 13-day period of the host species (Friedmann, 1929). At hatching, the young of the two species are virtually identical in appearance and weight. Friedmann noted that a newly hatched screaming cowbird weighed 2.4 g, versus 2.3 g for a just-hatched bay-winged. One newly hatched screaming cowbird lacked down on the femoral tract, whereas on a bay-winged chick this area was sparsely downy. Considering that the eggs of these species are nearly indistinguishable and that a single female typically lays several eggs in the same nest, the lack of egg-destruction behavior is not surprising. Young of both species are raised together in the nest, and there is no evidence of interspecies antagonism among nestlings (Friedmann, 1929).

Nestling period. The nestling periods of the screaming and bay-winged cowbirds are apparently the same, 12 days. At that stage the young of the two species are nearly identical in appearance. After leaving their nests, young bay-winged cowbirds are cared for by their parents for at least two additional weeks (Friedmann, 1919); it is likely that a similar period of postfledging dependency is typical of the screaming cowbird.

Population Dynamics

Parasitism rate. Fraga (1986, 1988) found an 87% parasitism rate for 79 nests of bay-winged cowbirds in Argentina; many of these same nests (24%) were additionally parasitized by shiny cowbirds. Friedmann (1929) noted that a local resident of Tucuman reported having found 66 nests of bay-winged cowbirds over a 20-year period, all of which had contained one or more eggs of the screaming cowbird. Screaming cowbirds often begin laying in the nests before the host bay-winged does, although such eggs are regularly evicted. Even if the bay-winged subsequently abandons its nest, screaming cowbirds may continue to deposit eggs in it (Hoy & Ottow, 1964).

Hatching and fledging success. Mason (1980) reported a relatively low hatching rate of less than 13% for screaming cowbird eggs in bay-winged nests and regarded the host-specific parasitic adaptations of the screaming cowbird as maladaptive. Fraga (1986) reported an overall egg-to-fledging success rate of 7.3% for screaming cowbird eggs in bay-winged nests, as compared with a success rate of 22.4% for the host species' eggs. Fraga (1984) observed that larvae of botflies and other ectoparasites are removed from the young of screaming cowbirds by their host-species nestmates and suggested that this might be a factor selecting for host-specific behavior on the part of the screaming cowbird. Evidently many eggs of the screaming cowbird are evicted by host bay-winged cowbirds, including all of those laid before its own clutch is begun, and sometimes also at least some of those that were deposited afterwards. The extremely large combined clutches that often develop, sometimes numbering more than 20 eggs, doubtless have a low rate of hatching success, but specific information is still lacking.

MAP 77. Historic range (filled) of shiny cowbird, plus the acquired range in the West Indies (shaded), showing chronology of northward breeding range expansion. Some recent U.S. sight records beyond Florida are also shown.

Host–parasite relations. Hoy and Ottow (1964) stated that bay-winged cowbirds regularly evict any eggs of screaming cowbirds that may have been laid before the start of their own clutch. Furthermore, when the nest-cup has been filled with eggs, the bay-winged hosts sometimes try to evict the eggs of the screaming cowbird, but may abandon the nest if unsuccessful. These authors also noted that when the bay-winged cowbirds selected the smallest available thornbird nests for laying their own eggs, they escaped parasitism. The same was true of a clutch laid in a woodpecker nest and of one laid in an apparently self-constructed nest. Fraga (1986) estimated that parasitism by shiny and screaming cowbirds reduced the breeding success by egg losses (39% of host eggs were destroyed or removed by the parasites) and by nestling competition (11% of the host nestlings died in parasitized nests). Evidently screaming cowbirds are more effective parasites of bay-winged cowbirds than are shiny cowbirds, at least in part because of the greater resemblance between the screaming and bay-wing nestling.

SHINY COWBIRD

(Molothrus bonariensis)

Other Vernacular Names: Glossy cowbird.

Distribution of Species (see map 77): Tropical and temperate South America south to Chile and southern Argentina. Also West Indies, where it was historically confined to the Lesser Antilles, but is now resident throughout and is currently colonizing southern Florida.

Subspecies

M. b. bonariensis: Eastern and southern Brazil, eastern Bolivia, Paraguay, Uruguay, and Argentina to Chubut; also central Chilean lowlands (Coquimbo to Valdivia).

M. b. riparius: Eastern Peru.

M. b. minimus: Brazil, Guianas, West Indies. Recently also reaching peninsular Florida, where it is now resident along the southern coast and expanding northward. Scattered occurrences elsewhere in the eastern USA from Texas to Maine.

M. b. venezuelensis: Eastern Colombia, Venezuela.

M. b. occidentalis: Southwestern Ecuador, western Peru.

M. b. aequatoralis: Western Ecuador, southwestern Colombia.

M. b. cabanisii: Panama, northern and western Colombia.

Measurements (mm)

7–8.5″ (18–21 cm)

M. b. bonariensis, wing, male avg. 114.5; tail, male avg. 82.6 (Friedmann, 1929). Wing:tail ratio 1:0.72.

M. b. cabanisii, wing, males 123.5–135, females 98.6–111.5. Tail, males 97–107.2; females 78.5–89.1 (Wetmore, 1984). Wing:tail ratio ~1:0.76.

M. b. occidentalis, wing, male ave. 109.6; tail, male avg. 83.5 (Friedmann, 1929). Wing: tail ratio 1:0.76.

M. b. minimus, wing, males 94–100; tail, male 75 (Friedmann, 1929). Wing:tail ratio ~1:0.77.

M. b. venezuelensis, wing, male avg. 112, tail, male avg. 87 (Friedmann, 1929). Wing:tail ratio 1:0.77.

Egg, avg. for species 23 × 19 (range 22–26 × 18–20) (Friedmann, 1929). Avg. of 302 *bonariarensis.* 22.7 × 18.1 (Mermoz & Reboreda, 1994). Avg. of 235 *minimus,* 20.65 × 16.46 (Wiley, 1988). Shape index 1.21 (= broad oval). Rey's index (for *minimus*) 1.54.

Masses (g)

Avg. of 479 males 38.7, of 670 females 31.9 (Dunning, 1993). Avg. of 21 nominate *bonariensis* males 55.5, of 31 females 45.6 (Mermoz & Reboreda, 1994). Avg. of 80 *minimus* males, 39.6 (Post et al., 1993). Five *minimus* males 38–41.5 (avg. 39.5), 7 females 29.5–33.5 (ffrench, 1991). Males of *minimus* 38–40, females 28–36 (Haverschmidt, 1968). Breeding females of *minimus* avg. of 18, 33.3 (Wiley, 1988). Estimated avg. egg mass 2.87 (*minimus*) to 4.9 (*cabanisii*): species mean 3.01 (Rahn et al., 1988). Egg:adult female mass ratio 9.4%.

Identification

In the field: This cowbird is notable for its lack of red eyes and a neck-ruff in both sexes, features that are typical of the similar bronzed cowbird (fig. 49). Instead, it has dark brownish black eyes and an entirely bluish black plumage (males), or is rather dull grayish brown above, grading to a paler, faintly streaked buff and grayish brown below (adult females). Immature individuals resemble adult females, but are more yellowish buff below, and the dorsal feathers have brownish buff edgings. They are also less streaked below than are adult females. Adult females resemble those of the brown-headed cowbird, but have slightly sharper (less robust) bills. The song of the male is a clear whistle, and when displaying it ruffles its feathers, touches its throat with its bill, and utters a prolonged "prro-prro-pro-TSLEE-yew." When circling a female in aerial display, the male may also utter a prolonged tinkling song.

In the hand: Neither sex of this species has a neck-ruff, hairlike feathers on the neck or upper back, nor toothlike projections on the inner vanes of the outer primaries, all of which are typical of bronzed cowbirds. The iris color in both sexes is brown, not red. Males differ from the brown-headed cowbird in lacking brown head coloration and instead are glossy purplish to violet-black throughout. Adult females resemble female brown-headed cowbirds in being generally dull brown, but the bill is less robust (under 10 mm high at its base where feathering begins), and they have a more yellowish superciliary stripe. Immature birds of both sexes resemble adult females but have even brighter

TABLE 32 Reported Host Species of the Tropical American Cowbirds[a]

Screaming Cowbird (44.5 g, mean host mass 107%)

Bay-winged cowbird, M (107%, O, Sp)

Shiny Cowbird (35 g, mean host mass 96%)

Rufous hornero, M (160%, S, Im)
Olive spinetail (37%, S, Im)
Short-billed canastero (51% S, Im)
Firewood-gatherer (117%, S, Im)
Collared antshrike (85%, O)
Black-tailed tityra (209%, C)
Short-tailed field tyrant (36%)
Yellow-browed tyrant (43%)
White-headed marsh tyrant (40%, S, Im)
Cattle tyrant, M (95%, C)
Fork-tailed tyrant, M (81%, O, Sp)
Tropical kingbird (106%, O, Bl)
Crowned slaty flycatcher (77%)
White-bearded flycatcher, M (O, Sp)
Great kiskadee (173%, S, Sp)
Yellow-bellied tyrannulet (21%)
White-rumped swallow (54%, C, Im)
Bicolored wren (120%, S)
Stripe-backed wren (66%)
Rufous-breasted wren (46%, S, Bl)
Superciliated wren
House wren, M (31%, C, Sp)
Long-tailed mockingbird (189%, O)
Chalk-browed mockingbird, M (207%, O, Sp)
Patagonian mockingbird (161%, O, Sp)
White-banded mockingbird (146%, O)
Rufous-bellied thrush (195%, O, Sp)
Creamy-bellied thrush (179%, O)
Masked gnatcatcher (O, Sp)
Rufous-browed peppershrike (81%, O, Sp)
Puerto Rican vireo (32%, O, Sp)
Black-whiskered vireo (51%, O, Sp)
Yellow warbler (27%, O, Sp)
Scrub blackbird (O, Sp)
Carib grackle (170%, O, Sp)
Greater Antillian grackle (210%, O, Bl)
Chestnut-capped blackbird, M (90%, P, Sp)
Yellow-shouldered blackbird, M (108%, O, Sp)
White-edged oriole
Black-cowled oriole, M (121%, P, Im)
Long-tailed meadowlark, M (320%, O, Bl)
Red-breasted blackbird, M (116%, O, Bl)
Bicolored conebill (30%, O, Bl)
Palm tanager (83%, O, Sp)
White-rumped tanager (83%, O, Sp)
Guira tanager (34%, O, Sp)
Sayaca tanager (91%, O, Sp)
Blue-and-yellow tanager (101%, O, Sp)
Silver-beaked tanager (79%, O, Sp)
Brazilian tanager, M (93%, O, Sp)
Grayish saltator (155%, O, Bl)
Colden-billed saltator (156%, O)
Streaked saltator (104%, O, Sp)
Cinereous finch, M (O)
Double-collared seedeater (31%, O, Sp)
Common diuca-finch, M (9%, O, Sp)
Ochre-breasted brush-finch (O)
Tumbes sparrow (O)
Rufous-collared sparrow, M (58%, O, Bl)
Grassland sparrow (48%, O, Im)
Long-tailed reed-finch
Black-and-rufous warbling-finch (54%)
Hooded siskin (O, Sp)

Bronzed Cowbird (~60 g; mean host mass 50%)

Green jay (126%, O, Bl)
Carolina wren (34%, C, Sp)
Plain wren (30%, S, Im.)
Bewick's wren (16%, C, Sp)
Orange-billed nightingale thrush (43%, O, Bl)
Northern mockingbird (78%, O, Bl)
Long-billed thrasher (112%, O, Sp)
Red-eyed vireo (27%, O, Sp)
Slaty vireo (20%, O, Sp)
Tropical parula (11%, O, Sp)
Golden-cheeked warbler (16%, O, Sp)
Scarlet-rumped tanager (51%, O, Bl)
Orchard oriole, M (32%, O, Sp)
Hooded oriole, M (39%, O, Sp)
Northern oriole, M (55%, P, Sp)
Black-headed oriole, M (68%, O, Sp)
Olive sparrow (38%, S, Im)
Rufous-sided towhee (65%, O, Sp)
Brown towhee (71%, O, Sp)
Song sparrow (33%, O, Bl)
Rufous-collared sparrow (33%, O, Bl)
Northern cardinal (72%, O, Bl)

(continued)

TABLE 32 (*continued*)

Red-crowned ant-tanager (52%, O, Bl)	Yellow-throated brush-finch (52%)
Summer tanager (45%, O, Sp)	White-eared ground-sparrow (69%, O, Im)
Flame-colored tanager (56%, O, Sp)	Prevost's ground-sparrow (45%, O, Bl)
Red-headed tanager (35%)	
Giant Cowbird (144 g; mean host mass 78%)	
Chestnut-headed oropendola, M (77%, P, Sp or Bl)	Montezuma oropendola, M (117%, P, Sp)
Russet-backed oropendola (123% P, Sp)	Yellow-rumped cacique, M (45%, P, Sp)
Crested oropendola, M (79%, P, Sp)	Red-rumped cacique, M (35%, P, Sp)
Green oropendola, M (111%, P, SP)	Green jay (41%, O, Bl)

[a]Species listed mainly after Friedmann & Kiff (1985), plus more recent references (e.g., Perez-Rivera, 1986; Cavalcanti & Pimentel, 1988). Only fostering hosts are shown for the bronzed and shiny cowbirds. Known major host species are identified by M. Mean adult host masses are usually shown as percentages of mean adult cowbird mass, but those for giant cowbirds and their hosts are based on females only. The giant cowbird's mean mass is uncertain (see text); using Smith's (1979) estimate (74 g), the host percentages would roughly double. Nest and egg types are coded for most species. O, open, cuplike; P, pensile or pendulous; and S, spherical or roofed. The egg types are Im, immaculate; Sp, spotted or streaked; and Bl, blotched. These trait summaries are not based on an exhaustive literature search, and some are incomplete.

yellowish superciliary stripes and buffy feather edging on the upperparts and are more or less streaked with buffy below. The skin of newly hatched nestlings is flesh-pink, and the upper mandible is slightly duskier than the lower one. Nestlings have deep red, orange-red or pinkish mouth linings and white to pale yellow mandibular flanges. Tufts of blackish down are present in newly hatched young, which helps to separate them from at least some host icterines (Mermoz & Reboreda, 1994). They are similar in appearance to those of the brown-headed cowbird (which is potentially sympatric in southern Florida), and effective distinguishing criteria remain to be established.

Habitats

This adaptable species occupies coastal mangroves, freshwater swamps, cultivated fields, pastures, recently deforested areas, and other partly wooded or open-canopy landscapes, especially where scattered shrubs or trees and livestock such as cattle are present. It occurs from sea level to about 3500 m, but is mostly found below 2000 m. In tropical to temperate climates. Based on radio-tagging and direct observations, Woodworth (1993) reported that females maintain breeding ranges but not defended territories in open-canopy forest interiors and range out about 4 km each day between breeding areas and surrounding foraging sites. Cowbirds densities were not found to be related to distances from forest edge, and thus in contrast to the brown-headed cowbird, there is no obvious "edge effect" that might help provide host protection from parasitism by nesting in forest interiors.

Host Species

A total of 63 probable biological hosts are listed in table 32, mostly on the basis of the summary by Friedmann & Kiff (1985), who have documented more than 200 host species. Sick (1993) listed nearly 60 host species from Brazil alone, but over much of southern Brazil the commonest host is the rufous-collared sparrow. In Trinidad at least 22 hosts are known, with several genera of icterines (*Agelaius, Sturneila*) and wrens (*Troglodytes*) especially important (ffrench, 1991).

Egg Characteristics

This species' eggs are quite variable in shape, from nearly spherical to almost elliptical, but on average are broad oval. In color, they range from white to whitish green in ground color, usually with markings (flecks, spots, blotches) of bright reddish, bright brown, or pale violet. However, some eggs may be entirely unmarked, and others are almost entirely a deep red color. In eastern Argentina, Uruguay, and parts of Brazil there are two distinct egg morphs (spotted versus immaculate), with few finely spotted intermediates present. Such dimorphism is not known in Venezuela or the West Indies, where the eggs are consistently freckled or spotted. There is evidently no clear trend toward regional host mimicry of particular host eggs and associated evolution of host-specific gentes, but differential egg-recognition and rejection behavior by at least one important host species (chalk-browed mockingbird) has been documented (Mason, 1986b). The shell mass (0.21–0.327 gm) and shell thickness (0.113–0.143 mm) are significantly greater than eggs of nonparasitic relatives (Rahn et al., 1988). Hoy & Ottow (1964) pointed out that although having a thick shell may be a great advantage to parasitic species, the presence of thickened shells among cowbirds is not necessarily the result of selection for adaptations favoring parasitism and instead may reflect the retention of an ancestral icterine trait associated with the construction of hanging nests and a resulting possible advantage in having strengthened eggshells.

Breeding Season

In Brazil this species has a lengthy breeding period, lasting about 6 months (Sick, 1993). Friedmann (1929) reported that the peak of the laying season in Argentina occurs during the second half of January and early February, but eggs have been found as early as mid-November and nestlings seen as late as early March. In the Lerma Valley of Argentina the overall cowbird breeding season extends from the end of October to early February, a span of about 4 months, but the most important nesting month for host species does not occur until March (Hoy & Ottow, 1964). In the Cauca Valley of Colombia the cowbird has a 9-month laying season (October–June), interrupting its breeding activity only during the dry season. (Kattan, 1993). In Trinidad, breeding has been recorded for nearly every month between May and January (Manolis, 1982; ffrench, 1991). However, Wiley (1988) found that in Puerto Rico the cowbirds were able to sustain their reproductive output throughout the egg-laying seasons of their major hosts (from late March through early August).

Breeding Biology

Nest selection, egg laying. This species is relatively nonselective in choosing its hosts and seems to deposit its eggs in any available nests of most small passerines nesting in its range. In Puerto Rico cowbirds locate host nests by furtive watching as well as by active searching and flushing of hosts. They closely monitor the status of host nests, and the peak in nest visits occurs on the host's first day of laying. Covered nests (domed or cavity nests) are as vulnerable to cowbird parasitism as open nests (Wiley, 1988). Eggs are laid at daily intervals; female cowbirds sometimes deposit more than one egg in a single nest, and several females frequently lay in the same nest (Friedmann, 1929). Some notably large clutches were reported by Miller (1917), who observed nests with as many as 37 cowbird eggs. He also found individual nests that had apparently been parasitized by as many as 12 and 13 separate females. One chalk-browed mockingbird nest mentioned by Hoy & Ottow (1964) had apparently been parasitized by 14 different females. Fraga (1978) reported that 29 parasitized nests of the rufous-collared sparrow found in eastern Argentina contained an average of 2.03 cowbird eggs and 2.14 sparrow eggs. A similar mean of 2.06 cowbird eggs was observed by King (1973) among parasitized nests found in northwestern Argentina. He (King, 1973) reported a mean of 2.29 sparrow eggs and 2.06 cowbird eggs in 17 parasitized nests as compared with 2.56 sparrow eggs in 9 unparasitized nests. Similarly, Sick & Ottow (1958) estimated a mean clutch size of 1.53 sparrow eggs and 1.84 cowbird eggs in 51 parasitized nests, as compared with 2.31 sparrow eggs in 32 unparasitized nests. It is likely that some of this host clutch size reduction results from egg destruction by visiting female cowbirds (Friedmann, 1929).

Mason (1986b) reported a wide range in host-species utilization, parasitism frequency, and parasitism intensity in two Argentine study areas. The chalk-browed mockingbird was found to be a favored host in both areas, with from 1–10 cowbird eggs present per parasitized nest, and a mean parasitism intensity of 2.64 cowbird eggs (in 98 parasitized nests) in the two sites collectively. Based on examination of ovarian follicies, Kattarr (1993) estimated a daily mean egg-laying rate of 0.66 egg. A mean of 3.2 eggs were estimated to be laid by each female cowbird per egg-laying cycle, followed by a nonlaying interval averaging 1.64 days. Over the remarkable 9-month breeding season typical of Colombia's Cauca Valley, it is possible that up to 120 eggs might thus be produced annually by a single female, representing an almost unbelievably high rate of fecundity for any wild bird.

Incubation and hatching. Friedmann (1929) reported an incubation period of 11–12 days, usually 12. Salvador (1984) reported an incubation period of 12–13 days when incubated by chalk-browed mockingbirds (which have an incubation period of 14–15 days). Fraga (1978) reported a 12-day period when incubated by rufous-collared sparrows, and a 11.5– to 12-day period under various other hosts. Mermoz & Reboreda (1994) reported a 11- to 13-day incubation period, with the cowbird chicks usually hatching before those of the host brown-and-yellow marshbirds, which have a 14- to 15-day incubation period.

Nestling period. Friedmann (1929) reported a usual 10-day nestling period, with the birds sometimes leaving the nest on the ninth day if frightened. Salvador (1984) reported a nestling period of 13–14 days under chalk-browed mockingbirds (whose corresponding fledging period was 14–15 days). Fraga (1978) reported a 12- to 13-day nestling period for birds reared by rufous-collared sparrow and 13–15 days with other host species.

Population Dynamics

Parasitism rate. Friedmann (1929) reported parasitism incidences for several host species, but most of these involved small sample sizes. He noted that the shiny cowbird's most common single host is probably the rufous-collared sparrow. He found 33 nests of this species, including 24 that were parasitized (72%), 5 that were unparasitized, and 4 that were empty. Sick & Ottow (1958) estimated a 61% parasitism rate among 93 nests of this species in eastern Argentina. Fraga (1978) reported a similar 72.5% rate among 40 Argentine nests of this species that were found prior to hatching. At the peak of the sparrow's breeding season, all of its nests found were parasitized. King (1973) similarly estimated an overall 66% parasitism rate among 50 rufous-collared sparrow nests in northwestern Argentina, and a 100% parasitism rate at the peak of the sparrow's breeding season. Studer & Vielliard (1988) reported a 100% parasitism rate (and no host chicks fledged) among 21 nests of the Forbes's blackbird in 1987; in 6 earlier years the parasitism rate had averaged 64%. In central Argentina, Salvador (1984) found an 88% parasitism rate for 92 nests of the chalk-browed mockingbird. This is a highly preferred host species that is an unusual host choice inasmuch as it is larger in mass than the cowbird and thus is a competitive advantage during the nestling period. Mason (1986a) also found a high incidence of parasitism (73.5%) for 68 nests of this mockingbird in eastern Argentina. Wiley & Wiley (1980) found a 33% parasitism rate among 87 nests of the yellow-hooded blackbird in Trinidad and Venezuela, and noted that small colonies sometimes receive 100% parasitism, leading to colony abandonment. Cruz et al. (1990) reported a 40.3% rate of parasitism among 377 yellow-hooded blackbird nests (those found before the nestling stage) in Trinidad. Fraga (1986, 1988) reported an overall parasitism rate of about 24% in sample of 79 nests of the bay-winged cowbird, which were also heavily parasitized (parasitism rate about 87%) by screaming cowbirds. A 74.3% parasitism rate among 74 nests of the brown-and-yellow marshbird was reported by Mermoz & Reboreda (1994). Wiley (1988) reported that 9 of 29 nesting species that he observed in Puerto Rico were parasitized, but he found no correlation between parasitism frequency and the host's relative abundance or its type of nest structure.

Hatching and fledging success. Salvador (1984) reported a 6.45% egg-to-fledging breeding success rate for 31 eggs, as compared with a 7.7% success rate for 39 chalk-browed mockingbird host eggs. Fraga (1986) reported that only one shiny cowbird fledged from 19 parasitized bay-winged cowbird nests, representing a 5% success rate. Of 59 cowbird eggs in 29 parasitized rufous-collared sparrow nests, only 10 (17%) hatched, and only four (6.8%) young fledged (Fraga, 1978). Mason (1986a) provided egg-to-fledging breeding success rates (= "survivorship estimates") relative to 15 host species, ranging from 78.3% for cowbirds in 15 parasitized nests of the rufous hornero to 5.3% for 45 nests of the rufous-collared sparrow. The largest sample, involving 59 nests of the chalk-browed mockingbird, produced a 16.8% rate of overall cowbird breeding success. Wiley (1985, 1988) identified six species as "high-quality" hosts in Puerto Rico (those fledging at least 55% of all cowbird chicks hatched), and five "low-quality" hosts that fledged lower percentages of cowbird chicks. The mean hatching success for four high-quality hosts was 39%, and their mean fledging success was 26%, representing an overall egg-to-fledging success rate of 10.1%. Mermoz & Reboreda (1994) reported that the primary factors affecting cowbird nesting success in their study were losses of eggs in multiple-parasitized nests, probably resulting from egg punctures made by other female cowbirds and the failure of some eggs to hatch.

Host–parasite relations. Cruz et al. (1990) judged that shiny cowbirds had a minimal adverse effect on reproductive success of yellow-hooded blackbirds in their Trinidad study, as a result of the species' colonial breeding behavior and effective joint nest defense by males, which reduced parasitism rates in centrally located nests. However, Wiley & Wiley (1980) judged that cowbirds could have serious effects on this species, including the abandonment of small colonies, because of high parasitism rates. Among the parasitized rufous-collared sparrow nests, 7 host young were fledged from 92 eggs (7.6%), as compared with 6 host young fledged from 35 eggs (17.1%) in unparasitized nests (Fraga, 1978). Fraga thus judged that shiny cowbirds probably do more harm to the reproductive efforts of rufous-collared sparrows than any nest predator. King (1973) estimated that the presence of a nestling cowbird reduced the sparrow's own productivity by a rate of about 0.4 fledgling per nest. All of the sparrow's mortality components were increased by the cowbird's presence, but the greatest effect was on egg mortality, presumably resulting from egg puncturing by the adult cowbirds. Egg pecking was also identified by Mason (1986a) as a source of host egg losses whereas Mermoz and Reboreda (1994) indicated that egg losses (either by direct cowbird removal or by host removal of punctured or cracked eggs) produced the greatest source of reduced host nesting success. Rejection behavior of foreign eggs occurs in some host species; the chalk-browed mockingbird accepts spotted cowbird eggs that are similar to its own, but selectively rejects immaculate eggs (Fraga, 1985; Mason, 1986b). This may help account for the pattern dimorphism in eggs (spotted vs. immaculate) laid by this cowbird in southern South America, although there is no current evidence that host-specific gentes exist in this species or that females laying different egg types select their hosts any differently from one another (Mason, 1986b).

BRONZED COWBIRD

(Molothrus aeneus)

Other Vernacular Names: Arment's cowbird (*armenti*). bronze-brown cowbird, glossy cowbird, lesser bronzed cowbird (*assimilis*). Miller's bronzed cowbird (*lovei*). red-eyed cowbird.

Distribution of Species (see map 78): From southern Texas, New Mexico and Arizona south through Central America to northern Colombia.

Subspecies

M. a. aeneus: Texas south through eastern

Central America to Panama. Slowly spreading eastward in Texas and now locally established in east–central Louisiana (New Orleans area).

M. a. lovei (=milleri): Arizona and New Mexico to west–central Mexico. Also reported from California.

M. a. assimilis: Southwestern Mexico.

M. a. armenti: Colombia.

Measurements (mm)

Females 7–7.5″ (18–19 cm); males 8″ (20cm)

M. a. aeneus, wing, males 117–122 (avg., 119), females 101–107.5 (avg. 104.5). Tail, males 82–88 (avg. 85), females 70–76.5 (avg. 73.5) (Friedmann, 1929). Wing:tail ratio 1:0.7.

M. a. assimilis, wing, males 105.2–111.8 (avg. 108.7). Tail, males 74.4–81.5 (avg. 77.5) (Ridgway, 1902). Wing:tail ratio 1:0.71.

Egg, avg. 23.11 × 18.29 (range 21–25 × 16.5–19 (Friedmann, 1929). Shape index 1.26 (= broad oval). Rey's index ~1.28.

Masses (g)

Avg. of 144 males 58.9, of 220 females 56.9 (Carter, 1986). Estimated egg weight 4.15 (*lovei*) to 4.85 (*assimilis*) (Rahn et al., 1988). Overall species' mean 4.51. Egg: adult female mass ratio 7.8%.

Identification

In the field: Both sexes of this species are shiny (males) to dull (females) black as adults, with bright red eyes. Adult males also exhibit a definite neck-ruff that is lacking or rudimentary in females and immature individuals (fig. 49). Immature birds also have brown, not red, eyes. Males utter various prolonged, thin, whistling notes, and when displaying before females males erect their back and rufflike neck feathers. During display the male also spreads his tail, arches his wings slightly, and bends his neck so that his bill touches the breast feathers. He may also flutter over, hover above, or circle around the female in low and undulating display flights. The male's advertising song is similar to that of the brown-headed cowbird, but is shorter and throatier. It consists of a guttural preliminary series of bubbling notes, followed by a squeaky and thin "ugh-gub-bub-te-pss-tseeee" whistle. A harsh and rasping "chuck" call is also uttered at times.

MAP 78. Summer breeding (hatched) and residential (filled) ranges of bronzed cowbird.

In the hand: Adults have a distinctive combination of red eyes, a soft and rather hairlike breast and neck plumage (with a well-developed neck-ruff in males) and a toothlike projection on each inner web of the second and third primaries. Adult females are blackish or sooty-brown, not glossy black like the males, and have only a poorly developed neckruff. Immature males are similarly dull sooty black to sooty brown, with paler feather edging on the underparts, and a rudimentary neck-ruff. Immature females are paler and grayer than young males and also have pale margins to the feathers of the underparts. Both sexes have longer bills than do brown-headed cowbirds and more closely resemble the Brewer's blackbird in bill shape. Newly hatched chicks resemble those of the brown-headed cowbird, having a orange-pink skin (which later becomes more greenish gray or brownish), greenish-blue eye-skin, yellowish bill and feet, a reddish gape, white mandibular flanges, and scattered mouse-gray down (Friedmann, 1929; Harrison, 1978). Carter (1986) described the mandibular flanges as cream-colored. Such white or cream-colored mandibular flanges may help separate nestling bronzed cowbirds from those of brown-headed cowbirds, which (in at least in sympatric populations) have yellow flanges.

Habitats

A variety of pastures, grasslands, woodland edges, or other areas with a combination of open areas and scattered trees are used by this species. It is mostly limited to low elevations in tropical to warm-temperate climates, but sometimes reaches altitudes of 1850 m.

Host Species

A total of 29 biological host species are listed in table 32, based mainly on the list provided by Friedmann & Kiff (1985). They noted that emberizine sparrows (16 species with 51 parasitism records) and icterines (11 species with 84 records) are evidently the most important of the 77 known hosts. The 29 biological hosts include 4 species of *Icterus* orioles, 3 species each of *Thyrothorus* wrens and *Piranga* tanagers, and 2 species each of *Pipilo* towhees and *Melozone* ground-sparrows. The streak-backed, hooded, and Audubon's orioles appear to be among the most frequently parasitized of all hosts, although the streak-backed is not yet known to be a fostering host. The Couch's kingbird is also now known to be a rare fostering host (Carter, 1986).

Egg Characteristics

The eggs have a ground color of light green or blue and lack darker surface markings. They are usually broad oval in shape, but some are subspherical. The eggs are more glossy than those of other cowbirds (Friedmann, 1929). The shell thickness (0.125–0.135 mm) and the shell mass (0.31–0.369 g) are both significantly greater than those of nonparasitic relatives (Rahn et al., 1988). These authors calculated that a 4.7 g cowbird egg, which falls within the observed mass range of this species, should have a breaking strength about 90% greater than that of a comparably sized nonparasitic icterine relative.

Breeding Season

Nine egg dates for Arizona range from May 30 to July 7, and 44 records for Texas are from April 1 to July 5, with half of the records occurring from May 12 to June 8 (Bent, 1958). Various records for Mexico and El Salvador encompass the 5-month period April to August (Friedmann et al., 1977; Friedmann & Kiff, 1985). In Costa Rica breeding extends from March or April to July (Stiles & Skutch, 1989). Breeding in Panama has been reported for late March (Wetmore, 1984), but no doubt is much more prolonged than this single record suggests.

Breeding Biology

Nest selection, egg laying. Both pendant and open-cup nests are used by this species. According to Friedmann (1929), eggs are laid at daily intervals, and the visiting cowbird neither removes any host eggs nor regularly pecks host eggs. However, Carter (1986) observed that pecking of host eggs (or those of other cowbirds) is a regular pattern of female behavior. Often only a single cowbird egg is placed in any one nest, although as many as seven eggs or young have been found (Friedmann, 1971; Carter, 1986), and there is one case of 14 eggs found in an apparent dump-nest (Friedman et al., 1977). Two females may also lay in the same nest, although this appears to be rather infrequent (Friedmann, 1929).

Incubation and hatching. The incubation period was reported by Friedmann (1929) as 12–13 days and by Carter (1986) as 10–12 days (avg. 11.0 days, $n = 4$).

Nestling period. Friedmann (1929) reported that the nestling period lasts about 11 days, which is followed by an additional postfledging dependency period of about 2 weeks. Carter found a 10- to 12-day fledging period (avg. 11.4 days, $n = 14$).

Population Dynamics

Parasitism rate. Carter (1986) found 100% parasitism rates for 12 green jay nests, 11 olive sparrow nests, and 10 northern cardinal nests, plus a 96% rate for 26 long-billed thrasher nests and a 71% rate for 35 northern mockingbird nests. Friedmann (1929) noted that both of two nests of the Audubon's oriole that he found in southern Texas were parasitized, as were seven of nine clutches (78%) present in the U.S. National Museum. Flood (1990) noted that each of four Audubon's oriole nests that he watched in eastern Mexico were visited at least once by cowbirds, and that at least one of these nests produced two fledged cowbird young. Additionally, 3 of 16 hooded oriole nests (19%) that Friedmann found in southern Texas were parasitized. Friedmann (1963) later added some additional records for both of these species. He also noted that cowbird eggs were found in only 2 of at least 150 nests of the Altamira oriole in Tamulipas, Mexico, although this oriole has been reported as a frequently parasitized species in the lowlands of El Salvador. All 10 nests of the yellow-winged cacique found by J. S. Rowley in Oaxaca were parasitized (Friedmann, 1971), although this seemingly highly tolerant species is still not known to be a fostering host.

Among hosts other than icterines, the song sparrow may be frequently victimized; 6 of 13 nests (46%) of this species found in Mexico City were parasitized by 13 eggs and 2 young cowbirds, an average parasitism intensity of 2.5 eggs or young per nest. Likewise, 9 of 11 nests (81%) of the rusty-crowned ground sparrow have been found to be parasitized, with up to 5 cowbird eggs per nest.

Hatching and fledging success. Among 13 parasitized nests in Texas, 5 nests fledged 1 cowbird, 4 fledged 2, and 4 fledged 3 (avg. 1.9 cowbirds per nest). In the same 13 nests, no hosts were fledged in 8, 2 nests fledged 1 host, 2 nests fledged 2, and 1 nest fledged 3 (avg. 0.7 hosts per nest) (Carter, 1986).

Host–parasite relations. Friedmann (1971) listed six species (three orioles, two sparrows, and the red-winged blackbird) that have served as common hosts to the eggs of bronzed and brown-headed cowbirds simultaneously. Carter (1986) observed simultaneous parasitism with the northern mockingbird, northern cardinal, and olive sparrow. However, Friedmann et al. (1977) suggested that sufficient differences in host choices occur between these cowbirds to help lessen competition for hosts in their limited areas of sympatric overlap. Friedmann (1963) noted, for example, that tyrant flycatchers, vireos, and wood warblers are used to a higher degree by brown-headed cowbirds than by bronzed cowbirds, and that, in contrast, orioles and emberizine sparrows are the primary hosts of the bronzed cowbird. Friedmann mentioned that this cowbird's eggs are similar in size and coloration to those of several genera of sparrows (*Melozone, Atlapetes,* and *Aimophila*). although this similarity appears to be fortuitous rather than the result of specific mimetic adaptation.

BROWN-HEADED COWBIRD

(Molothrus ater)

Other Vernacular Names: Buffalobird, common cowbird, dwarf cowbird (*obscurus*). Nevada brown-headed cowbird (*artemisiae*).

Distribution of Species (see map 79): Historically (pre-1850) mostly limited to the Great Plains, but now widespread throughout most of temperate North America. The breeding range currently extends east to the Atlantic coast, north to Great Slave Lake (in the Northwest Territories, Canada), west to the Pacific Coast, and south to central Mexico, the Gulf Coast, and southern Florida. Winters south to the Isthmus of Tehuantepec.

Subspecies

M. a. ater; Eastern Great Plains of North America east to the Atlantic Coast (Newfoundland to Florida).

M. a. artemisiae: Southeastern Alaska and western Canada (east to Ontario) south to

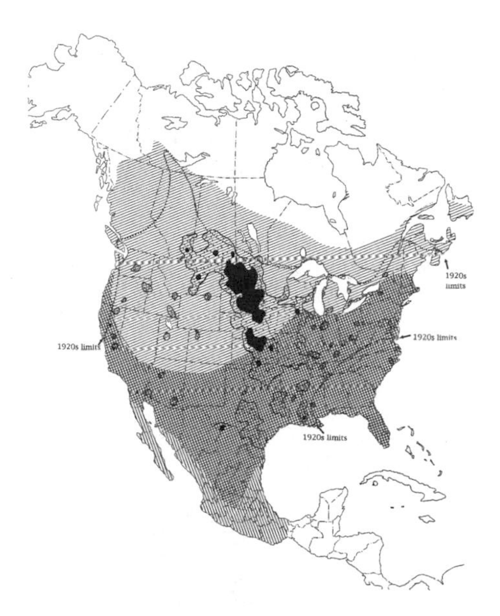

MAP 79. Summer breeding (lighter hatching), year-round residential (cross-hatched) and wintering (darker hatching) ranges of brown-headed cowbird. The dotted line shows the breeding range limits during the 1920s (after Friedmann, 1929). Enclosed dotted and linked areas, respectively, indicate moderate and dense breeding populations as of 1986–91 (after Lowther, 1995).

the eastern Sierras, the Great Basin, and the western Great Plains.

M. a. obscurus: California and New Mexico south to northern Baja and Guerrero, Mexico.

Measurements (mm)

7.5″ (19 cm)

M. a. artemisiae, wing, avg. of 283 males 105.9; 352 females 94.9 (Lowther, 1995).

M. a. ater, wing, males 105–116 (avg., 110.5), females 93.5–104.6 (avg., 101.1). Tail, males 70.1–80 (avg., 75.2), females 61.7–70 (avg., 66.8) (Ridgway, 1902). Wing:tail ratio 1:0.66–0.68.

M. a. obscurus, wing, avg. of 63 males 103.7, 35 females 93 (Lowther, 1993). Tail, avg. of 16 males 68.8, 15 females 62.2 (Ridway, 1902). Wing:tail ratio 1:0.66–0.67.

Egg, avg. of *ater* 21.24 × 16.42 (range 18.03–25.4 × 15.49–16.76), *artemisiae* 21.8 × 16.8; *obscurus* 19.3 × 14.99 (Lowther, 1993). Shape index 1.28–1.3 (= broad oval). Rey's index (*obscurus*) 1.51, (*ater*) 1.45, (*artemisiae*) 1.53.

Masses (g)

Avg. of nominate *ater:* 757 males 49, 692 females 38.8 (Dunning, 1993). Avg. of *artemisiae;* 232 adult ("after second year") males 27.5; 352 "after hatching year" females 37.6. Avg. of *obscurus;* 63 adult males 40.2, 35 yearling females 32 (Lowther, 1993). Avg. of various breeding-season male samples: *artemisiae* 44–47.3, *obscurus* 41–44.25, *ater* 51.3 (Rothstein et al., 1986). Avg. fresh egg weights 2.4 (*obscurus*), 3.13 (*ater*), 3.22 (*artemisiae*) (Rahn et al., 1988); overall species mean 2.9. Egg:adult female mass ratio, 7.5% (*obscurus*). 8.1% (*ater*), 8.6% (*artemisiae*): overall mean 8.1%.

Identification

In the field: The bicolored plumage of adult males (i.e., a brown head contrasting with an otherwise iridescent black body, wings, and tail) sets them apart from other cowbirds and "blackbirds" (fig. 49). Adult females are uniformly dark brown above, becoming olive-brown below, with rather obscure underpart streaking. Immature individuals resemble females, but are more distinctly streaked below. Like many other icterines, males assume bill-tilting display posture (fig. 50D) during hostile encounters, but additionally both sexes solicit preening from other birds (and perhaps reduce hostile responses) by assuming a silent head-down posture, with raised nape and crown feathers (fig. 50A, B). The male's breeding-season vocalizations include a squeaky, gurgling song uttered during the song-spread display, as the male spreads both wings and falls forward on his perch. This display is used both in courtship (toward females) and social dominance (toward other males)

FIGURE 50. Female (A) and male (B) brown-headed cowbirds in head-down (preening-invitation) display, plus males in low-level (C) and intense (D) bill-tilting displays (after photos in Selander & LaRue, 1961). Also shown is a male giant cowbird in head-down posture (E) (after a sketch in Orians, 1985), and a male bronzed cowbird nape-raising (F) (after a photo by J. Flynn).

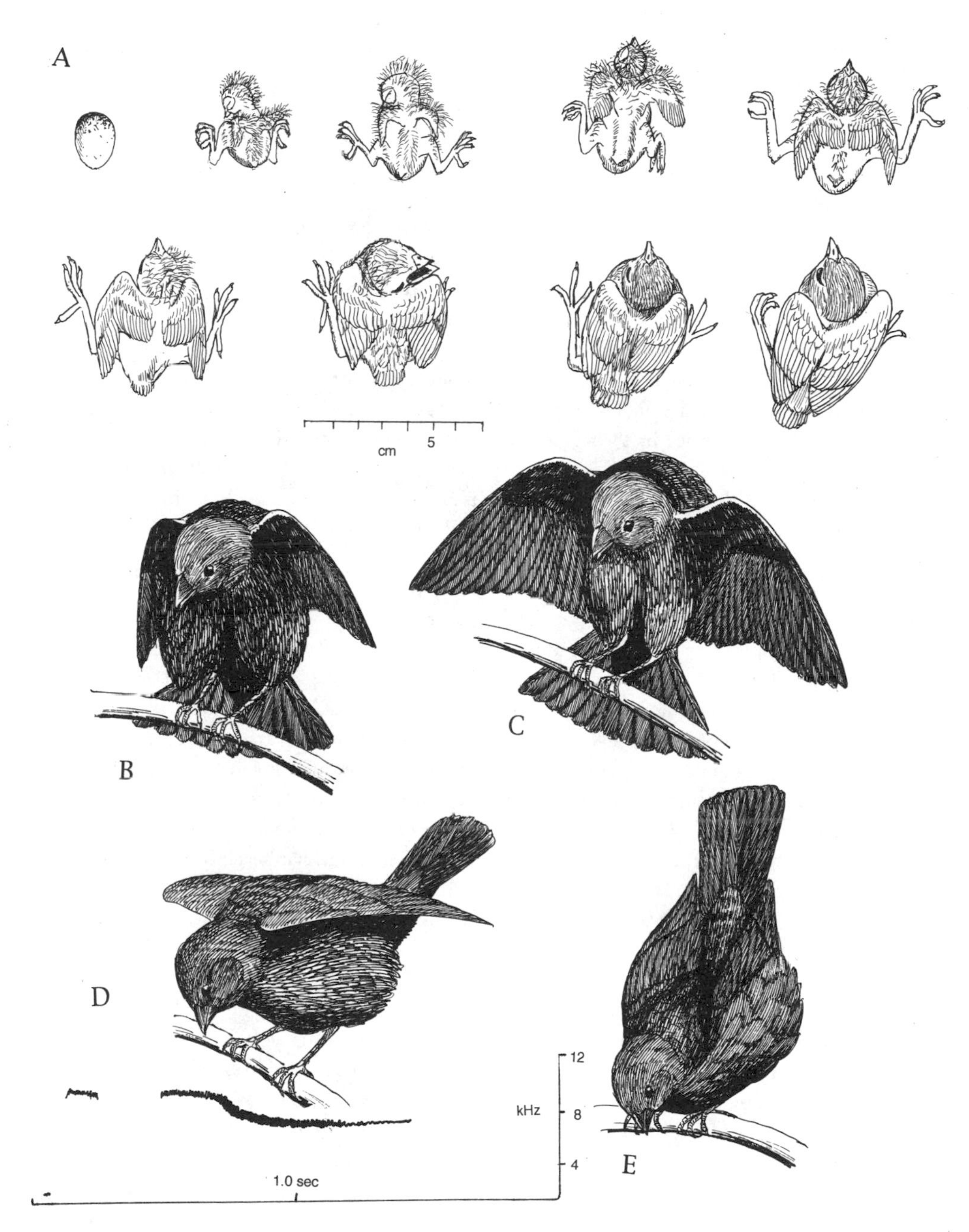

FIGURE 51. Egg and nestlings (shown at days 2–8 and day 10) of brown-headed cowbird (A, after photos in Norris, 1947). Also shown is a male brown-headed cowbird's song-spread display sequence (B–E, mainly after photos in Friedmann, 1929), and a sonogram (D) of the associated vocalization.

situations (fig. 51). Where this species occurs with the bronzed cowbird (as in southern Texas), both sexes of the latter species may be recognized by their larger size, longer and more massive bill shape, and reddish eyes. Females and juveniles of the bronzed cowbird are also considerably darker in plumage than those of the brown-headed cowbird. Shiny cowbirds are now increasingly found in company with brown-headed cowbirds (in southern Florida) and are similar in size and bill shape. Although male shiny cowbirds have iridescent head color that makes them readily recognizable, females of both species are similar in appearance and are best distinguished by their generally more uniform umber-brown (rather than tawny brown) plumages and their more uniform olive-brown underparts (rather than being distinctly streaked below with yellowish buff).

In the hand: All North American cowbirds are rather easily recognized by their short conical bills, as compared with the more elongated bills of typical "blackbirds." Both sexes of the brown-headed share with the shiny cowbird normal neck feathering (lacking a distinct neck-ruff) and normally shaped vanes of the outermost three primaries (rather than vanes that abruptly expand to form acute angles near their midpoints). Males are the only cowbirds with brown heads as adults, and females can be easily distinguished from adult males by their brownish overall color. Adult females differ from those of the shiny cowbird in having little or no streaking evident on their underparts, lacking a pale yellowish superciliary stripe, and showing generally less yellowish or tawny hues throughout. Immature birds of both sexes are more distinctly streaked than are adult females and also are somewhat spotted with brown and buff on the crown and especially on the flanks and underparts. Immature females tend to be somewhat paler overall than are young males.

Newly hatched young initially have a yellowish orange mouth lining, but this gape color becomes a deep red within a few days. The tongue is red, except for a light yellow rear edge. The mandibular (rictal) flanges at the base of the bill range from yellow (in *obscurus*) to white (in *artemisiae* and *ater*) with red rear edging, while the bill itself is dusky lemon, and the feet are light orange. The skin is likewise initially a light orange with a pinkish tinge, except for the more bluish-gray eye-skin, but becomes somewhat darker with increasing age. Some fairly long clumps of grayish (anteriorly) to white or buffy (posteriorly) down are also present at hatching on the head and upperparts (mainly along the supraorbital, dorsal and femoral feather tracts). Nestlings closely resemble those of the bronzed and shiny cowbirds (see previous account) and also closely resemble those of several host species (especially icterines), but tend to have more luxuriant down and more rapidly developing contour feathers than do most host chicks at comparable nestling stages.

Habitats

Primarily and historically associated with the grasslands of the Great Plains (note historic distribution on map 79), this species has progressively moved north, east, and west into areas of varied climates and habitats, especially into open woodlands, lumbered or burned-over forests, and forest edges or similar transitional brushy areas. The species has been able to occupy nearly every available non-heavily forested habitat imaginable, including many human-modified habitats. In general, fairly open woodlands or variously fragmented forest landscapes providing abundant forest-edge subhabitats are greatly favored. Using radio-tracking, Hahn (1994) found that during both of 2 years of observation, the host nests within a 1300-ha forest were more heavily parasitized (19–25% vs. 4–6%) than were those in a nearby old field, and that low- and ground-nesting forest species were parasitized more heavily (36% vs. 23%) than were mid-level and high-nesting species. Cowbirds penetrated all areas of the 1.6-km wide forest, suggesting that only in the largest, unfragmented forests are host species likely to be safe from cowbird parasitism.

Also using radio-tagging, Thompson (1993) estimated that female cowbirds traveled an average of 3.6 km from roosting sites to breeding areas (usually forest and shrub-sapling habitats) each morning and spent their afternoon hours foraging in shortgrass areas, croplands, or feedlots. In montane areas the species breeds to elevations as high as 2500 m and rarely has been seen at elevations approaching 3000 m. Wintering birds often associate with starlings, grackles, and blackbirds in massive roosts in the southern states; one such winter roost (Miller's Lake, in southern Louisiana) has estimated during recent winters to average 9.2 million birds, with a maximum record of 38.2 million during 1986 (Ortega, 1993). Such large numbers are hard to imagine, but if even close to accurate must make the brown-headed cowbird among the most abundant of North American passerine birds. However, an analysis of national Breeding Bird Survey data from 1966 to 1992 indicated a slight decline (averaging less than 1% annually) over that total period. The most marked declines have occurred since 1985 and are most evident from the Maritime Provinces south along the Appalachians, around the upper Great Lakes, and in the southern Great Plains. Stable or increasing populations occur in the southeastern states, in the Till Plains from Illinois east to Ohio, and from the Northern Great Plains west across the Great Basin (Peterjohn & Sauer, 1994).

Host Species

A total of 144 fostering hosts are listed in table 15, based on the most recent comprehensive summary (Friedmann & Kiff, 1985). Earlier summaries were provided by Friedmann (1929, 1963) and Friedmann et al. (1977). A list of about 80 additional victimized species (those that are known to have been parasitized by brown-headed cowbirds) but are not yet known to have actually fostered their chicks, was also provided by Friedmann & Kiff (1985). Minimum numbers of host records, based mostly but not entirely on these same authors, are indicated for those 44 species that are designated in table 15 as "frequent" fostering hosts (those having 20 or more host records). Nearly 90% of this species' total published records of parasitism (which now exceed 15,000) are attributable to these 44 species, and nearly 60% of the total are associated with only 20 of them. A comparison of host/parasite traits for brown-headed cowbirds and the common cuckoo is shown in table 8.

Egg Characteristics

Eggs of this species are broad oval, with a white to grayish white ground color and gray to brown freckling and spotting, especially toward the more rounded end. The surface is moderately glossy, and the eggs are similar in general appearance (color and patterning) to those of meadowlarks, but are considerably smaller. There is no evidence of interspecific egg mimicry, gentes development among females, or any special preference among females for parasitizing host species with similar egg traits. The eggs are typically about 30% larger than those of their most frequent hosts (tables 11 and 29), but the mean shape ratio of the cowbird (1:1.3) is nearly identical with the mean of the 20 most commonly exploited hosts, which average 1:1.35 (range 1:27–1.4). Mean eggshell thicknesses for various subspecies range from 0.107 mm to 0.135 mm, and mean shell masses from 0.185 g to 0.369 g. These figures represent significantly thicker and significantly heavier eggshells than those of various nonparasitic icterines (Rahn et al., 1988), and such traits presumably are associated with reduced probabilities of eggshell cracking or breakage during laying or egg manipulation by hosts or other laying cowbirds. In other parasitic *Molothrus* species and *Clamator*, this adaptation for increased resistance to breakage is correlated with multiple parasitism and the increased possibility of egg damage during multiple parasitism events (Brooker & Brooker, 1991). Both the increased shell thickness and rounded shell shape are important factors in increasing the puncture resistance of the eggs (Picman, 1989). There is no correlation between the volume of cowbird eggs and that of specific host-species eggs among 42 host species (Mills, 1987).

Breeding Season

The overall period when eggs have been found ranges from early April to August, with most of the records for May and June (Lowther, 1993), representing an approximate 60-day primary breeding period. Some representative spans of egg dates as summarized by Bent (1958) are: Alberta, May 24–July 1 (39 days); Ontario, May 15–July 1 (48 days); North Dakota, May 23–July 15 (54 days): Massachusetts, May 14–June 29 (47 days); Michigan, April 30–July 7 (68 days); Illinois, April 26–July 21 (87 days); Oklahoma, April 29–June 26 (59 days); Texas, April 7–July 2 (87 days); California, April 3–July 21 (110 days); Arizona, May 2–August 22 (111 days). The Canadian provinces thus have a maximum breeding season of only about 40–50 days, or about half as long as that typical of the southwestern states and California.

Breeding Biology

Nest selection, egg laying. Female cowbirds parasitize host species with a wide array of nest locations, nest structures, and egg traits (see tables 11 and 29). Whitehead et al. (1993) reported that among 1340 parasitized nests of more than 20 species in Indiana, parasitism levels are lower in forest interiors (~5%) than near exterior forest edges (~15%) or near clearcuts (~20%), and some individual species (such as Acadian flycatcher, worm-eating warbler, and ovenbird) that nest in peripheral as well as interior locations are more heavily parasitized near forest edges than in forest interiors. Similar edge-effect influences on parasitism rates have been reported for various woodland-breeding species by other workers (Brittingham & Temple, 1983; Robinson et al., 1993; Paton, 1994). Regarding specific habitat preferences, Peck and James (1987) found that hosts nesting in dry, open or semi-open sites within deciduous, mixed, or coniferous vegetation (shrubs or small trees) are preferred in Ontario. A general preference also exists for nests that are elevated (80% of 1925 records) and for those situated in living trees or shrubs (79%), especially deciduous woody species.

Other apparent habitat traits include a preference for selecting nests located in farmlands (54% of the host nests), including such agricultural subhabitats as overgrown fields and fence rows (27%), young conifer plantations and orchards (21%), and open fields (7%). Shrubby edge and wooded edge habitats collectively accounted for 27%. Residential habitats accounted for 14% of the selected host nests, and wetland habitats contributed 4%. Petit & Petit (1993) concluded that host breeding habitats (preferentially including deciduous forests, shrubby areas, and grasslands according to their classification) are more important than other ecological or life-history variables in influencing nest-selection tendencies among open-cup–nesting hosts. Host nest-placement traits were also judged important, but host life-history traits (clutch sizes, incubation periods, and nestling periods) were judged relatively unimportant. In Ontario, half of all parasitized nests had inner nest diameters of 3.87–7.6 cm. Likewise, half of the parasitized nests were situated 0.9–2.1 m above ground, although some affected nests were elevated as high as 19.8m (Peck & James, 1987). Nolan (1978) determined that prairie warbler nests that were situated no more than 1 m above ground were parasitized at a rate 10% below expectation (based on the vertical distribution of available nests), but those elevated 2–3 m above ground were parasitized at a rate that was 24% above random expectation.

Female cowbirds frequently but not invariably remove a host egg at about the time of egg laying (fig. 52). Nice (1937) estimated that removal of one or more host eggs occurred in 37% of the parasitized song sparrow nests she observed, and Payne (1992) estimated that parasitized indigo bunting nests have clutch sizes averaging 0.77–1.06 fewer bunting eggs than unparasitized clutches. Egg removal behavior is just as likely to occur on the day before the cowbird's egg laying as on the day of laying, and egg removal behavior is essentially restricted to those nests with at least two eggs already present (Hann, 1937). Of 96 instances in which female cowbirds had an opportunity to remove host prairie warbler eggs at about the time of their own laying, none were removed in 20% and only one was removed in 67% of the

FIGURE 52. Female brown-headed cowbird in a red-eyed vireo nest (A), and removing a host's egg (B). Gaping by a nestling cowbird is also shown (C). After photos by H. Harrison and A. D. DuBoise (in Bent, 1958).

cases. From two to four eggs were removed in the remaining 13% of the cases, but such multiple egg removal usually occurred only in those nests where host incubation had already begun. In most cases (82%) host egg removal occurred during the host's egg-laying period, and in the majority of cases (at least 73%) the host egg was removed at the cowbird's laying visit, or even 2 days before it laid (Nolan, 1978). Removal of one or more host eggs on the day before laying was observed in 7 of 11 cases of egg removal documented by Norris (1947); in 10 of these cases only a single egg was taken. In the majority (54%) of 212 Ontario parasitism incidents, the cowbird's eggs were deposited before any host eggs were present in the nest, in 38% they were laid during the host's own egg-laying period, and in the remaining cases they were introduced either after host incubation had begun or were laid in old or deserted nests. Eggs are usually laid about 10–15 minutes before sunrise, and generally only 20–40 seconds are spent at the host's nest (Norris, 1947; Nolan, 1978). Relative parasitism intensities (variations in numbers of eggs laid in individual host nests) for various hosts are summarized in table 5. As described earlier, such variations tend to follow Poisson distributions, indicating that each egg-laying event by cowbirds is essentially a random act; the presence or absence of other cowbird eggs does not influence the probability of additional depositions.

Female cowbirds are believed to lay an egg per day throughout an egg-laying cycle during which 1–7 eggs are deposited, after which a brief nonlaying interval may occur. Scott & Ankney (1980, 1983) estimated that in southern Ontario the total seasonal production of eggs per female may be about 40, which represents about 0.8 egg produced per day over the approximate 50-day laying season in Ontario. This figure also represents an annual female fecundity estimate nearly double that calculated by earlier workers (e.g., Payne, 1964, 1976a). However, Rothstein et al. (1986) similarly estimated that 48 eggs may be produced by a single female over a 67-day laying period, representing a mean laying rate of 0.72 egg per day. Twelve captive females had an average season-long egg production of 26.3 eggs each. Three of 24 captive females laid more than 40 eggs each in a single 89-day season, and one laid an egg per day for 67 consecutive days (Holford & Roby, 1993). Based on calculations by Scott & Ankney (1980), the average life expectancy for birds in their first to seventh year of life is only about 1.3 years, assuming an annual mortality rate of about 60% (Darley, 1971; Frankhauser, 1971), so most females probably achieve no more than 2 years of actual egg production during their lifetimes. However, there is a record of a banded cowbird surviving in the wild for a minimum of nearly 17 years (Klimkiewicz & Futcher, 1989).

Incubation and hatching. Judging from published reports, the incubation period is rather variable (e.g., 9–14 days reported by various observers), but this apparent variability probably reflects the varied times at which eggs are deposited in the host's nest (Peck & James, 1987; Lowther, 1993). Incubation periods for individual eggs have been determined to be as short as 11.2 days and as long as 12.6 days, with estimated means of 11.6 (10 eggs) (Norris, 1947) to 11.8 days (9 eggs) (Nolan, 1978). Among a sample of 19 cowbird eggs, 5 hatched from 1 to 4 days before the host's eggs, 10 hatched the same day, and 4 hatched from 1 to 5 days later than the host's (Norris, 1947). Newly hatched young of *ater* average 2.29 g (Nolan & Thompson, 1978), or about 75–80% of estimated mean fresh egg weight.

Nestling period. The usual nestling period is 9 days (fig. 51), although if frightened the chicks often depart the nest earlier, and sometimes the young fledge as late as day 10. The average of 11 nest-departure records was 8.7 days (Norris, 1947). The usual weight at the time of nest departure is 30–33 g, or 12–14 times greater than their hatching weight (Friedmann, 1929; Norris, 1947). Fledging is followed by an approximate 25- to 39-day period of dependency; in one study the mortality rate during this period was 47.6% (Woodward & Woodward, 1979).

Population Dynamics

Parasitism rate. Estimated nest parasitism rates are presented in table 10, and estimates of parasitism intensities are shown in table 26. Other estimated parasitism rates for numerous host species in various geographic regions have been provided by Elliott (1978) for Kansas, Mengel (1965) for Kentucky, Norris (1947) for western Pennsylvania, Bull (1974) for New York, and Peck & James (1987) for Ontario. Peck and James reported that the highest observed parasitism rates in Ontario (based on sample sizes of at least 50 host nests) were for the house finch (42%), followed by the purple finch (40%), red-eyed vireo (38%), chipping sparrow (32%), yellow-rumped warbler (31%), and yellow warbler (30%). One species (cedar waxwing) that has been previously categorized (Rothstein, 1976a,b) as a rejector nevertheless had 67 records of parasitism, and there were 90 records for the American goldfinch, which because of its specialized seed diet rarely raises cowbird chicks successfully. Among 86 known host species, the overall parasitism rate was 6.7% among 44,788 host nests, which represents the largest sample size yet available for any single geographic region.

Hatching and fledging success. Hatching and fledging success rates for this species are summarized in table 27. Young (1963) provided additional breeding success estimates from the literature for 36 host species, representing a total of 879 cowbird eggs. The hatching success rate for 795 of these eggs was 38%. The overall breeding (egg-to-fledging) success, considering all 36 host species, was 25%, but the success rate averaged about 10% higher among heavily exploited host species than for lightly parasitized ones. Norris (1947) estimated a similar overall 26.8% egg-to-fledging success for 108 cowbird eggs distributed among 14 host species in Pennsylvania. Elliott (1978) reported a much lower (8.3%) breeding success rate for 157 cowbird eggs in nests of 5 host species in Kansas. Mengel (1965) reported a still lower success rate among 25 host species in Kentucky; only 11 of 512 parasitized nests (2.1%), involving 6 host species, produced any fledged cowbirds. Over an 8-year study period, cowbirds successfully fledged young from 76 of 411 (18.5%) parasitized indigo bunting nests (Payne, 1992).

Host–parasite relations. Judging from estimates from the Breeding Bird Census data by Lown (1980), the density of breeding female brown-headed cowbirds averages approximately 3 birds per 100 available host nests, and this ratio had remained constant for the 4 decades previous to his study. Estimates of parasitism costs to hosts have been summarized for 10 separate studies in table 9. Similar earlier evaluations include those of Nice (1937), who estimated that for each fledged cowbird, one fewer song sparrow was produced. Norris (1947) similarly estimated a reduction of 0.89 host young (mean of 14 species) for each successfully parasitized nest. These host costs vary greatly, depending on the rate and intensity of parasitism and the degree to which the host species accepts or rejects the cowbird eggs. Like several other regularly exploited hosts, indigo buntings will abandon their nests if a cowbird egg is deposited before laying any of their own, but after that point will accept it. However, parasitized nests averaged 1.06 fewer bunting eggs than did nonparasitized nests. Overall nest success (percentage of nests fledging at least one bunting) was higher in nonparasitized nests (56.4%) than in singly parasitized nests (22.1%), and especially higher than in multiply parasitized nests (6.9%). However, among those indigo bunting pairs that succeeded in fledging a cowbird, the chances of fledging at least one bunting chick as well were not significantly reduced (Payne, 1992). In such species an "adaptive tolerance" strategy of dealing with cowbird eggs has been used, for in these hosts the associated costs of egg rejection (such as nest or clutch abandonment) are evidently greater than those of accepting the presence of such eggs (Petit, 1991). Discussion of these evolutionary aspects of cowbird parasitism were provided in chapters 1 and 5.

As mentioned earlier in this account, in spite of the recent expansion of the cowbird's range across much of temperate North America during the first half of this century, recent data from the Breeding Bird Survey suggest that this population surge is now over, and indeed since 1966 the national cowbird population has declined at an average rate of 0.9% per year (Peterjohn & Sauer, 1994). Recent land-use changes across much of the USA and southern Canada, such as changing farming methods and associated livestock, have evidently not been entirely to the species' benefit. Thus, although some areas may still have increasing or stable cowbird populations, in most areas their populations are declining, and so the impact of cowbird parasitism on sensitive host species is perhaps also past its peak.

GIANT COWBIRD

(Scaphidura oryzivora)

Other Vernacular Names: Rice grackle.

Distribution of Species (see map 80): Eastern Mexico south to southern Brazil, Paraguay, and Argentina, mainly in the tropical and subtropical zones.

Subspecies

S. o. impacificus: Veracruz south to Panama. Probably indistinguishable from *oryzivora* (Wetmore, 1984).

S. o. oryzivora: Panama, Trinidad, and South America.

MAP 80. Range of giant cowbird.

Measurements (mm)

Females 11″ (28 cm); males 13″ (33 cm)

S. o. impacificus, wing, males 177–203 (avg. 189, $n = 6$), female 160. Tail, males 133–152 (avg. 145.8, $n = 6$), female 119 (Ridgway, 1902). Wing:tail ratio 1:0.77.

S. o. oryzivora, wing, males 169.9–204.5 (avg. 191.1, $n = 10$), females 145.5–167 (avg. 155.2, $n = 10$). Tail, males 129.5–157 (avg. 145.4, $n = 10$), females 112.5–133 (avg. 119.2, $n = 10$) (Wetmore, 1984). Wing:tail ratio 1:0.76.

Egg, avg. of 10, 33.5 × 23.7 (ffrench, 1991).Range 30.2–35.3 × 25.1–28, Suriname (Haverschmidt, 1968). One Guatemalan egg 36.5 × 25.4 (Skutch, 1954). Shape index 1.41 (= oval). Rey's index 1.69.

Masses (g)

Avg. of 6 males, 219; of 5 females 162 (Dunning, 1993). One male 175, one female 144 (ffrench, 1991). One male 175, females 127–140 (Haverschmidt, 1968). Range (unsexed) 130–176 (Sick, 1993). Smith (1979) initially reported that males and females respectively weigh 120 and 74, which must have referred to nestling means. However, he later (1983) reported usual adult weights as 212 and 110, respectively, for males and females. Estimated egg weight 10.4; actual egg weights 9.7–12.6 (Haverschmidt, 1968). Egg:adult female mass ratio (assuming a mean female weight of 110) 9.45%.

Identification

In the field: The large size and gracklelike appearance of this species, together with its bright red eyes (males) to orange, brownish, or even yellow iris color (females, and perhaps also males in southern parts of the range) distinguish it from the local grackles, which have pale yellowish white eyes (fig. 53). Adult males also have a conspicuous neck-ruff that can be fully erected during social interactions, but is always somewhat apparent. Dur-

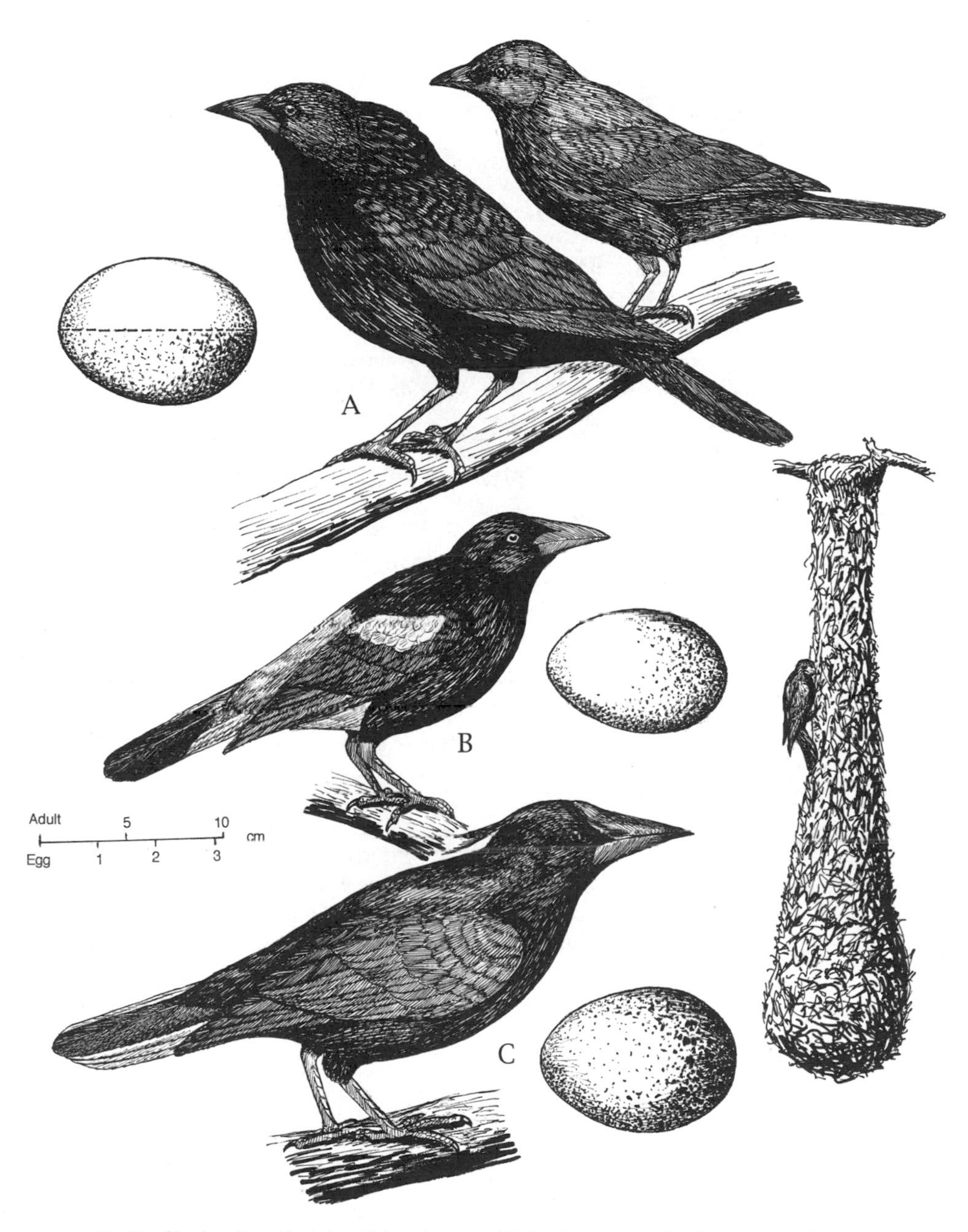

FIGURE 53. Profile sketches of adults of the giant cowbird (A) and two of its hosts, the yellow-rumped cacique (B) and the chestnut-headed oropendola (C). Also shown are their respective eggs and a female cowbird at an oropendola nest (after a photo by N. Smith).

ing courtship or territorial advertisement males often utter an unpleasant-sounding, ascending screech. They also utter a strident "jewli, chi, chi, chi, chi," while standing or strutting in a rather erect posture, erecting their neck-ruffs, and nearly touching the throat with the bill (see fig. 50). Females utter various nasal whistles. Females and immature males are smaller than adult males and have only a rudimentary neck-ruff but are otherwise similar in plumage. Both sexes fly in an undulating, woodpecker-like manner, folding their wings momentarily after every five to six wingstrokes. Their wing feathers produce an unusually loud and peculiar noise while in flight.

In the hand: The large size (wing at least 125 mm, usually 150 mm) immediately separates this species from the other cowbirds; it differs from others in its extremely large neck-ruff, which is most apparent in adult males. Adults also have a strongly convex culmen that becomes broad and rather flattened toward its dorsal base, producing a U-shaped forehead shield similar to that of an oropendola (see fig. 49). Adult females are much duller and are consistently smaller than adult males (wing <170 mm, tall <125 mm, culmen <35 mm, tarsus <45 mm). Juveniles of both sexes resemble adult females but may have paler bills. Immature males soon become more glossy and begin to develop their distinctive neck-ruffs. Nestlings are still poorly described, but apparently have a whitish skin that is partially obscured by relatively long and abundant grayish down on the head and upperparts, plus a white bill, white mandibular flanges, and pale yellow mouth lining (Friedmann, 1929), although a red mouth lining is typical of other icterine nestlings and would seem likely in this species too. By comparison, newly hatched cacique and oropendola nestlings are only sparsely downy (and entirely naked in at least one cacique), with pinkish skin and yellowish mandibular flanges. The mouth lining color is still poorly documented, but is red in at least one cacique (Skutch, 1954). After fledging, the cowbird's bill color gradually darkens, and by about 2 months of age only its tip has faint traces of white present (Friedmann, 1929).

Habitats

This adaptable species usually forages on the ground, in pastures, savannas, grasslands, and agricultural lands, often around large ungulates such as cattle. Generally it prefers open country to woodlands. However, it also forages along river banks or at woodland edges, where it is likely to encounter nesting colonies of oropendolas or other potential hosts. It ranges from sea level up to a maximum of about 2000 m in areas having tropical to subtropical climates and breeds to elevations of nearly 1700 m.

Host Species

Eight known host species are listed in table 32, based on the list provided by Friedmann & Kiff (1985) plus the account by Robinson (1988). Probably all of the larger troupials (oropendolas and caciques) breeding within the overall range of this species represent potential hosts. The larger species of oropendolas may be more effective in excluding females from their nests than are the considerably smaller caciques. Both groups of hosts are able to remove cowbird eggs from the nest by first impaling them with the lower mandible and then pushing them off the bill with the tongue (Smith, 1979).

Egg Characteristics

The eggs of this species are oval but seem to vary in both size and color. The ground color varies from white, bluish white, or pale green to gray. Blackish blotches, spots, or fine lines are variably present. Nearly all of the host species lay whitish, greenish, bluish, or grayish eggs that are distinctly marked with dark scrawls, speckles, spots, or larger blotches, but at times they may be nearly free of such markings. Smith (1968) reported that five recognizable egg and female morphs ("types") occurred in the Panama and Canal Zone population of giant cowbirds that he studied. Eggs deposited in colonies whose females tended to reject alien eggs mimicked those of host females. However, those eggs laid in colonies of more tolerant hosts were relatively nonmimetic (see "Nest selection, egg laying," below). Like the other parasitic icterines, the mean shell mass and shell thickness are significantly greater than those of nonparasitic relative (Rahn et al., 1988). According to Friedmann (1963), the eggs of this species are unusual in that they average smaller than those of their icterine hosts (range 30–39 × 19–25 mm; shape ratios 1:1.45–1.65 or oval to long oval, depending upon the species). However, giant cowbird eggs reported from Central America (Skutch, 1954), Suriname (Haverschmidt, 1968), and Trinidad (ffrench, 1991) appear to about as long and usually are even broader (30–35 × 25–28 mm, shape ratios 1:1.38–1.43) than their host's eggs. Perhaps marked regional or individual variation in egg size may exist, just as substantial variation in adult female weights may also be typical (see "Measurements").

Breeding Season

In Amazonian Peru the breeding season probably occurs from about July to December, with a peak in September, judging from the frequency of hostile interactions between the cowbird and two of its host species, which also are breeding then (Robinson, 1988). In Suriname the cowbird's breeding period extends from December to April (Haverschmidt, 1968). In Trinidad and Tobago its breeding season probably extends from December to June (actual records from January to May), which coincides with the nesting seasons of its two hosts, the crested oropendola and yellow-rumped cacique (ffrench, 1991). In Panama and elsewhere in Central America, host nesting activity may peak during March and April, but some birds may breed from as early as January or February to as late as August or even September (Skutch, 1965; Wetmore, 1984). Smith (1979) stated that host species' colonies that are not mutalistically associated with bees or wasps begin breeding at the start of the dry season (December–January), whereas those associated with these insects do not begin nesting until middle or late February, well into the dry season.

Breeding Biology

Nest selection, egg laying. Nearly all of the known host species (i.e., all the oropendolas and caciques) build pendulous nests up to 1 m or more in length, with a small opening near the roofed-over top (fig. 53). Smith (1968) stated that in Panama, five subpopulations of female giant cowbirds may be recognized with respect to their egg traits. Three of these subpopulations produce eggs that mimic the egg traits of the three Panamanian oropendolas, and the fourth has eggs that mimic those of caciques (of which three species occur in Panama). There is also a fifth (or "dumper") category of females that lay nonmimetic (generalized icterine) eggs. According to Smith, the four mimetic egg morphs vary both in size and color, depending on general egg mimicry requirements for the host species and even on the egg traits of individual host colonies. Smith reported that females laying mimetic eggs typically laid only a single egg in each host nest and usually deposited them in nests containing only a single host egg. However, "dumper" females laying the generalized type of egg sometimes laid as many as five eggs per nest (usually only two or three) and deposited their eggs in empty nests as well as in those with incomplete or even complete host clutches. Hosts in discriminator colonies ejected nonmimetic eggs from their nests almost immediately after they had been discovered, but those in nondiscriminator colonies accepted eggs showing a wide variety of colors, patterns, and sizes. An average of 1.17 parasitic eggs (in discriminator colonies) to 1.82 eggs (in nondiscriminator colonies) were deposited per nest. Host clutch sizes averaged 1.8 in nonparasitized nests and 1.27 in parasitized nests. This reduced host clutch size in parasitized nests evidently resulted from physiologically controlled clutch regulation among hosts rather than from direct egg destruction or removal by the parasitic females (Smith, 1968).

Incubation and hatching. Smith (1968) reported that the incubation period is 5–7 days less than its Panamanian host species (17–18 days for oropendolas and 16 days for caciques), which would place it in the range of 11–12 days, or no longer than those of the other much smaller cowbird species. A somewhat longer period would seem probable, given the fact that the cowbird's eggs are nearly the same size as those of the great-tailed and boat-tailed grackles, both of which have 13- to 14-day incubation periods (Skutch, 1954; Harrison, 1978).

Nestling period. No specific information exists. The fledging periods of the similar sized boat-tailed and great-tailed grackles are 20–23 days (Skutch, 1954; Harrison, 1978). The fledging periods for the host oropendolas and caciques is 30–37 days (Skutch, 1954; Smith, 1979), or about 4–5 weeks (ffrench, 1991).

Population Dynamics

Parasitism rate. Smith (1968) reported a mean 28% parasitism rate among 567 nests of discriminator colonies of oropendolas and caciques over 4 years of observation and a 73% parasitism rate among 935 nests of nondiscriminator colonies. The estimated collective mean overall parasitism rate for these two subgroups is 52.5% in a sample of 1502 nests.

Hatching and fledging success. Smith (1968) reported that 173 nests of icterine host species (mainly chestnut-headed oropendola and yellow-rumped cacique) averaged 111 fledglings of the giant cowbird per year (0.6 fledglings per nest) during 2 years of study. Among discriminator colonies, 666 cowbird eggs were laid in 567 nests (1.17 eggs per nest) over a 4-year period, of which 559 hatched (84%) and 433 fledged (Smith, 1968: table 2), producing a surprisingly high egg-to-fledging success rate of 65%. Among nondiscriminator colonies, 1708 cowbird eggs were laid in 935 nests (1.82 eggs per nest), of which 1263 hatched (74%) and 795 were fledged, representing an extremely high egg-to-fledging success rate of 74%. Added to the high parasitism rates reported by Smith, these data would suggest that the giant cowbird may be the most reproductively successful of all known avian brood parasites. Smith suggested that although the giant cowbird seemingly has a very high breeding success rate, it must also have a high postfledging mortality rate. By comparison, Robinson (1988) stated that the impact of giant cowbirds on their yellow-rumped cacique and russet-backed oropendola hosts was minimal. None of the 168 female caciques that he saw with fledglings was ever seen feeding a cowbird chick, and only 3 of 24 oropendola families were observed doing so.

Host–parasite relations. The remarkable mutualistic relationship between this cowbird and its icterine hosts in Panama that was discovered by Smith (1968) was briefly discussed earlier (chapter 5). In Smith's study area, 4-year average of 0.73–0.76 cowbird fledglings were produced per parasitized nest. Thus, given a mean clutch size of about 1.2–1.8 cowbird eggs per parasitized nest, the egg-to-fledging success rate for parasitic eggs must have been about 40–60%. This represents a breeding success rate substantially greater than that of many tropical birds having altricial young; for example, Skutch (1976) estimated a mean 29% egg-to-fledging success rate for 987 eggs of non-cavity nesters (including open, roofed and hanging nests) in Costa Rica. Smith similarly found that an average of 0.39–0.43 host young were fledged per nest over a 4-year period. Since host clutch-sizes averaged from 1.27 (in parasitized nests) to 1.8 (in unparasitized nests) eggs, and since host fledging success averaged 0.39–0.43 chick per nest, the overall host breeding success must have been in the range of about 25–30%, or about average for tropical altricial species.

The highest host fledging success (53–55%) observed by Smith occurred in nests containing two or three host chicks but no cowbirds, which were additionally protected from botflies by bees or wasps. Among parasitized nests placed in colonies unprotected by bees or wasps, the reduction in mean host clutch size (from 1.8 to 1.27 eggs) associated with the presence of cowbird eggs must be counterbalanced by a greatly improved nestling survival of host chicks in such parasitized nests to achieve an overall host benefit. In nests of such nondiscriminator colonies, the reproductive benefit to hosts resulting from the presence of a single cowbird chick was improved fledging by 0.34 host young per nest (rel-

Eastern crowned leaf-warbler feeding a nestling oriental cuckoo. After a photo in Kiyosu (1959).

ative to the 0.18 host young produced in all unparasitized nests), and in those with two cowbird chicks the relative improvement was 0.27 young per nest (see table 27).

On the other hand, unparasitized nests in wasp- or bee-protected colonies produced an average of 0.32 host young per nest in unparasitized nests, as compared with 0.25 young in singly parasitized nests and 0.2 in doubly parasitized nests. These differences represent a cost of parasitism of about 0.06 host young for each nestling parasite present under such conditions. Additionally, those host colonies using bees or wasps as an alternate antibotfly strategy had a shorter available nesting season, as well as a slightly lower overall rate of nestling survival, than did those colonies tolerating cowbird eggs and chicks in their nests.

More recently, Robinson (1988) reported that no cowbird chicks were known to fledge from colonies of yellow-rumped caciques and russet-backed oropendolas that he studied for 5 years in Peru. The actual rate of egg parasitism was not determined. Robinson suggested that the major difference in his and Smith's results was that, although 40–50% of the cacique and oropendola nests in his study area resulted in fledged young, these species managed to fledge only a single chick per nest and thus there could be no reproductive advantage in being parasitized. Similarly, Webster (1994) judged on the basis of host intolerance that the Montezuma oropendolas he studied in Costa Rica have not developed a mutualistic relationship with giant cowbirds, even though they also do not nest in association with social hymenopterans and thus have no apparent alternate defense against parasitism by botflies.

In another related study, Fleischer & Smith (1992) examined the supposed egg mimicry of the giant cowbird relative to two of its host taxa, using morphological and electrophoretic evidence. They observed significant, nonoverlapping differences between the eggs of the three species, although host discrimination studies are still needed to determine if functional egg mimicry exists.

Appendix A

GLOSSARY

Acceptor species. Those species that accept parasitic eggs in their nests. Acceptor species may include both unsuitable hosts (those whose foods or feeding methods are unsuited to the parasite) and fostering hosts (those who are able to foster parasites effectively).

Agonistic. "Struggling" behavior between antagonists, including attack–escape behavior and ritualized signals that are functionally associated with such behavior.

Allien. The egg or chick of a brood parasite when present in a host's nest.

Alloparental care. Parental care given to offspring other than one's own, including both nest-helping behavior involving kinship altruism and caring for unrelated offspring of brood parasites. *See also* altruism.

Allopatric. Populations or species occupying mutually exclusive ranges. *See also* sympatric.

Alloxenia (adj. alloxenic). Situation in which two brood parasites have different host species, presumably to reduce interspecific competition. *See also* homoxenia.

Altricial. Those species whose young are hatched in a helpless and sometimes nearly featherless state, unable to control their body temperatures or feed for themselves until they are nearly fledged. *See also* precocial.

Altruism Care-giving behavior performed for the benefit of others. Hosts of brood parasites are true (if unknowing) altruists, as neither kin selection or other types of natural selection can explain their behavior.

Alula. A group of small feathers located at the wrist that are associated with the first digit.

Anomalospiza. A monotypic genus (literally, "an anomalous sparrow") of African parasitic sparrows.

Atavistic. The retention or recurrence of an ancestral trait in an individual or a population.

Batesian mimicry. A type of mimicry in which a vulnerable or prey species (the mimic) resembles some better-protected species (the model), thereby improving its probability of survival. Egg mimicry and chick mimicry of brood parasites have some characteristics of Batesian mimicry.

Bill (or mandibular) flange. Soft, often colorful, outgrowths at the edges and bases of nestling bills that probably help stimulate direct parental feeding behavior toward the nestlings.

Biological host. Host species that hatches and tends a brood parasite's eggs and young. Synonymous with fostering and true host. *See also* hosts.

Biological parent. An adult bird that tends its own offspring. *See also* foster parent.

Breeding success. An estimate of reproductive efficiency, based on the percentage of eggs laid relative to the number of young that successfully fledge. *See also* fledging success, nesting.

Brood parasitism. The behavior of a species in which females lay their eggs in the nests of their own species (conspecific or intraspecific brood parasitism) or in those of other species (interspecific brood parasitism). Hypotheses advanced to account for the evolution of brood parasitism include dietary limitation, fortuitous egg laying, predation-risk spreading, and reproductive maladaptations models (q.v.).

Cacique. The vernacular name for a genus of medium-sized colonial icterines that build pendulous nests.

Cacomantis. A polytypic genus (literally, "an evil prophet") of Asian cuckoos.

Caliechthrus. A monotypic genus (literally, "an odious caller") of New Guinea cuckoos.

Call. Avian vocalizations that are typically short, acoustically simple, and are usually not seasonally or sexually restricted. They include juvenile begging calls, distress calls, and location calls, and adult alarm calls, aggressive calls, contact calls, and nest-site calls. *See also* song.

Cercococcyx. A polytypic genus (literally, "a tailed cuckoo") of long-tailed African cuckoos.

Choana (pl., choanae). The internal nostrils on the upper palate of a bird.

Chrysococcyx. A polytypic genus (literally, "a golden cuckoo") of Old World glossy cuckoos.

Clamator. A polytypic genus (literally, "a shouter") of crested cuckoos.

Clutch. The total number of eggs laid by a female during a laying cycle and normally incubated by her. In brood parasites "clutches" simply represent all the eggs laid during a female's egg-laying cycle.

Clutch parasitism. A synonym for nest or brood parasitism. Not in general use, but semantically preferable to either nest or brood parasitism, as nest or brood parasites might logically include actual ecto- and endoparasites.

Combassous. A vernacular name for the indigobirds of Africa. *See also* indigobirds; viduine finches; whydahs.

Commissure. The angular junction between the upper and lower mandibles of a bird. Commissural tubercles are distinctive commissural enlargements in nestlings that appear to provide species-specific stimuli for eliciting parental feeding.

Conspecific brood parasitism. Brood parasitism occurring within a species. Synonymous with intraspecific brood parasitism. *See also* dump-nesting; prehatching brood amalgamation.

Cooperative breeding. The situation in which two or more females (or pairs) cooperate in the incubation and rearing of their collective offspring.

Cowbirds. Member species (including five brood parasites) of *Molothrus* and *Scaphidura.* In the New World passerine family Fringillidae, that often associate with cattle or other large ungulates while foraging.

Crypsis. The evolution of concealing visual features, such as background mimicry, object mimicry, countershading, and disruptive patterning. *See also* mimicry.

Cuckoo. A general vernacular name applied to most members of the order Cuculiformes.

Cuculus. The type-genus of the Old World cuckoo family Cuculidae.

Culmen length. A straight-line measure of bill length, taken from the tip of the upper mandible to the base of the skull (total culmen) or to the edge of the forehead feathering (exposed culmen). The culmen is the upper-edge profile of the bill.

Dietary-limitations model. A hypothesis that those species whose adults have specialized diets unsuited for the feeding of their own offspring are predisposed toward evolving brood parasitism. Other dietary-related scenarios for brood-parasitism evolution involve those species whose required mobility and/or short optimum breeding periods (both limiting the amounts of food and durations of available time for feeding their young) make it more desirable to induce other species to rearing their young than to attempt it themselves.

Dromococcyx. A polytypic name (literally, "a running cuckoo") of a genus of New World ground cuckoos.

Dump-nesting (or egg-dumping). A situation in which two or more females lay eggs in a common nest, which subsequently may either be abandoned or incubated. Includes both intra- and interspecific egg combinations. *See also* prehatching brood amalgamation.

Edge-dependent species. Those species whose breeding habitat is related to the amount of available forest edge, rather than to the amount of forest interior. *See also* forest-interior species.

Egg parasitism. Often considered a synonym of brood parasitism, but also used to describe dump-nesting.

Egg-crypticity model. A hypothesis that a parasitic egg might gain added protection from removal through egg crypsis (by the egg's matching of the nest background and thus becoming less visible), rather than through mimicry of the host's eggs.

Egg-laying interval. The interval between the laying of successive eggs by an individual female during a laying cycle. The total eggs laid during such a cycle constitutes a clutch (q.v.).

Egg mimicry models. Hypotheses that attempt to account for the evolution of host-egg mimicry present in some brood parasites, based on the correspondingly reduced probabilities of (1) egg recognition and removal by hosts (host-discrimination model), (2) recognition and removal by other parasitic females (parasite-discrimination model, or (3) discovery of the clutch by predators because of egg conspicuousness (predation-reduction model).

Egg range. *See* home range.

Eggshell color. The rather uniform underlying "ground color" of an egg, which may have superficial darker markings (speckled, spotted, blotched, streaked, scrawled, scribbled).

Emarginate. Abruptly narrowing.

Eudynamis. A polytypic genus (literally, "a true koel") of Old World cuckoos. "Dynamis" is derived from Dunamene, one of the mythical Greek sea nymphs.

Estrildine finches. Members of the Old World passerine taxon Estrildinae, sometimes also called waxbills.

Eumelanin. A type of melanin responsible for gray, grayish black, and black hues in skin and skin derivatives. *See also* phaeomelanin.

Exploitative brood parasitism. Brood parasitism in which the parasite attains reproductive benefits, whereas the host endures reproductive costs. Nonexploitative brood parasitism may also occur, in which the host endures no reproductive costs and sometimes may even benefit.

Extra-pair copulations (EPC). Copulations by paired birds with other individuals, including both forced copulations ("rapes") and unforced or promiscuous copulations. *See also* kleptogamy.

Eye-ring. A bare, often brightly colored, area of skin around the eye; present in many cuckoos.

Facultative brood parasitism. Parasitism by a species that sometimes but not invariably reproduces in this manner. *See also* obligatory brood parasitism.

Fecundity. Relative reproductive rates, especially among females.

Fledging period. The interval occurring between hatching and initial flight of a baby bird; comparable to the nestling period in most altricial species. *See also* altricial, nestling period.

Fledging success. An estimate of reproductive efficiency, based on the percentage of a population's hatched young that survive to fledging. *See also* breeding success; hatching success.

Fledgling. A bird at the approximate age of fledging (initial flight) and relative independence from its parents.

Forest-interior species. Species whose breeding success depends on large areas of continuous (edge-free) forests. *See also* edge-dependent species.

Fortuitous egg-laying model. A hypothesis by Darwin that brood parasitism evolved as a result of chance laying of eggs in the active nests of others. Such eggs may have been increasingly accepted by hosts as adaptations facilitating brood parasitism (host mimicry, etc.) evolved. Related scenarios advanced as possible avenues to brood parasitism include the use of old nests, the takeover of nests of other species, and dump-nesting tendencies.

Fostering host. A species known to have hatched and nurtured parasitically laid eggs to fledging or independence. Synonymous with biological, suitable, and true hosts. *See also* hosts.

Foster parent. An adult that has assumed parental care of a chick other than or in addition to its own offspring. *See also* biological parent.

Fratricide. The killing of nest-mates (normally siblings) by nestlings. As broadly applied here, fratricide also includes the killing or elimination of host nestlings (and sometimes conspecifics) by first-hatched brood parasites, in addition to their destruction or ejection of any remaining eggs.

Gallery forest. Narrow forests associated with rivers or streams flowing through otherwise nonforested regions.

Gens (pl. gentes). A subpopulation of a brood parasite in which the females are genetically adapted to lay host-mimetic eggs and behaviorally predisposed (probably through imprinting effects) to parasitize a particular host species. *See also* individual-host specificity.

Grasp ejection. The elimination of foreign eggs from a nest by grasping them with the beak. *See also* puncture ejection.

Ground cuckoos. A vernacular name for a New World cuckoo family (Neomorphidae), which includes all of the obligatory brood parasites of the Americas.

Hatching success. An estimate of a population's breeding efficiency based on the percentage of its eggs known to hatch but that not necessarily fledge. *See also* breeding success; fledging success.

Hawk mimicry. The visual mimicry (in appearance and/or behavior) of hawks, particularly accipiters, by some parasitic cuckoos such as hawk cuckoos. Hawk mimicry may help prevent hosts from defending their nests, since hawklike females may intimidate them near their nests or hawklike males may distract hosts while the females visit their nests.

Hectare (ha). An area equal to 10,000 square meters, or 2.47 acres.

Hepatic morph. A rufous plumage phenotype of some female cuckoos that is rich in phaeomelanins. Also called "hepatic phase" or "rufous phase."

Heteronetta. A monotypic genus (literally, "a different duck") of the black-headed duck.

Hierococcyx. A subgenus (literally, "a hawk-cuckoo") of the Old World cuckoos.

Home range. The area occupied, but not necessarily defended, by an individual over a specified time period. Among brood parasites, female Eurasian cuckoos may have overlapping egg ranges comparable to their home ranges; males also have similar song ranges.

Homoxenia (adj. homoxenic). Situations in which two brood parasites share the same host species. *See also* alloxenia.

Host mimicry. Evolved phenotypic similarities between hosts and parasites. In avian brood parasites, such mimicry may include egg similarities (size, color, pattern), nestling similarities (gape or head colors, plumage similarities), and behavioral similarities of young or adults (postures, calls, songs). *See also* egg mimicry, mouth mimicry, vocal mimicry.

Hosts. Individuals or species that are victimized by brood parasites, but that do not necessarily rear their young. Even if they accept the parasitic eggs ("acceptors" or "willing" victims), they may be unsuitable hosts (their method of food presentation or the type of food provided may not suit the parasite). If they not only incubate the parasite's eggs but also effectively care for their young, host species may be described as biological, fostering, or true hosts (q.v.). Depending on their rate of parasitism, host species of host-generalist parasites may also be described as frequent, infrequent, occasional, or rare hosts; hosts of host-specific parasites may be described as primary versus marginal or accidental hosts.

Host-discrimination model. A hypothesis that egg mimicry evolved as a result of host discrimination, in which only those eggs most like the host are likely to be accepted and survive in host populations. By this scenario a brood parasite might evolve one or more egg types (morphs) that mimic specific primary host species or perhaps develop a less specific but still widely accepted egg type effective in parasitizing a variety of available and suitable hosts.

Host-generalist parasites. Brood parasites in which individual females may lay their eggs in the nests of a variety of host species, with no preference for those of a particular species. *See also* host-specific parasites; individual-host specificity.

Host-intolerant parasites. Brood parasites whose newly hatched young kill nest-mates or evict other eggs or young from the nest. *See also* host-tolerant parasites.

Host-specific parasites. Parasitic species whose host dependency is limited to a single species or a small group of closely related species. *See also* host generalists; individual-host specificity.

Host-tolerant parasites. Brood parasites whose young neither kill nor evict eggs or nest-mates from the nest but may cause their eventual starvation through food competition. *See also* host-intolerant parasites.

Icterines. An inclusive term for the New World passerine category (subfamily Icterinae, tribe Icterini) that includes cowbirds, blackbirds, oropendolas, caciques, etc.

Immature. *See* juvenile.

Imprinting. A unique type of age-dependent learning that usually occurs early in life, is irreversible, and whose functional expression may not be apparent until much later in life ("latent learning"). The learning of a host species' traits (such as its nests and/or nesting habitats) by nestling brood parasites, who when mature may use this information to locate potential host nests, is an example. *See also* gens; individual-host specificity.

Incubation period. The interval between the onset of incubation of an egg and hatching.

Indicator. The type genus (literally, "a guide") of the honeyguide family Indicatoridae.

Indigobirds. A vernacular name for those viduine finches (q.v.) whose breeding males have body plumages that are iridescent green, blue, or purple and lack ornamental and elongated tail feathers.

Individual-host specificity. Brood parasites whose host species are collectively variable within a parasite's population but are constant for individual parasitic females. The host is probably chosen through an imprinting-like attachment to it by the parasite while it is still a nestling. *See also* gens; host-generalist parasites; host-specific parasites; imprinting.

Interspecific brood parasitism. Brood parasitism occurring between, rather than within, species. *See also* conspecific brood parasitism.

Intraspecific brood parasitism. *See* Conspecific brood parasitism.

Juvenal. The first nondowny plumage that is acquired by unfledged birds (juveniles). The less precise term "immature" is used to refer to birds that are not yet sexually mature, whether or not they have already molted their juvenal plumage.

Juvenile. A young bird with most or all of the feathers of its first nondowny (juvenal) plumage.

Kin selection. A type of natural selection associated with differential reproductive success in producing genetically related descendants in a population, either by direct parentage (direct selection) or by improved production on nondescendant relatives (indirect selection).

Kleptogamy. The "stealing" of matings from other individuals' mates, whether by force ("forced copulations") or by overt participation. *See also* extra-pair copulations.

Kleptoparasitism. The stealing of food from other individuals.

Latilong. A geographic quadrant defined by lines or degrees of latitude and longitude.

Lore. The area between a bird's eye and the base of its bill, which is bare of feathers in some cuckoos.

Malar stripe. A moustachelike cheek stripe, present in some honeyguides and cuckoos.

Mass. An object's relative resistance to acceleration, as measured in grams. Proportional to weight, which is often used as a convenient but not wholly accurate synonym.

Mating system. The variations in evolved mating patterns of a taxon, such as seasonal monogamy, harem (simultaneous) polygyny, successive polygyny, etc.

Mechanical sounds. Sounds generated by feather vibration, wing-whirring, or other nonvocal means, whether occurring incidentally or serving as social signals.

Mellignomon. A polytypic genus (literally, "a honey-tracker") of African honeyguides.

Mellichneutes. A monotypic genus (literally, "a honey hunter") of the lyre-tailed honeyguide.

Microdynamis. A monotypic genus (literally, "a tiny koel") of New Guinea cuckoos.

Mimicry (= mimesis). Evolved similarities between two taxa, such as those existing between a brood parasite (mimic) and its host (model). Mimicry may include both structural and behavioral similarities that help increase the fitness of the parasite to its host. *See also* egg mimicry; hawk mimicry; host mimicry; mouth mimicry; vocal mimicry.

Miombe woodland. Woodlands in southern Africa dominated by *Brachystegia* trees.

Misocallus. A subgeneric name (literally, "a nest hater") sometimes applied to the black-eared cuckoo.

Molothrus. A polytypic genus (literally, "a glutton or greedy beggar") of New World cowbirds.

Monotypic. A taxon having only a single member at the next subsidiary taxonomic level (e.g., a genus having only a single species). *See also* polytypic.

Morph. A genetically controlled individual phenotype, within a polymorphic population having two or more such phenotypes. Sometimes called "phases."

Mortality rate. A measure of a population's rate of diminution by death or destruction of eggs, nestlings, or adults over some specified time interval.

Mouth mimicry. Signaling devices (color patterns, papillae, tubercles, etc.) on the palate, tongue, or mandibles of nestling parasites that adaptively mimic corresponding host signals and stimulate host-parent feeding behavior. *See also* bill flange.

Multiple parasitism. The simultaneous presence of two or more parasitic eggs in a host's nest, laid either by the same or (usually) different individual parasites. Multiple parasitism is more common in nest-sharing than nest-exclusive brood parasites.

Mutualism. An ecological interaction between two species in which both benefit, as opposed to those that are cost-free (neutralism) or unilaterally beneficial (exploitation).

Neotropical migrants. Migratory bird species seasonally present in the Neotropical region (Central and South America), especially those wintering in the Neotropics but breeding in the Nearctic region (North America south to northern Mexico).

Nest parasitism. The utilization of a nest built by another individual for egg laying, especially the active usurpation or takeover of a currently active nest of the same or a different species.

Nest structural types. Types of bird nests: open overhead (cup- or platformlike) or domed, with dorsal, lateral, or even ventral openings. Rarely nests are purselike, allowing limited entry. Nests may also be supported entirely from below (statant), from their sides (pensile), or from above (pendulous). Nest shape may greatly influence protection from brood parasites.

Nesting success. A measure of reproductive efficiency, based on the percentage of initiated nests that succeed in hatching at least one egg per clutch.

Nestling. A young bird still in the nest.

Nestling period. The interval between a chick's hatching and its leaving the nest. *See also* fledging period; nidicolous; nidifugous.

Nidicolous. Those species whose young remain in the nest until they are fledged or nearly fledged. *See also* altricial; nidifugous.

Nidifugous. Those species whose precocial young leave the nest shortly after hatching and often are able to fend for themselves well before they fledge. *See also* precocial; nidicolous.

Obligatory brood parasitism. Social parasites that never construct nests nor incubate their own eggs. *See also* facultative brood parasitism.

Oropendola. A collective vernacular name for a group of large, colonial-nesting icterines that build pendulous nests.

Oxylophus. A polytypic genus (literally, "sharp-crested") of crested cuckoos; often regarded as a subgenus within *Clamator.*

Pachycoccyx. A monotypic genus (literally, "a thick cuckoo") of the African thick-billed cuckoo.

Palatal patterns. *See* mouth mimicry.

Parasite-discrimination model. A hypothesis that egg mimicry in cuckoos evolved because parasitic eggs closely resembling host eggs have a reduced likelihood of subsequent recognition and removal during later visits to the same nest by other female cuckoos.

Parasites. Species or individuals that exist at the expense of other individuals or species ("hosts"), but do not usually cause their deaths. Social parasites are those that indirectly extract host energy, such as by stealing food (kleptoparasitism), matings (kleptogamy), or parental care (brood parasitism) from others and are thus distinct from true ecto- or endoparasites that directly extract energy from the host's body.

Parasitism costs. An estimate of the reduced fecundity of the host species resulting from the presence of the parasite, usually measured in terms of diminished rates of hatching and/or fledging success.

Parasitism intensity. A measure of the impact of brood parasitism, usually based on the number of parasitic eggs present in a single nest, but sometimes measured as the percentage of parasitic eggs present in a combined sample of host and parasite eggs. *See also* parasitism rate.

Parasitism pressure. A measure of brood parasitism impact on a host, based on the number of parasite species affecting that host. Equivalent to "species density" of parasitologists.

Parasitism rate. A measure of the impact of brood parasitism, usually based on the percentage of parasitized nests of a host population (equivalent to "parasitism prevalence" of parasitologists). Distinct from parasitism intensity (see above); these two statistics may differ considerably if multiple parasitism occurs commonly, but such differences are important only for host-tolerant parasites. Ideal maximum parasitism rates (from the parasite's standpoint) are those that barely allow the host to maintain its population indefinitely.

Passerid. A member of the Old World sparrow family Passeridae.

Phaeomelanin. A variety of melanin that produces rusty-brown hues, as distinct from the grayish-blacks produced by eumelanins. *See also* eumelanins.

Ploceid. A member of the Old World weaver finch family Ploceidae (here considered as a part of the enlarged family Passeridae).

Poisson distribution. A statistical distribution describing the occurrence of unlikely events in a large number of independent repeated trials.

Polygamy. A nonmonagamous mating system, such as polyandry (multiple male mates per female), polygyny (multiple female mates per male), and polygynandry (multiple mate-sharing in both directions). *See also* promiscuity.

Polytypic. A taxonomic group that has two or more members at the next subsidiary taxonomic level (e.g., a genus having two or more included species). *See also* monotypic.

Posthatching brood amalgamation. The fusion of broods of two or more females, usually of the same species.

Postocular. Located behind the eye, such as postocular eye stripes.

Prehatching brood amalgamation. The laying of eggs in a common nest by more than one female. *See also* dump-nesting.

Precocial. Species whose down-covered young are able to attain early independent thermoregulation and have motor and sensory abilities permitting rapid nest departure and self-care. *See also* altricial.

Predation-reduction model. A hypothesis that egg mimicry by brood parasites evolved because as the presence of a conspicuous egg in a host species' nest would expose that nest to a greater risk of predation.

Predation risk-spreading model. A hypothesis that brood parasitism may have evolved as an evolutionary strategy to spread the risks of nest predation and clutch or brood loss for an individual female.

Prodotiscus. A polytypic genus (literally, "a betrayer") of small African honeyguides.

Promiscuity. A mating system distinct from polygynandry in that it implies no individual pair-bonding or mate responsibilities beyond simple fertilization and may include polygynous promiscuity as well as polyandrous promiscuity. *See also* polygamy.

Proximate factors. Environmental or internal factors that trigger responses in an individual or species at the present time. *See also* ultimate factors.

Puncture ejection. The removal of foreign eggs by puncturing them with the beak and then removing them from the nest; also called "spiking." *See also* grasp-ejection.

Record of parasitism (ROP). An individual case or instance of nest parasitism.

Recruitment rate. A measure of population structure based on the percentage of current-year young in a species' population. This rate is estimated soon after the end of the breeding season and is equal to the species' fecundity rate less the collective rates of unreplaced egg losses and the mortality of prefledged young.

Rectrices (sing. rectrix). The collective tail feathers of a bird.

Rejector species. Potential host species that reject (usually by egg puncture ejection, covering-over, or nest abandonment) parasitic eggs. *See also* acceptor species.

Remiges (sing. remix). The collective flight feathers (primaries and secondaries) of a bird's wings.

Reproductive-maladaptations model. A hypothesis that brood parasitism resulting from breeding maladaptations (such as asynchrony between nest building and egg laying, or a gradual loss of brooding tendencies) that led to degenerative brood parasitism. This scenario contrasts with the progressive adaptations for and specialization in brood parasitism of most other models.

Rey's index. An index of relative eggshell strength, calculated as length × width (in mm) divided by eggshell weight (in mg). The lower the index (to a minimum of about 1:0 in brood parasites), the more the eggshell will resist puncture or breakage.

Rhamphomantis. A monotypic genus (literally, "a billed prophet") of New Guinea cuckoos.

Riparian. Refers to rivers, lakes, and their shorelines.

Scaphidura. A monotypic genus (literally, "a boatlike tail") of the giant cowbird.

Scythrops. A monotypic genus (literally, "an angry face") of the channel-billed cuckoo.

Search image. The learning of key characteristics associated with a particular host or prey species by an individual parasite or predator and its subsequent application in facilitating its searches for hosts or prey.

Shape index. A ratio based on the length of an egg relative to its width. Nearly spherical eggs have ratios approaching 1:1; highly elongated eggs may have ratios in excess of 1:1.5.

Site fidelity. A tendency of an individual to return to the same location (nest site, territory, etc.) in successive years.

Social parasitism. A term for social exploitative interactions that collectively include kleptogamy, kleptoparasitism, nest or nest-site stealing, and brood parasitism.

Song. A term for avian vocalizations that are prolonged and complex acoustically (often with both amplitude and frequency modulations). Songs are also often sex-limited and/or seasonally restricted in occurence, typically functioning as sexual and/or territorial advertisements. Song dialects are regional differences in songs typical of particular populations. *See also* calls; vocalizations.

Song range. *See* home range.

Specific-host parasites. *See* host-specific parasites.

Stiff-tailed ducks. A vernacular name for a group of diving duck species having unusually long, stiff rectrices.

Strategy. Behaviors or other adaptations that a species has evolved in maximizing its fitness.

Structural mimicry. Evolved visual similarities in the morphology (plumage, body shape, mouth patterns, etc.) between a parasite (mimic) and host (model).

Surniculus. A monotypic genus (literally, "a somber cuckoo") of Old World cuckoos.

Sympatry. Two populations with overlapping geographic distributions during the breeding season.

Syrinx (pl. syringes). The vocal organ of birds. In the groups considered here, the syrinx is usually located at the junction of the trachea and bronchi (tracheobronchial) but it also (as in some cuckoos) may be entirely bronchial. Vocalizations produced by the syrinx (calls and songs) may be modulated by the trachea and/or the oral cavity and esophagus. *See also* vocalizations.

Tapera. A monotypic genus (derived from a Tupi name for an animal with a ghostlike traits) of ground cuckoos.

Taxon (pl. taxa). A representative taxonomic group, such as a particular species or genus (which collectively are taxonomic categories). Taxonomy is the study, naming, and systematic classification of taxa.

Territoriality. Agonistic behavior related to localized social dominance (defense of a specific area), usually among conspecifics. Territories may be part of larger home ranges (those areas occupied but not necessarily defended). *See also* home range.

Tolerant host. A host that fledges more than 20% of a parasite's eggs. (Mayfield, 1985) *See also* true host.

Troupial. An English vernacular name for *Icterus icterus.* also sometimes used as a collective name for orioles, caciques, and oropendolas.

True host. Host species known to have hatched and reared a brood-parasite's young; synonymous with fostering and biological hosts. *See also* tolerant host; unsuitable host.

Ultimate factors. Environmental factors that during a species' past evolutionary history have shaped its present-day adaptations. *See also* proximate factors.

Unsuitable host. A host species that is known to hatch a parasite's eggs but, because of the manner of food presentation or type of food provided, the parasitic young cannot survive to fledging.

Vidua. A polytypic genus (literally "a widow") of African parasitic finches, family Passeridae.

Viduine finches. An inclusive term for the African passerine tribe Vidulni). *See also* whydahs.

Vocal mimicry. The mimicking of a host-species' vocalizations by its brood parasite, including both call mimicry (such as food-begging calls) and song mimicry (use of the host's species-specific vocalizations).

Vocalizations. Sounds generated by the syrinx and often modulated (in amplitude and/or frequency) by it and related respiratory-tract structures. *See also* mechanical sounds.

Waxbills. An inclusive term for various species of estrildine finches, especially those having a bill color resembling sealing wax. *See also* estrildine finch.

Whydahs. A variably inclusive (sometimes excluding indigobirds) vernacular term for some or all of the African viduine finches (Passeridae, Viduini). *See also* viduine finches.

Wing:tail ratio. A ratio of the wing length (measured from the wrist to the longest primary's tip) relative to the tail length (measured from the insertion of the middle rectrices to their tips); here expressed decimally, with the wing length designated as equal to 1.0.

Zygodactyl. The "yoke-toed" arrangement typical of cuckoos, in which the first and fourth toes are directed backward and the other two are pointed forward, as opposed to the usual avian arrangement of three toes forward and one backward (anisodactyl).

Appendix B

LATIN NAMES OF BIRDS MENTIONED IN THE TEXT

Latin nomenclature and adoption of English vernacular names generally follows Sibley & Monroe (1990), but some commonly encountered alternate vernacular names are cross-referenced. Names of brood parasites having individual text accounts are excluded from this list.

Abert's towhee *Pipilo aberti*
Abyssinian roller *Coracia abyssinica*
Abyssinian scimitarbill *Phoeniculus minor*
Abyssinian white-eye *Zosterops abyssinica*
Acadian flycatcher *Empidonax virescens*
African broadbill *Smithornis capensis*
African firefinch *Lagonosticta rubricata*
African golden-breasted bunting *Emberiza flaviventris*
African golden oriole *Oriolus auratus*
African golden weaver *Ploceus subaureus*
African paradise flycatcher *Terpsiphone viridis*
African pygmy kingfisher *Ceyx picta*
African quailfinch *Ortygospiza atricollis*
African red-eyed bulbul *Pycnonotus nigricans*
African reed warbler *Acrocephalus baeticatus*
African silverbill *Lonchura cantans*
Alder flycatcher *Empidonax alnorum*
Allied flycathcer warbler, *see* white-spectacled warbler
Altimira oriole *Icterus gularis*
American coot *Fulica americana*
American goldfinch *Carduelis tristis*
American redstart *Setophage ruticilla*
American robin *Turdus migratorius*
Amethyst sunbird *Nectarina amethystina*
Anambra waxbill *Estrilda poliopareia*
Anchieta's barbet *Stactolaema anchietae*
Anteater chat, *see* northern anteater-chat
Arctic warbler *Phylloscopus borealis*
Argentine blue-billed (or lake) duck *Oxyura vittata*
Arrow-marked babbler *Turdoides jardineii*

Ashy drongo *Dicrurus leucophaeus*
Ashy-gray wren warbler, *see* gray-breasted prinia
Ashy laughingthrush, *see* moustached laughingthrush
Ashy prinia (or wren-warbler) *Prinia socialis*
Asian fairy-bluebird *Irene puella*
Asian paradise-flycatcher *Terpsiphone paradisi*
Audubon's oriole *Icterus graduacauda*
Australasian magpie *Gymnorhina tibicen*
Australasian white-eyed pochard *Aythya australis*
Australian blue-billed duck *Oxyura australis*
Australian maned (or wood) duck *Chenonetta jubata*
Australian raven *Corvus coronoides*
Australian shelduck *Tadorna tadornoides*
Avadavit *Amandava amandava*
Azara's spinetail *Synallaxis azarae*
Azure-winged magpie *Cyanopica cyana*

Baglafecht weaver *Ploceus baglafecht*
Banded martin *Riparia cincta*
Bar-breasted firefinch *Lagonosticta rufopicta*
Bar-breasted honeyeater *Ramsayornis fasciatus*
Bar-headed goose *Anser indicus*
Bar-throated appalis *Apalis thoracica*
Bare-faced babbler *Turdoides gymnogenys*
Barn swallow *Hirundo rustica*
Barred antshrike *Thamnophilus doliatus*
Barrow's goldeneye *Bucephala islandica*
Bay-breasted warbler *Dendroica castanea*
Bearded scrub robin *Erythropygia quadrivirgata*
Beautiful sunbird *Nectarinia pulchella*
Bell minor *Manorina melanophrys*
Bell's vireo *Vireo bellii*
Bengalese finch *Lonchura striata*
Bennett's woodpecker *Campethera bennettii*
Bewick's wren *Thyromanes bewickii*
Bicolored honeycreeper *Conirostrium bicolor*
Bicolored wren *Campyiorhynchus griseus*
Black drongo *Dicrurus macrocercus*
Black flycatcher *Melaenornis pammelaina*
Black helmet shrike, *see* red-billed helmet shrike
Black scoter *Melanitta nigra*
Black sunbird *Nectarinia aspasia*
Black tit *Parus niger*
Black-bellied firefinch *Lagonosticta rara*
Black-bellied whistling duck *Dendrocygna autumnalis*
Black-billedcuckoo *Coccyzus erythropthalmus*
Black-billed magpie *Pica pica*
Black-browed warbler, *see* yellow-vented warbler
Black-capped chickadee *Parus atricapilla*
Black-capped social-weaver *Pseudonigrita arnaudi*
Black-capped vireo *Vireo atricapilla*
Black-cheeked waxbill *Estrilda erythronotos*
Black-chested prinia *Prinia flavicans*
Black-collared barbet *Lybius torquatus*
Black-collared starling (or myna) *Sturnus nigricollis*
Black-crowned night-heron *Nycticorax nycticorax*
Black-headed babbler, *see* dark-fronted babbler
Black-headed oriole *Icterus graduacauda*
Black-headed weaver *Ploceus melanocephalus*
Black-necked tailorbird, *see* dark-fronted tailorbird
Black-necked weaver *Ploceus nigricollis*
Black-and-red broadbill *Cymbirhynchus macrorhyncha*
Black-and-rufous warbling finch *Poospiza nigrorufa*
Black-rumped waxbill *Estrilda troglodytes*
Black-striped sparrow *Arremonops conirostris*
Black-tailed gnatcatcher *Polioptila melanura*
Black-tailed tityra *Tityra cayana*
Black-throated babbler, *see* gray-throated babbler
Black-throated blue warbler *Dendroica caerulescens*
Black-throated gray warbler *Dendroica nigrescens*
Black-throated green warbler *Dendroica virens*
Black-throated wattle eye *Platysteira peltata*
Black-throated firefinch *Lagonosticta larvata*
Black-throated weaver *Ploceus benghalensis*
Black-and-white warbler *Dendroica varia*
Black-winged bishop *Euplectes hordaeaceus*
Blackburnian warbler *Dendroica fusca*
Blackcap *Sylvia atricapilla*
Blackcap babbler *Turdoides reinwardtii*
Blackpoll warbler *Dendroica striata*
Bleating bush warbler *Camaroptera brachyura*
Blue grosbeak *Guira caerulea*
Blue magpie *Urocissa erythrorhyncha*
Blue whistling thrush *Myophonus caeruleus*
Blue-and-white fairywren *Malurus leucopterus*
Blue-and-white warbler *Cyanoptila cyanomelana*
Blue-and-yellow tanager *Thraupis bonariensis*

Blue-billed firefinch, *see* African firefinch
Blue-gray gnatcatcher *Polioptila caerulea*
Blue-throated flycatcher *Cyornis rubeculoides*
Blue-winged laughingthrush *Garrulax squamatus*
Blue-winged warbler *Vermivora pinus*
Blyth's leaf warbler *Phylloscopus reguloides*
Bobolink *Dolichonyx oryzivorus*
Bocage's weaver *Ploceus temporalis*
Boehm's bee-eater *Merops boehmi*
Boulder chat *Pinarornis plumosus*
Brambling *Fringilla montifringilla*
Brant *Branta bernicla*
Brazilian tanager *Ramphocelus bresilius*
Brewer's blackbird *Euphagus cyanocephalus*
Brewer's sparrow *Spizella breweri*
Broad-tailed thornbill *Acanthiza apicalis*
Bronze munia *Lonchura cucullata*
Bronze sunbird *Nectarinia kilimensis*
Brown babbler *Turdoides plebejus*
Brown bush warbler *Bradypterus luteoventris*
Brown creeper *Certhia americana*
Brown firefinch *Lagonosticta nitidula*
Brown grass warbler, *see* pectoral-patch cisticola
Brown hill warbler, *see* striated prinia
Brown illadopsis *Trichastoma fulvescens*
Brown shrike *Lanius cristatus*
Brown thornbill *Acanthiza pusilla*
Brown thrasher *Toxostoma rufum*
Brown towhee *Pipilo fuscus*
Brown twinspot *Clytospiza monteiri*
Brown wren babbler, *see* pygmy wren babbler
Brown-backed honeyeater *Ramsayornis modestus*
Brown-capped laughingthrush *Garrulax austeni*
Brown-cheeked fulvetta *Alcipe pollocephala*
Brown-headed gull *Larus ridibundus*
Brown-headed honeyeater *Melithreptus brevirostris*
Brown-hooded kingfisher *Halcyon albiventris*
Brown-necked raven *Corvus ruficollis*
Brown-throated wattle-eye *Platysteira cyanea*
Brown-and-yellow marshbird *Pseudolistes virescens*
Brownish-flanked bush-warbler *Cettia fortipes*
Buff-crowned foliage-gleaner *Philydor rufus*
Buff-rumped thornbill *Acanthiza reguloides*
Buff-spotted woodpecker *Campethera nivosa*
Buff-throated apalis *Apalis rufogularis*
Bufflehead *Bucephala albeola*
Bull-headed shrike *Lanius bucephalus*
Burchell's starling *Lamprotornis australis*
Bush warblers *Cettia* spp.

Cabanis' bunting *Emberiza cabanisi*
Caciques *Cacicus* and *Amblycercus* spp.
Canada goose *Branta canadensis*
Canada warbler *Wilsonia canadensis*
Canvasback *Aythya valisineria*
Cape anteater chat, *see* southern anteater-chat
Cape batis *Batis capensis*
Cape crombec *Sylvietta rufescens*
Cape bulbul *Pycnonotis capensis*
Cape robin chat *Cossypha caffra*
Cape rock thrush *Monticola rupestris*
Cape rook *Corvus capensis*
Cape sparrow *Passer melanurus*
Cape wagtail *Motacilla capensis*
Cape weaver *Ploceus capensis*
Cardinal, *see* Northern cardinal
Cardinal woodpecker *Dendropicos fuscescens*
Carib grackle *Quiscalus lugubris*
Carmine bee-eater *Merops nubius*
Carolina chickadee *Parus carolinensis*
Carolina wren *Thyromanes ludovicianus*
Carrion crow *Corvus corone*
Cattle tyrant *Machetornis rixosus*
Cedar waxwing *Bombycilla cedrorum*
Cerulean warbler *Dendroica cerulea*
Chalk-browed mockingbird *Mimus saturninus*
Chaplin's barbet *Lybius chaplini*
Chattering cisticola *Cisticola anonymus*
Chestnut munia *Lonchura malacca*
Chestnut sparrow *Passer eminibey*
Chestnut teal *Anas castanea*
Chestnut weaver *Ploceus rubiginosus*
Chestnut-bellied (or red-billed) helmet shrike *Prionops caniceps*
Chestnut-bellied rock thrush *Monticola rufiventris*
Chestnut-bellied starling *Spreo pulcher*
Chestnut-capped blackbird *Agelaius ruficapillus*
Chestnut-collared longspur *Calcarius ornatus*

Chestnut-crowned laughingthrush *Garrulax erythrocephalus*
Chestnut-crowned warbler *Siecercus castaniceps*
Chestnut-fronted helmet-shrike *Prinops scopifrons*
Chestnut-headed (or Wagler's) oropendola *Psarocolius wagleri*
Chestnut-rumped thornbill *Acanthiza uropygialis*
Chestnut-sided warbler *Dendroica pennsylvanica*
Chiffchaff *Phylloscopus colybyta*
Chimango *Milvago chimango*
Chinese babax *Babas lanceolatus*
Chinspot batis *Batis molitor*
Chipping sparrow *Spizella passerina*
Chorister robin-chat *Cossypha dichroa*
Chotoy spinetail *Synallaxis (Schoeniophylax) phryganophila*
Chough, *see* red-billed chough
Cinciodes *Cinclodes* spp.
Cinereous finch *Piezorhina cinerea*
Cinnamon-chested bee-eater *Merops oreobates*
Cisticolas *Cisticola* spp.
Clay-colored sparrow *Spizella pusilla*
Cliff swallow *Hirundo pyrrhonota*
Codrington's indigobird, *see* variable indigobird
Collared antshrike *Sakesphorus bernardi*
Collared sunbird *Anthreptes collaris*
Comb duck *Sarkidiornis melanotos*
Common babbler *Turdoides caudatus*
Common bulbul *Pycnonotus barbatus*
Common diuca-finch *Diuca diuca*
Common eider *Somateria mollissima*
Common goldeneye *Bucephala clangula*
Common grenadier *Uraeginthus granatina*
Common iora *Aegithina tiphia*
Common jery, *see* northern jery
Common kestrel *Falco tinnunculus*
Common mallard, *see* mallard
Common merganser *Mergus merganser*
Common moorhen *Gallinula chloropus*
Common myna *Acridotheres tristis*
Common pochard *Aythya ferina*
Common raven *Corvus corax*
Common redstart *Phoenicurus phoenicurus*
Common shelduck *Tadorna tadorna*
Common stonechat *Saxicola torquata*
Common tailorbird *Orthotomus sutorius*
Common thornbird *Phacellodomus rufifrons*
Common waxbill *Estrilda astrild*
Common yellowthroat *Geothlypis trichas*
Copper sunbird *Nectarinia cuprea*
Coscoroba swan *Coscoroba coscoroba*
Cotton pygmy goose *Nettapus coromandelianus*
Couas *Coua* spp.
Couch's kingbird *Tyrannus melanicholicus*
Creamy-bellied thrush *Turdus amaurochalinus*
Crescent-chested babbler *Stachyris melanothorax*
Crested barbet *Trachyphonus vaillantii*
Crested malimbe *Malimbus malimbicus*
Crested oropendola *Paroscolius decumanus*
Crimson sunbird *Aethopyga siparaja*
Crimson-breasted boubou (or gonolek) *Laniarius atrococcineus*
Crimson-rumped waxbill *Estrilda rhodopyga*
Croaking cisticola *Cisticola natalensis*
Crowned slaty flycatcher *Empidonax aurantioatrocristatus*

Dark firefinch *Lagonosticta rubricata*
Dark-fronted babbler *Rhopocichla atriceps*
Dark-necked tailorbird *Orthotomus atrogularis*
Dark-eyed junco *Junco hyemails*
Dartford warbier *Sylvia undata*
Desert cisticola *Cisticola aridula*
Dickcissel *Spiza americana*
Double-collared seedeater (or finch) *Sporophila caerulescens*
Drab-breasted bamboo tyrant *Hemitriccus diops*
Drongos *Dicrurus* spp.
Dunnock, *see* hedge accentor
Dusky alseonax *Musicapa adusta*
Dusky flycatcher *Empidonax oberholseri*
Dusky flycatcher (African), *see* dusky alseonax
Dusky sunbird *Nectarinia fusca*
Dwarf cuckoo *Cuculus pumilus*

Eared pygmy tyrant *Myiornis auricularis*
Earthcreepers *Upucerthia* spp.
Eastern bearded scrub robin *Erythropygia quadrivirgata*
Eastern black-browed warbler, *see* golden-spectacled warbler
Eastern bluebird *Sialia sialis*

Eastern crowned leaf warbler *Phylloscopus cornatus*
Eastern kingbird *Tyrannus tyrannus*
Eastern meadowlark *Sturneila magna*
Eastern phoebe *Sayornis phoebe*
Eastern wood-pewee *Contopus virens*
Egyptian goose *Alopochen aegypticus*
Emperor fairywren *Malurus cyanocephalus*
Emperor goose *Anser canagica*
Eurasian blackbird *Turdus merula*
Eurasian jackdaw *Corvus monedula*
Eurasian jay *Garrulus glandarius*
Eurasian sparrowhawk *Accipiter nisus*
Eurasian wigeon *Anas penelope*
European robin *Erithacus rubecula*
Eurasian skylark *Alauda arvensis*
European starling *Sturnus vulgaris*
Evening grosbeak *Coccothrautes vespertinus*
Eye-ringed flatbill *Rhynchocyclus brevirostris*
Eyebrowed wren babbler *Napothera epilidota*

Fawn-breasted waxbill *Estrilda paludicola*
Fan-tailed raven *Corvus rhipidurus*
Familiar chat *Cercomela familaris*
Fork-tailed drongo *Dicrurus adsimilis*
Fairy gerygone (or warbler) *Geregone palperosa*
Ferruginous white-eye *Aythya nyroca*
Field sparrow *Spizella pusilla*
Figbird *Sphecotheres viridis*
Fiji bush-warbler *Cettla ruficapilla*
Firewood-gatherer *Anumbius annumbi*
Fiscal shrike *Lanius collaris*
Five-striped sparrow *Amphispiza quinquestriata*
Flame-colored tanager *Piranga bidentata*
Forbes' blackbird *Curaeus forbesi*
Forest raven *Corvus tasmanicus*
Fork-tailed tyrant (or flycatcher) *Tyrannus savana*
Fox sparrow *Passerella iliaca*
Fulvous whistling duck *Dendrocygna bicolor*

Gadwall *Anas strepera*
Garden bulbul *Pycnonotus barbatus*
Garden warbler *Sylvia borin*
Garganey *Anas querquedula*
Goldbreast, *see* zebra waxbill
Golden bush robin *Tarsiger chrysaeus*
Golden-backed weaver *Ploceus jacksoni*
Golden-bellied gerygone *Gerygone sulphurea*
Golden-billed saltator *Saltator aurantirostris*
Golden-cheeked warbler *Dendroica chrysoparia*
Golden-crowned kinglet *Regulus satraps*
Golden-spectacled warbler *Siecercus burkii*
Golden-tailed woodpecker *Campethera abingoni*
Golden-tailed warbler *Vermivora chrysoptera*
Goosander, *see* common merganser
Gouldian finch *Chloebia gouidiae*
Gould's sunbird *Aethopyga gouldiae*
Graceful prinia *Prinia gracilis*
Grace's warbler *Dendroica graciae*
Grasshopper sparrow *Ammodramus savannarum*
Gray catbird *Dumatella carolinensis*
Gray duck, *see* Pacific black duck
Gray fantail *Rhipidura fuliginosa*
Gray flycatcher *Empidonax wrightii*
Gray gerygone *Gergone igata*
Gray sibia *Heteropohasia gracilis*
Gray teal *Anas gibberifrons*
Gray tit-flycatcher *Myioparus plumbeus*
Gray vireo *Vireo vicinior*
Gray wagtail *Motacilla cinerea*
Gray woodpecker *Dendropicos goertae*
Gray-backed cameroptera *Cameroptera brachyura*
Gray-bellied wren *Tesia cyaniventer*
Gray-breasted prinia *Prinia hodgsonii*
Gray-cheeked tit-babbler *Macronous flavicollis*
Gray-headed canary-flycatcher *Culicapa ceylonensis*
Gray-headed kingfisher *Halcyon leucocephala*
Gray-headed sparrow *Passer griseus*
Gray-hooded warbler *Seicercus xanthoschistus*
Gray-sided laughingthrush *Garrulax caerulatus*
Gray-throated babbler *Stachyris nigriceps*
Gray-throated barbet *Gymnobucco bonapartei*
Grayish saltator *Saltator coerulescens*
Great kiskadee *Pitangus sulphurata*
Great reed warbler *Acrocephalus arundinaceus*
Great-tailed grackle *Quiscalus mexicanus*
Greater blue-eared starling *Lamprotornis chalybaeus*
Greater double-collared sunbird *Nectarinia afra*
Greater necklaced laughingthrush *Garrulax pectoralis*
Greater red-breasted meadowlark, *see* long-tailed meadowlark

Greater roadrunner *Geococcyx californianus*
Greater scaup *Aythya marila*
Greater swamp warbler *Acrocephalus rufescens*
Greater thornbird *Phacellodomus ruber*
Green barbet *Stactolaema olivacea*
Green crombec *Sylvietta virens*
Green jay *Cyanocorax yncas*
Green oropendola *Psarocolius viridus*
Green wood-hoopoe *Phoeniculus purpureus*
Green white-eye *Zosterops virens*
Green-backed camaroptera *Camaroptera brachyura*
Green-backed twinspot *Lagonosticta nitidula*
Green-headed sunbird *Nectarinia verticalis*
Green-winged pytilia *Pytilia melba*
Greenfinch *Chloris chloris*
Greylag goose *Anser anser*
Groove-billed ani *Crotophaga sulcirostris*
Grosbeak-weaver *Amblyospiza albifrons*

Hardhead, *see* Australisian white-eyed pochard
Harlequin duck *Histrionicus histrioncus*
Hartlaub's babbler *Turdoides hartlaubi*
Hedge accentor *Prunella modularis*
Helmated friarbird *Philemon buceroides*
Hermit thrush *Catharus guttatus*
Hermit warbler *Dendroica occidentalis*
Heuglin's masked weaver *Ploceus heuglini*
Heuglin's robin chat, *see* white-browned robin-chat
Hildebrandt's starling *Spreo hildebrandti*
Hill myna *Gracula religiosa*
Hill prinia *Prinia atrogularis*
Himalayan barwing, *see* rusty-fronted barwing
Himalayan whistling thrush, *see* blue whistling-thrush
Holub's golden weaver *Ploceus xanthops*
Hooded crow *Corvus corone*
Hooded merganser *Mergus cucullatus*
Hooded oriole *Icterus cucullatus*
Hooded siskin *Spinus magellanicus*
Hooded warbler *Wilsonia citrina*
Hoopoe *Upupa epops*
Horned lark *Eremophila alpestris*
Horneros *Furnarius* spp.
Horsfield's babbler *Trichastoma separia*
House crow *Corvus splendens*
House finch *Carpodacus mexicanus*
House sparrow *Passer domesticus*
House wren *Troglodytes aedon*
Hutton's vireo *Vireo huttoni*

Indian blue robin (or chat) *Lucinia brunnea*
Indian gray thrush *Turdus unicolor*
Indian myna *Acridotheres tristis*
Indigo bunting *Passerina cyanea*
Inornate warbler *Phylloscopus inornatus*

Jackdaw, *see* Eurasian jackdaw
Jameson's firefinch *Lagonosticta rhodopareia*
Japanese bush warbler *Cettia diphone*
Japanese robin *Erithacus akahige*
Java sparrow *Lonchura orizivora*
Jungle babbler *Turdoides striatus*
Jungle crow *Corvus levaillantii*

Karoo scrub wren *Cercotrichas coryphaeus*
Kentucky warbler *Oporornis formosus*
King eider *Somateria spectabilis*
Kirtland's warbler *Dendroica kirtlandii*
Knysna woodpecker *Campethera notata*
Kurrichane thrush *Turdus libonyanus*

Lapland longspur (or bunting) *Calcarius lapponicus*
Large crowned warbler, *see* western crowned warbler
Large gray babbler *Turdoides malcolmi*
Large scrub wren *Sericornis nouhyusi*
Large-billed crow *Corvus macrorhynchus*
Large-billed gerygone *Gerygone magnirostris*
Large-billed scrubwren *Sericornis magnirostris*
Large-billed warbler *Gerygone magnirostris*
Lark sparrow *Chondestes grammacus*
Laughingthrushes *Garrulus* spp.
Lazuli bunting *Passerina amoena*
Leaden flycatcher *Myiagra rubecula*
Leaf warblers *Phylloscopus* spp.
LeConte's sparrow *Ammodramus leconteii*
Lemon-breasted flycatcher *Microeca flavigaster*
Lesser goldfinch *Spinus psaltria*
Lesser ground cuckoo *Morococcyx erythropygus*
Lesser masked weaver *Ploceus intermedius*
Lesser necklaced laughingthrush *Garrulax moniligera*
Lesser scaup *Aythya affinis*

Lesser short wing *Brachipteryx leucophrys*
Lesser whistling duck *Dendrocygna javanica*
Limpkin *Aramus guarauna*
Lincoln's sparrow *Melospiza lincolni*
Lineated pytillia *Pytilla lineata*
Linnet *Carduelis cannabina*
Little bee-eater *Merops pusillus*
Little crow *Corvus bennetti*
Little friarbird *Philemone citreogularis*
Little green bee-eater *Merops orientalis*
Little minivet *Pericrocotus peregrinus*
Little sparrowhawk *Accipiter minullus*
Little spider hunter *Arachnothera longirostris*
Long-billed thrasher *Toxostoma longirostre*
Long-tailed duck, *see* oldsquaw
Long-tailed meadowlark *Sturnella loyca* (= *Pezites militaris* of Freidmann & Kiff, 1985)
Long-tailed mockingbird *Mimus longicaudatus*
Long-tailed reed finch *Donacospiza albifrons*
Long-tailed shrike *Lanius schach*
Long-tailed starling *Lamprotornis mevesii*
Long-tailed wagtail *Motacilla clara*
Long-tailed wren babbler, *see* tawny-breasted wren-babbler
Louisiana waterthrush *Seiurus motacilla*
Lucy's warbler *Vermivora luciae*

Maccoa duck *Oxyura maccoa*
MacGillivray's warbler *Oporornis tolmiei*
Madagascar bee-eater *Merops superciliosa*
Madagascar coucal *Centropus toulou*
Madagascar cisticola *Cisticola cherina*
Madagascar paradise flycatcher *Terpsiphone mutata*
Madagascar swamp warbler *Acrocephalus newtoni*
Magnolia warbler *Dendroica magnolia*
Magpie, *see* clack-billed magpie
Magpie lark *Grallina cyanoleuca*
Magpie munia *Lonchura fringilloides*
Maguari stork *Eluxenura maguari*
Malachite sunbird *Nectarina famosa*
Mallard *Anas platyrhynchos*
Malkohas *Phaenicophaeus* spp.
Manchurian bush-warbler *Cettia canturians*
Mangrove gerygone (or warbler) *Gerygone levigaster*
Marbled teal *Marmaronetta angustirostris*
Mariqua sunbird *Nectarinia mariquensis*
Marsh warbler *Acrocephalus palustris*
Masked duck *Oxyura dominica*
Masked firefinch, *see* black-throated firefinch
Masked gnatcatcher *Ptiloptila dumicola*
Masked weaver *Ploceus vellatus*
Meadow pipit *Anthus pratensis*
Melodious blackbird, *see* scrub blackbird
Mocking chat *Thamnolaea cinnamomeiventris*
Mockingbird, *see* northern mockingbird
Montane white-eye *Zosterops poliogastra*
Montezuma oropendola *Gymnostinops montezuma*
Mourning dove *Zenaida macroura*
Mourning warbler *Oporornis philadelphia*
Mouse-colored sunbird *Nectarinia veroxii*
Moussier's redstart *Phoenicurus moussieri*
Moustached laughingthrush *Garrulax cineraceus*
Muscovy duck *Cairina moschata*
Musk duck *Biziura lobata*

Naked-faced barbet *Gymnobucco calvus*
Nashville warbler *Vermivora ruficapilla*
Natal robin chat *Cossypha natalensis*
Nepal fulvetta *Alcippe nepalensis*
New Zealand shelduck *Tadorna variegata*
Newton's sunbird *Nectarinia newtoni*
Noisy friarbird *Philemone corniculatus*
North American black duck *Anas rubripes*
North American wood duck *Aix sponsa*
Northern anteater chat *Myrmecocichla aethiops*
Northern cardinal *Cardinalis cardinalis*
Northern jery *Neomixis tenella*
Northern masked weaver *Passer taeniopterus*
Northern mockingbird *Mimus polyglottos*
Northern oriole *Icterus galbula*
Northern parula *Parula americana*
Northern shrike *Lanius excubitor*
Northern shoveler *Anas clypeata*
Northern waterthrush *Seiurus novaeboracensis*
Nubian woodpecker *Campethera nubica*

Ochre-breasted brush-finch *Atlapetes simirufus*
Ochre-faced tody-flycatcher *Todirostrum plumbeiceps*
Oldsquaw *Clangula hyemalis*
Old World ground cuckoos *Carpococcyx* spp.

Olive sparrow *Arremonops rufivirgatus*
Olive spinetail *Cranioleuca obsoleta*
Olive sunbird *Nectarinia olivacea*
Olive thrush *Turdus olivaceus*
Olive woodpecker *Dendropicos griseocephalus*
Olive-backed oriole *Oriolus sagittatus*
Olive-backed pipit *Anthus hodgsoni*
Olive-backed sunbird *Nectarinia jugularis*
Olive-backed tailorbird *Orthotomus sepium*
Olive-bellied sunbird *Nectarinia chloropygia*
Olive-capped coucal *Centropus ruficeps*
Olive-sided flycathcer *Contopus borealis*
Orange minivet *Pericrocotus flammeus*
Orange-billed nightingale-thrush *Catharus aurantirostris*
Orange-breasted waxbill, *see* zebra waxbill
Orange-cheeked waxbill *Estrilda melpoda*
Orange-crowned warbler *Vermivora celata*
Orange-winged pytillia *Pytillia afra*
Orchard oriole *Icterus sprurus*
Orioles (Old World) *oriolus* spp. and (New World) *Icterus* spp.
Oropendolas *Psarocolius* & *Gymnostinops* spp.
Ovenbird *Seiurus aurocapillus*

Pacific black (or gray) duck *Anas superciliosa*
Painted bunting *Passerina ciris*
Pale flycatcher *Bradornis pallidus*
Pale-breasted spinetail *Synallaxis albescens*
Pale-footed bush warbler *Cettia pallidipes*
Pale-winged starling *Onychognathus nabouroup*
Pallas' warbler *Phylloscopus proregulus*
Palm warbler *Dendroica palmarum*
Paradise flycatchers *Terpsiphone* spp.
Patagonian mockingbird *Mimus patagonicus*
Pectoral-patch cisticola *Cisticola brunnescens*
Peruvian meadowlark, *see* greater red-breasted meadowlark
Peters's twinspot *Hypargos niveoguttatus*
Phainopepla *Phainopepla nitens*
Philadelphia vireo *Vireo philadelphia*
Pied babbler *Turdoides bicolor*
Pied barbet *Tricholaema leucomelas*
Pied bushchat (or stonechat) *Saxicola caprata*
Pied crow *Corvus albus*
Pied currawong *Strepera graculina*
Pied starling *Spreo bicolor*
Pied wagtail (African) *Motacilla aguimp*
Pied wagtail (British) *Motacilla alba yarrelli*
Pied water tyrant *Fluvicola pica*
Pine siskin *Cardulis pinus*
Pine warbler *Dendroica pinus*
Pink-eared duck *Malacorhynchus membranaceus*
Piping cisticola *Cisticola fulvicapillus*
Pipipi *Mohaua novaeseelandiae*
Plain ant-vireo *Dysithamnus mentalis*
Plain prinia (or wren-warbler) *Prinia inornata*
Plain wren *Thryothorus modestus elatus*
Plain-crowned spinetail *Synallaxis guianensis*
Plum-colored starling, *see* violet-backed starling
Plumbeous water-redstart *Rhyacornis fuliginosus*
Prairie warbler *Dendroica discolor*
Prevost's ground-sparrow *Melozone biarcuatum*
Prinias *Prinia* spp.
Pririt batis *Batis pririt*
Prothonotary warbler *Protonotaria citrea*
Purple finch *Carpodacus purpureus*
Purple grenadier *Uraeginthus ianthinogaster*
Purple sunbird *Nectarinia asiatica*
Purple-crowned fairy-wren *Malurus coronatus*
Pygmy sunbird *Anthreptes platurus*
Pygmy wren-babbler *Pnoepyga pusilla*

Quail finch, *see* African quailfinch

Racket-tailed drongo *Dicrurus paradisea*
Rattler cisticola *Cisticola chiniana*
Raven, *see* common raven
Red bishop *Euplectes orix*
Red wattlebird *Anthochaera carunculata*
Red-backed scrub wren *Cercotrichas leucophrys*
Red-backed shrike *Lanius collurio*
Red-backed wren *Malurus melanocephalus*
Red-billed chough *Pyrrhocorax pyrrhocorax*
Red-billed firefinch *Lagonosticta senegala*
Red-billed (or Retz's) helmet-shrike *Prionops retzi*
Red-billed leothrix *Leothrix lutea*
Red-billed pytilia *Pytilia lineata*
Red-breasted blackbird *Leistes militaris*

Red-breasted merganser *Mergus serrator*
Red-breasted nuthatch *Sitta canadensis*
Red-breasted shrike, *see* crimson-breasted gonolek
Red-capped robin *Petroica goodenovii*
Red-capped robin chat, *see* Natal robin-chat
Red-cheeked cordon-bleu *Uraeginthus bengalus*
Red-chested sunbird *Nectarinia erythroceria*
Red-collared widowbird *Euplectes ardens*
Red-crested pochard *Netta rufina*
Red-crowned ant-tanager *Habia rubica*
Red-eyed thornbird *Phacellodomus erythropthalmus*
Red-eyed vireo *Vireo olivaceus*
Red-faced cisticola *Cisticola erythrops*
Red-faced liocichla *Liocichla phoenicea*
Red-faced pytilia *Pytilia hypogrammica*
Red-flanked bluetail *Tarsiger cyanurus*
Red-fronted barbet *Tricholaema diademata*
Red-fronted coot *Fulica rufifrons*
Red-gartered coot *Fulica armillata*
Red-headed laughingthrush, *see* chestnut-crowned laughingthrush
Red-headed tanager *Piranga erythrocephala*
Red-headed weaver *Anaplectes rubriceps*
Red-rumped cacique *Cacicus haemorrhous*
Red-rumped waxbill *Estrilda charmosyna*
Red-shouldered glossy starling *Lamprotornis chalybeus*
Red-vented bulbul *Pycnonotus cafer*
Red-winged backbird *Agelaius phoeniceus*
Red-winged glossy starling *Lamprotornis nitens*
Red-winged pytilia *Pytilia phoenicoptera*
Red-winged starling *Onychognathus morio*
Redhead *Aythya americana*
Redthroat *Sericornis (Pyrrholaemus) brunneus*
Reed bunting *Emberiza schoeniclus*
Reed warbler *Acrocephalus scirpaceus*
Restless flycathcer *Myiagra inquieta*
Richard's pipit *Anthus richardi*
Robin, *see* Specific types
Rock pipit *Anthus (spinoletta) petrosus*
Rock wren *Salpinctes obsoletus*
Rose robin *Petroica rosea*
Rose-breasted grosbeak *Pheucticus ludovicianus*
Roseate spoonbill *Ajaia ajaia*
Ross's goose *Anser rossi*
Rosy-billed pochard (or rosybill) *Netta peposaca*
Ruby-crowned kinglet *Regulus calendula*
Ruddy duck *Oxyura jamaicensis*
Ruddy shelduck *Tadorna ferruginea*
Rufescent prinia *Prinia rufescens*
Rufous ant thrush *Neocossyphus fraseri*
Rufous bush robin *Cercotrichas galactotes*
Rufous fantail *Rhipidura rufifrons*
Rufous hornero *Funarius rufus*
Rufous grass warbler, *see* winding cisticola
Rufous whistler *Pachycephala rufiventris*
Rufous-and-white wren *Thryothorus rufalbus*
Rufous-bellied gerygone *Gerygone dorsalis*
Rufous-bellied niltava *Niltava sundara*
Rufous-bellied thrush *Turdus rufiventris*
Rufous-breasted spinetail *Synallaxis erythrothorax*
Rufous-breasted wren *Thryothorus rutilus*
Rufous-breasted wryneck *Jynx ruficollus*
Rufous-browed peppershrike *Cyclarhis gujanesis*
Rufous-chested swallow *Hirundo semirufa*
Rufous-chinned laughingthrush *Garrulax rufigularis*
Rufous-collared sparrow *Zonotrichia capensis*
Rufous-fronted babbler *Stachyris rufifrons*
Rufous-necked laughingthrush *Garrulax ruficollis*
Rufous-sided gerygone *Gerygone dorsalis*
Rufous-sided towhee *Pipilo erythropthalmus*
Rufous-vented laughingthrush *Garrulax gularis*
Rufous-winged sparrow *Aimophila carpalis*
Ruppell's glossy (or long-tailed) starling *Lamprotornis purpuropterus*
Ruppell's robin chat *Cossypha semirufa*
Russet-backed oropendola *Psarocolius angustifrons*
Rusty-cheeked simitar babbler *Pomotorhinus erythrogenys*
Rusty-crowned ground-sparrow *Melozone kieneri*
Rusty-fronted barwing *Actinodura egertoni*

Sao Tome' weaver *Ploceus sanctaethomae*
Satin flycathcer *Myiagra cyanoleuca*
Savannah sparrow *Passerculus sandwichensis*
Sayaca tanager *Thraupis sayaca*
Scaly thrush *Zoothera dauma*
Scaly weaver *Sporopipes squamifrons*
Scaly-breasted wren babbler *Pnoepyga albiventer*
Scarlet robin *Petroica multicolor*

Scarlet tanager *Piranga olivacea*
Scarlet-chested sunbird *Nectarina senegalensis*
Scarlet-rumped tanager *Ramphocelus passerinii*
Scimitarbill *Phoeniculus cyanomelas*
Scissor-tailed flycathcer *Muscivora forficata*
Scrub blackbird *Dives warszewiczi*
Seaside sparrow *Ammospiza maritima*
Sedge warbler *Acrocephalus schoenobaenus*
Senegal coucal *Centropus senegalensis*
Senegal firefinch, *see* red-billed firefinch
Sharpe's akalat *Sheppardi sharpei*
Shortwings *Brachpteryx* spp.
Short-billed canastero *Asthenes baeri*
Short-tailed field-tyrant *Muscigralla brevicauda*
Siberian blue robin *Lucinia cyanae*
Silver-beaked tanager *Ramphocelus carbo*
Silver-crowned friarbird *Philemon argenticeps*
Silver-eared mesia *Leiothrix argentaurus*
Singing cisticola *Cisticola cantans*
Singing honeyeater *Lichenostomus virescens*
Slender-billed babbler *Argya longirostris*
Slender-billed weaver *Ploceus pelzelni*
Small niltava *Niltava macgrigoriae*
Small, spider hunter, *see* little spider hunter
Small wren babbler, *see* eyebrowed wren-babbler
Smew *Mergus albellus*
Smooth-billed ani *Crotophaga ani*
Snow goose *Anser caerulescens*
Solitary vireo *Vireo solitaria*
Sombre greenbul *Andropadus importunus*
Song sparrow *Melospiza melodea*
Song thrush *Turdus musicus*
Sooty-fronted spinetail *Synallaxis frontalis*
Sooty-headed bulbul *Pycnonotus aurigaster*
Soulimanga sunbird *Nectarinia sovimanga*
South African cliff swallow *Hirundo spilodera*
South Island robin *Miro australis*
South Island tomtit *Petroica dannefaerdi*
Southern anteater chat *Myrmecocichla formicivora*
Southern boubou *Laniarus ferruginea*
Southern brown-throated weaver *Passer xanthopterus*
Southern masked weaver *Ploceus velatus*
Southern pochard *Netta erythropthalma*
Southern puffback *Dryscopus cubla*
Southern rufous sparrow *Passer motitensis*
Southern screamer *Chauna torquata*
Southern whiteface *Aphelocephala leucopsis*
Sparrowhawk *Accipiter nisus*
Speckled warbler *Sericornis (Chhonicola) sagittatus*
Spectacled weaver *Ploceus ocularis*
Speke's weaver *Ploceus spekei*
Spider hunters *Arachnothera* spp.
Spinetails *Cranioleuca* spp.
Spix's spinetail *Synallaxis spixi*
Splendid fairywren *Malurus splendens*
Splendid starling *Lamprotornis splendidus*
Spot-breasted laughingthrush *Garrulax merulina*
Spot-throated babbler *Pellorneum albiventre*
Spotted flycatcher *Muscicapa striata*
Spotted forktail *Enicurus maculatus*
Spotted rail *Paradirallus maculatus*
Spur-winged goose *Plectropterus gambensis*
Starred robin *Pogonocichia stellata*
Stonechat *Saxicola torquata*
Streak-backed oriole *Icterus pustulatus*
Streaked fantail-warbler, *see* zitting cisticola
Streaked laughingthrush *Garrulax lineatus*
Streaked saltator *Saltator albicollis*
Streaked scrub-warbler *Scotocerca inquieta*
Streaked spider hunter *Arachnothera magna*
Streaked weaver *Ploceus manyar*
Streaky seedeater *Poliospiza striolatus*
Striated laughingthrush *Garrulax striatus*
Striated grassbird *Megalurus palustris*
Striated prinia *Prinia criniger*
Striated thornbill *Acanthiza lineata*
Stripe-backed wren *Campylorhynchus nuchalis*
Stripe-crowned spinetail *Cranioleuca pyrrhophia*
Striped grass warbler, *see* croaking cisticola
Striped kingfisher *Halcyon chelicuti*
Striped tit-babbler *Macronous gularis*
Strong-footed bush warbler, *see* brownish-flanked bush-warbler
Stub-footed bush warbler *Cettia squameiceps*
Summer tanager *Piranga rubra*
Superb blue fairywren *Malurus cyaneus*
Superciliated wren *Thryothorus superciliaris*
Swainson's warbler *Limnothlypis swainsonii*
Swallow-tailed bee-eater *Merops hirundineus*
Swamp sparrow *Melospiza georgiana*

Swee waxbill *Estrilda melanotis*
Swynnerton's robin *Swynnertonia swynnertoni*

Tabora cisticola *Cisticola angusticauda*
Tailorbirds *Orthotomus* spp.
Tasmanian thornbill *Acanthiza ewingii*
Tawny-breasted wren-babbler *Spelaeornis longicaudatus*
Tawny-flanked prinia *Prinia subflava*
Tennessee warbler *Vermivora peregrina*
Thick-billed reed warbler *Acrocephalus aedon*
Tickell's thrush, *see* Indian gray thrush
Tinkling camaroptera *Camaroptera rufilata*
Tinkling cisticola *Cisticola tinniens*
Tody flycatchers *Todirostrum* spp.
Tody tyrants *Myizetetes* spp.
Torresian crow *Corvus orru*
Townsend's warbler *Dendroica townsendi*
Tree pipit *Anthus trivialis*
Tree swallow *Tachycineta bicolor*
Tristam's warbler *Sylvia deserticola*
Tropical boubou *Laniarus aethiopicus*
Tropical kingbird *Tyrannus melancholicus*
Tropical parula *Parula pitiayumi*
Troupial *Icterus icterus*
Tufted duck *Aythya fuligula*
Tullberg's woodpecker *Campethera tullbergi*
Tumbes sparrow *Rhynchospiza stolzmanni*

Variable sunbird *Nectarinia venusta*
Variegated fairywren *Malurus lamberti*
Veery *Catharus fuscenscens*
Verdin *Auriparus flaviceps*
Verdita flycatcher *Eumyias thalassina*
Vermilion flycatcher *Pyrocephalus rubinus*
Vesper sparrow *Poecetes grammacus*
Vieillot's black weaver *Ploceus nigerrimus*
Village weaver *Ploceus cucullatus*
Violet-backed starling *Cinnyricinclus leucogaster*
Violet-eared waxbill, *see* common grenadier
Virginia's warbler *Vermivora virginiae*

Wagler's oropendola, *see* chestnut-headed oropendola
Warbling vireo *Vireo gilvus*
Wattle-eye, *see* brown-throated wattle-eye
Western crowned warbler *Phylloscopus occipitalis*
Western flycatcher *Empidonax difficilis*, including *occidentalis*
Western kingbird *Tyrannus verticalis*
Western tanager *Piranga ludoviciana*
Western thornbill *Acanthiza inornata*
Western wood pewee *Contopus sordidulus*
White helmet-shrike *Prionops plumatus*
White wagtail *Motacilla alba*
White-banded mockingbird *Mimus triurus*
White-bearded flycatcher *Conopias inornatus*
White-breasted sunbird *Nectarinia talatala*
White-browed coucal *Centropus superciliosus*
White-browed fantail *Rhipidura aureola*
White-browed robin chat *Cossypha heuglini*
White-browed scrub robin *Erythropygia leucophrys*
White-browed scrubwren *Sericornis frontalis*
White-browed sparrow weaver *Plocepasser mahali*
White-crowned forktail *Enicurus leschenaulti*
White-crested laughingthrush *Garrulax leucolophus*
White-crowned sparrow *Zonotrichia leucophrys*
White-crowned starling *Spreo albicapillus*
White-eared ground sparrow *Melozone leucotis*
White-eared honeyeater *Lichenostomus leucotis*
White-edged oriole *Icterus graceannae*
White-eyed vireo *Vireo griseus*
White-faced ground-sparrow, *see* Prevost's ground-sparrow
White-faced ibis *Plagadis falcinellus*
White-fronted bee-eater *Merops bullockoides*
White-fronted chat *Ephthianura albifrons*
White-headed barbet *Lybius leucocephalus*
White-headed duck *Oxyura leucocephala*
White-headed marsh-tyrant *Arundinicola leucocephala*
White-naped honeyeater *Melithreptus lunatus*
White-plumed honeyeater *Lichenostomus penicillatus*
White-rumped minor *Manorina flavigula*
White-rumped swallow *Tachycineta leucorrhoa*
White-shouldered fairywren *Malurus alboscapulatus*
White-spectacled warbler *Siecercus affinis*
White-tailed robin *Cinclidium leucurum*
White-throated babbler *Turdoides gularis*
White-throated gerygone (or warbler) *Gerygone olivacea*
White-throated robin chat *Cossypha humeralis*
White-throated silverbill (or munia) *Lonchura malabarica*

White-throated sparrow *Zonotrichia albicollis*
White-throated swallow *Hirundo albigulis*
Whitehead *Mahoua albicilla*
Whitethroat *Sylvia communis*
Whyte's barbet *Lybius whytii*
Willow flycatcher *Empidonax traillii*
Willow warbler *Phylloscopus trochilus*
Williewagtail, *Rhipidura leucophrys*
Wilson's warbler *Wilsonia pusilla*
Winding cisticola *Cisticola galactotes*
Wing-snapping cisticola *Cisticola ayresii*
Winter wren *Troglodytes troglodytes*
Wood thrush *Hylocichla mustelina*
Wood warbler *Phylloscopus sibilatrix*
Woodchat shrike *Lanius senator*
Worm-eating warbler *Helmintheros vermivorus*
Wren, *see* winter wren
Wren grass warbler, *see* zitting cisticola
Wrentit *Chamaea fasciata*

Yellow bishop *Euplectes capensi*
Yellow thornbill *Acanthiza nana*
Yellow wagtail *Motacilla flava*
Yellow warbler *Dendroica petechia*
Yellow white-eye *Zosterops senegalensis*
Yellow-backed sunbird, *see* scarlet sunbird
Yellow-bellied eremomela *Eremomela icteropygialis*
Yellow-bellied prinia (or wren-warbler) *Prinia flaviventris*
Yellow-bellied tyrannulet *Pseudocolopteryx flaviventris*
Yellow-billed cuckoo *Coccyzus americanus*
Yellow-billed shrike *Corvinella corvina*
Yellow-breasted babbler *see* striped tit-babbler
Yellow-breasted chat *Icteria virens*
Yellow-breasted laughingthrush, *see* rufous-vented laughingthrush
Yellow-browed ground warbler, *see* gray-bellied wren
Yellow-browed tyrant *Satrapa icterophrys*
Yellow-chinned spinetail *Certhiaxis cinnamomea*
Yellow-eyed canary *Serinus mozambicus*
Yellow-faced honeyeater *Lichenostomus chrysops*
Yellow-hooded blackbird *Agelaius icterocephalus*
Yellow-rumped cacique *Cacicus cela*
Yellow-rumped tinkerbird *Pogoniulus bilineatus*
Yellow-rumped thornbill *Acanthiza chrysorrhoa*
Yellow-rumped warbler *Dendroica coronata*
Yellow-shouldered blackbird *Agelaius xanthomus*
Yellow-spotted petronia *Petronia pyrgita*
Yellow-tinted honeyeater *Lichenostomus flavescens*
Yellow-throated brush-finch *Atlapetes albinucha*
Yellow-throated petronia *Petronia superciliaris*
Yellow-throated scrubwren *Sericornis citreogularis*
Yellow-throated vireo *Vireo flavifrons*
Yellow-throated warbler *Dendroica dominica*
Yellow-tufted honeyeater *Lichenostomus melanops*
Yellow-vented warbler *Phylloscopus cantantor*
Yellow-whiskered greenbul *Andropadus latirostris*
Yellow-winged cacique *Cacicus melanicterus*
Yellow-winged pytilia *see* red-faced waxbill
Yellowbill *Ceuthmochares aereus*
Yellowhammer *Emberiza citrinella*
Yellowhead *Mohous ochrocephala*
Yuhinas *Yuhina* spp.

Zebra waxbill *Estrilda subflava*
Zitting cisticola *Cisticola juncidis*

REFERENCES

Ali, S. & S. D. Ripley. 1983. Handbook of the birds of India and Pakistan. Compact Edition. Delhi: Oxford University Press.

Amadon, D. 1942. Birds collected by the Whitney South Sea Expedition. Notes on some non-passerine genera. *Amer. Mus. Novitates* 1176:1–21.

Amat, J. A. 1991. Effects of red-crested pochard nest parasitism on mallards. *Wilson Bull.* 103:501–3.

______. 1993. Parasitic laying in red-crested pochard nests. *Ornis. Scand.* 24:65–70.

American Ornithologists' Union. 1983. Check-list of North American birds. Sixth ed. Lawrence, Kansas: Allen Press.

Andersson, M. 1984. Brood parasitism within species. In: Producers and Scroungers (C. J. Barnard, ed.) pp. 195–228. London: Chapman & Hall.

______, & M. O. G. Ericksson. 1984. Nest parasitism in goldeneyes *Bucephala clangula:* some evolutionary aspects. *Am. Nat.* 120:1–16.

Arias de Reyna, L., & S. J. Hidalgo. 1982. An investigation into egg-acceptance by azure-winged magpies and host-recognition by great spotted cuckoo chicks. *Anim. Behav.* 30:819–23.

______, M. Corvillo, & I. Aguilar. 1982. Reproduccion dei crialo *Clamator glandarius* en Sierra Morena Central. *Donanva Acta Verg.* 9:177–93 [English summary].

Baker, E. C. S. 1913. The evolution of adaptation in parasitic cuckoo's eggs. *Ibis.* 10(3):384–98.

______. 1923. Cuckoos' eggs and evolution. *Proc. Zool. Soc. Lond.* 277–94.

______. 1942. Cuckoo problems. London: Witherby.

Balatski, N. N. 1994. (On egg identification in cuckoos in the territory of the USSR.) In: Modern ornithology, 1992 (E. N. Kurochkin, ed.) pp. 31–46. Russian Acad. of Sci. & menzbir Ornith. Soc. Moscow: Nauka Publ. [In Russian, English summary.]

Baldamus, E. 1853. Neue Belträge zur Fortpflanzungsgeschichte des Europaisch Kuckucks. *Naumannia* 307–26.
Bannerman, D. 1949. The birds of tropical West Africa. Vol. 7. Edinburgh: Oliver & Boyd.
Becking, J. H. 1981. Notes on the breeding of Indian cuckoos. *J. Bombay Nat. Hist. Soc.* 78:201–31.
______. 1988. The taxonomic status of the Madagascan cuckoo *Cuculus poliocephalus rochii* and its occurrence on the African mainland, including southern Africa. *Bull. Br. Ornithol. Club* 108:195–206.
______, & D. W. Snow. 1985. Brood-parasitism. In: A dictionary of birds (B. Campbell & E. Lack, eds.) pp. 57–70. Calton, UK: T. & A. D. Poyser.
Beddard, F. E. 1885. On the structural characters and classification of the cuckoos. *Proc. Zool. Soc. Lond.* pp. 168–87.
Benson, C. W., & C. R. S. Pitman. 1964. Further breeding records from Northern Rhodesia (no. 4). *Bull. Br. Ornithal. Club* 84:54–60.
______, R. K. Brooke, R. J. Dowsett, & M. P. S. Irwin. 1971. The birds of Zambia. London: Collins.
______, & M. P. S. Irwin. 1972. The thick-billed cuckoo *Pachycoccyx audeberti* (Schlegel)(Aves: Cucuiidae). *Arnoldia* (Rhodesia) 5(33):1–22.
Bent, A. C. 1940. Life histories of North American cuckoos, goatsuckers, hummingbirds and their allies. *U. S. Natl. Mus. Bull.* 176:1–506.
______. 1953. Life histories of North American wood warblers. *U. S. Natl. Mus. Bull.* 203:1–734.
______. 1958. Life histories of North American blackbird, orioles, tanagers and allies. *U. S. Natl. Mus. Bull.* 211:1–549.
Berger, A. J. 1951. The cowbird and certain host species in Michigan. *Wilson Bull.* 63:26–34.
______. 1960. Some anatomical characters of the Cuculidae and Musophagidae. *Wilson Bull.* 72:26–34.
Biswas, B. 1951. Notes on the taxonomic status of the Indian plaintive cuckoo. *Ibis* 93:596–8.
Blaise, M. 1965. Contribution á l'etude de la reproduction du coucou gris *Cuculus canorus* dans ie nord-est de la France. *L'Oiseau et R.F.O.* 35:87–116.
Blankespoor, G. W., J. Oolman, & C. Uthe. 1982. Eggshell strength and cowbird parasitism of red-winged blackbirds. *Auk* 99:363–65.
Bowen, B. S., & A. D. Kruse. 1994. Brown-headed cowbird parasitism of western meadowlarks in North Dakota [abstract]. North Am. Ornithol. Conf., June 21–26, 1994, Missoula, Mont.
Brazil, M. A. 1991. The birds of Japan. Washington, DC: Smithsonian Inst. Press.
Brehm, A. E. 1853. Zur Fortpflanzungsgeschichte des *Cuculus glandarius. J. f. Ornith.* 144–45.
Briggs, S. V. 1991. Intraspecific nest parasitism in maned ducks. *Emu* 91:230–35.
Briskie, J. V., & S. G. Sealy. 1989. Changes in nest defense against a brood parasite over the breeding cycle. *Ethology* 89:61–67.
______, ______, & K. A. Hobson. 1992. Behavioral defenses against avian brood parasitism in sympatric and allopatric populations. *Evol.* 46:334–40.
Brittingham, M. C., & S. A. Tample. 1983. Have cowbirds caused forest songbirds to decline? *BioScience* 33:31–35.
Brooke, M. de L. & N. B. Davies. 1987. Recent changes in host usage by cuckoos *Cuculus canorus* in Britain. *J. Anim. Ecol.* 56:873–83.
______, & ______. 1991. A failure to demonstrate host imprinting in the cuckoo (*Cuculus canorus*) and alternate hypotheses for the maintenance of egg mimicry. *Ethology* 89:154–66.
Brooker, M. G., & L. C. Brooker. 1986. Identification and development of the nestling cuckoos *Chrysococcyx basalis* and *C. lucidus plagosus* in Western Australia. *Aust. Wildl. Res.* 13:197–202.
______, & ______. 1989a. The comparative breeding behavior of two sympatric cuckoos, Horsfield's bronze-cuckoo *Chrysococcyx basalis* and the shining bronze-cuckoo *C. lucidus.* in Western Australia: a new model for the evolution of egg morphology and host specificity in avian brood parasites. *Ibis* 133:528–47.
______, & ______. 1989b. Cuckoo hosts in Australia. *Aust. Zool. Rev.* 2:1–67.

______, & ______. 1991. Eggshell strength in cuckoos and cowbirds. *Ibis* 133:406–13.
______, ______, & C. Rowley. 1988. Egg deposition by the bronze-cuckoos *Chrysococcyx basalis* and *Ch. lucidus*. *Emu* 88:107–9.
Brosset, A. 1976. Observations sur le parasitisme de la reproduction du coucou emeraud au Gabon. *Oiseau et R. F. O.* 46:201–08.
Brown, B. T. 1994. Rates of brown-headed cowbird parasitism on riparian passerines in Arizona. *J. Field Ornith.* 65:160–68.
Brown, C. R, and M. B. Brown. 1989. Behavioral dynamics of intraspecific brood parasitism in colonial cliff swallows. *Anim. Behav.* 37:777–96.
Brown, J. L. 1978. Avian communal breeding systems. *Annu. Rev. Ecol. Syst.* 9:123–56.
Bull, J. 1974. Birds of New York state. Garden City, NY: Doubleday.
Burhans, D. E. 1993. Dawn nest arrivals and aggression in two cowbird hosts: early bird catches the worm [abstract]. North American Research Workshop on the Ecology and Management of Cowbirds, Nov. 4–5, 1993, Austin, Texas.
Carello, C. 1993. Comparison of rates of cowbird brood parasitism between two red-winged blackbird nesting sites [abstract]. North American Research Workshop on the Ecology and Management of Cowbirds, Nov. 4–5, 1993, Austin Texas.
Carter, M. D. 1986. The parasitic behavior of the bronzed cowbird in south Texas. *Condor* 88:11–25.
Cavalcanti, R. B., M. R. Lemes & R. Cintra. 1991. Egg losses in communal nests of the guira cuckoo. *J. f. Ornith.* 62:177–80.
______, & T. M. Pimentel. 1988. Shiny cowbird parasitism in central Brazil. *Condor* 90:40–43.
Chalton, D. O. 1991. Development of a diederik cuckoo chick in a spectacled weaver nest. *Ostrich* 62:84–85.
Chance, E. 1922. The cuckoo's secret. London: Sedgwick & Jackson.
______. 1940. The truth about the cuckoo. London: Country Life.
Chapin, J. 1917. The classification of the weaver-birds. *Bull. Am. Mus. Nat. Hist.* 37:243–80.
______. 1939. The birds of the Belgian Congo. II. *Bull Am. Mus. Nat. Hist.* 75:1–632.
Cheesman, R. E. & W. L. Sclater. 1935. On a collection of birds from north-western Abyssinia. *Ibis* 13(5):594–622.
Chisholm, A. H. 1935. Bird wonders of Australia. Sydney: Angus & Robertson.
______. 1973. Cuckoos are very resolute. *Aust. Bird Watcher* 5:49–54.
Clark, K. L., & R. J. Robertson. 1981. Cowbird parasitism and evolution of anti-parasite strategies in the yellow warbler. *Wilson Bull.* 93:249–58.
Clawson, R. L., G. W. Hartman, & L. H. Fredrickson. 1979. Dump nesting in a Missouri wood duck population. *J. Wildl. Mgmt.* 43:347–55.
Coates, B. J. 1985. The birds of Papua New Guinea, vol. 1. Non-passerines. Alderley, Australia: Dove Publ.
Colebrook-Robjent, J. F. R. 1984. The breeding of the didric cuckoo *Chrysococcyx caprius* in Zambia. *Proc. Pan-African Ornithol. Congr.* 5:765–77.
Cott, H. B. & C. W. Benson. 1970. The palatability of birds, mainly based on observations of a tasting panel in Zambia. *Ostrich* (Suppl.) 8:375–84.
Courtney, J. 1967. The juvenile food-begging call of some fledgling cuckoos—vocal mimicry or vocal duplication by natural selection. *Emu* 67:154–57.
Craig, A. J. F. K. 1982. Breeding success of a red bishop colony. *Ostrich* 53:182–8.
Cramp, S. (ed.). 1985. Handbook of the birds of Europe, the Middle East and North Africa, vol. 4. Terns to woodpeckers. Oxford: Oxford University Press.
Cronin, E. W., & P. W. Sherman. 1977. A resource-based mating system: the orange-rumped honeyguide, *Indicator indicator*. *Living Bird* 15:5–32.
Crouther, M. M. 1985. Some breeding records of the common koel. *Aust. Bird Watcher* 11:49–56.

Crouther, M. M., & M. J. Crouther. 1984. Observations on the breeding of figbirds and common koels. *Corella* 28:89–92.

Cruz, A., T. D. Manolis & R. W. Andrews. 1990. Reproductive interactions of the shining cowbird *Molothrus bonariensis* and the yellow-headed blackbird *Agelaius icterocephalus* in Trinidad. *Ibis* 132:436–44.

______, & J. W. Wiley. 1989. The decline of an adaptation in the absence of a presumed selection pressure. *Evolution* 43:55–62.

Darley, J. A. 1968. The social organization of breeding brown-headed cowbirds (Ph.D. dissertation). London, Ontario: Univ. Western Ontario.

______. 1971. Sex ratio and mortality in the brown-headed cowbird. *Auk* 88:560–66.

Darwin, C. 1872. The origin of species by means of natural selection, 6th ed. London: J. Murray.

Davies, N. B. 1992. Dunnock behaviour and social evolution. Oxford: Oxford Univ. Press.

______, & M. de L. Brooke. 1988. Cuckoos versus reed warblers: adaptations and counteradaptations. *Anim. Behav.* 36:262–84.

______ & ______. 1989a. An experimental study of co-evolution between the cuckoo, *Cuculus canorus,* and its hosts. I. Host egg discrimination. *J. Anim. Ecol.* 58:207–24.

______ & ______. 1989b. An experimental study of co-evolution between the cuckoo, *Cuculus canonus,* and its hosts. II. Host egg markings, chick discrimination and general discussion. *J. Anim. Ecol.* 58:225–36.

______. 1940a. Social nesting habits of *Guira guira. Auk* 57:472–84.

Davis, D. E. 1940b. Social nesting habits of the smooth-billed ani. *Auk* 57:179–218.

______. 1941. Social nesting habits of *Crotophaga major. Auk* 58:179–83.

______. 1942. The phylogeny of social nesting habits in the Crotophaginae. *Q. Rev. Biol.* 17:115–34.

Dawkins, R., & J. R. Krebs. 1979. Arms races between and within species. *Proc. Roy. Acad. Lond.* B 205:489–511.

DeCapita, M. E. 1993. Brown-headed cowbird control on Kirtland's warbler nesting areas in Michigan. [Abstract] North American Research Workshop on the Ecology and Management of Cowbirds, Nov. 4–5, 1993, Austin, Texas.

Dean, W. R. J., I. A. W. Macdonald, & C. J. Vernon. 1974. Possible breeding record of *Cercococcyx montanus. Ostrich* 45:188.

De Geus, D. W., & L. B. Best. 1991. Brown-headed cowbirds parasitize loggerhead shrikes: first records for family Laniidae. *Wilson Bull.* 103:504–6.

Deignan, H. G. 1945. The birds of northern Thailand. *U.S. Natl. Mus. Bull.* 186:1–616.

Deignan, H. G., & B. Amos. 1950. Notes on some forms of the genus *Chalchites* Lesson. *Emu* 49:167–8.

Delacour, J., & P. Jabouille. 1931. Les Oiseaux de l'Indochine Francaise, vol. 2. Paris: Exp. Col. Int.

Dewar, D. 1907. An enquiry into the parasitic habits of the Indian koel. *J. Bombay Nat. Hist. Soc.* 17:765–82.

Dhindsa, M. A. 1983a. Intraspecific nest parasitism in two species of Indian weaverbirds, *Ploceus benghalensis* and *P. manyar. Ibis* 125:243–45.

Diamond, J. M. 1972. Avifauna of the eastern highlands of New Guinea. *Pub. Nutt. Ornith. Club* 12:1–438.

Dickinson, E. C., R. S. Kennedy, & K. C. Parkes. 1991. The birds of the Philippines: an annotated checklist. B.O.U. checklist no. 12. Tring: British Museum.

Dowsett-Lemaire, F. 1983. Scaly-throated honeyguide *Indicator variegatus* parasitizing olive woodpeckers *Dendropicos griseocephalus* in Malawi. *Bull. Br. Ornithol. Club* 103:71–76.

Duftfy, A. M., Jr. 1982. Movements and activities of radio-tagged brown-headed cowbirds. *Auk* 99:319–27.

______. 1994. Vocalizations and brown-headed cowbird behavior. *J. f. Ornith.* 135:463.

______, A. R. Goldsmith, & J. C. Wingfield. 1987. Prolactin secretion in a brood parasite, the brown-headed cowbird. *J. Zool. Lond.* 212:669–75.

Dunning, J. B., Jr. (ed.). 1993. CRC handbook of avian body masses. Baton Rouge,LA: C.R.C. Press.

Dyer, M. I., J. Pinowski, & B. Pinowska. 1977. Population dynamics. In: Granivorous birds in ecosystems: their evolution, populations, energetics, adaptations, impact and control (J. Pinowski & S. C. Kendeigh, eds.) pp. 53–105. Cambridge, UK: Cambridge Univ. Press.

Eadle, J. M. 1991. Constraint and opportunity in the evolution of brood parasitism in waterfowl. *Proc. 20th Inter. Ornith. Congress, Christchurch, New Zealand.* vol. II: 1031–40.

______, F. R. Kehoe & T. D. Nudds. 1988. Pre-hatch and post-hatch and brood amalgamation in North American Anatidae: a review of hypotheses. *Can. J. Zool.* 66:1709–21.

Earle, R. A. 1986. The breeding biology of the South African cliff swallow. *Ostrich* 57:138–56.

Edwards, G., E. Hosking. & S. Smith. 1949. Reactions of some passerine birds to a stuffed cuckoo. *Brit. Birds* 42:13–19.

______. 1950. Reactions of some passerine birds to a stuffed cuckoo. II. A detailed study of the willow warbler. *Brit. Birds* 43:144–50.

Elliott, P. F. 1978. Cowbird parasitism on the Kansas tallgrass prairie. *Auk* 95:161–67.

______. 1980. Evolution of promiscuity in the brown-headed cowbird. *Condor* 82:138–41.

Emlen, J. T. 1957. Display and mate selection in the whydahs and bishop birds. *Ostrich* 28:202–13.

Emlen, S. T., & P. H. Wrege. 1986. Forced copulations and interspecific parasitism: Two costs of social living in white-throated bee-eaters. *Ecology* 71:2–29.

Erhlich, P., D. S. Dobkin, & D. Wheye. 1988. The birder's handbook: A field guide to the natural history of North American birds. New York: Simon & Schuster.

Erickson, R. C. 1948. Life history and ecology of the canvas-back *(Nyroca valisneria)* in southeastern Oregon (Ph.D. dissertation). Ames: Iowa State Univ.

Fankhauser, D. P. 1971. Annual adult survival rates of blackbirds and starlings. *Bird-banding* 42:36–42.

ffrench, R. 1991. A guide to the birds of Trinidad & Tobago. 2nd. ed. Wynnewood, Pa.:Livingstone Press.

Flood, N. J. 1990. Aspects of the breeding biology of Audubon's oriole. *J. Field Ornithol.* 61:290–302.

Folkers, K. L., & P. E. Lowther. 1985. Responses of nesting red-winged blackbirds and yellow warblers to brown-headed cowbirds. *J. Field. Ornith.* 56:175–77.

Ford, J. 1963. Breeding behavior of the yellow-tailed thornbill in south-western Australia. *Emu* 63:185–200.

______. 1982. Hybridization and migration in an Australian population of little and rufous-breasted bronze-cuckoos. *Emu* 81:209–222.

Fraga, R. M. 1978. The rufous-collared sparrow as a host of the shiny cowbird. *Wilson Bull.* 90:271–84.

______. 1979. Differences between nestlings and fledgling of screaming and baywinged cowbirds. *Wilson Bull.* 90:151–54.

______. 1984. Bay-winged cowbirds (*Molothrus badius*) remove ectoparasites from their brood parasite, the screaming cowbird (*M. rufoaxillaris*). *Biotropica* 16:223–26.

______. 1986. The bay-winged cowbird (*Molonthrus badius*) and its brood parasites: interactions, co-evolution and comparative efficiency (Ph.D. dissertation). Santa Barbara: Univ. of Calif.

______. 1988. Nest sites and breeding success of bay-winged cowbirds (*Molothrus badius*). *J. f. Ornith.* 129:175–83.

Fleischer, R. F. & N. G. Smith. 1992. Giant cowbird eggs in the nests of two icterid hosts: the use of morphology and electrophoretic variants to identify individuals and species. *Condor* 94:572–78.

Freeman, S., D. F. Gori, & S. Rohwer. 1990. Red-winged blackbirds and brown-headed cowbirds: some aspects of a host-parasite relationship. *Condor* 92:336–40.

Friedmann, H. 1929. The cowbirds: a study in the biology of social parasitism. Springfield, Ill: C. C. Thomas

______. 1993. A contribution to the life history of the crespin or four-winged cuckoo, *Tapera naevia. Ibis* 13 (3):532–38.

______. 1948. The parasitic cuckoos of Africa. *Wash. Acad. Sci. Monogr.* 1:1–204.

______. 1949. Additional data on victims of parasitic cowbirds. *Auk* 66:154–63.

______. 1955. The honey-guides. *Bull. U.S. Natl. Mus.* 208:1–292.

______. 1960. The parasitic weaverbirds. *Bull. U.S. Natl. Mus.* 223:1–196.

______. 1963. Host relations of the parasitic cowbirds. *Bull U.S. Natl. Mus.* 233:1–273.

______. 1964. Evolutionary trends in the avian genus *Clamator. Smithson. Misc. Coll.* 146(4):1–127.

______. 1968. The evolutionary history of the avian genus *Chrysococcyx. Bull. U.S. Natl. Mus.* 265:1–137.

______, & L. F. Kiff. 1985. The parasitic cowbirds and their hosts. *Proc. West. Found. Zool.* 2:226–304.

______, ______, & S. I. Rothstein. 1977. A further contribution to knowledge of the host relations of the parasitic cowbirds. *Smithson. Contrib. Zool.* 235:1–75.

von Frisch, O. 1969. Die Entwicklung des Häherkuckucks im Nest der Wirtsvögel und seine Nachzucht in Gefangenschaft. *Z. Tierpsychol* 26:641–50 [English summary].

______. 1973. Ablenkungsmonöver bei der Eiablage des Häherkuckucks (*Clamator glandarius*). *J. f. Ornith.* 114:129–31 [English summary].

______, & H. von Frisch. 1967. Beobachtungen zur Brutbiologie under Jungentwicklung des Häherkuckucks. *Z. Tierpsychol.* 24:129–36 [English summary].

Frith, H. J. (ed.) 1977. Reader's Digest complete book of Australian birds, revised ed. Sydney: Reader's Digest Services.

Fry, C. H., S. Keith & E. K. Urban. 1988. The birds of Africa, vol. 3. London: Academic Press.

Gaston, A. J. 1976. Brood parasitism by the pied crested cuckoo *Clamator jacobinus. J. Anim. Ecol.* 45:331–48.

Gärtner, K. 1981. Das wegnehmen von Wirtsvogeleiern durch den Kuckuck. *Ornith. Mittl.* 33:115–31.

Gates, J. E., & L. W. Gysel. 1978. Avian nest dispersion and fledging success in a field-forest ecotone. *Ecology* 59:871–83.

Gibbons, D. W. 1986. Brood parasitism and cooperative nesting in the moorhen, *Gallinula chloropus. Behav. Ecol. Sociobiol.* 19:221–32.

Gill, B. J. 1982. Notes on the shining bronze-cuckoo (*Chrysococcyx lucidus*) in New Zealand. *Notornis* 29:215–27.

______. 1983. Brood parasitism by the shiny cuckoo *Chrysococcyx lucidus* at Kaikoura, New Zealand. *Ibis* 125:40–55.

Ginn, P. J., W. G. McIlleron, & P. le S. Milstein. 1989. The complete book of Southern African birds. Cape Town: Struik Winchester.

Glue, D., & R. Morgan. 1972. Cuckoo hosts in British habitats. *Bird Study* 19:187–92.

______. 1984. Cuckoo hosts in Britain. *BTO News* 134:5.

Glutz, U. N. von Blotzheim, & K. M. Bauer (eds.). 1980. Handbuch der Vögel Mitteleuropas. Band 9. Wiesbaden: Akademische Verlagsgesellschaft.

Gochfeld, M. 1979. Brood parasite and host coevolution: interactions between shiny cowbirds and two species of meadowlarks. *Am Nat.* 113:855–70.

Goddard, M. T., & S. Marchant. 1981. The parasitic habits of the channel-billed cuckoo *Scythrops novaehollandiae* in Australia. *Aust. Birds* 17:65–72.

Goldwasser, S., D. Gaines, & S. R. Wilbur. 1980. The least Bell's vireo in California, a de facto endangered race. *Am. Birds* 34:742–45.

Goodwin, D. 1982. Estrildid finches of the world. Ithaca, N. Y.: Cornell Univ. Press.

Gosper, D. 1964. Observations on the breeding of the koel. *Emu* 64:39–41.

Gowaty, P. A. 1984. Cuckoldry: the limited scientific usefulness of a colloquial terms. *An. Behav.* 32:924–5.

______, & A. A. Karlin. 1984. Multiple maternity and paternity in single broods of apparently monogamous eastern bluebirds (*Sialia sialis*). *Behav. Ecol. & Sociobiol.* 15:91–95.

Grice, D., & J. P. Rogers.1965. The wood duck in Massachusetts. Final rpt., W-19-R. Westboro: Mass. Div. Fisheries & Wildlife.

Guilford, T., & A. F. Read. 1990. Zahavian cuckoos and the evolution of nestling discrimination by hosts. *An. Behav.* 39:600–01.

Haguchi, H., & S. Sato. 1984. An example of character release in host selection and egg colour of cuckoos *Cuculus* spp. in Japan. *Ibis* 126:398–404.

Hahn, D. C. 1994. Parasitism at the landscape level: do cowbirds prefer forests? [abstract]. North American Research Workshop on the Ecology and Management of Cowbirds, Nov. 4–5, Austin, Texas.

Hahn, H. W. 1941. The cowbird at the nest. *Wilson Bull.* 53:211–21.

Hall, B. P. (ed.) 1974. Birds of the Harold Hall Australian Expedition, 1962–70. Publ. No. 745. London: British Museum.

______. & R. E. Moreau. 1970. An atlas of speciation in African passerine birds. London: British Museum.

Hamilton, W. J., & G. H. Orians. 1965. Evolution of brood parasitism in altricial birds. *Condor* 67:361–82.

Handford, P., & M. A. Mares. 1985. The mating systems of ratities and tinamous: an evolutionary perspective. *Biol. J. Linn. Soc.* 25:77–104.

Hann, H. W. 1937. Life history of the oven-bird in southern Michigan. *Wilson Bull.* 49:145–237.

Harmes, K. E., D. Beletsky, & G. H. Orians. 1991. Conspecific nest parasitism in three species of New World blackbirds. *Condor* 93:967–74.

Harrison, C. 1978. A field guide to the nests, eggs and nestlings of North American birds. Glasgow: Collins.

Harrison, T. P., & F. D. Hoeniger (eds.). 1972. The fowles of heaven, or history of birds, by E. Topsell. Austin: U. of Texas Press.

Haverschmidt, F. 1955. Beobachtungen on *Tapera naevia* und ihren Wirtsvögeln. *J. f. Ornith.* 96:337–43.

______. 1961. Der Kuckuck *Tapera naevia* und seine Wirte in Surinam. *J. f. Ornith.* 102:353–59.

______. 1968. Birds of Surinam. Edinburgh: Oliver & Boyd.

Heinroth, O., & M. Heinroth. 1926–28. Die Vogel Mitteleuropas. 4 vols.

Herricks, F. H. 1910. Life and behavior of the cuckoo. *J. Exp. Zool.* 9:169–233.

Hill, R. A. 1976. Sex ratio and sex determination of immature brown-headed cowbirds. *Bird-Banding* 47:112–14.

Höhn, E. O. 1959. Prolactin in the cowbird's pituitary in relation to avian parasitism. *Nature* 184:20–30.

Holford, K. C., & D. D. Roby. 1993. Factors limiting fecundity of brown-headed cowbirds. *Condor* 95:336–45.

Hoover, J. P., & M. C. Brittingham. 1993. Regional variation in cowbird parasitism of wood thrushes. *Wilson Bull.* 105:228–38.

Howell, A. B. 1914. Cowbird notes. *Auk* 21:250–51.

Hoy, G., & J. Ottow. 1964. Biological and oological studies of the molothrine cowbirds (Icteridae) of Argentina. *Auk* 81:186–203.

Hunter, H. C. 1961. Parasitism of the masked weaver *Ploceus valatus arundinaceus. Ostrich* 32:55–63.

Hussain, S. A., & S. Ali. 1984. Some notes on the ecology and status of the orange-rumped honeyguide *Indicator xanthonotus* in the Himilayas. *J. Bombay Nat. Hist. Soc.* 80:564–74.

Inskipp, C., & T. Inskipp. 1991. A guide to the birds of Nepal. Washington, DC: Smithsonian Inst. Press.

Jackson, N., & W. Roby. 1992. Fecundity and egg-laying patterns of captive yearling brown-headed cowbirds. *Condor* 94:585–90.

Jackson, W. M. 1992. Relative importance of parasitism by *Chrysococcyx* cuckoos versus conspecific nest parasitism in the northern masked weaver *Ploceus taeniopterus. Ornis Scand.* 23:203–06.
Jensen, R. A. C. 1966. Genetics of cuckoo egg polymorphism. *Nature* 209:827.
______, & C. F. Clinning. 1974. Breeding biology of two cuckoos and their hosts in South West Africa. *Living Bird* 13:5–50.
______, & M. K. Jensen. 1969. On the breeding biology of southern African cuckoos. *Ostrich* 40:163–81.
______, & C. J. Vernon. 1970. On the biology of the didric cuckoo in southern Africa. *Ostrich* 41:237–46.
Johnsgard, P. A. 1973. Grouse and quails of North America. Lincoln: Univ. of Nebraska Press.
______. 1979. Birds of the Great Plains: Breeding species and their distribution. Lincoln: Univ. of Nebr. Press.
______. 1993. Arena birds: sexual selection and behavior. Washington, DC: Smithsonian Inst. Press.
______, & M. Carbonell. 1996. Ruddy ducks and other stifftails, their behavior and biology. Norman: Univ. of Oklahoma Press.
Johnson, A. W. 1967. The birds of Chile, vol. 2. Buenos Aires: Johnson.
Johnson, R. G., & S. A. Temple. 1990. Nest predation and brood parasitism of tallgrass prairie birds. *J. Wildl. Manage.* 54:106–11.
Jones, J. M. B. 1985. Striped crested cuckoo parasitizing arrow-marked babbler. *Honeyguide* 31:170–71.
Jourdain, F. C. R. 1925. A study of parasitism in the cuckoos. *Proc. Zool. Soc. London* 1925:639–67.
Joyner, D. 1975. Nest parasitism and brood-related behavior of the ruddy duck (*Oxyura jamaicensis*) (Ph.D. dissertation). Lincoln: Univ of Nebraska.
Kattan, G. H. 1993. Extraordinary annual fecundity of shiny cowbirds at a tropical locality, and its energy trade-off. Abstract, North American Research Workshop on the Ecology and Management of Cowbirds, Nov. 4–5, 1993, Austin, Texas.
Kendra, P. E., R. R. Roth, & D. W. Tallamy. 1988. Conspecific brood parasitism in the house sparrow. *Wilson Bull.* 100:80–90.
Kiff, L. F., & A. W. Williams. 1978. Host records for the striped cuckoo from Costa Rica. *Wilson Bull.* 90:138–9.
King, A. P. 1979. Variables affecting parasitism in the North American cowbird *(Molothrus ater)* (Ph.D. dissertation). Ithaca, NY: Cornell Univ.
King, J. R. 1973. Reproductive relationships of the rufous-crowned sparrow and shiny cowbird. *Auk* 90:19–34.
Klaas, E. E. 1975. Cowbird parasitism and nesting success in the eastern phoebe. *Occas. Pap. Mus. Nat. Hist. Univ. Kansas* 41:1–18.
Klimkiewilcz, M. K., & A. G. Futcher. 1989. Longevity records of North American birds. Supplement I. *J. Field Ornithol.* 60:469–94.
Kuroda, N. 1966. A note on the problem of hawk-mimicry in cuckoos. *Japan J. Zool.* 1966:173–81.
Lack, D. 1963. Cuckoo hosts in England. *Bird Study* 10:185–203.
______. 1966. Population studies of birds. Oxford: Oxford Univ. Press.
______. 1968. Ecological adaptations for breeding in birds. London:Methuen.
Lamba, B. S. 1963. The nidification of some common Indian birds—Part 1. *J. Bombay Nat. Hist. Soc.* 60:121–33.
______. 1975. The Indian crows: a contribution to their breeding biology, with notes on brood parasitism on them by the Indian koel. *Record. Zool. Surv. Indian* 71(1–4):183–300.
Landgren, O. 1990. Guide to the birds of Madagascar. New Haven, Conn.: Yale Univ. Press.
Lanyon, S. M. 1992. Interspecific brood parasitism in blackbirds: a phylogenetic perspective. *Science* 255:77–79.
Laskey, A. R. 1950. Cowbird behavior. *Wilson Bull.* 62:157–74.

Linz, G. M., & S. B. Bolin. 1982. Incidence of brown-headed cowbird parasitism on red-winged blackbirds. *Wils. Bull.* 94:93–5.

Liversidge, R. 1971. The biology of the jacobin cuckoo *Clamator jacobinus. Proc. 3rd Pan-Afr. Ornith. Congr. Ostrich (Suppl.)* 8:117–37.

Lokemoen, J. T. 1966. Breeding ecology of the redhead duck in western Montana. *J. Wildl. Manage.* 30:668–81.

______. 1991. Brood parasitism among waterfowl nesting on islands and peninsulas in North Dakota. *Condor* 93:340–45.

Lombardo, M. P. 1988. Evidence of intraspecific brood parasitism in the tree swallow. *Wilson Bull.* 100:126–28.

Lotem, A., H. Nakamura & Z. Zahavi. 1992. Rejection of cuckoo eggs in relation to host age: a possible evolutionary equilibrium. *Behav. Ecol.* 3:128–32.

______. 1995. Constraints on egg discrimination and cuckoo-host co-evolution. *Anim. Behav.* 49:1185–209.

Lown, B. A. 1980. Reproductive success of the brown-headed cowbird: a prognosis based on Breeding Bird Census data. *Am Birds* 34:15–17.

Lowther, P. E. 1995. Brown-headed cowbird (*Molothrus ater*). In: The Birds of North America, no. 47. Washington, DC. Am. Ornithol. Union.

Lyon, B. E. 1993. Conspecific brood parasitism as a flexible female reproductive tactic in American coots. *Anim. Behav.* 46:911–28.

Maclean, G. L. 1984. Roberts' birds of southern Africa, 5th ed. Cape Town: Trustees of the John Voelcker Bird Book Fund.

MacLeod, J. G. R., and M. Hallack. 1956. Some notes on the breeding of Klaas's cuckoo. *Ostrich* 27:2–5.

Mackworth-Praed, C. W., & G. H. B. Grant. 1962. Birds of eastern and northern Africa. London: Longmans.

______, & ______. 1973. Birds of west central and western Africa. vol. 2. London: Longmans.

Makatsch, W. 1955. Der Brutparasitismus in der Vogelweit. Redebeul: Neumann.

Mal'chevsky, A. S. 1987. Kukuschka i eyo vospitateli. [The cuckoo and its hosts.] Leningrad: Leningrad Univ. Press.

Manolis, T. D. 1982. Host relationships and reproductive strategies of the shiny cowbird in Trinidad and Tobago (Ph.D. dissertation). Boulder: Univ. of Col.

Marchant, S. 1974. Analyses of nest records of the willie wagtail. *Emu* 74:149–60.

Marchetti, K. 1992. Costs to host defense and the persistence of parasitic cuckoos. *Proc. R. Soc. Lond. B* 248:41–45.

Mark, D., & B. J. Stutchbury. 1994. Response of forest-interior songbirds to the threat of cowbird parasitism. Anim. Behav. 47:275–80.

Marshall, A. J. 1931. Notes on cuckoos. *Emu* 30:298–99.

Mason, P. 1980. Ecological and evolutionary aspects of host selection in cowbirds (Ph.D. dissertation). Austin: Univ. of Texas.

______. 1986a. Brood parasitism in a host generalist, the shiny cowbird. 1. The quality of different species as hosts. *Auk* 103:52–60.

______. 1986b. Brood parasitism in a host generalist, the shiny cowbird. 2. Host selection. *Auk* 103:61–69.

May, R. M., & S. K. Robinson. 1985. Population dynamics of avian brood parasites. *Am. Nat.* 126:475–94.

Mayfield, H. 1965. Chance distribution of cowbird eggs. *Condor* 67:257–63.

Mayr, E. 1932. Notes on the bronze cuckoo *Chalcites lucidus* and its subspecies. *Am. Mus. Novitates* 520:1–9.

______. 1942. Notes on the genus *Cacomantis. Am. Mus. Novitates* 1176:15–21.

______. 1944. The birds of Sumba and Timor. *Bull. Am. Mus. Nat. Hist.* 83:123–84.

______, & A. L. Rand. 1937. Results of the Archbold Expedition. 14. Birds of the 1933–34 Papuan Expedition. *Bull Am. Mus. Nat. Hist.* 73:1–248.

McCabe, R. A. 1991. The little green bird: ecology of the willow flycatcher. Madison, Wis.: Rusty Rock Press.

McGeen, D. S. 1972. Cowbird-host relationships. *Auk* 89:360–80.

McGilp, J. N. 1929. Should cuckoos' eggs be destroyed? *Emu* 28:298.

McKinney, F. 1985. Primary and secondary male reproductive strategies of dabbling ducks. In: Avian monogamy (P. A. Gowaty & D. W. Mock, eds.) pp. 68–82. AOU Monographs 37:1–121.

______, S. R. Derrickson & P. Mineau. 1983. Forced copulation in waterfowl. *Behaviour* 86:250–94.

McLachlin, G. R., & R. Liversidge. 1957. Roberts' birds of South Africa, revised ed. Cape Town: Trustees of the South African Bird Book Fund.

McLean, I. G. 1982. Whitehead breeding, and parasitism by long-tailed cuckoos. *Notornis* 29:156–8.

McLean, I., & J. R. Waas. 1987. Do cuckoo chicks mimic the begging calls of their hosts? *Anim. Behav.* 35:1896–98.

Medway, Lord & D. R. Wells. 1976. Birds of the Malay Peninsula, Vol. 5. London:H. F. & G. Witherby.

Mengel, R. 1965. The birds of Kentucky. *Ornithol. Monogr.* 3:1–587.

Menon, G. K., & R. V. Shah. 1979. Adaptive features in juvenile plumage pattern of the Indian koel *Eudynamys scolopaceus:* host mimesis and hawk-pattern. *J. Yamashina Inst. Ornithol.* 11:87–95.

Mermoz, M. E., & J. C. Reboreda. 1994. Brood parasitism of the shiny cowbird, *Molothrus bonariensis,* on the brown-and-yellow marshbird *Pseudolistes virescens. Condor* 96:716–21.

Miles, D. B. 1986. A record of brown-headed cowbird (*Molothrus ater*) parasitism of rufous-crowned sparrows (*Aimophila ruficeps*). *Southwestern Nat.* 31:253–254.

Miller, L. 1917. Field notes on *Molothrus. Bull. Am. Mus. Nat. Hist.* 1917:579–92.

Miller, R. S. 1932. Some remarks on the nesting of the brown-backed honeyeater. *Emu* 32:110–12.

Mills, A. M. 1987. Size of host egg and egg size in brown-headed cowbird. *Wilson Bull.* 99:490–91.

Moksnes, A., & Røyskaft, E. 1987. Cuckoo host interactions in Norwegian mountain areas. *Ornis. Scand.* 18:168–72.

______, & ______. 1989. Adaptations of meadow pipits to parasitism by the common cuckoo. *Behav. Ecol. Sociobiol.* 24:25–30.

______, & ______. 1996. Egg-morphs and host preference in the common cuckoo (*Cuculus canorus*): an analysis of cuckoo and host eggs from European museum collections. *J. Zool. Lond.* 236:625–49.

______, ______, & T. Tyssee. 1995. On the evolution of blue cuckoo eggs in Europe. *J. Avian Biol.* 26:13–19.

______, ______, ______, L. Korsnes, H. M. Lampe, & H. C. Pedersen. 1991a. Behavioral responses of potential hosts toward artificial cuckoo eggs and dummies. *Behaviour* 116:64–85.

______, ______, & A. T. Braa. 1991b. Rejection behavior by common cuckoo hosts toward artificial brood parasite eggs. *Auk* 108:348–54.

Moksnes, A., E. Roskraft, A. T. Braa, L. Korsnes, H. M. Lampe, & H. C. Pedersen. 1991. Behavioral responses of potential hosts toward artificial cuckoo eggs and dummies. *Behaviour* 116:64–89.

Morel, M.-Y. 1973. Contribution á l'etude dynamique de la population de *Lagonosticta senegala* L. (estrildides) à Richard-Toll (Senegal). Interrelations avec le parasite *Hypochera chalybeata* (Müller) (viduines) *Mem. Mus. Nat. d' Hist. Nat.,* Ser. A (Zool.) 78:1–156.

Morton, E., S., & S. M. Farabaugh. 1979. Infanticide and other adaptations of the nestling striped cuckoo *Tapera naevia. Ibis* 121:212–13.

Mountfort, G. 1958. Portrait of a wilderness: the story of the Coto Donaña expedition. London: Hutchinson.

______, & I. J. Ferguson-Lees. 1961. The birds of the Coto Donaña. *Ibis* 103a:86–109.

Mundy, P. J. 1973. Vocal mimicry of their hosts by nestlings of the great spotted cuckoo and striped crested cuckoo. *Ibis* 115:602–4.

______, & A. W. Cook. 1977. Observations on the breeding of the pied crow and great spotted cuckoo in northern Nigeria. *Ostrich* 48:72–84.

Nakamura, H. 1990. Brood parasitism by the cuckoo *Cuculus canorus* in Japan and the start of new parasitism on the azure-winged magpie. *Japan J. Ornithol.* 39:1–18.

______. 1994. Population dynamics and mating system of the common cuckoo. [Abstract]. *J. f. Ornith.* 135:466.

de Naurois, R. 1979. The emerald cuckoo of Sao Tome and Principe Islands (Gulf of Guinea). *Ostrich* 50:88–93.

Neufeldt, I. 1966. Life history of the Indian cuckoo, *Cuculus micropterus micropterus* Gould, in the Soviet Union. *J. Bombay Nat. Hist. Soc.* 63:399–419.

Neunteufel, A. 1951. Observaciones sobre del *Dromococcyx pavoninus* Pelzein y el parasitismo de los cuculides. *Hornero* 9:288–90.

Neunzig, R. 1929. Zum Brutparasitismus der Viduinen. *J. f. Ornith.* 77:1–21.

Newman, G. A. 1970. Cowbird parasitism and nesting success of lark sparrows in southern Oklahoma. *Wilson Bull.* 82:304–9.

Newton, A. 1896. A dictionary of birds. London: A. & C. Black.

Nice, M. M. 1937. Studies in the life history of the song sparrow, Vol. 2. Trans Linn. Soc. N. Y., 1937: 1–246.

______. 1957. Nesting success in altricial birds. *Auk* 74:305 21.

Nicolai, J. 1964. Der Brutparasitismus der Viduinae also ethologische Problem. *Z. Tierpsychol.* 21:129–204.

______. 1967a. Rassen und Artbildung in der Viduengattung *Hypochera*. *J. f. Ornith.* 108:309–19.

______. 1967b. Die isolierte Frühmauser der Farbmerkmale der Kopgefieders bei *U. granatinus* (L.) and *U. inathinogaster* Reichw. (Estrildidae). *Z. Tierpsychol.* 25:854–61.

______. 1968. Die Schabelfarbung als potentialler Isolationsfactor zwischen *Pytilia phaenicoptera* Swainson und *Pytilia lineata* Heuglin. *J. f. Ornith.* 109:450–61.

______. 1969. Beobachtungen an Paradieswitwen (*Steganura paradisaea* L., *Steganura obtusa*) Chapin, und der Strowitwe (*Tetraenura fischeri* Reichenow) in Ostafrika. *J. f. Ornith.* 110:421–47.

______. 1972. Zwel neue *Hypochera* Arten aus West Afrika (Ploceidae, Viduinae). *J. f. Ornith.* 113:229–40.

______. 1974. Mimicry in parasitic birds *Sci. Am.* 231(4):92–98.

______. 1977. Der Rotmaskenastrild (*Pytilia hypogrammica*) als Wirt der Togo-Paradiswitwe *(Steganura togoensis)*. *J. f. Ornith.* 118:175–88.

______. 1989. Brutparasitismus der Glanzwitwe (*Vidua hypocherina*). *J. f. Ornith.* 130:423–34.

Nolan, V., Jr. 1978. The ecology and behavior of the prairie warbler. *Ornith. Monogr.* 26:1–595.

______, & C. F. Thompson. 1978. Egg volume as a predictor of hatchling weight in the brown-headed cowbird. *Wilson Bull.* 90:353–8.

Norris, R. T., 1947. The cowbirds of Preston Frith. *Wilson Bull.* 59:83–103.

North, A. J. 1895. Oological notes. *Proc. Linn. Soc. N.S.W.* 10:215–17.

Oatley, T. B. 1970. Robin hosts of the red-chested cuckoo in Natal. *Ostrich* 41:232–36.

______. 1980. Eggs of two cuckoo genera in one nest, and a new host for emerald cuckoo. *Ostrich* 51:126–27.

Ohmart, R. D. 1973. Observations on the breeding adaptations of the roadrunner. *Condor* 75:140–149.

Ølen, I. J., A. Moksnes, & E. Røskaft. 1995. Evolution of variation in egg color and marking pattern in European passerines: adaptations in a coevolutionary arms race with the cuckoo, *Cucius canorus. Behav. Ecol.* 6:166–74.

Oliver, W. R. B. 1935. New Zealand birds. Wellington: A. H. & A. W. Reed.

Olsen, H. 1958. Opdraet af atlasfinken. *Stuefuglene* 33:125–28.
Orians, G. 1985. Blackbirds of the Americas. Seattle: Univ. of Wash. Press.
______, E. Røskaft, & L. D. Beletsky. 1989. Do brown-headed cowbirds lay their eggs at random in the nests of red-winged blackbirds? *Wilson Bull.* 101:599–605.
Oring, L. 1982. Avian mating systems. In: Avian Biology, Vol. VI (D. S. Farner & J. R. King, eds.), pp. 1–92. New York: Academic Press.
Ortego, B. 1993. Brown-headed cowbird population trends at a large winter roost in southwest Louisiana from 1974–1992 [abstract]. North American Research Workshop on the Ecology and Management of Cowbirds, Nov. 4–5, 1993, Austin, Texas.
Owen, J. H. 1933. The cuckoo in the Felsted district. *Rep. Felsted School Sci. Soc.* 33:25–39.
Parkenham, R. H. W. 1939. Field notes on the birds of Zanzibar and Pemba. *Ibis* 14 (3):522–54.
Parker, S. A. 1981. Prolegomenon to further studies in the *Chrysococcyx "malayensis"* group. *Zool. Verhandl. Rijkmus Nat. Hist. Leiden* 187:1–57.
Parkes, K. C. 1960. Notes on some non-passerine birds from the Philippines. *Ann. Carn. Mus.* 35:331–40.
Paton, W. C. 1994. The effect of edge on avian nest success: how strong is the evidence? *Conserv. Biol.* 8:17–26.
Payne, R. B. 1965. Clutch size and numbers of eggs laid by brown-headed cowbirds. *Condor* 67:44–60.
______. 1967a. Gonadal responses of brown-headed cowbirds to long daylengths. *Condor* 69:289–97.
______. 1967b. Interspecific communication signals in parasitic birds. *Am. Nat.* 101:363–76.
______. 1970. The mouth markings of juvenal *Vidua regia* and *Uraeginthus granatinus. Bull. Br. Ornithol. Club.* 90:16–18.
______. 1971. Paradise whydahs *Vidua paradisaea* and *V. obtusa* of southern and eastern Africa, with notes on differentiation of the females. *Bull. Br. Ornith. Club* 91:66–76.
______. 1973a. Behavior, mimetic songs and song dialects, and relationships of the parasitic indigobirds *(Vidua)* of Africa. *Ornithol. Monogr.* 11:1–333.
______. 1973b. Individual histories and the clutch size and numbers of eggs of parasitic cuckoos. *Condor* 75:414–38.
______. 1973c. Vocal mimicry of the paradise whydahs *(Vidua)* and response of female whydahs to the songs of their hosts and their mimics. *Animal Behav.* 21:762–71.
______. 1976a. The clutch size and numbers of eggs of brown-headed cowbirds: effects of latitude and breeding season. *Condor* 78:337–42.
______. 1976b. Song mimicry and species relationships among the west African pale-winged indigobirds. *Auk* 93:25–38.
______. 1977a. Clutch size, egg size, and the consequences of single vs multiple parasitism in parasitic finches. *Ecology* 58:500–13.
______. 1977b. The ecology of brood parasitism in birds. *Annu. Rev. Ecol. Syst.* 8:1–28.
______. 1980. Behavior and songs in hybrid parasitic finches. *Auk* 94:118–34.
______. 1982. Species limits in the indigobirds (Ploceidae, *Vidua*) of West Africa: mouth mimicry, song mimicry, and description of new species. *Misc. Publ. Mus. Zool. Univ. Mich.* 162:1–96.
______. 1985. The species of parasitic finches in West Africa. *Malimbus* 7:103–13.
______. 1990. Song mimicry by the village indigobird *(Vidua chalybeata)* of the red-billed firefinch (*Lagonosticta senegala*). *Vogelwarte* 35:321–28.
______. 1991. Female and first-year male plumages of paradise whydahs (*Vidua interjects*). *Bull Br. Ornithol. Club* 111:95–100.
______. 1992. Indigo bunting. In: The birds of North America, no. 4 (A. Poole et al., eds.). Washington, DC: Am. Ornithol. Union.
______, & K. Payne. 1967. Cuckoo hosts in southern Africa. *Ostrich* 38:135–43.

______, & ______. 1977. Social organization and mating success in a local song population of village indigobird, *Vidua chalybeata. Z. Tierpsychol.* 45:113–73.

______, & L. L. Payne. 1994. Song mimicry and species associations of West African indigobirds *Vidua* with quail-finch *Ortygospiza atricollis,* goldbreast *Amadava subflava* and brown twinspot *Clytospiza monteiri. Ibis* 136:291–304.

______, ______, & M. E. D. Nhlane. 1992a. Song mimicry and species status of the green widowfinch *Vidua codringtoni. Ostrich* 63:86–97.

______, ______, ______, & K. Hustler. 1992b. Species status and distribution of the parasitic indigobirds *Vidua* in east and central Africa. *Proc. VIII Pan-African Ornithol. Cong., Bujumbura,* pp. 40–50.

Peck, G. K. & R. D. James. 1987. Breeding birds of Ontario: nidiology and distribution, Vol. 2. Passerines. Toronto: Royal Ontario Museum.

Perez-Rivera, R. A. 1986. Parasitism by the shiny cowbird in the interior parts of Puerto Rico. *J. f. Ornith.* 57:99–104.

Perrins, C. 1967. The short apparent incubation period in the cuckoo. *Br. Birds* 60:51–52.

Peterjohn, B. G., & J. R. Sauer. 1994. Temporal and geographic patterns in population trends of brown-headed cowbirds [abstract]. North American Research Workshop on the Ecology and Management of Cowbirds, Nov. 4–5, 1993, Austin, Texas.

Peters, J. L. 1940. Check-list of the birds of the world. Cambridge: Harvard Univ. Press.

Petit, L. J. 1991. Adaptive tolerance of cowbird parasitism by prothonotary warblers: a consequence of nest-site limitations? *Anim. Behav.* 41:425–32.

______, & D. R. Petit. 1993. Host selection by cowbirds in North America: adaptations to life history traits or ecological opportunism? [abstract]. North American Workshop on the Ecology and Management of Cowbirds, Nov. 4–5, 1993, Austin, Texas.

Petrie, M., & A. P. Möller. 1991. Laying eggs in others' nests: intraspecific brood parasitism in birds. *Trends Ecol. Evol.* 6:315–20.

Phillips, W. W. A. 1948. Cuckoo problems of Ceylon. *Spolia Zeylandica* 25:45–60.

Picman, J. 1989. Mechanism of increased puncture resistance of eggs of brown-headed cowbirds. *Auk* 106:577–83.

Pitman, C. R. S. 1957. On the egg of the African cuckoo *Cuculus canorus gularis* Stephens. *Bull. Br. Ornith. Club* 77:138–9.

Post, W., A. Cruz, & D. B. McNair. 1993. The North American invasion pattern of the shiny cowbird. *J. Field Ornithal.* 64:32–41.

______, & J. W. Wiley. 1976. The yellow-shouldered blackbird—present and future. *Am. Birds* 30:13–20.

______, & ______. 1977. Reproductive interactions of the shiny cowbird and the yellow-shouldered blackbird. *Condor* 76:176–84.

Powell, A. 1979. Cuckoo in the nest. *Wildfowl News* (Wildfowl and Wetlands Trust Newsletter) 81:15–16.

Preston, F. W. 1948. The cowbird (*Molothrus ater*) and the cuckoo (*Cuculus canorus*). *Ecology* 29:115–16.

Pycraft, W. F. 1910. A history of birds. London: Methuen.

Rahn, H., L. Curran-Everett, & D. T. Booth. 1988. Eggshell differences between parasitic and non-parasitic Icteridae. *Condor* 90:962–64.

Rand, A. L. 1941. Results of the Archbold Expedition. No. 32. *Am. Mus. Novitates* 1102:1–15.

______. 1942a. Results of the Archbold Expedition. No. 42. *Bull. Am. Mus. Nat. Hist.* 79:289–366.

______. 1942b. Results of the Archbold Expedition. No. 43. *Bull Am. Mus. Nat. Hist.* 79:425–515.

______, & E. T. Gilliard. 1969. Handbook of New Guinea birds. Garden City, N.Y.: Natural History Press.

______, & D. S. Rabor. 1960. Birds of the Philippine Islands. *Fieldiana (Zool).* 35:222–441.
Reade, W. & E. Hosking. 1967. Nesting birds, eggs and fledglings. London: Blandford Press.
Reed, R. A. 1968. Studies of the diederick cuckoo *Chrysococcyx caprius* in the Transvaal. *Ibis* 110:321–31.
______. 1969. Notes on the red-crested cuckoo in the Transvaal. *Ostrich* 40:1–4.
Rey, E. 1892. Altes und Neues aus dem Houshalte des Kuckucks. Leipzig: Freese.
Richardson, M., & R. W. Knapton. 1994. Regional use, selection and nesting success of wood ducks, *Aix sponsa* using nest boxes in Ontario. *Can. Field-Nat.* 107:293–303.
Ridgway, R. 1902. Birds of North and Middle America. Tanagriidae to Mniotiltidae Pt. 2. *Bull. U. S. Natl. Mus.* 50:1–834.
______. 1916 (Idem.) Cuculidae to Columbidae. Pt. 7. *Bull. U. S. Natl. Mus.* 50:1–543.
Roberts, T. J. 1991. The birds of Pakistan. 2 vol. Oxford: Oxford Univ. Press.
Robertson, R. J., & R. F. Norman. 1976. Behavioral defenses to brood parasitism by potential hosts of the brown-headed cowbird. *Condor* 78:167–73.
______, & ______. 1977. The function and evolution of aggressive host behavior toward the brown-headed cowbird (*Molothrus ater.*) *Can. J. Zool.* 55:508–18.
Robinson, S. K. 1988. Foraging ecology and host relationships of giant cowbirds in southeastern Peru. *Wilson Bull.* 100:224–35.
Robinson, S., J. P. Hoover, & R. Jack. 1993. Patterns of variation in cowbird parasitism levels in fragmented Illinois landscapes [abstract]. North American Research Workshop on the Ecology and Management of Cowbirds, Nov. 4–5, 1993, Austin, Texas.
Rohwer, S., & C. D. Spaw. 1988. Evolutionary lag versus bill-shape constraints: a comparative study of the acceptance of cowbird eggs by old hosts. *Evol. Ecol.* 2:27–36.
Romagnano, L., A. S. Hoffenberg, & H. W. Power. 1990. Intraspecific brood parasitism in the European starling. *Wilson Bull.* 102:279–91.
Røskaft, E. & A. Moksnes. 1994. Host preferences in the common cuckoo and anti-parasite host defense. Abstract. *J. f. Ornith.* 135:467.
Rothstein, S. I. 1970. An experimental investigation of the defenses of the hosts of the brown-headed cowbird (*Molothrus ater*) (Ph.D. dissertation). New Haven, Conn.: Yale Univ.
______. 1975a. Evolutionary rates and host defenses against avian brood parasitism. *Am. Nat.* 109:151–76.
______. 1975b. An experimental and teleonomic investigation of avian brood parasitism. *Condor* 77:50–71.
______. 1975c. Mechanisms of avian egg recognition: do birds know their own eggs? *Anim. Behav.* 23:268–78.
______. 1976a. Cowbird parasitism of the cedar waxwing and its evolutionary implications. *Auk* 93:498–509.
______. 1976b. Experiments on defenses cedar waxwings use against cowbird parasites. *Auk* 93:675–91.
______. 1977a. Cowbird parasitism and egg recognition of the northern oriole. *Wilson Bull.* 89:21–32.
______. 1977b. The preening invitation or head-down display of parasitic cowbirds: 1. Evidence for intraspecific occurrence. *Condor* 79:13–23.
______. 1982. Mechanisms of avian egg recognition: which egg parameters elicit responses by rejecter species? *Behav. Ecol. Sociobiol.* 11:229–39.
______. 1990. A model system for coevolution: avian brood parasitism. *Ann. Rev. Ecol. Syst.* 21:481–508.
______, D. A. Yokel & R. C. Fleisher. 1986. Social dominance, mating and spacing, female fecundity, and vocal dialects in captive and free-ranging brown-headed cowbirds. In: Current Ornithology (R. F. Johnston, ed.) pp. 127–85. New York: Plenum Press.
Rowan, M. K. 1983. Doves, parrots, louries & cuckoos of southern Africa. London: Croom Helm.
Royama, T. 1963. Cuckoo hosts in Japan. *Bird Study* 10:201–2.
Salter, B. E. 1978. A note on the channel-bill cuckoo, *Bird Observer* 559.

Salvador, S. A. 1982. Estudio de parasitismo del Crespin, *Tapera naevia. Hist. Nat.* 2:65–70 [English summary]. (Aves: Icteridae).

______. 1983. Parasitismo de cria del renegrido (*Molothrus bonariensis*) en Villa Maria, Cordoba, Argentina (Aves: Icteridae). [English summary]. *Hist. Nat.* 3:149–58.

______. 1984. Estudio de parasitismo de cria del renegrido (*Molothrus bonariensis*) en calandria (*Mimus saturninus*) en Villa Maris, Cordoba [English summary]. *El Hornero* 12:141–49.

Schönwetter, M. 1967–1984. Handbuch der Oologie. 3 Vols. Berlin: Akademische-Verlag.

Schulze-Hagen, K. 1992. Parasitierung und Brutverluste durch den Kuckuck (*Cuculus canorus*) bei Teich- und Sumpfröhrsanger (*Acrocephalus scirpaceus, A. palustris*) in Mittel- und Westeuropa. [English summary]. *J. f. Ornith.* 133:237–49.

Scott, D. M. 1991. The time of day of egg laying by the brown-headed cowbird and other icterines. *Can. J. Zool.* 69:2093–99.

______, & C. D. Ankney. 1980. Fecundity of the brown-headed cowbird in southern Ontario. *Auk* 97:677–83.

______, & ______. 1983. The laying cycle of brown-headed cowbirds: passerine chickens? *Auk* 100:583–92.

Sealy, S. G., K. A. Hobson, & J. V. Briskie. 1989. Responses of yellow warblers to experimental intraspecific brood parasitism. *J. Field Ornithol.* 60:224–9.

Seaton, C. 1962. The yellow-breasted sunbird as a host to the rufous-breasted bronze-cuckoo. *Emu* 62:174–6.

Selander, R. K, & C. J. La Rue, Jr. 1961. Interspecific preen invitation display of parasitic cowbirds. *Auk* 78:473–504.

Semel, B., P. W. Sherman, & S. M. Byers. 1988. Effects of brood parasitism and nest box placement on wood duck breeding ecology. *Condor* 90:920–30.

Serle, W. 1957. A contribution to the ornithology of the eastern region of Nigeria. *Ibis* 99:628–85.

Shaw, P. 1984. Social behaviour of the pin-tailed whydah *Vidua macroura* in northern Ghana. *Ibis* 126:463–73.

Sibley, C. G., & B. L. Monroe, Jr. 1990. Distribution and taxonomy of birds of the world. New Haven, Conn.: Yale University Press.

Sick, H. 1953. Zur Kenntnis der brasilianischen Lerchenkuckucke *Tapera* und *Dromococcyx. Bonn Zool Beitr.* 4:305–26.

______. 1993. Birds in Brazil. Princeton, N.J.: Princeton Univ. Press.

______, & J. Ottow. 1958. Vom brasilianischen Kuhvogel, *Molothrus bonariensis,* und sienen Wirten, besonders dem Ammerfinken, *Zonotrichia capensis. Bonn. Zool. Beitr.* 9:40–62.

Skead, C. J. 1951. Notes on honeyguides in southeastern Cape Province, South Africa. *Auk* 68:52–62.

______. 1952. Cuckoo studies on a South African farm (Part II). *Ostrich* 23:2–15.

______. 1962. Jacobin crested cuckoo *Clamator jacobinus* (Boddaert) parasitising the fork-tailed drongo *Dicrurus adsimilis* (Beckstein). *Ostrich* 33:72–3.

______. 1975. Ecological studies of four estrildines in the central Transvaal. *Ostrich* suppl. 11:1–54.

Skutch, A. F. 1954. Life histories of Central American birds. *Pacific Coast Avifauna* 31:1–448.

______. 1976. Parent birds and their young. Austin: Univ. of Texas Press.

______. 1987. Helpers at the nest: A worldwide survey of cooperative breeding and related behavior. Iowa City: Univ. of Iowa Press.

Slud, P. 1964. The birds of Costa Rica: distribution and ecology. *Bull. Am. Mus. Nat. Hist.* 128:1–420.

Smith, J. N. M., & P. Arcese. 1994. Brown-headed cowbirds and an island population of song sparrows: A 16-year study. *Condor* 96:916–34.

Smith, N. G. 1968. The advantage of being parasitized. *Nature* 219:690–94.

______. 1979. Alternate responses by hosts to parasites which may be helpful or harmful. In: Host-parasite interactions (B. B. Nickol, ed.) pp. 7–15. New York: Academic Press.

______. 1983. *Zarhynchus wagleri.* In: Costa Rican natural history (D. H. Janzen, ed.) pp. 614–16. Chicago: Univ. of Chicago Press.

Smith, S., & E. Hosking. 1955. Birds fighting. London:Faber & Faber, Ltd.

Smithe, F. B. 1966. The birds of Tikal. Garden City; N.Y.: Natural History Press.

Smythies, B. E. 1953. The birds of Burma. 2nd. ed. Edinburgh: Oliver & Boyd.

Snow, D. W. 1968. An atlas of speciation in African non-passerine birds. London: British Museum (Natural History).

Soler, M. 1990. Relationships between the great spotted cuckoo *Clamator glandarius* and its corvid hosts in a recently colonized area. *Ornis Scand.* 21:212–23.

______, & Møller, A. P. 1990. Duration of sympatry and coevolution between the great spotted cuckoo and its magpie host. *Nature* 343:748–50.

______, J. J. Palomino, J. G. Martinez, & J. J. Soler. 1994. Activity, survival, independence and migration of great spotted cuckoos. *Condor* 96:802–5.

______, J. J. Soler, J. G. Martinex, & A. P. Møller. 1994. Micro-evolutionary change in host response to a brood parasite. *Behav. Ecol. Sociobiol.* 35:295–301.

______, ______, ______, & ______. 1995. Magpie host manipulation by great spotted cuckoos: evidence for an avian Maffia? *Evol.* 49:770–75.

Sorenson, M. D. 1991. The functional significance of parasitic egg laying and typical nesting in redhead ducks: an analysis of individual behavior. *Anim. Behav.* 42:771–96.

Sorenson, M. D. 1993. Parasitic egg-laying in canvasbacks: frequency, success and individual behavior. *Auk* 110: 57–69.

Southern, H. N. Mimicry in cuckoos' eggs. In: Evolution as a process (J. Huxley, A. C. Hardy, & E. B. Ford, eds.) pp. 219–32. London: Allen & Unwin.

Spaw, C. D., & S. Rohwer. 1987. A comparative study of eggshell thickness in cowbirds and other passerines. *Condor* 89:307–18.

Steyn, P. 1973. Some notes on the breeding biology of the striped cuckoo. *Ostrich* 44:163–69.

______, & W. W. Howells. 1975. Supplementary notes on the breeding biology of the striped cuckoo. *Ostrich* 46:258–60.

Stewart, B. G., W. A. Carter & J. D. Tyler. 1988. Third known nest of the slaty vireo *Vireo brevipennis* (Vireonidae) in Colima, Mexico. *Southwest. Nat.* 33:252–3.

Stiles, G. F., & A. F. Skutch. 1989. A guiode to the birds of Costa Rica. Ithaca: Cornell Univ. Press.

Stresemann, E. 1931. Zur Ornithologie von Halmahera und Batjan. *Orn Montsb.* 1931:167–71.

______, & V. Stresemann. 1961. Die Handswingen-mauser der Kuckucke (Cuculidae). *J. f. Ornith.* 102:317–52.

______, & ______. 1969. Die Mauser der Schopfkuckucke (*Clamator*). *J. f. Ornigh.* 110:192–204.

Studer, A., & J. Vielliard. 1988. Premiers donnees etho-ecologiques sur l'icteride bresilien *Curaeus forbesi* (Sclater, 1886)(Aves, Passeriformes) [English summary]. *Rev. Suisse Zool.* 95:1063–77.

Swynnerton, C. F. M. 1916. On the colouration of the mouths and eggs of birds. *Ibis* 10(4):264–94.

Tarboton, W. 1975. African cuckoo parasitising fork-tailed drongo. *Ostrich* 46:186–88.

______. 1986. African cuckoo: the agony and ectasy of being a parasite. *Bokmakierie* 38:109–11.

Teather, K. L., & R. J. Robertson. 1986. Pair bonds and factors influencing the diversity of mating systems in brown-headed cowbirds. *Condor* 88:63–69.

Teuschl, Y., B. Taborsky & M. Taborsky. 1994. Habitat imprinting and egg mimicry in European cuckoos. Abstract. *J. f. Ornith.* 135:137.

Thompson, F. R. 1993. Temporal and spatial patterns of breeding brown-headed cowbirds in the midwestern United States [abstract]. North American Research Workshop on the Ecology and Management of Cowbirds, Nov. 4–5, 1993, Austin, Texas.

Thompson, M. 1966. Birds from north Borneo. *U. of Kansas, Mus. Nat. Hist* 17(8):377–433.

Traylor, M. A. 1966. Relationships of the combassous (sub-genus *Hypothera*). *Ostrich* (Suppl.) 6:57–74.
______. 1968. Family Ploceidae, subfamily Viduinae. In: Checklist of birds of the world, vol. 14. (R. A. Paynter, Jr., ed.) pp. 390–7. Cambridge: Harvard Univ. Press.
Trine, C. L. 1993. Multiple parasitism by brown-headed cowbirds in wood thrush nests: Effects on fledging production of parasite and host.[Abstract] North American Research Workshop on the Ecology and Management of Cowbirds, Nov. 4–5, 1993, Austin, Texas.
Uyehara, J. G. 1993. How do female cowbirds search for nests? [abstract]. North American Research Workshop on the Ecology and Management of Cowbirds, Nov, 4–5, 1993, Austin, Texas.
Valverde, J. A. 1971. Notas sobre la biologia de reproduccion del crialo *Clamator glandarius* (L.). *Ardeola* (numero especial), pp. 591–647.
Verhrencamp, S. L. 1976. The evolution of communal nesting in the groove-billed ani (Ph.D. dissertation). Ithaca, N.Y.: Cornell Univ.
______. 1977. Relative fecundity and parental effort in communally nesting anis, *Crotophaga sulcirostris. Science* 197:403–5.
Vernon, C. I. 1964. The breeding of the cuckoo-weaver (*Anomalospiza imberis* (Cabanis)) in southern Rhodesia. *Ostrich* 35:260–63.
Vernon, C. J. 1984. The breeding biology of the thick-billed cuckoo. In: Proc. 5th. Pan-African Ornithol. Congress, Johannesburg, pp. 825–40.
Victoria, J. K. 1972. Clutch characteristics and egg discrimination ability of the African village weaverbird *Ploceus cucullatus. Ibis* 114:367–76.
Voipio, P. 1953. The hepaticus variety and the juvenile plumage of the cuckoo. *Ornis Fenn.* 30:97–117.
Walkinshaw, L. H. 1949. Twenty-five eggs apparently laid by a cowbird. *Wilson Bull.* 61:82–5.
______. 1961. The effect of parasitism by the brown-headed cowbird on *Empidonax* flycatchers in Michigan. *Auk* 78:266–8.
______. 1983. Kirtland's warbler: the natural history of an endangered species, *Bull. Cranbrook Inst. Sci.* 58:1–207.
Weatherhead, P. J. 1989. Sex ratios, reproductive success and impact of brown-headed cowbirds. *Auk* 106:358–66.
Webster, M. S. 1994. Interspecific brood parasitism of Montezuma oropendolas by giant cowbirds: parasitism or mutualism? *Condor* 96:794–98.
Weigmann, C., & J. Lamprecht. 1991. Intraspecific nest parasitism in bar-headed geese, *Anser indicus. Anim. Behav.* 41:677–88.
Weller, M. W. 1959. Parasitic egg laying in the redhead (*Aythya americana*) and other North American Anatidae. *Ecol. Monogr.* 29:333–65.
______. 1967. Notes on plumages and weights of the black-headed duck, *Heteronetta atricapilla. Condor* 69:133–45.
______. 1968. The breeding biology of the parasitic black-headed duck, *Living Bird* 7:169–208.
West, M. J., A. P. King & D. H. Eastzer. 1981. Validating the female bloassay of cowbird (*Molothrus ater*) song: relating differences in song potency to mating success. *Anim. Behav.* 29:490–501.
Wetherbee, D. K., & N. S. Wetherbee. 1961. Artificial incubation of eggs of various bird species and attributes of neonates. *Bird-banding* 113:339–46.
Wetmore, A. 1968. The birds of the Republic of Panama. Pt. 2. *Smithsonian Misc. Coll.* 150(2):1–605.
______. 1984. The birds of the Republic of Panama. Pt. 4. *Smithsonian Misc. Coll.* 150(4):1–670.
White, C. M. N., & M. D. Bruce. 1986. The birds of Wallacea (Sulawesi, the Moluccas and Lesser Sunda Islands, Indonesia): an annotated checklist, London: British Ornithologists' Union Checklist No. 7.
Whitehead, D. R., D. E. Winslow, M. S. Koukol, B. Ford & G. M. Greenberg. 1993. Landscape pattern and cowbird parasitism in the forest of south-central Indiana [abstract]. North American

Research Workshop on the Ecology and Management of Cowbirds, Nov. 4–5, 1993, Austin, Texas.

Wiens, J. A. 1965. Nest parasitism of the dickcissel by the yellow-billed cuckoo in Marshall County, Oklahoma. *Southerwestern Nat.* 10:142.

Wiley, J. W. 1985. Shiny cowbird parasitism in two avian communities in Puerto Rico. *Condor* 87:167–76.

______. 1988. Host selection by the shiny cowbird. *Condor* 90:289–303.

______, & M. S. Wiley. 1980. Spacing and timing in the nesting ecology of a tropical blackbird: comparison of populations in different environments. *Ecol. Monogr.* 50:153–78.

Williams, J. G., & G. S. Keith. 1962. A contribution to our knowledge of the parasitic weaver, *Anomalospiza imberbis. Bull. Br. Ornithol. Club* 82:141–2.

Wilson, E. O. 1971. The insect societies. Cambridge: Harvard Univ. Press.

Woodsworth, B. L. 1993. Ecology and behavior of the shiny cowbird (*Molothrus bonariensis*) in a dry subtropical forest [abstract]. North American Research Workshop on the Ecology and Management of Cowbirds, Nov. 4–5, 1993, Austin, Texas.

Woodward, P. W., & J. W. Woodward. 1979. Survival of fledgling brown-headed cowbirds. *Bird-banding* 50:66–68.

Wyllie, I. 1981. The cuckoo. London: Batsford.

Yamauchi, A. 1995. Theory of evolution of nest parasitism in birds. *Am. Nat.* 145:434–56.

Yokel, D. A. 1986. Monogamy and brood parasitism: an unlikely pair. *Anim. Behav.* 34:1348–58.

______. 1987. Sexual selection and the mating system of the brown-headed cowbird (*Molothrus ater*) in eastern California (Ph.D. dissertation) University of California, Santa Barbara.

______. 1989. Intrasexual aggression and the mating behavior of brown-headed cowbirds: their relation to population densities and sex ratios. *Condor* 91:43–51.

______, & S. I. Rothstein. 1991. The basis for female choice in an avian brood parasite. *Behav. Ecol. Sociobiol.* 29:39–45.

Yom-Tov, Y. 1980. Intraspecific nest parasitism in birds. *Biol. Rev.* 55:93–108.

______, G. M. Dunnet, & A. Anderson. 1974. Intraspecific nest parasitism in the starling *Sturnus vulgaris. Ibis* 116:87–90.

Young, A. D. 1963. Breeding success of the cowbird. *Wilson Bull.* 75:115–22.

______, & R. D. Titman. 1988. Intraspecific nest parasitism in red-breasted mergansers. *Can. J. Zool.* 66:2454–8.

Zahavi, A. 1979. Parasitism and nest predation the parasitic cuckoos. *Am. Nat.* 113:157–59.

Zimmerman, J. L. 1966. Polygyny in the dickcissel. *Auk* 83:534–46.

Zimmerman, J. L. 1983. Cowbird parasitism of dickcissels in different habitats and at different nest densities *Wils. Bull.* 95:7–22.

Zuniga, J. M., & T. Redondo. 1992. No evidence for variable duration of sympatry between the great spotted cuckoo and its magpie host. *Nature* 359:410–11.

INDEX

This index includes the English vernacular names of all bird species that are mentioned in the text, plus the generic and specific epithets of the 85 species of brood parasites having individual descriptive text entries (shown by italics). Complete text indexing for these species is limited to the entries associated with their English vernacular names; the entries associated with Latin specific and generic epithets index only the primary descriptive accounts for these taxa. The appendices are not indexed.